Systems Engineering and Control Systems

Systems Engineering and Control Systems

Edited by **Brian Maxwell**

NY RESEARCH PRESS
New York

Published by NY Research Press,
23 West, 55th Street, Suite 816,
New York, NY 10019, USA
www.nyresearchpress.com

Systems Engineering and Control Systems
Edited by Brian Maxwell

International Standard Book Number: 978-1-63238-503-1 (Hardback)

Printed in the United States of America.

Contents

Preface

Systems engineering is mainly concerned with design and management of complex engineering systems. It tries to incorporate the requirements specified by the end users in the initial steps of system design. This discipline overlaps with control engineering, software engineering, industrial engineering and project management. Control systems are generally used for controlling and regulating the behavior of other devices. There are different kinds of control systems available such as on-off control, logic control, linear control, PID control, etc. This book elucidates new techniques and their applications in a multidisciplinary approach. It presents researches and studies performed by experts across the globe. In this book, using case studies and examples, constant effort has been made to make the understanding of the difficult concepts of systems engineering and control systems as easy and informative as possible, for the readers.

All of the data presented henceforth, was collaborated in the wake of recent advancements in the field. The aim of this book is to present the diversified developments from across the globe in a comprehensible manner. The opinions expressed in each chapter belong solely to the contributing authors. Their interpretations of the topics are the integral part of this book, which I have carefully compiled for a better understanding of the readers.

At the end, I would like to thank all those who dedicated their time and efforts for the successful completion of this book. I also wish to convey my gratitude towards my friends and family who supported me at every step.

Editor

1

Synchronization of the fractional-order chaotic system via adaptive observer

Ruoxun Zhang* and Jingbo Gong

College of Teacher Education, Xingtai University, Hebei province, 054001, People's Republic of China

The means to design the observer for a class of fractional-order chaotic systems is investigated. A novel Lyapunov function is proposed and a robust adaptive observer is designed to synchronize a given fractional-order chaotic system. The constructed observer could guarantee the error of state converges to zero asymptotically. Simulation results demonstrate the effectiveness and robustness of the proposed scheme.

Keywords: adaptive observer; synchronization; fractional-order nonlinear system

PACS: 05.45.Gg; 05.45.Xt

1. Introduction

In the last 30 years, fractional calculus has attracted attention of many physicists and engineers. Notable contributions have been made to both the theory and applications of fractional differential equations (Bagley & Calico, 1991; Duarte & Macado, 2002; Heaviside, 1971; Ichise, Nagayanagi, & Kojima, 1971; Linares, Baillot, Oustaloup, & Ceyral, 1996; Mandelbrot & Van Ness, 1968; Oustaloup, 1995; Podlubny, 1999b; Sun, Abdlwahad, & Onaral, 1984; and references therein). Also, fractional differential equations have recently proved to be valuable tools in modeling of many physical phenomena in various fields of science and engineering.

Recently, studying fractional-order chaotic systems has become an active research field. Synchronization of fractional-order chaotic systems starts to attract increasing attention due to its potential applications in secure communication and control processing. Some approaches have been proposed to achieve chaos synchronization in fractional-order chaotic systems (Li & Deng, 2006; Li, Yu, & Luo, 2012; Liu, 2013; Lu, 2006a, 2006b; Qi, Yang, & Zhang, 2010; Wang, Wang, & Niu, 2011; Wang, Zhang, Lin, & Zhang, 2011; Wang, Zhang, & Wang, 2013; Wu, Lu, & Shen, 2009; Wu, Zhang, & Yang, 2011; Zhang & Yang, 2011a, 2011b; Zhang & Yang, 2012a, 2012b). However, the lack of the extension of the existing adaptive observers for fractional order systems is sensible.

In this paper, a novel adaptive observer is presented to solve the problem of state reconstruction for fractional-order nonlinear systems. It is shown that the proposed observer guarantees that the state estimation errors are convergent to zero. The Lyapunov approach is utilized to analyze the stability of the estimation error system. It ought to be mentioned that the proposed observer is very simple and constructive for practical applications. Moreover, utilizing fractional calculus, a new Lyapunov function is proposed for the error dynamics when the fractional-order observer is applied.

The rest of the paper is organized as follows: In Section 2, some basic concepts of fractional calculus is described and its properties are discussed. In Section 3, a novel adaptive observer is presented and a stability analysis of the fractional-order error system is given. In Section 4, the adaptive observer scheme has been tested via numerical simulations and the corresponding results are presented. Finally, some concluding remarks are drawn in Section 5.

2. Preliminaries and definitions

In this section, we introduce the definition of Reimann–Liouville fractional integration and derivative. The α th-order Reimann–Liouville fractional integration of function $f(t)$ with respect to t and the terminal value t_0 is given by

$$ {}_{t_0}D_t^{-\alpha} f(t) = \frac{1}{\Gamma(\alpha)} \int_{t_0}^{t} \frac{f(\tau)}{(t-\tau)^{1-\alpha}}\, \mathrm{d}\tau, \qquad (1)$$

and the Reimann–Liouville definition of α th-order fractional derivative is given by

$$ {}_{t_0}D_t^{\alpha} f(t) = \frac{1}{\Gamma(m-\alpha)} \frac{\mathrm{d}^m}{\mathrm{d}t^m} \int_{t_0}^{t} \frac{f(\tau)}{(t-\tau)^{\alpha-m+1}}\, \mathrm{d}\tau, \qquad (2)$$

where m is the first integer which is larger than α, i.e. $m-1 \le \alpha < m$ and Γ is the Gamma function.

*Email: xtzhrx@126.com

The material presented in the sequel is based on the aforementioned definitions of fractional differentiation and integration. Two properties of Reimann–Liouville fractional integration and derivative are given as follows (Podlubny, 1999a):

Property 1

$$ {}_{t_0}D_t^{\beta}\, {}_{t_0}D_t^{-\alpha} f(t) = {}_{t_0}D_t^{\beta-\alpha} f(t), \quad (\alpha > 0, \beta > 0) \tag{3} $$

Property 2 (Leibniz's rule)

$$ \begin{aligned} {}_0D_t^{\alpha} f^2(t) &= \sum_{k=0}^{\infty} \frac{\Gamma(1+\alpha)}{\Gamma(1+k)\Gamma(1-k+\alpha)} {}_0D_{t,}^{k} f(t) {}_0D_t^{\alpha-k} f(t) \\ &= f(t)\, {}_0D_t^{\alpha} f(t) + \sum_{k=1}^{\infty} \frac{\Gamma(1+\alpha)}{\Gamma(1+k)\Gamma(1-k+\alpha)} \\ &\quad {}_0D_t^{k} f(t) {}_0D_t^{\alpha-k} f(t). \end{aligned} \tag{4} $$

3. Main results

Considering a class of fractional-order chaotic systems described by

$$ \begin{aligned} {}_0D_t^{\alpha} x &= Ax + Bf(x), \\ y &= Cx \end{aligned} \tag{5} $$

where $0 < \alpha < 1$, $x \in R^n$ is the state vector, $y \in R^m$ is the out vector, $A \in R^{n\times n}$, $B \in R^{n\times q}$, and $C \in R^{m\times n}$ are known matrices, and $f(\cdot) \in R^q$ is a nonlinear function vector.

The adaptive observer for system (5) is constructed as follows:

$$ {}_0D_t^{\alpha} \hat{x} = A\hat{x} + Bf(\hat{x}) + \tfrac{1}{2} kB(y - C\hat{x}). \tag{6} $$

To study the synchronization between systems (5) and (6), some necessary assumptions must be made as follows:

ASSUMPTION 1 *The nonlinear function vector $f(x)$ satisfies Lipschitz conditions*:

$$ \| f(x) - f(\hat{x}) \| < l_f \| x - \hat{x} \| = l_m \| e \|, \tag{7} $$

where l_f is a Lipschitz constant, $\| \cdot \|$ denotes 2norm.

Remark 1 Based on the boundedness of chaotic systems, Assumption 1 is reasonable.

ASSUMPTION 2 *Suppose that the pair (A, C) is observable and the pair (A, B) is controllable. Further, there exists a constant vector $L \in R^{n\times m}$ to make $(A - LC)^{\mathrm{T}}P + P(A - LC) = -Q$, $B^{\mathrm{T}}P = C$, where $P = P^{\mathrm{T}}$ and $Q = Q^{\mathrm{T}}$ are two positive definite matrices.*

Let the error signals be

$$ e = x - \hat{x}, \quad \tilde{k} = k^* - k, $$

where k^* is a constant to be determined. From systems (5) and (6), the error system is as below:

$$ {}_0D_t^{\alpha} e = (A - LC)e + B(f(x) - f(\hat{x})) - \tfrac{1}{2} kBCe + LCe. \tag{8} $$

It is obvious that the synchronization between systems (5) and (6) is achieved if and only if the error system (8) is asymptotically stable at the origin.

Now, we give our main result.

THEOREM 1 *Suppose that Assumptions* 1 and 2 *hold, then the response system* (6) *can asymptotically synchronize the drive system* (5) *if the following conditions are satisfied*:

The adaptation law k is chosen as

$$ \dot{k} = c_k \| y - C\hat{x} \|^2, \tag{9} $$

where c_k is positive number to be chosen suitably.

Proof From Assumption 2, there exist two positive definite matrices $P = P^{\mathrm{T}}$ and $Q = Q^{\mathrm{T}}$ such that the following equation holds:

$$ (A - LC)^{\mathrm{T}}P + P(A - LC) = -Q, \quad B^{\mathrm{T}}P = C. \tag{10} $$

Now, consider a Lyapunov candidate function

$$ V = 2\, {}_0D_t^{\alpha-1}\, e^{\mathrm{T}}Pe + \frac{1}{2c_k}(k^* - k)^2. \tag{11} $$

By using (7), (9), and (10) and the Properties 1 and 2, the derivative of V with respect to t is given by

$$ \begin{aligned} \dot{V} &= 2D^1\, {}_0D_t^{\alpha-1} e^{\mathrm{T}}Pe - (k^* - k)\dot{k}/c_k \\ &= 2\, {}_0D_t^{\alpha} e^{\mathrm{T}}Pe - (k^* - k)\dot{k}/c_k \\ &= e^{\mathrm{T}}P\, {}_0D_t^{\alpha} e + ({}_0D_t^{\alpha} e)^{\mathrm{T}}Pe + \gamma - (k^* - k) \| y - C\hat{x} \|^2 \\ &= e^{\mathrm{T}}[(A - LC)^{\mathrm{T}}P + P(A - LC)]e + 2e^{\mathrm{T}}PB[f(x) - f(\hat{x})] \\ &\quad + 2e^{\mathrm{T}}PLCe + \gamma - ke^{\mathrm{T}}PBCe - (k^* - k)||Ce||^2 \\ &= e^{\mathrm{T}}[(A - LC)^{\mathrm{T}}P + P(A - LC)]e + 2e^{\mathrm{T}}C^{\mathrm{T}}[f(x) - f(\hat{x})] \\ &\quad + 2e^{\mathrm{T}}PLCe - ke^{\mathrm{T}}C^{\mathrm{T}}Ce + \gamma - (k^* - k) \| Ce \|^2 \\ &= e^{\mathrm{T}}[(A - LC)^{\mathrm{T}}P + P(A - LC)]e + 2e^{\mathrm{T}}C^{\mathrm{T}}[f(x) - f(\hat{x})] \\ &\quad + 2e^{\mathrm{T}}PLCe + \gamma - k^* \| Ce \|^2, \end{aligned} $$

where $D^1 = \mathrm{d}/\mathrm{d}t$, $\gamma = 2\sum_{k=1}^{\infty}[\Gamma(1+\alpha)/\Gamma(1+k)\Gamma(1-k+\alpha)]({}_0D_t^{k} e)^{\mathrm{T}}P\, {}_0D_t^{\alpha-k} e$. Since $B^{\mathrm{T}}P = C$, and $({}_0D_t^{k} e)^{\mathrm{T}} \quad P\, {}_0D_t^{\alpha-k} e = ({}_0D_t^{k} e)^{\mathrm{T}}PBB^{m}(BB^{\mathrm{T}})^{-1}\, {}_0D_t^{\alpha-k} e = ({}_0D_t^{k} Ce)^{\mathrm{T}}B^{\mathrm{T}}(BB^{\mathrm{T}})^{-1}\, {}_0D_t^{\alpha-k} e$

we suppose

$$ \gamma \le 2c_\gamma \| e \| \cdot \| Ce \| \le \frac{c_\gamma^2}{\varepsilon_\gamma} \| Ce \|^2 + \varepsilon_\gamma \| e \|^2, \tag{12} $$

in which c_γ is a positive constant, and ε_γ is a suitable positive constant.

By Assumption 1, we get the following inequalities:

$$2e^{\mathrm{T}}C^{\mathrm{T}}[f(x)-f(\hat{x})] \leq 2l_{\mathrm{f}} \parallel Ce \parallel \cdot \parallel x-\hat{x} \parallel \leq \frac{l_{\mathrm{f}}^2}{\varepsilon_{\mathrm{f}}} \parallel Ce \parallel^2 + \varepsilon_{\mathrm{f}} \parallel e \parallel^2 . \quad (13)$$

Notice that

$$2e^{\mathrm{T}}PLCe = 2e^{\mathrm{T}}PLB^{\mathrm{T}}Pe \leq 2 \parallel L^{\mathrm{T}}Pe \parallel \cdot \parallel B^{m}Pe \parallel \leq \frac{1}{\varepsilon_{\mathrm{l}}} \parallel B^{m}Pe \parallel^2 + \varepsilon_{\mathrm{l}} \parallel L^{\mathrm{T}}Pe \parallel^2 = \frac{1}{\varepsilon_{\mathrm{l}}} \parallel B^{\mathrm{T}}Pe \parallel^2 + \varepsilon_{\mathrm{l}}\lambda_{\max}(PLL^{\mathrm{T}}P) \parallel e \parallel^2,$$

where ε_{f} and ε_{l} are two suitable positive constants.

Then, we get

$$\dot{V} \leq e^{\mathrm{T}}[(A-LC)^{\mathrm{T}}P + P(A-LC) + (\varepsilon_{\gamma} + \varepsilon_{\mathrm{f}} + \varepsilon_{\mathrm{l}}\lambda_{\max} (PLL^{\mathrm{T}}P)I_n]e + \left(\frac{c_{\gamma}^2}{\varepsilon_{\gamma}} + \frac{l_{\mathrm{f}}^2}{\varepsilon_{\mathrm{f}}} + \frac{1}{\varepsilon_{\mathrm{l}}} - k^*\right) \parallel Ce \parallel^2 .$$

Let

$$k^* = \frac{c_{\gamma}^2}{\varepsilon_{\gamma}} + \frac{l_{\mathrm{f}}^2}{\varepsilon_{\mathrm{f}}} + \frac{1}{\varepsilon_{\mathrm{l}}},$$

then,

$$\dot{V} \leq e^{\mathrm{T}}[(A-LC)^{\mathrm{T}}P + P(A-LC) + (\varepsilon_{\gamma} + \varepsilon_{\mathrm{f}} + \varepsilon_{\mathrm{l}}\lambda_{\max}(PLL^{\mathrm{T}}P))I_n]e.$$

From the above inequality, ε_{γ}, ε_m, and ε_{l} can be chosen to be small enough such that $\dot{V} < 0$ for all nonzero error. Therefore, the error system (7) is asymptotically stable at the origin, i.e. the response system (6) can asymptotically synchronize the drive system.

4. Simulation

In this section, in order to show the effectiveness of the proposed scheme in the preceding section, a numerical example on the fractional-order chaotic Chen system will be provided.

The fractional-order chaotic Chen system (Li & Chen, 2004)

$$\begin{aligned} D_t^{\alpha}x_1 &= a(x_2 - x_1) \\ D_t^{\alpha}x_2 &= dx_1 - x_1x_3 + cx_2 \\ D_t^{\alpha}x_3 &= x_1x_2 - bx_3. \end{aligned} \quad (14)$$

When $\alpha = 0.93$, $a = 35$, $b = 3$, $c = 28$, and $d = -7$, the fractional-order Chen system behaves chaotically. Figure 1 shows the chaotic trajectory of (14). Rewrite the fractional-order chaotic Chen system as follows:

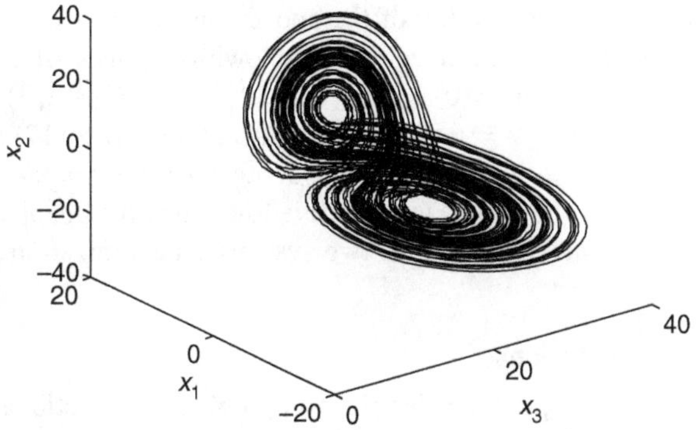

Figure 1. Chaotic attractors in the fractional-order Chen chaotic system with $\alpha = 0.93$.

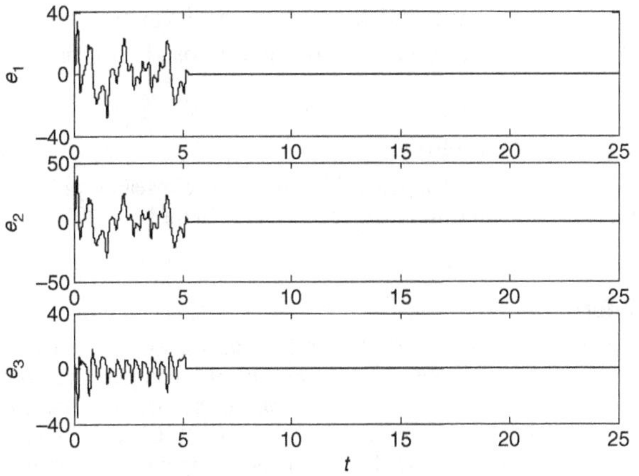

Figure 2. Synchronization errors e_1, e_2, e_3 of two fractional-order Chen chaotic systems.

$$D_t^{\alpha}x = Ax + Bf(x), \quad (15)$$

where

$$A = \begin{pmatrix} -35 & 35 & 0 \\ -7 & 28 & 0 \\ 0 & 0 & -3 \end{pmatrix}, \quad x = \begin{pmatrix} x_1 \\ x_2 \\ x_3 \end{pmatrix}, \quad B = \begin{pmatrix} 0 & 0 \\ 1 & 0 \\ 0 & 1 \end{pmatrix},$$

$$f(x) = \begin{pmatrix} -x_1x_3 \\ x_1x_2 \end{pmatrix}.$$

The output vector is $y = Cx$. Matrix C is chosen

$$C = \begin{pmatrix} 0 & 1 & 0 \\ 0 & 0 & 1 \end{pmatrix}.$$

Clearly, the pair (A, C) is observable. Let system (15) be a master system. Applying the results in Section 3, an adaptive observer is designed as follows:

$$D_t^{\alpha}\hat{x} = A\hat{x} + Bf(\hat{x}) + u, \quad (16)$$

$$u = \tfrac{1}{2}kB(y - C\hat{x}), \quad \dot{k} = c_k \parallel y - C\hat{x} \parallel^2 . \quad (17)$$

To confirm the validity of the above conclusion, we give numerical simulation with the following choices of the initial conditions: $x(0) = (0.2, 0.5, 6)^T, \hat{x}(0) = (1, -2, 3)^T$, and $k(0) = 0$, $\alpha = 0.93$. When an adaptive part u (17) is added into the system at the time of 5th, the error $e = x - \hat{x}$ curves are shown in Figure 2. It is found that the adaptive part can quickly render the two systems synchronization.

5. Conclusions

In this paper, the synchronization problem for fractional-order chaotic systems is investigated. An adaptive observer-based slave system is designed to synchronize a given chaotic master system. Based on the Lyapunov stability theorem, the global synchronization between the master and slave systems is ensured. Simulation results demonstrate the effectiveness and robustness of the proposed scheme.

Acknowledgements

The present work is supported by the science research project of Hebei higher education institutions under Grant No. z2012021.

References

Bagley, R. L., & Calico R. A. (1991). Fractional-order state equations for the control of viscoelastically damped structures. *Journal of Guidance, Control, and Dynamics, 14*, 304–311.

Duarte, F. B. M., & Macado, J. A. T. (2002). Chaotic phenomena and fractional-order dynamics in the trajectory control of redundant manipulators. *Nonlinear Dynamics, 29*, 315–342.

Heaviside, O. (1971). *Electromagnetic theory*. New York: Chelsea.

Ichise, M., Nagayanagi, Y., & Kojima, T. (1971). An analog simulation of non integer order transfer functions for analysis of electrode processes. *Journal of Electroanalytical Chemistry and Interfacial Electrochemistry, 33*, 253–265.

Li, C. G., & Chen, G. R. (2004). Chaos in the fractional order Chen system and its control. *Chaos, Solitions & Fractals, 22*, 549–554.

Li, C. L., Yu, S. M., & Luo, X. S. (2012). Fractional-order permanent magnet synchronous motor and its adaptive chaotic control. *Chinese Physics B, 21*, 100506.

Li, C. P., & Deng, W. H. (2006). Chaos synchronization of fractional-order differential systems. *International Journal of Modern Physics B, 20*, 791–803.

Linares, H., Baillot, Ch., Oustaloup, A., & Ceyral, Ch. (1996, July). *Generation of a fractal ground: Application in robotics*. International Congress in IEEE-SMC CESA'96 IMACS Multiconf., Lille.

Liu, J. G. (2013). A novel study for impulsive synchronization of fractional-order chaotic systems. *Chinese Physics B, 22*, 060510.

Lu, J. G. (2006a) Nonlinear observer design to synchronize fractional-order chaotic systems via a scalar transmitted signal. *Physica A, 359*, 107–118.

Lu, J. G. (2006b). Synchronization of a class of fractional-order chaotic systems via a scalar transmitted signal. *Chaos, Solitions & Fractals, 27*, 519–525.

Mandelbrot, B., & Van Ness, J.W. (1968). Fractional Brownian motions, fractional noises and applications. *SIAM Review, 10*, 422–437.

Oustaloup, A. (1995). *La Derivation Non Entiere: Theorie, Synthase et Applications*. Paris: Editions Hermes.

Podlubny, I. (1999a). *Fractional differential equations*. New York: Academic Press.

Podlubny, I. (1999b). Fractional-order systems and PI $\lambda D\mu$ controllers. *IEEE Transactions on Automatic Control, 44*, 208–213.

Qi, D. L., Yang, J., & Zhang, J. L. (2010). The stability control of fractional order unified chaotic system with sliding mode control theory. *Chinese Physics B, 19*, 100506.

Sun, H. H., Abdelwahad, A. A., & Onaral, B. (1984). Linear approximation of transfer function with a pole of fractional order. *IEEE Transactions on Automatic Control, 29*, 441–444.

Wang, D. F., Zhang, J. Y., & Wang, X. Y. (2013). Synchronization of uncertain fractional-order chaotic systems with disturbance based on fractional terminal sliding mode controller. *Chinese Physics B, 22*, 040507.

Wang, M. J., Wang, X. Y., & Niu, Y. J. (2011). Projective synchronization of a complex network with different fractional order chaos nodes. *Chinese Physics B, 20*, 010508.

Wang, X. Y., Zhang, Y. L., Lin, D., & Zhang, N. (2011). Impulsive synchronisation of a class of fractional-order hyperchaotic systems. *Chinese Physics B, 20*, 030506.

Wu, C. J., Zhang, Y. B., & Yang, N. N. (2011). The synchronization of a fractional order hyperchaotic system based on passive control. *Chinese Physics B, 20*, 060505.

Wu, X. J., Lu, H. T., & Shen, S. L. (2009). Synchronization of a new fractional-order hyperchaotic system. *Physics Letters A, 373*, 2329–2337.

Zhang, R. X., & Yang, S. P. (2011a). Adaptive stabilization of an incommensurate fractional-order chaotic system via a single state controller. *Chinese Physics B, 20*, 090512.

Zhang, R. X., & Yang, S. P. (2011b). Adaptive synchronization of fractional-order chaotic systems via a single driving variable. *Nonlinear Dynamics, 66*, 831–837.

Zhang, R. X., & Yang, S. P. (2012a). A single adaptive controller with one variable for synchronization of fractional-order chaotic systems. *Chinese Physics B, 21*, 080505.

Zhang, R. X., & Yang, S. P. (2012b). Modified adaptive controller for synchronization of incommensurate fractional-order chaotic system. *Chinese Physics B, 21*, 030505.

2

A nonlinear oscillator with strange attractors featured Sinai-Ruelle-Bowen measure

Fengjuan Chen*, Shujiao Jin and Liqun Zhou

College of Mathematics, Physics and Information Engineering, Zhejiang Normal University, Jinhua, Zhejiang 321004, People's Republic of China

This paper studies a class of Duffing oscillator with a forcing parameter ε. We obtain Hénon-like attractors, rank one attractors, and periodic sinks as ε changes. Hénon-like attractors and rank one attractors are chaotic in the sense of SRB measures, while periodic sinks represent stable dynamics with a basin of positive Lebesgue measure. As $\varepsilon \to 0$, three attractors construct a dynamical pattern repeating with certain period. Through numerical simulations, we observe three attractors perfectly as well as the dynamical pattern.

Keywords: homoclinic tangle; heteroclinic tangle; Hénon-like attractor; rank one attractor; SRB measure

1. Introduction

Duffing oscillator is known as a simple model displaying rich nonlinear dynamics, such as homoclinic tangles and horseshoes. In general, for an equation with homoclinic solution, periodic perturbation often leads to the intersection of stable and unstable manifolds. Therefore, Melnikov method is an efficient tool to detect horseshoes. However, it is possible for stable and unstable manifolds pulling apart. In this case, Melnikov method fails. Are there any important dynamics in this situation? What methods can be used to clear the important dynamics? How to deal with the intersection and pulling apart of stable and unstable manifolds together? Recently, Wang and Oksasoglu (2011) provided a complete theory on two-dimensional homoclinic tangles. With a new return map, they proved Hénon-like attractors and periodic sinks in homoclinic tangles beyond horseshoes. Further, for pulling apart, they obtained rank one attractors.

This paper studies a special class of Duffing oscillator, concentrating on its strange attractors as the forcing parameter changes. First, we present homoclinic solutions and heteroclinic solutions in unperturbed equation. Then, with double homoclinic tangle theory (Wang, 2009), we obtain Hénon-like attractors, rank one attractors, and periodic sinks, along with two dynamical patterns. With heteroclinic tangle theory (Chen, Oksasoglu, & Wang, 2013), we show Hénon-like attractors, periodic sinks, and transient tangles, also with a dynamical pattern. During this course, horseshoes are always participants, though we cannot observe it in numerical simulations.

2. Dynamics of perturbed homoclinic solutions

Although homoclinic tangles was discovered earlier by Poincaré (1899), the overall dynamics are far from understood. In Wang and Oksasoglu (2011), the authors presented a systematic study on homoclinic tangles for equation:

$$\begin{aligned}\frac{\mathrm{d}x}{\mathrm{d}t} &= f(x,y) + \varepsilon P(x,y,t),\\ \frac{\mathrm{d}y}{\mathrm{d}t} &= g(x,y) + \varepsilon Q(x,y,t),\end{aligned} \tag{1}$$

where $f(x,y)$, $g(x,y)$, $P(x,y,t)$, $Q(x,y,t)$ are $C^r (r \geq 3)$ functions, and $P(x,y,t+T) = P(x,y,t)$, $Q(x,y,t+T) = Q(x,y,t)$ for a constant $T > 0$.

Assume that the unperturbed equation, that is, $\varepsilon = 0$ in Equation (1), has a dissipative saddle point O connected with a homoclinic solution $\ell(t)$. If O is non-resonant, then a new return map R was derived in Wang and Oksasoglu (2011). Let Σ^- be a section, and (θ, x) be variables on Σ^-. Denote $(\theta_1, x_1) = \mathcal{R}(\theta, x)$. Then

$$\begin{aligned}\theta_1 &= \theta + a - \frac{\omega}{\beta}\ln F(\theta,x,\varepsilon) + \mathcal{O}_{\theta,x,h}(\varepsilon),\\ x_1 &= bF^{-\alpha/\beta}(\theta,x,\varepsilon),\end{aligned} \tag{2}$$

where a, b are constants, $\alpha < 0, \beta > 0$ are two eigenvalues of saddle O, and

$$F(\theta,x,\varepsilon) = \mathcal{W}(\theta) + kx + E(\theta,\varepsilon) + \mathcal{O}_{\theta,x,h}(\varepsilon). \tag{3}$$

*Corresponding author. Email: fjchen@zjnu.cn

In Equation (3), $\mathcal{W}(\theta)$ is a Melnikov function defined as

$$\mathcal{W}(\theta) = \int_{-\infty}^{+\infty} [P(\ell(t), t+\theta), Q(\ell(t), t+\theta)] \cdot \tau_{\ell(t)}^{\perp} \mathrm{e}^{-\int_0^t \mathbb{E}_\ell(s)\,\mathrm{d}s}\,\mathrm{d}t, \tag{4}$$

k is a small constant, and $E(\theta,\varepsilon)$, $\mathcal{O}_{\theta,x,h}(\varepsilon)$ are small error terms. In Equation (4), $\tau_{\ell(t)}^{\perp}$ is a vector perpendicular to the tangent vector of $\ell(t)$ at time t, and $\mathbb{E}_\ell(t)$ is the expansion rate of solutions in the neighborhood of $\ell(t)$.

From Equation (2), the domain of $\mathcal{R}$ depends on the image of $F(\theta, x, \varepsilon)$, which is related to the zeros of $\mathcal{W}(\theta)$. Denote

$$M = \max_{\theta\in S^1} \mathcal{W}(\theta), \quad m = \min_{\theta\in S^1} \mathcal{W}(\theta), \tag{5}$$

where $S^1 = \mathbb{R}/\{nT\}$ for $n \in \mathbb{Z}$.

THEOREM 2.1 (Wang, 2011; Wang & Oksasoglu, 2011)

(i) *If $\mathcal{W}(\theta)$ is a Morse function satisfying $m < 0 < M$, then Hénon-like attractors, periodic sinks, and horseshoes of infinitely symbols occur in the space of ε. Moreover, as $\varepsilon \to 0$, there is a dynamical pattern repeating with period $\mathrm{e}^{\beta T}$, where β is the unstable eigenvalue of saddle O.*
(ii) *If $m > 0$, then rank one attractors occur for large ω, where ω is the frequency of perturbation.*

Remark 1 In Theorem 2.1(ii), $m > 0$ implies the pulling apart of stable and unstable manifolds. At this time, $\mathcal{R}$ defines on the full Σ^-. It is a rank one map admitting rank one attractors (Wang & Oksasoglu, 2008) characterized by SRB measures (Young, 2002; Benedicks & Young, 1993).

Denote two Melnikov functions by $\mathcal{W}^+(\theta)$ and $\mathcal{W}^-(\theta)$. Let

$$\begin{aligned} M^+ &= \max_{\theta\in S^1} \mathcal{W}^+(\theta), \quad m^+ = \min_{\theta\in S^1} \mathcal{W}^+(\theta), \\ M^- &= \max_{\theta\in S^1} \mathcal{W}^-(\theta), \quad m^- = \min_{\theta\in S^1} \mathcal{W}^-(\theta). \end{aligned} \tag{6}$$

THEOREM 2.2 (Wang, 2009; Wang & Oksasoglu, 2010) *Assume that $\mathcal{W}^+(\theta)$ and $\mathcal{W}^-(\theta)$ are two Morse functions.*

(i) *If $m^+, m^- < 0 < M^+, M^-$, then Equation (1) shows the mixture of two homoclinic tangles. It contains Hénon-like attractors, rank one attractors, and periodic sinks.*
(ii) *If $m^+ < 0 < M^+, m^- > 0$ (or $m^- < 0 < M^-$, $m^+ > 0$), then Equation (1) shows one homoclinic tangle and one rank one attractor. One-sided Hénon-like attractors, one-sided rank one attractors, and one-sided periodic sinks are created.*
(iii) *If $m^+ < 0 < M^+, M^- < 0$ (or $m^- < 0 < M^-$, $M^+ < 0$), then Equation (1) shows one tangle mixed with one rank one attractor. Hénon-like attractors, rank one attractors, and periodic sinks, including two-sided and one-sided, are permitted.*
(iv) *If $m^+ > 0, m^- > 0$, then Equation (1) shows two rank one attractors.*
(v) *If $m^+ > 0, M^- < 0$ (or $m^- > 0, M^+ < 0$ or $M^+ < 0, M^- < 0$), then Equation (1) shows only one rank one attractor.*

As $\varepsilon \to 0$, each case shows a repetitive dynamical pattern.

Remark 2 In Theorem 2.2, one-sided attractor means that the attractor is only on the left side or right side of saddle O. If it is on both sides, we call it two-sided attractor.

The setting of Theorems 2.1 and 2.2 are homoclinic solutions. Intuitively, heteroclinic cycle, formed by heteroclinic solutions, looks like a homoclinic loop. Therefore, perturbation to heteroclinic cycle also leads to complicated dynamics (Chen et al. 2013). Precisely, given two saddles O and O_*, connecting with two heteroclinic solutions $\ell(t)$ and $\ell_*(t)$ in unperturbed equation of (1). Let $\mathcal{W}(\theta)$, $\mathcal{W}_*(\theta)$ be two Melnikov functions with four extrema M, m, and M_*, m_*. We have

THEOREM 2.3 (Chen et al. 2013) *Suppose $\mathcal{W}(\theta)$ and $\mathcal{W}_*(\theta)$ are two Morse functions satisfying $m < 0 < M$, $m_* < 0 < M_*$, and O, O_* are both dissipative and non-resonant saddles. Then, Hénon-like attractors, periodic sinks, and horseshoes of infinitely many symbols occur for Equation (1) as $\varepsilon > 0$ changes.*

Remark 3 For homoclinic tangles, there is a well-defined dynamical pattern repeating itself periodically. However, for heteroclinic tangles, the repetitive dynamical pattern depends on two unstable eigenvalues. If they are rationally related, we have a repetitive pattern. If not, the repetitive pattern disappears.

3. Strange attractors in a class of duffing oscillator

In this section we study a nonlinear oscillator

$$\ddot{x} + (\alpha + \gamma x^2)\dot{x} + \beta x + \delta x^3 + \varepsilon[\eta x\dot{x} + (x^2-1)^2 \cos\omega t] = 0, \tag{7}$$

where $\alpha > 0$, γ, β, δ, and $\varepsilon > 0$, η, ω are parameters. We are interested in strange attractors for every $\varepsilon > 0$.

In history, many studies were carried out for Equation (7) on some parameters. For example, when $\varepsilon = 0$, Holmes and Rand (1980) studied its phase portrait and bifurcation set in (α, β) space. When $\gamma = \eta = 0, \varepsilon \neq 0$, Holmes (1979) discussed strange attractors with Poincaré map only for

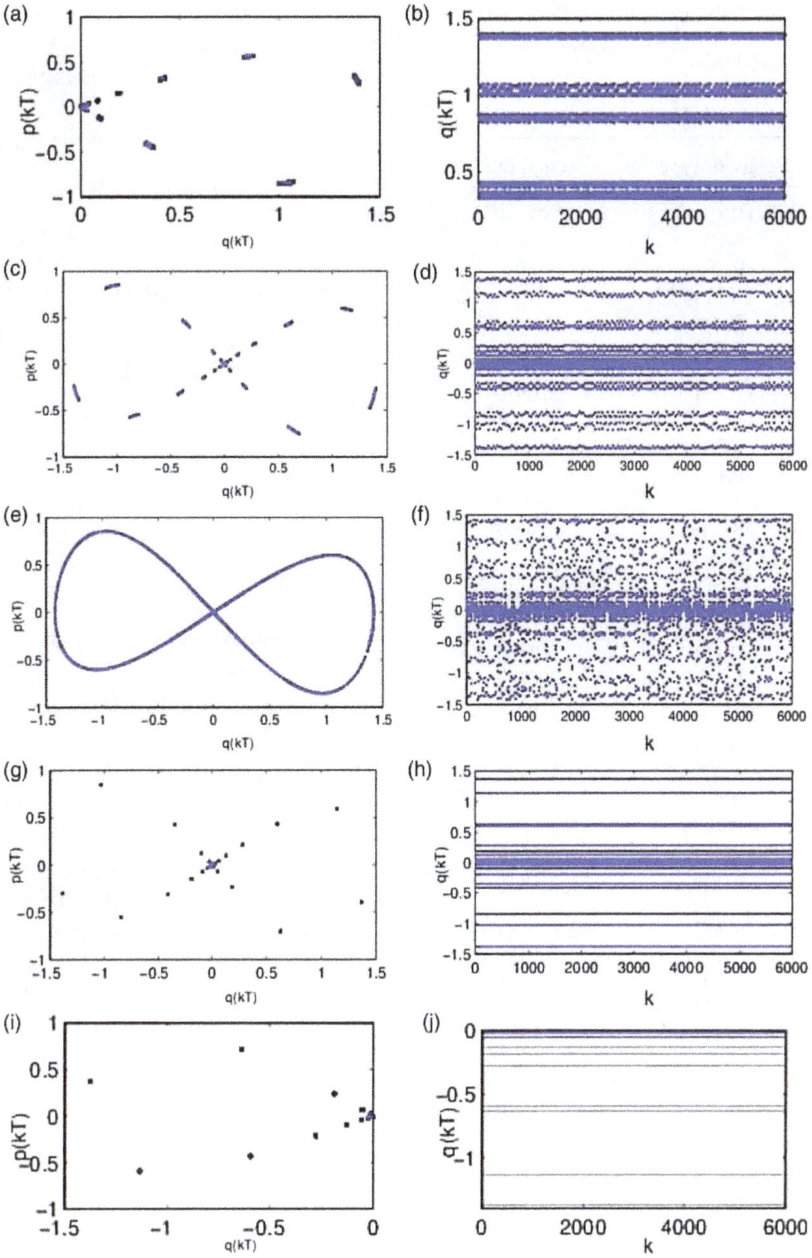

Figure 1. Case (i) of Equation (9). (a) One-sided Hénon-like attractor at $\varepsilon = 2.3 \times 10^{-4}$; (b) Time evolution of (a); (c) Two-sided Hénon-like attractor at $\varepsilon = 1.577 \times 10^{-4}$; (d) Time evolution of (c); (e) Two-sided rank one attractor at $\varepsilon = 3.2 \times 10^{-4}$; (f) Time evolution of (e); (g) Two-sided periodic sink at $\varepsilon = 3.3 \times 10^{-4}$; (h) Time evolution of (g); (m) One-sided periodic sink at $\varepsilon = 2.2 \times 10^{-4}$; and (n) Time evolution of (m).

certain ε. According to what we know, few studies are on strange attractors for every ε.

Write $\dot{x} = y$, Equation (7) becomes

$$\begin{aligned} \dot{x} &= y, \\ \dot{y} &= -\beta x - \delta x^3 - (\alpha + \gamma x^2)y \\ &\quad - \varepsilon[\eta xy + (x^2 - 1)^2 \cos \omega t]. \end{aligned} \tag{8}$$

We consider two cases for Equation (8). One case is $(\beta, \delta) = (-1, 1)$. In this case, we present dynamics of double homoclinic tangles. The other case is $(\beta, \delta) = (1, -1)$. This case leads to the heteroclinic tangles.

3.1. *Homoclinic tangles*

For $(\beta, \delta) = (-1, 1)$, Equation (8) simplifies to

$$\begin{aligned} \dot{x} &= y, \\ \dot{y} &= x - x^3 - (\alpha + \gamma x^2)y - \varepsilon[\eta xy + (x^2 - 1)^2 \cos \omega t]. \end{aligned} \tag{9}$$

Table 1. The dynamical pattern of Equation (9, case i).

$\alpha = 0.5, \gamma = -0.620831624826$
$\eta = 0.0, \omega = 2\pi$
Theoretical multiplicity $e^{\lambda_2} \approx 2.1832$

ε	Behavior type	Actual ratio
3.4×10^{-4}	HL(2)	–
3.3×10^{-4}	S(2)	–
3.2×10^{-4}	R(2)	–
2.3×10^{-4}	HL(1)	–
2.2×10^{-4}	S(1)	–
2.1×10^{-4}	R(2)	–
1.577×10^{-4}	HL(2)	2.1560
1.552×10^{-4}	S(2)	2.1263
1.477×10^{-4}	R(2)	2.1666
1.057×10^{-4}	HL(1)	2.1760
1.006×10^{-4}	S(1)	2.1869
1.000×10^{-4}	R(2)	2.1000
7.231×10^{-5}	HL(2)	2.1809
7.114×10^{-5}	S(2)	2.1816
6.775×10^{-5}	R(2)	2.1801
4.823×10^{-5}	HL(1)	2.1916
4.588×10^{-5}	S(1)	2.1927
4.586×10^{-5}	R(2)	2.1805
3.313×10^{-5}	HL(2)	2.1826
3.260×10^{-5}	S(2)	2.1822
3.104×10^{-5}	R(2)	2.1827
2.215×10^{-5}	HL(1)	2.1774
2.103×10^{-5}	S(1)	2.1816
2.101×10^{-5}	R(2)	2.1828
1.518×10^{-5}	HL(2)	2.1825
1.494×10^{-5}	S(2)	2.1821
1.422×10^{-5}	R(2)	2.1828
1.019×10^{-5}	HL(1)	2.1737
9.633×10^{-6}	S(1)	2.1831
9.629×10^{-6}	R(2)	2.1820
6.955×10^{-6}	HL(2)	2.1826
6.844×10^{-6}	S(2)	2.1829
6.517×10^{-6}	R(2)	2.1820
4.661×10^{-6}	HL(1)	2.1862
4.415×10^{-6}	S(1)	2.1818
4.411×10^{-6}	R(2)	2.1830

First, we need a dissipative saddle and homoclinic solution for the unperturbed equation

$$\begin{aligned} \dot{x} &= y, \\ \dot{y} &= x - x^3 - (\alpha + \gamma x^2)y. \end{aligned} \tag{10}$$

PROPOSITION 3.1 *For sufficiently small $\alpha > 0$, there is a γ_α such that Equation (10) has a saddle point $O(0,0)$ and two homoclinic solutions.*

Proof $O(0,0)$ is an equilibrium point of Equation (10) with eigenvalues

$$\lambda_1 = \frac{-\alpha - \sqrt{\alpha^2+4}}{2}, \quad \lambda_2 = \frac{-\alpha + \sqrt{\alpha^2+4}}{2}.$$

So, for $\alpha > 0$, we have $\lambda_1 < 0, \lambda_2 > 0$, and hence $O(0,0)$ is a saddle. Since $\lambda_1 + \lambda_2 = -\alpha < 0$, $O(0,0)$ is a dissipative saddle. The non-resonant is guaranteed by the irrational $\sqrt{\alpha^2+4}$.

For small $\alpha > 0$, we rewrite Equation (10) as

$$\begin{aligned} \dot{x} &= y, \\ \dot{y} &= x - x^3 - \alpha(1 + \gamma_\alpha x^2)y, \end{aligned} \tag{11}$$

where $\gamma_\alpha = \gamma/\alpha$. Equation (11) is an auto-perturbation of equation

$$\begin{aligned} \dot{x} &= y, \\ \dot{y} &= x - x^3. \end{aligned} \tag{12}$$

By simple computation, Equation (12) has two homoclinic solutions: $\ell_1(t) = \{(a_1(t), b_1(t)) : t \in \mathbb{R}\}$ and $\ell_2(t) = \{(-a_1(t), -b_1(t)) : t \in \mathbb{R}\}$, where

$$a_1(t) = \sqrt{2}\,\mathrm{sech}t, \; b_1(t) = -\sqrt{2}\mathrm{sech}t \cdot \tanh t.$$

The standard Melnikov function on $\ell_1(t)$ is

$$M_1 = \int_{-\infty}^{\infty} (1 + \gamma_\alpha a_1^2(t)) b_1^2(t)\,\mathrm{d}t = \frac{4}{3} + \frac{16}{15}\gamma_\alpha. \tag{13}$$

Thus, we have $M_1 = 0$ at $\gamma_\alpha = -\frac{5}{4}$, and a new homoclinic solution arises for Equation (10). Denote this homoclinic solution as $\tilde{\ell}_1(t) = \{(\tilde{a}_1(t), \tilde{b}_1(t)) : t \in \mathbb{R}\}$. Note that $\tilde{\ell}_2(t) = \{(-\tilde{a}_1(t), -\tilde{b}_1(t)) : t \in \mathbb{R}\}$ is also a homoclinic solution for Equation (10). They are all homoclinic to $O(0,0)$. This completes the proof of Proposition 3.1. ■

In the following, we fix α, γ_α as in Proposition 3.1, and consider Equation (9):

$$\begin{aligned} \dot{x} &= y, \\ \dot{y} &= x - x^3 - \alpha(1 + \gamma_\alpha x^2)y - \varepsilon[\eta x y + (x^2-1)^2 \cos \omega t]. \end{aligned} \tag{14}$$

Comparing Equation (14) with Equation (1), we have $P(x,y,t) = 0$, $Q(x,y,t) = -[\eta x y + (x^2-1)^2 \cos \omega t]$. Therefore, the Melnikov function (4) on $\tilde{\ell}_1(t)$ is

$$\mathcal{W}_1(\theta) = A_1\eta + \sqrt{B_1^2 + C_1^2} \cos(\omega\theta + \varphi_1), \tag{15}$$

where, for $(x,y) \in \tilde{\ell}_1(t)$,

$$\begin{aligned} A_1 &= \int_{-\infty}^{\infty} \frac{x y^2 e^{-\int_0^t \mathbb{E}_{\tilde{\ell}_1}(s)\,ds}}{\sqrt{y^2 + [x - x^3 - \alpha(1+\gamma_\alpha x^2)y]^2}}\,\mathrm{d}t, \\ B_1 &= \int_{-\infty}^{\infty} \frac{(x^2-1)^2 y e^{-\int_0^t \mathbb{E}_{\tilde{\ell}_1}(s)\,ds} \cos \omega t}{\sqrt{y^2 + [x - x^3 - \alpha(1+\gamma_\alpha x^2)y]^2}}\,\mathrm{d}t, \\ C_1 &= \int_{-\infty}^{\infty} \frac{(x^2-1)^2 y e^{-\int_0^t \mathbb{E}_{\tilde{\ell}_1}(s)\,ds} \sin \omega t}{\sqrt{y^2 + [x - x^3 - \alpha(1+\gamma_\alpha x^2)y]^2}}\,\mathrm{d}t, \\ \tan \varphi_1 &= \frac{C_1}{B_1}, \end{aligned} \tag{16}$$

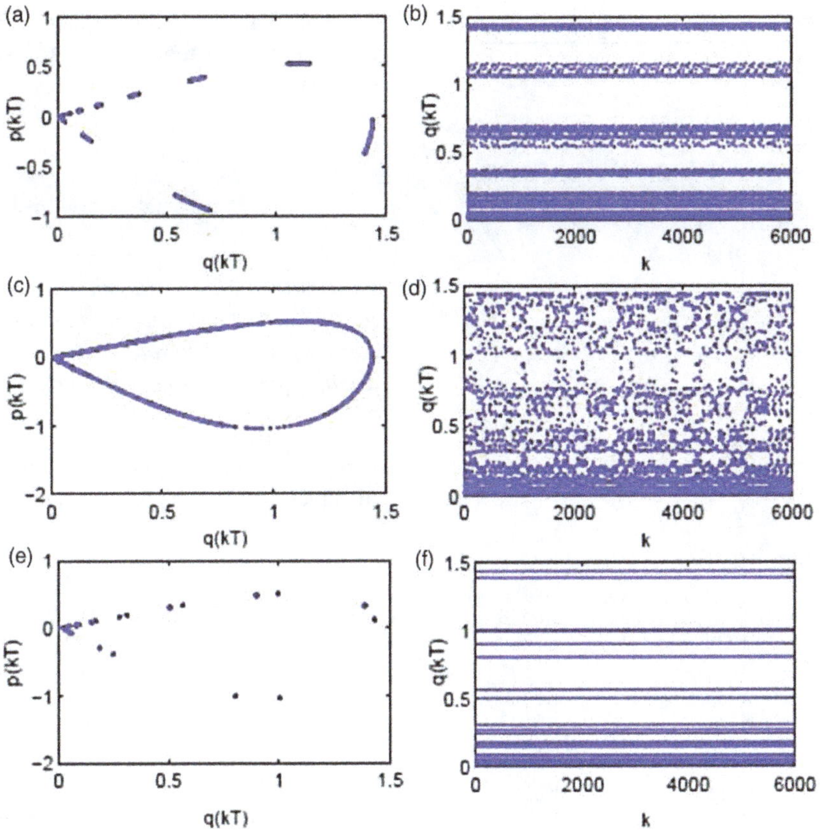

Figure 2. Case (v) of Equation (9). (a) One-sided Hénon-like attractor at $\varepsilon = 5.8 \times 10^{-3}$; (b) Time evolution of (a); (c) One-sided rank one attractor at $\varepsilon = 4.2 \times 10^{-3}$; (d) Time evolution of (c); (e) One-sided periodic sink at $\varepsilon = 5.1 \times 10^{-3}$; and (f) Time evolution of (e).

and

$$\mathbb{E}_{\tilde{\ell}_1}(t) = \tau_{\tilde{\ell}_1}^{\perp}(t)\begin{pmatrix} 0 & 1 \\ 1-3x^2-2\alpha\gamma_\alpha xy & -\alpha(1+\gamma_\alpha x^2)\end{pmatrix}\tilde{\tau}_{\ell_1}^{\perp}(t),$$

with

$$\tau_{\tilde{\ell}_1}^{\perp}(t) = \frac{(x-x^3-\alpha(1+\gamma_\alpha x^2)y,\ -y)}{\sqrt{y^2+[x-x^3-\alpha(1+\gamma_\alpha x^2)y]^2}}.$$

Since $\tilde{\ell}_2(t) = -\tilde{\ell}_1(t)$, $t \in \mathbb{R}$, the Melnikov function on $\tilde{\ell}_2(t)$ is

$$\mathcal{W}_2(\theta) = -A_1\eta + \sqrt{B_1^2+C_1^2}\cos(\omega\theta+\varphi_1). \tag{17}$$

From Equations (15) and (17), the four extrema of $\mathcal{W}_1(\theta), \mathcal{W}_2(\theta)$ are

$$M_1 = A_1\eta + \sqrt{B_1^2+C_1^2},$$

$$m_1 = A_1\eta - \sqrt{B_1^2+C_1^2},$$

$$M_2 = -A_1\eta + \sqrt{B_1^2+C_1^2},$$

$$m_2 = -A_1\eta - \sqrt{B_1^2+C_1^2}.$$

With Theorem 2.2, only two cases (i): $m_1 < 0 < M_1, m_2 < 0 < M_2$, and (v): $m_1 > 0, M_2 < 0$ or $m_2 > 0$, $M_1 < 0$ hold for real η. Solve case (i), we have

$$-\frac{\sqrt{B_1^2+C_1^2}}{A_1} < \eta < \frac{\sqrt{B_1^2+C_1^2}}{A_1}. \tag{18}$$

Solve case (v) of $m_1 > 0, M_2 < 0$, we have

$$\eta > \frac{\sqrt{B_1^2+C_1^2}}{A_1}, \tag{19}$$

and from $m_2 > 0, M_1 < 0$, we have

$$\eta < -\frac{\sqrt{B_1^2+C_1^2}}{A_1}. \tag{20}$$

With Theorem 2.2(i), for η in Equation (18), Equation (9) exhibits mixture of two homoclinic tangles. Consequently, both one-sided and two-sided Hénon-like attractors, rank one attractors, and periodic sinks occur for Equation (9), see Figure 1. The corresponding dynamical pattern is listed in Table 1, which repeats itself with period $e^{\lambda_2} \approx 2.1832$. The symbols "HL(1), HL(2)" stand for respectively one-sided and two-sided Hénon-like attractors. The same meaning also applies to symbols "S(1), S(2), R(1), R(2)." For η in Equation (19) or Equation (20), Theorem 2.2(v) tells us

Table 2. The dynamical pattern of Equation (9, case v).

$\alpha = 1.0, \gamma = -1.219038499970$
$\eta = 0.3, \omega = 2\pi$
Theoretical multiplicity $e^{\lambda_2} \approx 1.85527$

ε	Behavior type	Actual ratio
5.8×10^{-3}	HL(1)	–
5.1×10^{-3}	S(1)	–
4.4×10^{-3}	HL(1)	–
4.2×10^{-3}	R(1)	–
3.8×10^{-3}	S(1)	–
3.7×10^{-3}	R(1)	–
3.6×10^{-3}	HL(1)	–
3.4×10^{-3}	S(1)	–
3.3×10^{-3}	HL(1)	–
3.2×10^{-3}	S(1)	–
3.167×10^{-3}	HL(1)	1.8314
2.782×10^{-3}	S(1)	1.8332
2.389×10^{-3}	HL(1)	1.8418
2.283×10^{-3}	R(1)	1.8397
2.051×10^{-3}	S(1)	1.8528
1.999×10^{-3}	R(1)	1.8509
1.958×10^{-3}	HL(1)	1.8396
1.847×10^{-3}	S(1)	1.8408
1.819×10^{-3}	HL(1)	1.8142
1.746×10^{-3}	S(1)	1.8328
1.707×10^{-3}	HL(1)	1.8553
1.499×10^{-3}	S(1)	1.8559
1.288×10^{-3}	HL(1)	1.8548
1.230×10^{-3}	R(1)	1.8561
1.104×10^{-3}	S(1)	1.8578
1.099×10^{-3}	R(1)	1.8189
1.054×10^{-3}	HL(1)	1.8577
9.949×10^{-4}	S(1)	1.8565
9.832×10^{-4}	HL(1)	1.8501
9.403×10^{-4}	S(1)	1.8569
9.195×10^{-4}	HL(1)	1.8564
8.054×10^{-4}	S(1)	1.8612
6.929×10^{-4}	HL(1)	1.8589
6.629×10^{-4}	R(1)	1.8555
5.947×10^{-4}	S(1)	1.8564
5.916×10^{-4}	R(1)	1.8577
5.678×10^{-4}	HL(1)	1.8563
5.359×10^{-4}	S(1)	1.8565
5.296×10^{-4}	HL(1)	1.8565
5.063×10^{-4}	S(1)	1.8572

only one rank one attractor for Equation (9). Therefore, one-sided Hénon-like attractors, one-sided rank one attractors, and one-sided periodic sinks emerge expectedly, see Figure 2. Table 2 is the corresponding dynamical pattern.

3.2. Heteroclinic tangles

For $(\beta, \delta) = (1, -1)$, Equation (8) is changed to

$$\dot{x} = y,$$
$$\dot{y} = -x + x^3 - (\alpha + \gamma x^2)y - \varepsilon[\eta xy + (x^2 - 1)^2 \cos \omega t]. \tag{21}$$

To apply Theorem 2.3 to Equation (21), we need two heteroclinic solutions together with two saddles for unperturbed equation

$$\begin{aligned} \dot{x} &= y, \\ \dot{y} &= -x + x^3 - (\alpha + \gamma x^2)y. \end{aligned} \tag{22}$$

Proposition 3.2 *For sufficiently small $\alpha > 0$, there is a γ_α such that Equation (22) has two heteroclinic solutions associated with two saddles $O_1(-1, 0)$ and $O_2(1, 0)$, which are dissipative and non-resonant.*

Proof $O_1(-1, 0)$ and $O_2(1, 0)$ are two equilibrium points of Equation (22), sharing the same eigenvalues

$$\lambda_1 = \frac{-\alpha - \gamma - \sqrt{(\alpha + \gamma)^2 + 8}}{2},$$
$$\lambda_2 = \frac{-\alpha - \gamma + \sqrt{(\alpha + \gamma)^2 + 8}}{2}.$$

For $\alpha + \gamma > 0$, we have $\lambda_1 < 0, \lambda_2 > 0$. So $O_1(-1, 0)$ and $O_2(1, 0)$ are both saddles. Since $\lambda_1 + \lambda_2 = -(\alpha + \gamma) < 0$, they are simultaneously dissipative. Moreover, they are both non-resonant due to the irrational $\sqrt{(\alpha + \gamma)^2 + 8}$.

For small $\alpha > 0$, Equation (22) is also equivalent to

$$\begin{aligned} \dot{x} &= y, \\ \dot{y} &= -x + x^3 - \alpha(1 + \gamma_\alpha x^2)y, \end{aligned} \tag{23}$$

which perturbs from equation

$$\begin{aligned} \dot{x} &= y, \\ \dot{y} &= -x + x^3. \end{aligned} \tag{24}$$

Clearly, Equation (24) has two heteroclinic solutions with symmetry: $\ell(t) = \{(a(t), b(t)) : t \in \mathbb{R}\}$ and $\ell_*(t) = \{(-a(t), -b(t)) : t \in \mathbb{R}\}$, where

$$a(t) = \tanh\left(\frac{\sqrt{2}}{2}t\right), \quad b(t) = \frac{\sqrt{2}}{2}\operatorname{sech}^2\left(\frac{\sqrt{2}}{2}t\right).$$

The standard Melnikov function on $\ell(t)$ is

$$M = \int_{-\infty}^{\infty} (1 + \gamma_\alpha a^2(t))b^2(t)\,dt = \frac{2\sqrt{2}}{3} + \frac{2\sqrt{2}}{15}\gamma_\alpha. \tag{25}$$

So $M = 0$ at $\gamma_\alpha = -5$, and thus a new heteroclinic solution arises for Equation (23). Denote this heteroclinic solution as $\tilde{\ell}(t) = \{(\tilde{a}(t), \tilde{b}(t)) : t \in \mathbb{R}\}$. Easy to check that

$\tilde{\ell}_*(t) = \{(-\tilde{a}(t), -\tilde{b}(t)) : t \in \mathbb{R}\}$ is also a heteroclinic solution for Equation (23). They are both heteroclinic to saddles $O_1(-1,0)$ and $O_2(1,0)$. This confirms Proposition 3.2. ■

For Equation (21), two Melnikov functions on $\tilde{\ell}(t), \tilde{\ell}_*(t)$ are

$$\begin{aligned} \mathcal{W}(\theta) &= A\eta + \sqrt{B^2 + C^2}\cos(\omega\theta + \varphi), \\ \mathcal{W}_*(\theta) &= -A\eta + \sqrt{B^2 + C^2}\cos(\omega\theta + \varphi), \end{aligned} \tag{26}$$

where, for $(x,y) \in \tilde{\ell}(t)$,

$$\begin{aligned} A &= \int_{-\infty}^{\infty} \frac{xy^2 e^{-\int_0^t \mathbb{E}_{\tilde{\ell}}(s)\,ds}}{\sqrt{y^2 + [-x + x^3 - \alpha(1+\gamma_\alpha x^2)y]^2}}\,dt, \\ B &= \int_{-\infty}^{\infty} \frac{(x^2-1)^2 y e^{-\int_0^t \mathbb{E}_{\tilde{\ell}}(s)\,ds}\cos\omega t}{\sqrt{y^2 + [-x + x^3 - \alpha(1+\gamma_\alpha x^2)y]^2}}\,dt, \\ C &= \int_{-\infty}^{\infty} \frac{(x^2-1)^2 y e^{-\int_0^t \mathbb{E}_{\tilde{\ell}}(s)\,ds}\sin\omega t}{\sqrt{y^2 + [-x + x^3 - \alpha(1+\gamma_\alpha x^2)y]^2}}\,dt, \\ \tan\varphi &= \frac{C}{B}, \end{aligned} \tag{27}$$

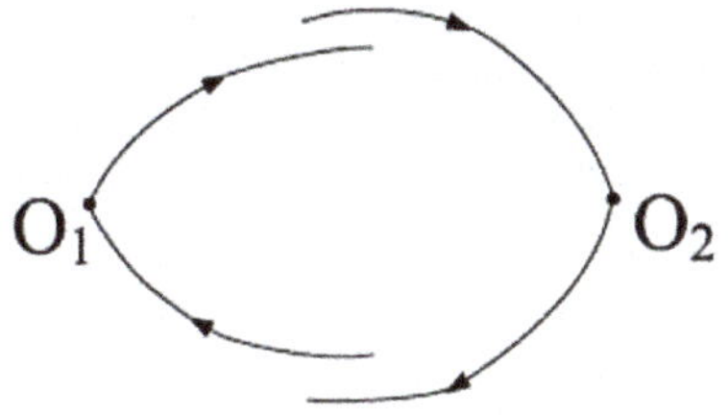

Figure 3. Case (v) of Equation (21).

Table 3. The dynamical pattern of Equation (21, case i).

$\alpha = -0.5, \gamma = 2.429857626335$
$\eta = 0, \omega = 2\pi$
Theoretical multiplicity $e^{\lambda_2} \approx 2.0793$

ε	Behavior type	Actual ratio
3.7×10^{-3}	Hénon-like attractors	–
3.6×10^{-3}	Periodic sinks	–
3.5×10^{-3}	Transient tangles	–
1.772×10^{-3}	Hénon-like attractors	2.0880
1.749×10^{-3}	Periodic sinks	2.0583
1.669×10^{-3}	Transient tangles	2.0971
8.386×10^{-4}	Hénon-like attractors	2.1130
8.276×10^{-4}	Periodic sinks	2.1133
7.896×10^{-4}	Transient tangles	2.1137
3.960×10^{-4}	Hénon-like attractors	2.1177
3.900×10^{-4}	Periodic sinks	2.1221
3.728×10^{-4}	Transient tangles	2.1180
1.864×10^{-4}	Hénon-like attractors	2.1245
1.839×10^{-4}	Periodic sinks	2.1207
1.753×10^{-4}	Transient tangles	2.1266
8.710×10^{-5}	Hénon-like attractors	2.1401
8.595×10^{-5}	Periodic sinks	2.1396
8.175×10^{-5}	Transient tangles	2.1443

and

$$\mathbb{E}_{\tilde{\ell}}(t) = \tau_{\ell}^{\perp}(t)\begin{pmatrix} 0 & 1 \\ 3x^2 - 1 - 2\alpha\gamma_\alpha xy & -\alpha(1+\gamma_\alpha x^2) \end{pmatrix}\tilde{\tau}_{\ell}^{\perp}(t),$$

with

$$\tau_{\ell}^{\perp}(t) = \frac{(-x + x^3 - \alpha(1+\gamma_\alpha x^2)y,\ -y)}{\sqrt{y^2 + [-x + x^3 - \alpha(1+\gamma_\alpha x^2)y]^2}}.$$

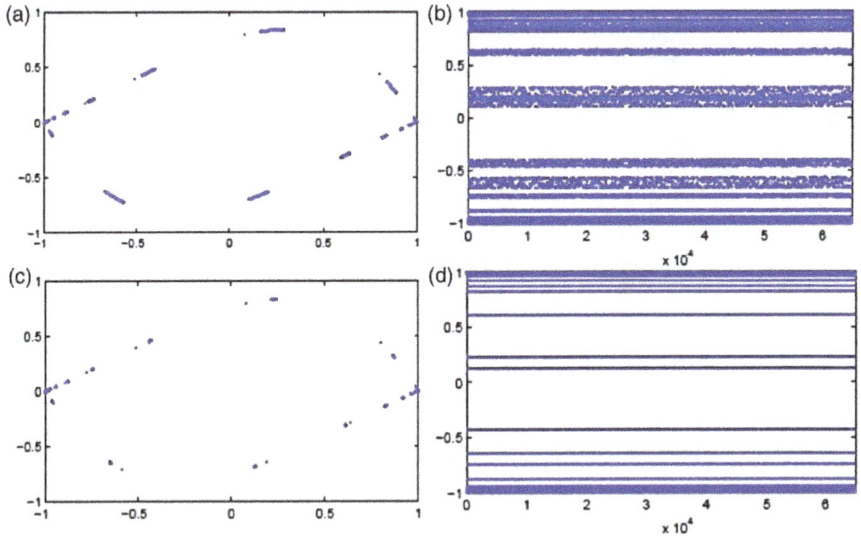

Figure 4. Case (i) of Equation (21). (a) Hénon-like attractor at $\varepsilon = 8.710 \times 10^{-5}$; (b) Time evolution of (a); (c) Periodic sink at $\varepsilon = 3.900 \times 10^{-5}$; and (d) Time evolution of (c).

The four extrema of $\mathcal{W}(\theta), \mathcal{W}_*(\theta)$ are $M = A\eta + \sqrt{B^2 + C^2}$, $m = A\eta - \sqrt{B^2 + C^2}$, $M_* = -A\eta + \sqrt{B^2 + C^2}$, $m_* = -A\eta - \sqrt{B^2 + C^2}$. With the same reason, only two cases hold for real η: (i) $m < 0 < M, m_* < 0 < M_*$; (v) $m > 0, M_* < 0$ or $m_* > 0, M < 0$. However, for heteroclinic tangles, there is no return map corresponding to case (v), see Figure 3. Therefore, only case (i) make sense. Without loss of generality, we assume $A > 0$. Then case (i) implies

$$-\frac{\sqrt{B^2 + C^2}}{A} < \eta < \frac{\sqrt{B^2 + C^2}}{A}. \tag{28}$$

Take η satisfying Equation (28), from Theorem 2.3 we have Hénon-like attractors and periodic sinks, see Figure 4. Horseshoes are there, but we can not visualize it in numerical simulations. Table 3 is a dynamical pattern for Equation (21) when $\alpha = -0.5, \gamma = 2.42985626335$, $\eta = 0$.

4. Conclusions

In this paper, Hénon-like attractors, rank one attractors, and periodic sinks are obtained in a class of Duffing oscillator. Among the three attractors, Hénon-like attractors and rank one attractors are chaotic in the sense of the SRB measure. All the three attractors are organized in an invariant pattern that repeats itself periodically with respect to the forcing magnitude ε as $\varepsilon \to 0$.

Acknowledgements

The authors thank Prof. Jibin Li and Prof. Fangyue Chen for their encouragement and generous support. This paper is supported by a NSFC grant (No. 11171309).

References

Benedicks, M., & Young, L. S. (1993). Sinai-Bowen-Ruelle measure for certain Hénon maps. *Inventiones Mathematicae, 112*, 514–576.

Chen, F. J., Oksasoglu, A., & Wang, Q. D. (2013). Heteroclinic tangles in time-periodic equations. *Journal of Differential Equations, 254*, 1137–1171.

Holmes, P. (1979). A nonlinear oscillator with a strange attractor. *The Philosophical Transactions of the Royal Society of London*, Series A, *292*, 419–448.

Holmes, P., & Rand, D. (1980). Phase portraits and bifurcations of the non-linear oscillator: $\ddot{x} + (\alpha + \gamma x^2)\dot{x} + \beta x + \delta x^3 = 0$. *Journal Non-linear Mechanics, 15*, 449–458.

Poincaré, H. (1899). Les *Méthodes* Nouvelles de la *Mécanique Céleste*. 3 vols. Paris: Gauthier-Villars.

Wang, Q. D. (2009). *Periodically forced double homoclinic loops to a dissipative saddle*. Unpublished manuscript.

Wang, Q. D. (2011). *Dynamics of non-autonomously perturbed homoclinic solutions*. Unpublished manuscript.

Wang, Q. D., & Oksasoglu, A. (2008). Rank one chaos: theory and applications. *International Journal of Bifurcation Chaos, 18*, 1261–1319.

Wang, Q. D., & Oksasoglu, A. (2010). Periodic occurrence of chaotic behavior of homoclinic tangles. *Physica D: Nonlinear Phenomena, 239*, 387–395.

Wang, Q. D., & Oksasoglu, A. (2011). Dynamics of homoclinic tangles in periodically perturbed second-order equations. *Journal of Differential Equations, 250*, 710–751.

Young, L. S. (2002). What are SRB measure, and which dynamical systems have them? *Journal of Statistical Physics, 108*, 733–754.

3

Numerical analysis and circuit realization of the modified LÜ chaotic system

Guoqing Huang[a,b*] and Zuozun Cao[b]

[a]*School of Civil Engineering and Architecture, Nanchang University, Nanchang, People's Republic of China;* [b]*School of Science, Nanchang University, Nanchang, People's Republic of China*

A novel three-dimensional autonomous chaotic system from the LÜ chaotic system is given. By using the theoretical analysis and numerical simulation, we provide an insight into the dynamic properties and characterizations of this system, such as Hopf bifurcation. In particular, we are interested in focusing on the dependence of varying parameters on chaos with the help of some chaos indicators including the fast Lyapunov indicator, small alignment indexes and Lyapunov exponent. It is shown that growing the parameter c leads to the extent of chaos. Finally, a chaotic electronic circuit is designed for the realization of the chaotic attractor with aim of Multisim software, and it gives almost the same rules of types of orbits as numerical ones by an alternating value of a circuital resistor.

Keywords: fast Lyapunov indicator; small alignment; circuit realization

1. Introduction

Chaos with exponential sensitivity of initial conditions is a typical nonlinear phenomenon in nonintegrable dynamical systems. This phenomenon shows that two identical systems have different trajectories as time evolves when initial points of these systems are slightly different. Chaotic dynamics in classical physics has been intensively investigated within the mathematics, science and engineering communities for about 40 years. As well known, since Lorenz (1963) discovered a simple three-dimensional smooth autonomous chaotic system, the investigation of chaotic behavior has attracted great attention. From then on, many researchers have proposed and analyzed the novel three-dimensional chaotic systems such as Rössler systems (Rössler, 1979), the Chua's circuit (Cafagna & Grassi, 2003), the Chen system (Chen & Ueta, 1999) and the LÜ system (Lü & Chen, 2002).

There are many ways for the identification of chaotic orbits from regular ones in the classical systems (Contopoulos, 2002). Each of them has its advantages and disadvantages in quantifying the regular or chaotic nature of orbits in the nonlinear systems. As well known, the Lyapunov exponents (LEs) (Benettin, Galgani, & Strelcyn, 1976) are frequently used for the measuring the average exponential deviation of two nearby trajectories. As an efficient method to detect regular from chaotic orbits, the LEs are applicable to a phase space with any dimension, but quite a long integration time is often needed to get a reliable value of LEs in a multidimensional system. There are also other qualitative methods for the nonlinear systems, such as the power spectra, Poincaré sections (Hénon & Heils, 1964), Smaller Alignment Index (SALI) (Skokos, 2001; Skokos, Bountis, & Antonopoulos, 2007), fast Lyapunov indicators (FLIs) (Froeschlé, Lega, & Gonczi, 1997), etc. As very fast tools to find chaos, both the SALI and the FLI change following completely different time rates for different orbits thus allowing them to detect between the ordered and chaotic case. The SALI and the FLI are still suitable for discussing dissipative system as in Huang and Wu (2012).

One main aim of the present paper is to use numerical approaches to study the dynamical properties of a new three-dimensional nonlinear system. Because the LÜ system consists of quadratic nonlinearities, while the new system has exponent nonlinearities, the latter shows more complex dynamic characteristics than the former. As another purpose, the FLI and the SALI are tested to be very fast, efficient tools to distinguish chaotic orbits from regular ones in the dissipative system, which are commonly used in the conservative Hamiltonian systems and first applied to treat the strongly dissipative system in our previous work in Huang and Wu (2012). Finally, chaotic dynamics have also been used in engineering and experimental applications (Matsumoto, Chua, & Kobayashi, 1986; Qi, Wyk, Wyk, & Chen, 2009). Some dissipative systems can be performed as electronic oscillator circuits, and signals are seen from an oscilloscope or a digital signal processor. Immediately, the nature of chaotic or regular orbits is clearly shown by the experimental observations.

*Corresponding author. Email: huanggq@ncu.edu.cn

The rest of this paper is organized in the following manner. In Section 2, we construct a three-dimensional dynamical system based on the LÜ system, and analyze the structure of equilibria and hopf bifurcation in the system. Meantime in Section 3, we also study the transitions from regular orbits to chaotic ones varying parameters by chaos indicators such as the LEs, the FLIs and the SALI. In Section 4, experiment circuit has been built for implementing the novel system. Finally, Section 5 summarizes the conclusions.

2. Numerical investigations

The LÜ Chaos system is written as $\mathrm{d}x/\mathrm{d}t = a(y - x), \mathrm{d}y/\mathrm{d}t = cy - xz, \mathrm{d}z/\mathrm{d}t = -bz + xy$. With xy replaced by e^{xy}, a novel simple three-dimensional autonomous system is shown with an exponent nonlinear term as follows:

$$\frac{\mathrm{d}x}{\mathrm{d}t} = a(y - x),$$
$$\frac{\mathrm{d}y}{\mathrm{d}t} = cy - xz,$$
$$\frac{\mathrm{d}z}{\mathrm{d}t} = -bz + e^{xy}, \tag{1}$$

where $x(t), y(t), z(t)$ are the state vector and $a = 36, b = 3, c$ are positive constants. The new system (1) consists of one quadratic and one exponent term. We obtain phase portraits of individual orbits from various directions varying parameter that concern ordered or chaotic orbits. As a single parameter c increases, the system has interesting complex dynamical behavior. By setting parameter $c = 25$ in Figure 1 emerges the two irregular attractors in the novel three-dimensional smooth system, namely the so-called two-wing irregular attractors, which is similar to that of the LÜ system. For the case $c = 5$, we can watch phase portraits with respect to projections on the $y - z$ plane in Figure 2. It is clear to show that the solution is a regular attractor. The simulation results are obtained by using the four-order Runge–Kutta method with the step length taken as 0.01 and initial conditions given with $(x, y, z) = [1, -1, 1]$.

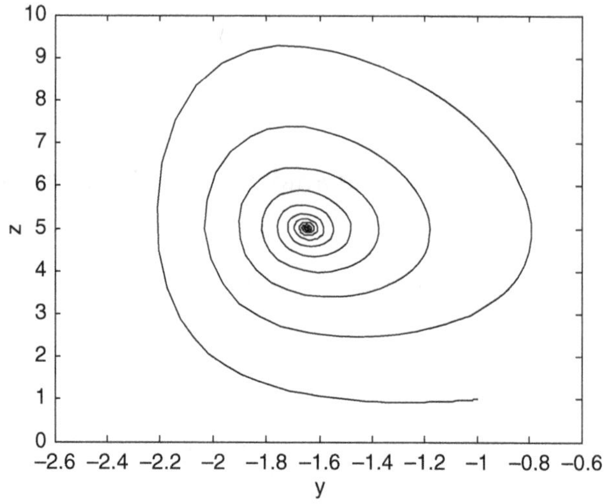

Figure 2. Phase portraits $(y - z)$ of the regular orbit with $c = 5$.

The equilibria as particular solutions of the system meet the following algebraic equations:

$$a(y - x) = 0, \quad cy - xz = 0, \quad -bz + e^{xy} = 0. \tag{2}$$

Obviously, we solve these nonlinear algebraic equations and obtain three equilibriums:

$$s_0 = \left[0, 0, \frac{1}{b}\right], \quad s_{1,2} = \left[\pm\sqrt{\ln(bc)}, \pm\sqrt{\ln(bc)}, c\right]. \tag{3}$$

To study the stability of the equilibrium points, the system (1) is linearized at s_1 and the Jacobianb matrix is obtained:

$$\begin{pmatrix} -a & a & 0 \\ -z_1 & c & -x_1 \\ y_1 e^{x_1 y_1} & x_1 e^{x_1 y_1} & -b \end{pmatrix}. \tag{4}$$

The equilibrium points excluding s_0 should depend on the parameter c. We discuss them according to three different values of c. For the first case $c = 5$, we obtain two

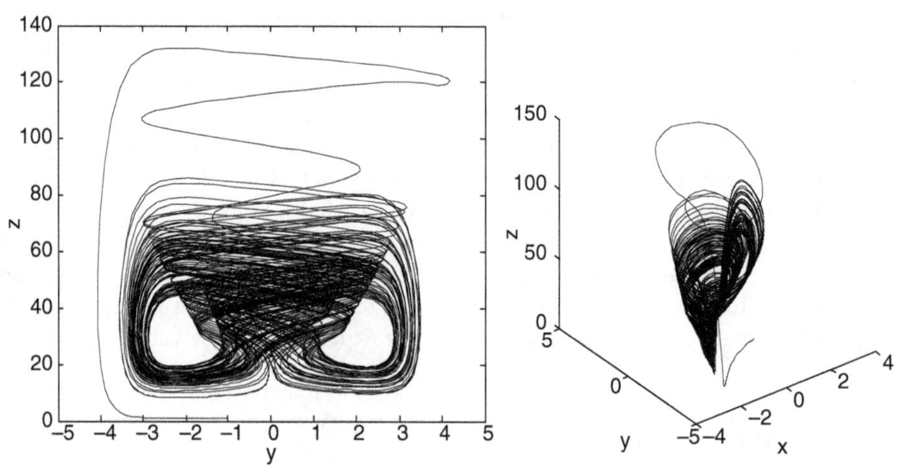

Figure 1. Phase portraits of the chaotic orbit with $c = 25$.

other nontrivial equilibria $s_{1,2} = [\pm 1.6456, \pm 1.6456, 5]$. The equilibrium s_1 is a stable node-focus point because three eigenvalues of Equation (4) are $[-32.6510, -0.6745 \pm 9.4403i]$ with $i = \sqrt{-1}$. In addition, the eigenvalues at the equilibrium s_2 are the same as those at the equilibrium s_1 and also show that the equilibrium s_2 has the same property. This is because the system is symmetric with respect to the $x-y$ plane, in other words, it remains invariant under the transformation $(x,y,z) \to (-x,-y,z)$. As far as the second case $c = 25$, two nontrivial equilibria are $s_{1,2} = [\pm 2.0779, \pm 2.0779, 25]$. The equilibrium s_1 is an unstable node-focus point because there are three eigenvalues $[7.5845 \pm 27.2350i, -29.1691]$. For the case $c = 7.31$, we obtain two nontrivial equilibria, $s_{1,2} = [\pm 1.7572, \pm 1.7572, 7.31]$. In this case, we obtain the characteristic equation of the Jacobian matrix in the following form:

$$\lambda^3 + B_1\lambda^2 + B_2\lambda + B_3 = 0, \qquad (5)$$

with $B_1 = 31.6918, B_2 = 153.8443$, and $B_3 = 4875.6038$. It is easy to find that $B_1 > 0, B_2 > 0, B_3 > 0$ and $B_1B_2 - B_3 \approx 0$. This ensures that Equation (5) has one pair of approximate pure imaginary eigenvalues $\Gamma_{1,2} = 0.0009 \pm 12.4034 \approx \pm 12.4034$ and other negative eigenvalues $\Gamma_3 = 31.6918$. On the one hand, all the eigenvalues of the characteristic equations have negative real parts which verify that the system is stable when $c < 7.31$, on the other hand, there are always one eigenvalue with positive real part as $c > 7.31$. Namely, the stability of the existing equilibrium changes from being stable to unstable as the parameter c grows to span the critical value of 7.31. Thus, we can determine that a Hopf bifurcation occurs at $c = 7.31$. Our analysis of the Hopf bifurcation is just based on that one pair of purely imaginary eigenvalues and all other eigenvalues containing negative real parts are sufficient conditions for the emergence of a Hopf bifurcation at a critical parameter value (Arrowsmith & Place, 1990).

3. Distinguishing orbits by different chaotic indicators

There are various methods to distinguish between chaotic and ordered orbits. Now we propose three methods to deal with our problems.

3.1. Lyapunov exponents

In classical physics, LEs, as a common chaos indicator to distinguish whether a system is chaotic or regular, have been widely used to measure the chaoticity of orbits in the nonlinear dynamical system. They calculate the rate of exponential divergence between neighboring trajectories in the phase space, precisely, if the motion is ordered, the corresponding LEs are all negative, otherwise. If the motion is chaotic, the largest LE is strictly positive. There are two different methods for numerically calculating LEs.

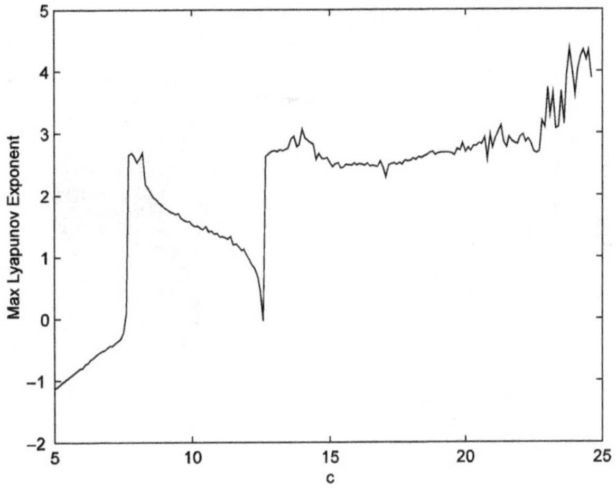

Figure 3. Lyapunov spectra with the variations of c.

One rigorous method is to use the tangent vector from the solution of the variational equations of the system. Another less rigorous method is the so-called two-particle method (Tancredi, Sanchez, & Roig, 2001) using the deviation vector between two nearby trajectories in place of the tangent vector. All LEs of the three-dimensional system (1) can be attained, where initial conditions are $(x,y,z) = (1,-1,1)$. In the case $c = 5$, there are three negative LEs $(\lambda_1, \lambda_2, \lambda_3) = (-1.2217, -1.2592, -13.6291)$. This sufficiently tests that the system is Lyapunov stable and its attractor is a stable fixed point. For $c = 25$, there are one positive and three negative LEs $(\lambda_1, \lambda_2, \lambda_3) = (3.8993, -1.7070, -16.5162)$. The onset of such a positive LE confirms that the system is dynamically unstable and chaotic.

An ordered and chaotic behavior can be checked with the LEs method varying system parameters. There is a sudden varying from regular behaviors to chaotic ones when the parameter c passes 7.31 as shown in Figure 3. In addition, a bifurcation is very convenient to search for an abrupt change of a qualitatively different solution (such as the structure of attractors) for a nonlinear system when a control parameter is smoothly changed. With the help of the bifurcation, a period doubling, quadrupling, etc., and the onset of chaos can be found. Let the parameter c vary in the interval [2.0, 10.0] and give the initial condition $(1,-1,1)$, the bifurcation diagram versus c is proposed to show how system changes with increasing value of parameter c displayed in Figure 4. The bifurcation is good to show the transition to chaos as the parameter spans 7.31. The above ways offer enough dynamical information that $c = 7.31$ is a threshold value from regular to chaotic motion.

3.2. Fast Lyapunov indicators

It is well known that the time necessary to reach a limit value, either of the length of any tangential vector or of the angle between tangent vectors, is taken as an indicator of stochasticity for a nonlinear dynamical system. Following

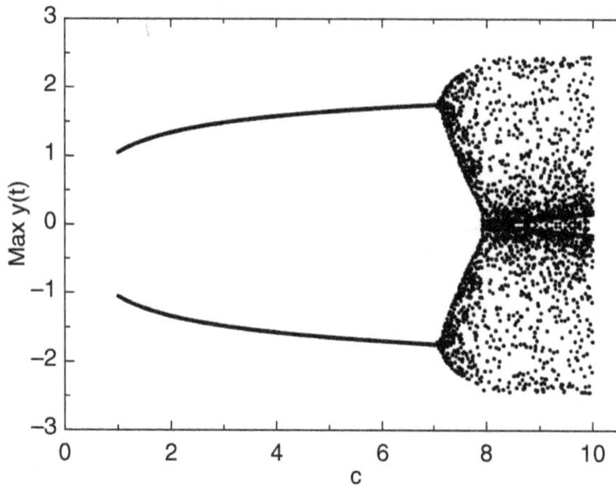

Figure 4. Bifurcation diagrams of max y versus c.

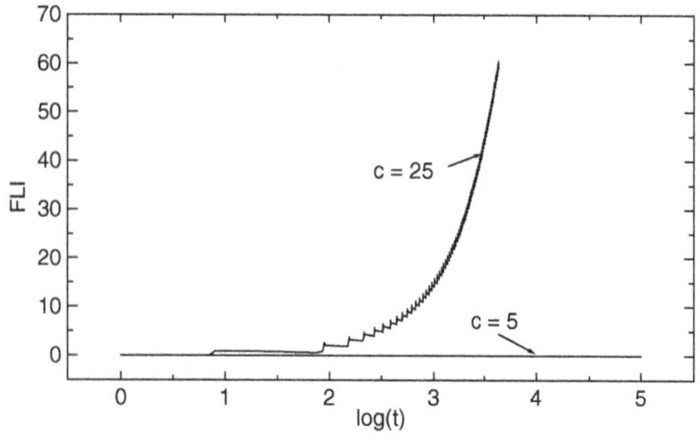

Figure 5. The FLIs versus $\log(t)$.

this general idea, FLI, as a simple, qualitative index, was first introduced by Froeschlé et al. and then modified by Froeschlé and Lega (2000). Given any threshold, the FLI will reach the value fast for a chaotic orbit, and slowly for a regular orbit. Conversely, in the same time span, the indicators will show different values for a regular and a chaotic nature of orbit with completely different time rates. More specifically, the logarithm of the length of a tangent vector method of inspecting the dynamics of a nonlinear physical evolves exponentially for a chaotic orbit, grows only polynomially for a regular motion. In addition, this index was further developed as the two-particle method (Wu & Xie, 2007, 2008) in which the equations of motion must be solved two times, and the renormalization technique was used to calculate their FLI within a sufficiently long time span. As shown from Figure 5 that FLIs change with the two different values of parameter c. The variations of FLIs are entirely different for the two case. The FLIs are almost invariant for $c = 5$, but some grow exponentially for $c = 25$. As a consequence, the orbit is regular for the former, but chaotic for the latter. The result is consistent with the one given by the the Hopf bifurcation and LEs. It has proved successful that FLI succeeds in detecting the nature of orbits

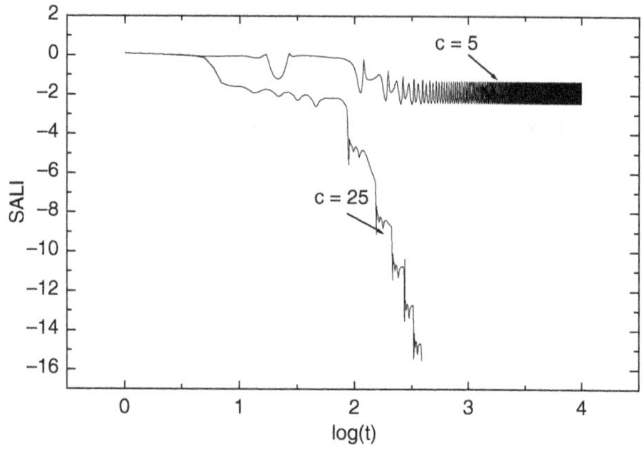

Figure 6. The SALIs versus $\log(t)$.

faster than the computation of the LEs due to less or no renormalization.

3.3. Small Alignment index

The construction of SALI (Skokos, 2001)also stems from the idea of the calculation of two LEs in any degree system. Two deviation vectors at the same point in the tangent space converge to the direction of the eigenvector which corresponds to the maximal Lyapunov characteristic exponent (LCE) if the Gram–Schmidt orthogonalization is not used. In general, any two randomly chosen initial tangent vectors will become aligned with the most unstable direction and the angle between them will rapidly converge to zero. The speed of the convergence is completely of a opposite nature for chaotic and regular orbits, namely the misalignment between the two tangent vectors tends rapidly to zero for chaotic orbits, while it shows small fluctuations around non-zero values for ordered orbits and so it clearly detects between the two types of orbits. The SALI method has already been confirmed to be an efficient and quick indicator of chaoticity for diagnosing between chaotic and ordered motion independent of the dimensions of the system. In our opinion, the indicator is still suitable for a dissipative system because the speed of the SALI converging to zero is strongly linked to the properties of trajectories but does not at all depend on the conservative or dissipative nature of the system. The SALI also converges exponentially to zero for chaotic orbits in dissipative systems, while it exhibits small fluctuations around non-zero values for ordered ones. These facts can be confirmed in Figure 6. The SALI grows to a constant for $c = 5$. This is a regular behavior of the orbit. But the case $c = 25$ means the appearance of chaos because their SALIs are fast close to the value -16 before $t = 10,000$. The result is in conformity to one given by the above other methods.

4. Circuit realization of the chaotic system

There are some common methods of the circuit realization of the chaotic system: one of them is to apply the piece-wise

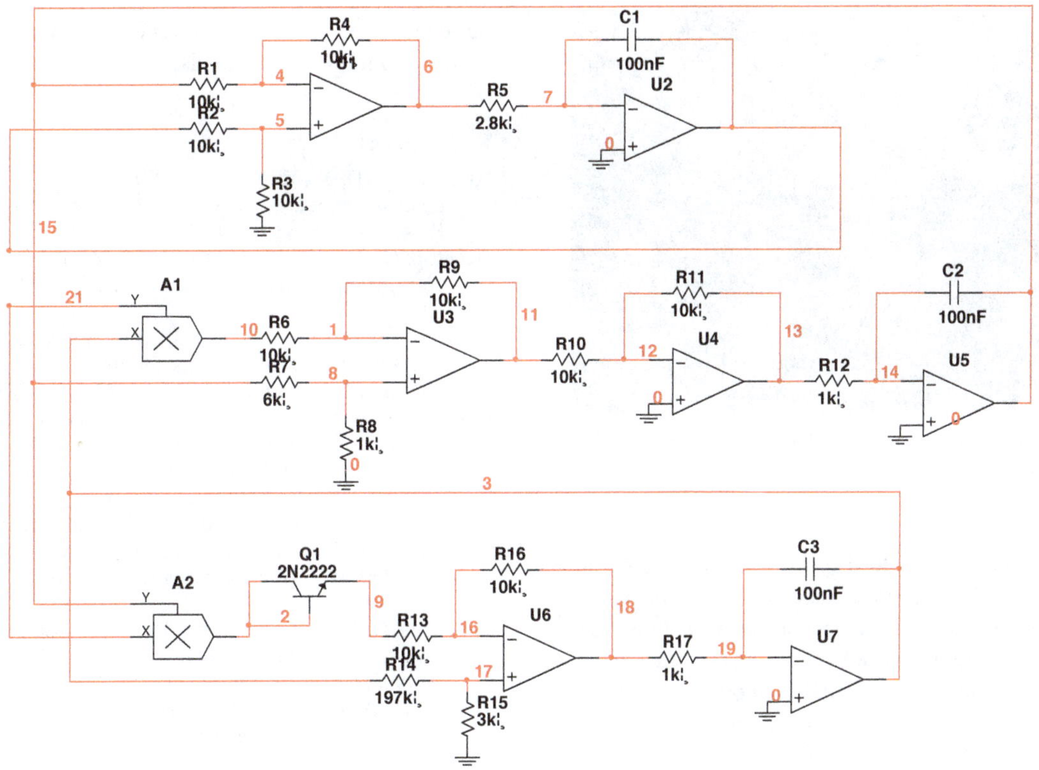

Figure 7. Circuit diagram of the system.

linear function to replace the quadratic, cubic term or other nonlinear term. Other of methods is to use the switch function to substitute the nonlinear term, respectively. The present system is implemented by applying analog multiplier to replace the quadratic product term and by using the triode emitter junction to substitute exponent term, without influencing the system original nonlinear nature. In a practical circuit implementation concerning signals needs to be observed, the strength of the experimental circuit signals is changed to some scale to that of the original circuit signals so that the implementation can be obtained without any difficulty. Of course, the replacement of chaotic variables does not change system properties. As shown in Figure 7, the voltages C_1, C_2, and C_3 are used as U_{C_1}, U_{C_2}, and U_{C_3}, respectively. The operational amplifiers and its electronic circuitry perform the basic operations of addition, subtraction and integration. In the light of the characters of ideal op-amp (virtual short and virtual open) and the Kirchhoff's current and voltage laws, the corresponding circuit equation can be described as

$$\frac{dU_{C_1}}{dt} = \frac{R_4}{R_1R_5C_1}U_{C_2} - \frac{R_3(R_1+R_4)}{(R_2+R_3)R_1R_5C_1}U_{C_1},$$
$$\frac{dU_{C_2}}{dt} = \frac{R_8R_{11}(R_6+R_9)}{(R_7+R_8)R_{10}R_{12}R_6C_2}U_{C_2} - \frac{R_{11}R_9}{R_6R_{10}R_{12}C_2}U_{C_1}U_{C_3},$$
$$\frac{dU_{C_3}}{dt} = \frac{R_{14}(R_{13}+R_{16})}{(R_{14}+R_{15})R_{13}R_{17}C_3}U_{C_2} - \frac{R_{16}I_s}{R_{17}C_3}e^{U_{C_1}U_{C_2}/V_T}, \tag{6}$$

where $V_T = KT \approx 26\,\text{mV}$ is the thermal voltage and $I_s \approx 200\,\mu\text{A}$ is the reverse saturation leakage current. In the later analysis, related electronic devices are as follows. The resistors are following as: $R_1, R_2, R_3, R_4, R_6, R_8 = 10\,\text{k}\Omega$; $R_9, R_{10}, R_{11}, R_{13}, R_{16} = 10\,\text{k}\Omega$; $R_5 = 2.8\,\text{k}\Omega$; $R_7 = 6\,\text{k}\Omega$; $R_{12}, R_{17} = 1\,\text{k}\Omega$; $R_{14} = 197\,\text{k}\Omega$ and $R_{15} = 3\,\text{k}\Omega$. Meantime, set the capacitors $C_1, C_2, C_3 = 100\,\text{nF}$. Two values of the parameter c will be given by changing value resistors R_7. We get circuital simulation viewpoints in Figures 8

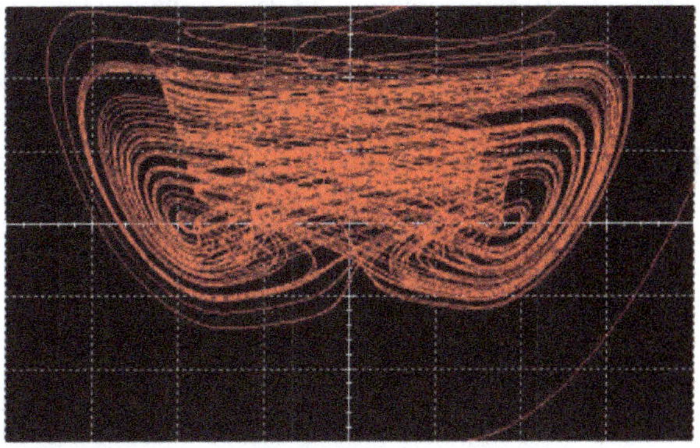

Figure 8. The experiment observations $(y - z)$ of chaotic orbit with $c = 25$.

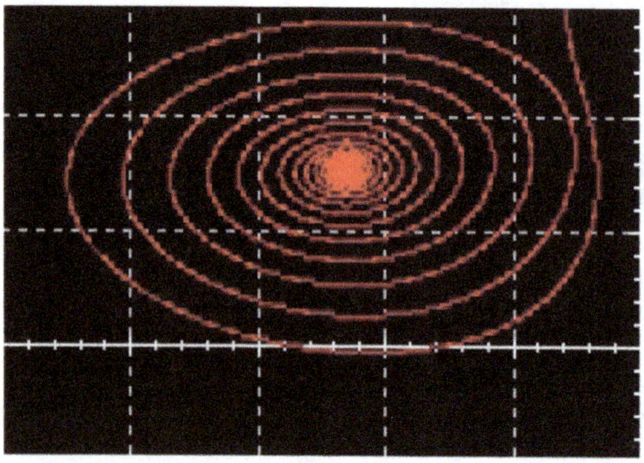

Figure 9. The experiment observations $(y - z)$ of regular trajectory with $c = 5$.

and 9, which correspond to the numerical counterparts of Figures 1 and 2. Setting parameter $c = 25$ with the resistors $R_7 = 6\,\mathrm{k}\Omega$, two chaotic attractors appear. As an illustration, Figure 8 is the same as Figure 1, and it can be shown that the experimental results and computational ones are consistent with for the chaotic motions. For the case $c = 5$ with the resistors $R_7 = 38.8\,\mathrm{k}\Omega$, we get phase portraits with respect to projections on different planes in Figure 9. However, for the chaos case, there is a slight difference between the experimental views and the numerical views. Generally speaking, the numerical views are clearer than the experimental ones because experimental initial conditions are never known perfectly.

5. Conclusions

A new modified LÜ chaotic system has been investigated with exponential terms. Some basic properties of this system have been discussed in terms of chaotic attractors, equilibria, and eigenvalues of the Jacobian matrices. Bifurcations, LEs are used to find the dependence of the transitivity from order to chaos on changing parameter c. As a result, they get the same results that $c = 7.31$ is a threshold value from a regular dynamic to an irregular one. Namely, under some necessary conditions for the occurrence of chaotic motions, increasing the parameter c always leads to the strength of chaos. With the help of SALIs and FLI, we can explore two different types of orbits (regular or chaotic), as the dynamical parameter c varies. In addition, an electronic circuitry is designed for the realization of the chaotic attractor, identifying experiment results with computer simulations. The new chaotic systems can be regarded as information sources that naturally produce digital communication signals.

References

Arrowsmith, D. K., & Place, C. M. (1990). *An introduction to dynamical systems*. New York, NY: Cambridge University Press.

Benettin, G., Galgani, L., & Strelcyn, J.-M. (1976). Kolmogorov entropy and numerical experiments. *Physical Review A, 14*, 2338–2345.

Cafagna, D., & Grassi, G. (2003). New 3D-scroll attractors in hyperchaotic Chua's circuits forming a ring. *International Journal of Bifurcation and Chaos, 13*, 2889–2903.

Chen, G., & Ueta, T. (1999). Yet another chaotic attractor. *International Journal of Bifurcation and Chaos, 9*(7), 1465–1466.

Contopoulos, G. (2002). *Order and chaos in dynamical astronomy*. Berlin: Springer Verlag.

Froeschlé, C., & Lega, E. (2000). On the structure of symplectic mappings. The fast Lyapunov indicator: A very sensitive tool. *Celestial Mechanics and Dynamical Astronomy, 78*, 167–195.

Froeschlé, C., Lega, E., & Gonczi, R. (1997). Fast Lyapunov indicators. Applications to asteroidal motion. *Celestial Mechanics and Dynamical Astronomy, 67*, 41–62.

Hénon, M., & Heils, C. (1964). The applicability of the third integral of motions: Some numerical experiments. *Astronomical Journal, 69*, 73–79.

Huang, G.-Q., & Wu, X. (2012). Analysis of new four-dimensional chaotic circuits with experimental and numerical methods. *International Journal of Bifurcation and Chaos, 22*(2), 1250042 (13 pp.).

Lorenz, E. N. (1963). Deterministic nonperiodic flow. *Journal of Atmospheric Sciences, 20*, 130–141.

Lü, J., & Chen, G. (2002). A new chaotic attractor coined. *International Journal of Bifurcation and Chaos, 12*(3), 659–661.

Matsumoto, T., Chua, L. O., & Kobayashi, K. (1986). Hyperchaos: Laboratory experiment and numerical confirmation. *IEEE Transactions on Circuits and Systems, 33*, 1143–1147.

Qi, G. Y., Wyk, M. A., Wyk, B. J., & Chen, G. R. (2009). A new hyperchaotic system and its circuit implementation. *Chaos, Solitons and Fractals, 40*, 2544–2549.

Rössler, O. E. (1979). An equation for hyperchaos. *Physics Letters A, 71*(2,3), 155–156.

Skokos, C. (2001). Alignment indices: A new, simple method for determining the ordered or chaotic nature of orbits. *Journal of Physics A, 34*, 10029–10043.

Skokos, Ch., Bountis, T., & Antonopoulos, Ch. (2007). Geometrical properties of local dynamics in Hamiltonian systems: The Generalized Alignment Index (GALI) method. *Physica D, 231*, 30–54.

Tancredi, G., Sanchez, A., & Roig, F. (2001). A comparison between methods to compute Lyapunov exponents. *Astronomical Journal, 121*, 1171–1179.

Wu, X., & Xie, Y. (2007). Revisit on "Ruling out chaos in compact binary systems". *Physical Review D, 76*(6), 124004D.

Wu, X., & Xie, Y. (2008). Resurvey of order and chaos in spinning compact binaries. *Physical Review D, 77*(10), 103012.

4

Decay rate constrained stability analysis for positive systems with discrete and distributed delays

Jun Shen and James Lam*

Department of Mechanical Engineering, The University of Hong Kong, Pokfulam Road, Hong Kong

This paper is concerned with the decay rate constrained exponential stability analysis for continuous-time positive systems with both time-varying discrete and distributed delays. A necessary and sufficient condition is first given to ensure that a positive system with distributed delay is exponentially stable and satisfies a prescribed decay rate. Furthermore, by exploiting the monotonicity of the trajectory of a constant delay system and comparing the trajectory of the time-varying delay system with that of the constant delay system, the results are extended to positive systems with both bounded time-varying discrete delays and distributed delays.

Keywords: distributed delays; exponential stability; positive systems; time-delay systems

1. Introduction

The study of positive systems has attracted tremendous attention in the recent years. The fact that many practical models involve quantities which are intrinsically nonnegative naturally gives rise to such systems whose state variables and output signal are always constrained in the first orthant whenever both the initial condition and the input signal are nonnegative. This stimulates a variety of works on the issue of dynamic systems under positivity constraint. Positive systems have applications in a wide range of disciplines involving systems biology (de Jong 2002), pharmacokinetics (Jacquez 1985) and ecology (Caswell 2001). A typical example is the mathematical modeling of compartmental networks (Haddad, Chellaboina, & Hui, 2010), which captures the exchange of nonnegative quantities of materials among compartments and the environment with conservation of mass of materials. Along with abundant practical applications, the mathematical theory of positive systems is originated from the well-known Perron–Frobenius theorem (Berman & Plemmons 1994), which arises in the analysis of nonnegative matrices. In the past decade, a wealth of literature has been devoted to the analysis and synthesis of positive systems, just to name a few, we refer the readers to Kaczorek (2002), Kaczorek (2008), Liu (2009), Haddad et al. (2010), Li, Lam, and Shu (2010), Feng, Lam, Li, and Shu (2011), Kaczorek (2011), Ait Rami and Tadeo (2007), Ait Rami and Napp (2012), Ait Rami (2011), Zhao, Zhang, Shi, and Liu (2012), and Zhao, Zhang, and Shi (2013) and the references therein.

On the other hand, as in other dynamic systems, time delay is often encountered in the analysis of positive systems. Many works have been reported on the stability analysis of positive delay systems in the literature. The stability of both linear and nonlinear positive systems with constant delays was studied in Haddad and Chellaboina (2004) and it is pointed out that asymptotic stability can be preserved regardless of the magnitude of delays. It is further shown in Liu, Yu, and Wang (2009, 2010) that both discrete- and continuous-time positive systems with bounded time-varying delays are asymptotically stable as long as the corresponding delay-free systems are asymptotically stable. Moreover, it is found that similar results also hold for positive switched systems with unbounded time-varying delays in Liu and Dang (2010). Nevertheless, exponential stability analysis of positive delay system has not received much attention until quite recently. Decay rate constrained output-feedback stabilization is solved in Feng et al. (2011) via iterative linear matrix inequality approach by a modified cone complementarity linearization method (El Ghaoui, Oustry, & Ait Rami, 1997). In Zhu, Li, and Zhang (2012) and Zhu, Meng, and Zhang (2013), it has been shown that although bounded discrete delay has no impact on the asymptotic stability of positive systems, it does affect the decay rate of the state trajectory. The positivity and exponential stability for linear and nonlinear systems with both discrete and distributed delays were studied in Ngoc (2013). Recently, the characterization of the L_∞-gain of positive systems with

*Corresponding author. Email: james.lam@hku.hk

discrete and distributed delays was given in Shen and Lam (2013a, b).

Motivated by the above discussion, in this paper, we address the decay rate constrained stability analysis problem for positive systems with both bounded time-varying discrete delays and distributed delays. The results obtained in this paper can be regarded as extensions of the asymptotic stability analysis for positive systems with bounded time-varying discrete delays in Liu et al. (2010) and the exponential stability analysis for positive systems with both time-varying discrete and distributed delays in Ngoc (2013). It is worth mentioning that the approach employed in this paper is different from that used in either Liu et al. (2010) or Ngoc (2013). Note that the technical proof in Liu et al. (2010) is extremely complicated. In fact, utilizing the methods in this paper, one can give an alternative proof for the results in Liu et al. (2010) and Ngoc (2013). The advantage of our approach lies in two aspects. Firstly, for the constant delay case, we make use of a Lyapunov–Krasovskii functional, which can be applied to nonlinear positive system as well as switched positive systems with average dwell time switching signal like those considered in Zhao et al. (2012) and Zhao et al. (2013). For the time-varying delay case, we provide a very simple proof, which only needs to look into the monotonicity of the constant delay system and compare the trajectory of the time-varying delay system with that of the constant delay system. This would be useful in the stability analysis of switched positive system such as that investigated in Liu and Dang (2010). In our opinion, generally, distributed delays play a role similar to bounded discrete delays in the analysis of positive systems.

2. Notations and preliminaries

In this section, we introduce some elementary notations and lemmas on positive systems and nonnegative matrices, which will be needed in the following sections. All the matrices, if their dimensions are not explicitly stated, are assumed to be compatible for algebraic operations. x_i denotes the ith entry of a column vector $x \in \mathbb{R}^n$. $\mathbf{1}$ denotes a column vector with each entry equals 1. A real matrix $A \in \mathbb{R}^{m\times n}$ with all of its entries nonnegative is called a nonnegative matrix and is denoted by $A \succeq 0$ and $A \in \mathbb{R}_+^{m\times n}$. A square matrix $A \in \mathbb{R}^{n\times n}$ with all its off-diagonal entries nonnegative is called Metzler and is denoted by $\mathbb{M}^n$. For two matrices $A = [a_{ij}], B = [b_{ij}] \in \mathbb{R}^{m\times n}$, $A \succeq B$ (respectively, $A \succ B$) means that $a_{ij} \geq b_{ij}$ (respectively, $a_{ij} > b_{ij}$) for $i = 1, 2, \ldots, m$ and $j = 1, 2, \ldots, n$. The ∞-norm of a column vector $x \in \mathbb{R}^n$ is defined as $\|x\|_\infty = \max_{i=1,2,\ldots,n} |x_i|$. $\mathbb{C}([-h, 0], \mathbb{R}^n)$ is the Banach space of all vector-valued continuous functions defined on $[-h, 0]$ with norm $\|\phi\| \triangleq \max_{t\in[-h,0]} \|\phi(t)\|_\infty$. Given a vector-valued continuous function $x(t) : [-h, \infty) \to \mathbb{R}^n$, for every $t \geq 0$, $x_t \in \mathbb{C}([-h, 0], \mathbb{R}^n)$ is defined by $x_t(\theta) = x(t+\theta)$, $\theta \in [-h, 0]$.

The following lemma gives a necessary and sufficient condition to ensure that a Metzler matrix is Hurwitz via linear programming.

Lemma 1 (Farina & Rinaldi 2000). *A Metzler matrix $A \in \mathbb{M}^n$ is Hurwitz if and only if one of the following conditions is satisfied:*

(1) there exists a column vector $p \succ 0$, such that $p^{\mathrm{T}}A \prec 0$.
(2) there exists a column vector $q \succ 0$, such that $Aq \prec 0$.

3. Positive systems with distributed delays

Consider the following continuous-time linear system with pure distributed delays:

$$\dot{x}(t) = Ax(t) + \int_{-h}^{0} A_h(s)x(t+s)\,\mathrm{d}s, \quad t \geq 0,$$
$$x(s) = \phi(s), \quad s \in [-h, 0], \tag{1}$$

where $x(t) \in \mathbb{R}^n$ represents the state vector; $\phi(\cdot) \in \mathbb{C}([-h, 0], \mathbb{R}^n)$ is the initial condition; $h > 0$ is a constant; $A_h(s) \in \mathbb{C}([-h, 0], \mathbb{R}^{n\times n})$.

The definition and characterization of the positivity of system (1) are presented below.

Definition 1 (Farina & Rinaldi 2000) *System (1) is called (internally) positive if for all initial condition $\phi(s) \succeq 0$ ($s \in [-h, 0]$), the state trajectory $x(t) \succeq 0$ for all $t \geq 0$.*

Lemma 2 (Ngoc, 2013, Theorem II.2) *System (1) is (internally) positive if and only if A is Metzler and $A_h(s)$ is nonnegative for all $s \in [-h, 0]$.*

The definition of exponential stability with a given decay rate is given in the following.

Definition 2 *Given $\alpha > 0$, system (1) is said to be exponentially stable with decay rate α if there exists a constant $M > 0$, such that for any initial condition $\phi(\cdot) \in \mathbb{C}([-h, 0], \mathbb{R}^n)$ satisfying that $\phi(s) \succeq 0$ ($s \in [-h, 0]$), the solution of system (1) satisfies that $\|x(t; \phi)\|_\infty \leq M\|\phi\|\mathrm{e}^{-\alpha t}$ for all $t \geq 0$.*

Remark 1 It is well known that all vector norms $\|\cdot\|_p$ ($p \in [1, \infty]$) defined on $\mathbb{R}^n$ are equivalent, that is, for any $p_1 \neq p_2$, there exists constants $c_1, c_2 > 0$, such that $c_1\|x\|_{p_2} \leq \|x\|_{p_1} \leq c_2\|x\|_{p_2}$ for any $x \in \mathbb{R}^n$. Therefore, without loss of generality, one can employ ∞-norm to define the exponential stability with a prescribed decay rate, which is more convenient for the analysis of positive systems.

In the following theorem, based on the Lyapunov–Krasovskii functional as well as a simple transformation, we give a necessary and sufficient condition under which positive system (1) is exponentially stable with a given decay rate.

THEOREM 1 *Suppose that system* (1) *is positive. Then, for any given* $\alpha > 0$, *system* (1) *is exponentially stable with decay rate* α *if and only if Metzler matrix* $\alpha I + A + \int_{-h}^{0} \mathrm{e}^{-\alpha s} A_h(s)\,\mathrm{d}s$ *is Hurwitz.*

Proof (Sufficiency) Construct a Lyapunov–Krasovskii functional:

$$V(x_t) = \lambda^{\mathrm{T}} \mathrm{e}^{\alpha t} x(t) + \lambda^{\mathrm{T}} \int_{-h}^{0} A_h(s) \int_{s+t}^{t} \mathrm{e}^{\alpha(\theta - s)} x(\theta)\,\mathrm{d}\theta\,\mathrm{d}s,$$

where $\lambda \succ 0$. Taking derivative of $V(x_t)$ along the trajectory of system (1) yields that

$$\begin{aligned}\dot{V}(x_t) &= \lambda^{\mathrm{T}} \mathrm{e}^{\alpha t} \left((\alpha I + A) x(t) + \int_{-h}^{0} A_h(s) x(t+s)\,\mathrm{d}s \right) \\ &\quad + \lambda^{\mathrm{T}} \mathrm{e}^{\alpha t} \left(\left(\int_{-h}^{0} \mathrm{e}^{-\alpha s} A_h(s)\,\mathrm{d}s x(t) \right. \right. \\ &\quad \left. - \int_{-h}^{0} A_h(s) x(t+s)\,\mathrm{d}s \right) \\ &= \lambda^{\mathrm{T}} \mathrm{e}^{\alpha t} \left(\alpha I + A + \int_{-h}^{0} \mathrm{e}^{-\alpha s} A_h(s)\,\mathrm{d}s \right) x(t).\end{aligned}$$

By Lemma 1, $\alpha I + A + \int_{-h}^{0} \mathrm{e}^{-\alpha s} A_h(s)\,\mathrm{d}s$ is a Hurwitz matrix implies that there exists a column vector $\lambda \succ 0$, such that $\lambda^{\mathrm{T}}(\alpha I + A + \int_{-h}^{0} \mathrm{e}^{-\alpha s} A_h(s)\,\mathrm{d}s) \prec 0$. Therefore, $\dot{V}(x_t) \leq 0$ for all $t \geq 0$, which implies that $V(x_t) \leq V(x_t)|_{t=0} = \lambda^{\mathrm{T}} x(0) + \lambda^{\mathrm{T}} \int_{-h}^{0} A_h(s) \int_{s}^{0} \mathrm{e}^{\alpha(\theta - s)} x(\theta)\,\mathrm{d}\theta\,\mathrm{d}s$ for all $t \geq 0$. It is obvious that there exists $M > 0$ dependent on h and $\max_{s \in [-h,0]} A_d(s)$, such that $V(x_t) \leq \|\lambda\|_\infty M \|\phi\|$ for all $t \geq 0$. Also note that $V(x_t) \geq \lambda^{\mathrm{T}} \mathrm{e}^{\alpha t} x(t)$ and thus for all $t \geq 0$, we have

$$\begin{aligned}\min_{i=1,2\ldots,n} \{\lambda_i\} \mathrm{e}^{\alpha t} \|x(t)\|_\infty &\leq \min_{i=1,2,\ldots,n} \{\lambda_i\} \mathrm{e}^{\alpha t} \mathbf{1}^{\mathrm{T}} x(t) \\ &\leq \lambda^{\mathrm{T}} \mathrm{e}^{\alpha t} x(t) \leq M \|\lambda\|_\infty \|\phi\|.\end{aligned}$$

Then, it follows that $\|x(t)\|_\infty \leq (M\|\lambda\|_\infty / \min_{i=1,2\ldots,n} \{\lambda_i\}) \mathrm{e}^{-\alpha t} \|\phi\|$ for all $t \geq 0$, which reveals that system (1) is exponentially stable with decay rate α. This completes the sufficiency part.

(Necessity) Suppose that there exists a constant $M > 0$, such that for any initial condition $\phi(s) \succeq 0$ ($s \in [-h, 0]$), the state trajectory of system (1) satisfies that $\|x(t;\phi)\|_\infty \leq M \mathrm{e}^{-\alpha t} \|\phi\|$. For any constant ϵ satisfying that $\alpha > \epsilon > 0$, we define $\beta \triangleq \alpha - \epsilon > 0$ and $y(t) \triangleq \mathrm{e}^{\beta t} x(t)$, then $y(t)$ is the solution of the following system:

$$\begin{aligned}\dot{y}(t) &= \beta \mathrm{e}^{\beta t} x(t) + \mathrm{e}^{\beta t} \dot{x}(t) \\ &= (\beta I + A) y(t) + \int_{-h}^{0} \mathrm{e}^{-\beta s} A_h(s) y(t+s)\,\mathrm{d}s. \quad (2)\end{aligned}$$

Note that $\|y(t)\|_\infty = \mathrm{e}^{\beta t} \|x(t)\|_\infty \leq M \mathrm{e}^{(\beta - \alpha)t} \|\phi\| = M \mathrm{e}^{-\epsilon t} \|\phi\|$. Therefore, system (2) is exponentially stable with decay rate ϵ. By (Ngoc, 2013, Theorem III.1), we have $\beta I + A + \int_{-h}^{0} \mathrm{e}^{-\beta s} A_h(s)\,\mathrm{d}s$ is a Hurwitz matrix. It is well known that the spectrum of a matrix is continuously varying with respect to the variations of its entries and so is the spectral abscissa. For sufficiently small ϵ, we can deduce that $\alpha I + A + \int_{-h}^{0} \mathrm{e}^{-\alpha s} A_h(s)\,\mathrm{d}s$ is a Hurwitz matrix, which completes the necessity part. ■

Remark 2 Unlike (Ngoc 2013), Lyapunov method is utilized here for exponential stability analysis, which can easily take into account decay rate of the state trajectory. It would also be useful in the stability analysis of switched positive systems such as those considered in Zhao et al. (2012) and Zhao et al. (2013). In addition, this approach has potential applications in the analysis of nonlinear positive systems when the nonlinearity fulfills a Lipschitz condition.

4. Extension to positive systems with time-varying discrete and distributed delays

In this section, we aim to extend the results in the last section to positive systems with both time-varying discrete and distributed delays. Let us consider the following continuous-time linear system with both time-varying discrete and distributed delays:

$$\dot{x}(t) = Ax(t) + A_\tau x(t - \tau(t)) + \int_{-h(t)}^{0} A_h(s) x(t+s)\,\mathrm{d}s, \quad (3)$$

where $x(t) \in \mathbb{R}^n$ stands for the state vector; $\tau(t) \leq \tau$ and $h(t) \leq h$ where $\tau, h > 0$ are constants; $\phi(\cdot) \in \mathbb{C}([-\max\{h, \tau\}, 0], \mathbb{R}^n)$ is the initial condition; $A_h(s) \in \mathbb{C}([-h, 0], \mathbb{R}^{n \times n})$.

The positivity of system (3) is characterized in the following lemma.

LEMMA 3 (Ngoc 2013) *For all delays* $\tau(t)$ *and* $h(t)$ *satisfying that* $\tau(t) \leq \tau$ *and* $h(t) \leq h$, *system* (3) *is positive if and only if* A *is Metzler,* A_τ *is nonnegative and* $A_h(s)$ *is nonnegative for all* $s \in [-h, 0]$.

In the following, we always assume that A is Metzler, A_τ is nonnegative and $A_h(s)$ is nonnegative for all $s \in [-h, 0]$. In order to analyze the state trajectory of system (3), we first study the following positive systems with constant discrete delays and distributed delays over a fixed interval:

$$\dot{x}(t) = Ax(t) + A_\tau x(t-\tau) + \int_{-h}^{0} A_h(s)x(t+s)\,\mathrm{d}s. \quad (4)$$

The stability analysis of system (4) can be performed in a manner similar to that of system (1).

THEOREM 2 *Given $\alpha > 0$, positive system (4) is exponentially stable with decay rate α if and only if $\alpha I + A + A_\tau e^{\alpha\tau} + \int_{-h}^{0} e^{-\alpha s} A_h(s)\,\mathrm{d}s$ is a Hurwitz matrix.*

Proof The sufficiency can be proved by constructing the following Lyapunov–Krasovskii functional:

$$V(x_t) = \lambda^{\mathrm{T}} e^{\alpha t} x(t) + \lambda^{\mathrm{T}} A_\tau \int_{t-\tau}^{t} e^{\alpha(\theta+\tau)} x(\theta)\,\mathrm{d}\theta + \lambda^{\mathrm{T}} \int_{-h}^{0} A_h(s) \int_{s+t}^{t} e^{\alpha(\theta-s)} x(\theta)\,\mathrm{d}\theta\,\mathrm{d}s.$$

The necessity can be proved by following a line similar to the proof of Theorem 1 and hence the detailed proof is omitted here. ■

Now we are in the position to investigate the decay rate constrained exponential stability of system (3). Before moving on, the following lemma is needed for further development. The result directly follows from the linearity and the positivity of system (3).

LEMMA 4 *Suppose that $x(t;\phi_1)$ and $x(t;\phi_2)$ are state trajectories of system (3) with initial condition ϕ_1 and ϕ_2, respectively. Then, $\phi_1(s) \preceq \phi_2(s)$ for $s \in [-\max\{h,\tau\},0]$ implies that $x(t;\phi_1) \preceq x(t;\phi_2)$ for $t \geq 0$.*

The following two lemmas aim to give a monotonic property of system (4) with a particular constant initial condition.

LEMMA 5 *Suppose that $\lambda \succ 0$ satisfies that $(A + A_\tau + \int_{-h}^{0} A_h(s)\,\mathrm{d}s)\lambda \prec 0$. Then, the state trajectory of system (4) with initial condition $\phi(s) \equiv \lambda$ $(s \in [-\max\{h,\tau\},0])$ satisfies that $x(t) \preceq \lambda$ for all $t \geq 0$.*

Proof Let $e(t) \triangleq \lambda - x(t)$, then $e(t)$ satisfies that

$$\dot{e}(t) = Ae(t) + A_\tau e(t-\tau) + \int_{-h}^{0} A_h(s)e(t+s)\,\mathrm{d}s - (A + A_\tau + \int_{-h}^{0} A_h(s)\,\mathrm{d}s)\lambda. \quad (5)$$

Noting that $e(s) = 0$ for $s \in [-\max\{h,\tau\},0]$ and system (5) is positive, it follows that $e(t) \succeq 0$ for all $t \geq 0$ by regarding $-(A + A_\tau + \int_{-h}^{0} A_h(s)\,\mathrm{d}s)\lambda$ as a nonnegative input. This implies that $x(t) \preceq \lambda$ for all $t \geq 0$, which completes the proof. ■

LEMMA 6 *Suppose that $\lambda \succ 0$ satisfies that $(A + A_\tau + \int_{-h}^{0} A_h(s)\,\mathrm{d}s)\lambda \prec 0$. Then, the state trajectory of system (4) with initial condition $\phi(s) \equiv \lambda$ $(s \in [-\max\{h,\tau\},0])$ is monotonically non-increasing, that is, $x(t_1) \succeq x(t_2)$ for any $t_2 > t_1 \geq 0$.*

Proof Given any constant $c > 0$, define $e(t) \triangleq x(t) - x(t+c)$, then $e(t)$ satisfies that

$$\dot{e}(t) = Ae(t) + A_\tau e(t-\tau) + \int_{-h}^{0} A_h(s)e(t+s)\,\mathrm{d}s. \quad (6)$$

Note that the initial condition $e(s) = x(s) - x(s+c) = \lambda - x(s+c) \succeq 0$ for $s \in [-\max\{h,\tau\},0]$ by Lemma 5. Then $e(t) \succeq 0$ holds due to the positivity of the error system (6). Therefore, $x(t) \succeq x(t+c)$ for all $t \geq 0$, which completes the proof. ■

Based on the monotonicity of constant delay system (4) with constant initial condition $\phi(s) \equiv \lambda$ $(s \in [-\max\{h,\tau\},0])$, we compare the state trajectories of systems (3) and (4) under this initial condition.

LEMMA 7 *Suppose that $\lambda \succ 0$ satisfies that $(A + A_\tau + \int_{-h}^{0} A_h(s)\,\mathrm{d}s)\lambda \prec 0$ and that $x_1(t)$ and $x_2(t)$ are the state trajectories of systems (3) and (4) with the same initial condition $\phi(s) \equiv \lambda$ $(s \in [-\max\{h,\tau\},0])$, respectively. Then $x_1(t) \preceq x_2(t)$ for all $t \geq 0$.*

Proof Let $e(t) \triangleq x_2(t) - x_1(t)$, then $e(t)$ is the solution of system

$$\dot{e}(t) = Ae(t) + A_\tau e(t-\tau(t)) + \int_{-h(t)}^{0} A_h(s)e(t+s)\,\mathrm{d}s + A_\tau(x_2(t-\tau) - x_2(t-\tau(t))) + \int_{-h}^{-h(t)} A_h(s)x_2(t+s)\,\mathrm{d}s. \quad (7)$$

By Lemma 6, $x_2(t-\tau) - x_2(t-\tau(t)) \succeq 0$ for all $t \geq 0$. Noting that $e(s) = 0$ for $s \in [-\max\{h,\tau\},0]$ and the error system (7) is positive, it follows that $e(t) \succeq 0$ for all $t \geq 0$ by regarding $A_\tau(x_2(t-\tau) - x_2(t-\tau(t))) + \int_{-h}^{-h(t)} A_h(s)x_2(t+s)\,\mathrm{d}s$ as a nonnegative input. Therefore, $x_2(t) \succeq x_1(t)$ for all $t \geq 0$, which completes the proof. ■

In the light of Lemmas 4 and 7, we are ready to state the main theorem of this section.

THEOREM 3 *Given $\alpha > 0$, positive system (3) is exponentially stable with decay rate α for all delays $\tau(t)$ and $h(t)$ satisfying that $\tau(t) \leq \tau$ and $h(t) \leq h$ if and only if positive system (4) is exponentially stable with decay rate α, or equivalently, $\alpha I + A + A_\tau e^{\alpha\tau} + \int_{-h}^{0} e^{-\alpha s} A_h(s)\,\mathrm{d}s$ is a Hurwitz matrix.*

Proof Necessity holds trivially. In the following we prove the sufficiency part. Suppose that positive system (4) is exponentially stable with decay rate α and that $x_1(t;\phi)$ and $x_2(t;\phi)$ are the state trajectories of systems (3) and (4) with the same initial condition ϕ. Then, from (Ngoc, 2013, Theorem III.1), it can be concluded that there exists a column vector $\lambda \succ 0$, such that $(A + A_\tau + \int_{-h}^{0} A_h(s)\,\mathrm{d}s)\lambda \prec 0$. For any initial condition ϕ, one can define $c = \|\phi\| / \min_{j=1,2,\ldots,n}\{\lambda_j\}$, then it is obvious that $c\lambda \succeq \phi(s)$ for $s \in [-\max\{h,\tau\}, 0]$. For initial condition $\phi \equiv c\lambda$, by Lemma 7, we can conclude that $x_1(t;c\lambda) \preceq x_2(t;c\lambda)$ for all $t \geq 0$. By Lemma 4, this implies that $x_1(t;\phi) \preceq x_1(t;c\lambda) \preceq x_2(t;c\lambda)$ for all $t \geq 0$. Since system (4) is exponentially stable with decay rate α, there exists constant $M > 0$, such that $\|x_2(t;c\lambda)\|_\infty \leq M\mathrm{e}^{-\alpha t}c\|\lambda\|_\infty = (M\|\lambda\|_\infty / \min_{j=1,2,\ldots,n}\{\lambda_j\})\mathrm{e}^{-\alpha t}\|\phi\|$. This further implies that $\|x_1(t;\phi)\|_\infty \leq \|x_2(t;c\lambda)\|_\infty \leq (M\|\lambda\|_\infty / \min_{j=1,2,\ldots,n}\{\lambda_j\})\mathrm{e}^{-\alpha t}\|\phi\|$. Hence, system (3) is exponentially stable with decay rate α, which completes the proof. ∎

The above theorem can be easily extended to the multiple discrete delay case, and hence we only present the results in the following corollary while the detailed proof is omitted.

COROLLARY 1 *Suppose that A is Metzler, A_{τ_i}, $i = 1,2,\ldots,N$, are all nonnegative and $A_h(s)$ is nonnegative for all $s \in [-h, 0]$. Consider the following continuous-time linear delay system:*

$$\dot{x}(t) = Ax(t) + \sum_{i=1}^{N} A_{\tau_i} x(t - \tau_i(t)) + \int_{-h(t)}^{0} A_h(s)x(t+s)\,\mathrm{d}s. \tag{8}$$

System (8) is exponentially stable with decay rate $\alpha > 0$ for all delays $\tau_i(t)$ and $h(t)$ satisfying that $\tau_i(t) \leq \tau_i$ ($i = 1,2,\ldots,N$) and $h(t) \leq h$, if and only if one of the following conditions holds:

(1) $\alpha I + A + \sum_{i=1}^{N} A_{\tau_i}\mathrm{e}^{\alpha\tau_i} + \int_{-h}^{0} \mathrm{e}^{-\alpha s}A_h(s)\,\mathrm{d}s$ is a Hurwitz matrix.

(2) there exists a column vector $\lambda \succ 0$, such that $(\alpha I + A + \sum_{i=1}^{N} A_{\tau_i}\mathrm{e}^{\alpha\tau_i} + \int_{-h}^{0} \mathrm{e}^{-\alpha s}A_h(s)\,\mathrm{d}s)\lambda \prec 0$.

5. Numerical example

Let us consider positive delay system (3) with the following system matrices:

$$A = \begin{bmatrix} -5 & 1 \\ 0.8 & -6 \end{bmatrix}, \quad A_\tau = \begin{bmatrix} 0.3 & 0.4 \\ 0.2 & 0.5 \end{bmatrix},$$

$$A_h(s) = \begin{bmatrix} 0 & -0.1s \\ -0.1s^3 & -0.1s \end{bmatrix}, \quad s \in [-3, 0].$$

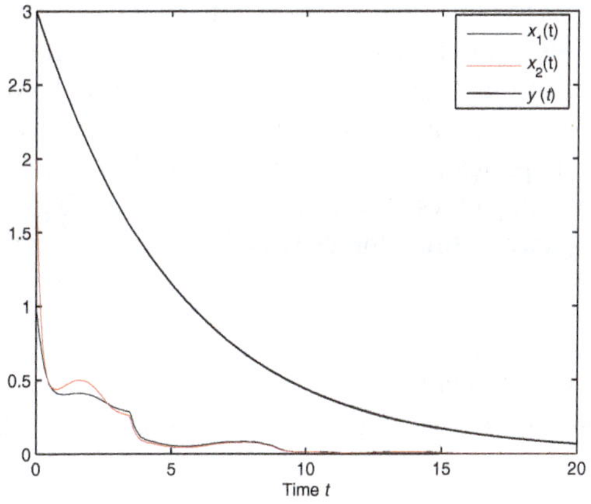

Figure 1. State trajectory of system (3).

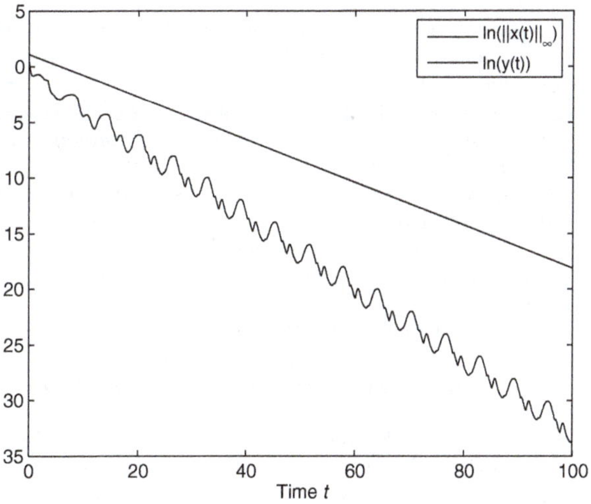

Figure 2. Time evolution of $\ln(\|x(t)\|_\infty)$.

The delays are given as $\tau(t) = 4 + 2\sin t$ and $h(t) = 2 + \sin t$, respectively. According to Theorem 3, one can easily check through linear programming that the decay rate that can be achieved is $\alpha = 0.192$. Given initial conditions $\phi(s) = [1 - \sin(0.1t)\;\; 1 + \cos(0.1t)]^{\mathrm{T}}$ ($t \in [-6, 0]$), the state trajectory of system (3) is depicted in Figure 1. One can observe that the state of system (3) satisfies that $\|x(t)\|_\infty \leq y(t) = 3\mathrm{e}^{-0.192t}$. Therefore, it holds that $\ln(\|x(t)\|_\infty) \leq \ln(y(t)) = \ln 3^{-0.192t}$, which can be verified from Figure 2.

6. Conclusions

In this paper, the decay rate constrained exponential stability problem of positive systems with both time-varying discrete and distributed delays has been investigated. An explicit characterization has been given to ensure that a positive system with distributed delays is exponentially stable and satisfies a prescribed decay rate. The results have been further extended to positive systems with both bounded time-varying discrete delays and distributed delays. In fact, this paper has provided a very simple alternative proof for the results in Liu et al. (2010) and Ngoc (2013). It is worth noting that our approach only relies on the monotonicity and positivity of the system, hence it is flexible and also has potential applications in the analysis of switched positive systems with delays. Applications of the methods proposed in this paper to these topics would be our future research directions.

Acknowledgements

This work was partially supported by GRF HKU 7140/11E.

References

Ait Rami, M. (2011). Solvability of static output-feedback stabilization for LTI positive systems. *Systems & Control Letters, 60*(9), 704–708.

Ait Rami, M., & Napp, D. (2012). Characterization and stability of autonomous positive descriptor systems. *IEEE Transactions on Automatic Control, 57*(10), 2668–2673.

Ait Rami, M., & Tadeo, F. (2007). Controller synthesis for positive linear systems with bounded controls. *IEEE Transactions on Circuits and Systems II: Express Briefs, 54*(2), 151–155.

Berman, A., & Plemmons, R. J. (1994). *Nonnegative matrices.* SIAM: Philadelphia, PA.

Caswell, H. (2001). *Matrix population models: Construction, analysis and interpretation.* Sunderland, MA: Sinauer Associates.

El Ghaoui, L., Oustry, F., & Ait Rami, M. (1997). A cone complementarity linearization algorithm for static output-feedback and related problems. *IEEE Transactions on Automatic Control, 42*(8), 1171–1176.

Farina, L., & Rinaldi, S. (2000). *Positive linear systems: Theory and applications.* New York, NY: Wiley-Interscience.

Feng, J., Lam, J., Li, P., & Shu, Z. (2011). Decay rate constrained stabilization of positive systems using static output feedback. *International Journal of Robust and Nonlinear Control, 21*(1), 44–54.

Haddad, W. M., & Chellaboina, V. (2004). Stability theory for nonnegative and compartmental dynamical systems with time delay. *Systems & Control Letters, 51*(5), 355–361.

Haddad, W. M., Chellaboina, V. S., & Hui, Q. (2010). *Nonnegative and compartmental dynamical systems.* Princeton, NJ: Princeton University Press.

Jacquez, J. (1985). *Compartmental analysis in biology and medicine.* Ann Arbor, MI: University of Michigan Press.

de Jong, H. (2002). Modeling and simulation of genetic regulatory systems: A literature review. *Journal of Computational Biology, 9*(1), 67–103.

Kaczorek, T. (2002). *Positive 1D and 2D systems.* London: Springer-Verlag.

Kaczorek, T. (2008). Fractional positive continuous-time linear systems and their reachability. *International Journal of Applied Mathematics and Computer Science, 18*(2), 223–228.

Kaczorek, T. (2011). Positive linear systems consisting of n subsystems with different fractional orders. *IEEE Transactions on Circuits and Systems I: Regular Papers, 58*(6), 1203–1210.

Li, P., Lam, J., & Shu, Z. (2010). H_∞ positive filtering for positive linear discrete-time systems: An augmentation approach. *IEEE Transactions on Automatic Control, 55*(10), 2337–2342.

Liu, X. (2009). Constrained control of positive systems with delays. *IEEE Transactions on Automatic Control, 54*(7), 1596–1600.

Liu, X., & Dang, C. (2010). Stability analysis of positive switched linear systems with delays. *IEEE Transactions on Automatic Control, 56*(7), 1684–1690.

Liu, X., Yu, W., & Wang, L. (2009). Stability analysis of positive systems with bounded time-varying delays. *IEEE Transactions on Circuits and Systems II: Express Briefs, 56*(7), 600–604.

Liu, X., Yu, W., & Wang, L. (2010). Stability analysis for continuous-time positive systems with time-varying delays. *IEEE Transactions on Automatic Control, 55*(4), 1024–1028.

Ngoc, P. H. A. (2013). Stability of positive differential systems with delay. *IEEE Transactions on Automatic Control, 58*(1), 203–209.

Shen, J., & Lam, J. (2013a). L_∞-gain analysis for positive systems with distributed delays. *Automatica.* Advance online publication. doi:10.1016/j.automatica.2013.09.037

Shen, J., & Lam, J. (2013b). On ℓ_∞ and L_∞ gains for positive systems with bounded time-varying delays. *International Journal of Systems Science.* Advance online publication. doi:10.1080/00207721.2013.843217

Zhao, X., Zhang, L., & Shi, P. (2013). Stability of a class of switched positive linear time-delay systems. *International Journal of Robust and Nonlinear Control, 23*(5), 578–589.

Zhao, X., Zhang, L., Shi, P., & Liu, M. (2012). Stability of switched positive linear systems with average dwell time switching. *Automatica, 48*(6), 1132–1137.

Zhu, S., Li, Z., & Zhang, C. (2012). Exponential stability analysis for positive systems with delays. *IET Control Theory & Applications, 6*(6), 761–767.

Zhu, S., Meng, M., & Zhang, C. (2013). Exponential stability for positive systems with bounded time-varying delays and static output feedback stabilization. *Journal of the Franklin Institute, 350*(3), 617–636,

5

The analysis of the dynamic behavior of the electro-optical bistable systems

Zhang Sheng-Hai*, Yang Hua and Zhao Zhen-Hua

Institute of Science, Information Engineering University, Zhengzhou 450001, People's Republic of China

In this paper, research on the dynamic behaviors of the electro-optical bistable system, especially for its chaotic dynamic behavior, was carried out. The dynamic equation is resolved by the method of the diagrammatized mode. The state of the electro-optical bistable system with changing system parameter is analyzed in detail, and the concrete positions of the bifurcation points are calculated. The system can generate tangent bifurcation and period double bifurcation as the systemic parameter changes, thus the system can generate chaos through period double bifurcation; on the other hand, the system can generate intermittent chaos.

Keywords: electro-optical bistable system chaos; stability; diagrammatized mode; periodic window

PACS: 02.30.Oz; 02.60.Cb; 05.45.-a; 05.45.Pq

1. Introduction

Li (2006) and Szöke, Daneu, Goldhar, and Kurnit, (1969) advanced the optical bistable system for the first time in 1969. Because it was very important in all optical communications, optical switch, optical detect, optical logic gate, optical information storage, optical computer and so on, the optical bistable system received broad attention consequently (Chen, Shen, Jiang, & Shi, 2007; Hua & Lu, 2000; Jin, Chen, Huang, & Jiang, 2009; Wang & Yan, 2009; Yang, Dai, & Zhang, 1994). Optical bistable system includes all optical bistable and hybrid bistable systems (Shen, 2000), and there are many works on the acousto-optical bistable system and the hybrid bistable system (Liu, Wang, Gao, & Lu, 1999; Lv, Li, & Cao, 2004; Réal & Claude, 1985; Zhang, Pan, Luo, & Gao, 2008; Zhang, Zhai, Zhang, & Liu, 2001), but there are few on the electro-optical bistable system. Gao, Narducci, Schulman, Squicciarini, and Yuan (1983) did research on the dynamic characters of the electro-optical bistable system, because this system's structure is simple and can be realized easily in experiment, and aroused the interest of many researchers. Up to now, some researchers have made some works on system chaos with change in the input intensity levels (Hopf, Kaplan, Rose, & Sanders, 1986; Niu, Ma, & Wang, 2009; Zhang, 1997; Zhang, Li, Zheng, Jiang, & Gao, 1998), but there is no work on system chaos caused by the system parameter as we know. In this paper, the stability of the electro-optical bistable system with change in the system parameter is analyzed, and the relation between the periodic windows and intermittent chaos was discovered.

2. The electro-optical bistable system

Figure 1 shows the electro-optical bistable system. LP is the polaroid, A is an amplifier and PLZT is the nonlinear medium made by Pb-based lanthanum-doped zirconate titanate. A/D and D/A stand for the equipments that made the analogue signal to digital signal and digital to analogue, respectively. The light outputs from the He–Ne laser enter into the electro-optic modulator, which is composed of the polaroid LP and PLZT. It is converted into the electrical signal by a photoelectric detector. The signal is delayed by the delay device and feedback to PLZT nonlinear medium. The refractive index is changed nonlinearly, which causes the change in output light intensity.

The dynamic behavior of the electro-optical bistable system can be described by the following equation (Niu et al., 2009; Zhang et al., 1998):

$$\frac{dx(t)}{dt} + x(t) = I\frac{\{1 - k\cos[x(t-\tau_d) + \theta]\}}{2}, \quad (1)$$

where I and $x(t)$ represent the input and output intensity levels, I always represents bifurcation parameter researched in chaos control (Niu et al., 2009; Zhang et al., 1998), $x(t)$ is proportional to the feedback voltage, τ_d is the effective delay time in feedback loop, θ is the extinct parameter and k is the modulation depth of the device, which represents the bifurcation parameter in this paper. Both τ_d and the time variable t are scaled to the natural response time of the electro-optical bistable system.

*Corresponding author. Email: ccstshz@163.com

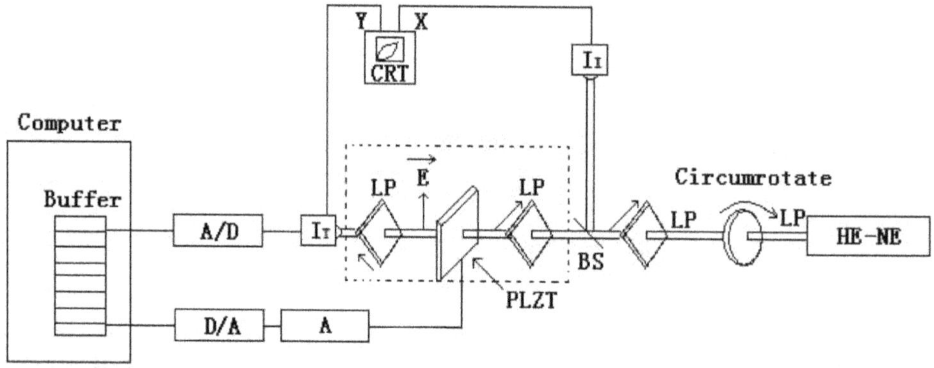

Figure 1. The electro-optical bistable system.

The dynamic equation of the electro-optical bistable system (1) can be expressed as the dimensionless iterate equation at extended delayed:

$$x_{n+1} = I\frac{[1 - k\cos(x_n + \theta)]}{2}, \tag{2}$$

where n represents the iterate times and k *is* the bifurcation parameter.

3. Analysis for the stability of the electro-optical bistable system

In this paper, the parameter I and θ are setas 2 and π, for other values, the results are similar.

3.1. Period one state

When the system is in period one state, and let $x_{n+1} = x_n$, Equation (2) can be expressed as the following equation, x^* is the steady-state solution:

$$x^* = 1 + k\cos x^*. \tag{3}$$

If the stability of the electro-optical bistable system is analyzed with a diagrammatized mode, Equation (3) can be switched as the following equation:

$$\begin{aligned} y &= x^*, \\ y &= 1 + k\cos x^*. \end{aligned} \tag{4}$$

Figure 2 shows the steady period one-state solution of quation (4), we find the points of intersection between the beeline $y = x^*$ and the curves $y = 1 + k\cos x^*$ are steady-state solutions. On the other hand, for Equation (2), if there is a very small deviation ε_0 at x^* for $n = 0$ and ε_k at x^* for $n = k$, then, if $|dy/dx^*| > 1$, $\varepsilon_k \to \infty$ when $k \to \infty$, so, if $|dy/dx^*| > 1$, the steady-state solution is not stable; if $|dy/dx^*| < 1$, $\varepsilon_k \to 0$ when $k \to \infty$, so, the steady-state solution is stable and if $|dy/dx^*| = 1$, the points of k are critical points (bifurcation points) (Liu, 1994). Obviously, $k = \pm 1/\sin x^*$ accords with the conditions. The critical points are also steady-state solutions, so x^* accords with the following equation:

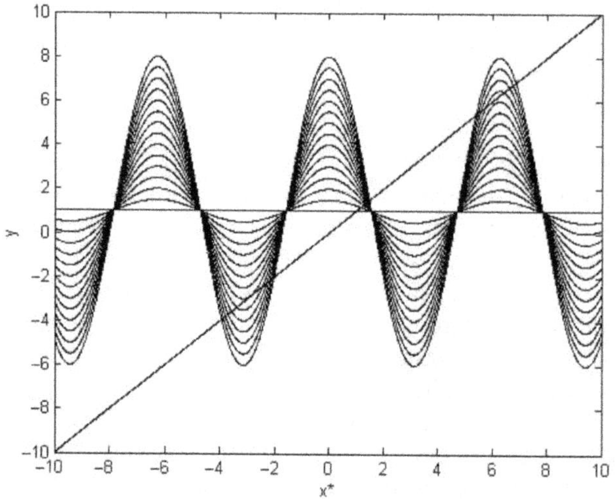

Figure 2. Steady period one-state solution diagram for the diagrammatized mode.

$$k = \frac{x^* - 1}{\cos x^*} = \frac{\pm 1}{\sin x^*}. \tag{5}$$

It can also be written as

$$x^*\sin x^* - \sin x^* \mp \cos x^* = 0. \tag{6}$$

We obtain the values of x^* from Equation (6) through the method of numerical calculation, and substituting these values into Equation (5), we can obtain the values of k. At the points $k = 1.04$, the system bifurcates from period one state to period two state.

In Figure 2, there are tangent points between $y = x^*$ and $y = 1 + k\cos x^*$ at different places. From Figure 3(a) one can find that $y = x^*$ and $y = 1 + 4.05\cos x^*$ have a point of intersection when x^* is close to 1.5. It implies there is a steady-state solution. By numerically solving $x^* = 1 + 4.05\cos x^*$, we can obtain the accurate value $x^* = 1.4576$. When $k = 4.05$ and $x^* = 1.4576$, the slope of $y = 1 + 4.05\ \cos x^*$ equals $-4.0241(|-4.0241| > 1)$,

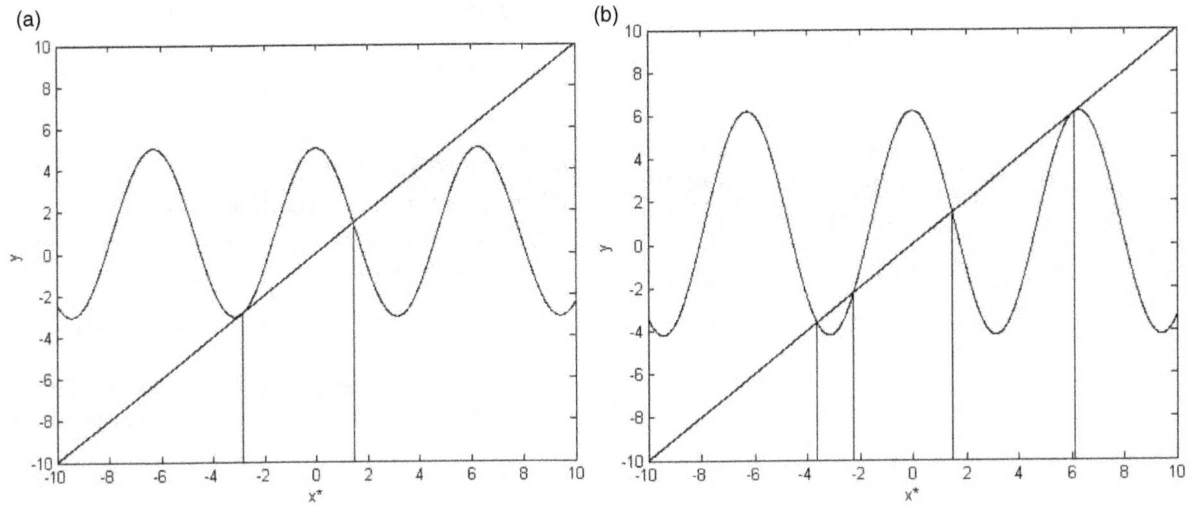

Figure 3. (a) Steady period one-state solution diagram for the diagrammatized mode at $k = 4.05$ and (b) steady period one-state solution diagram for the diagrammatized mode at $k = 5.2$.

which means that this is an unstable point. When x^* is close to -3, $y = x^*$ and $y = 1 + 4.05\ \cos\ x^*$ are tangential, the slope of $y = 1 + 4.05\ \cos\ x^*$ equals 1. It infers that this point is a critical point (bifurcation point). If $k < 4.05$, there is no point of intersection. It means there is no steady-state solution; if $k > 4.05$, the slope's absolute value of $y = 1 + 4.05\ \cos\ x^*$ is less than 1. That is to say, there is a steady-state solution. As $k < 4.05$, a slit will be made between $y = x^*$ and $y = 1 + k\ \cos\ x^*$. In the slit, the difference of evolvement is very small. It is an approximate periodic movement, but with an increase in the iterative times, the value of x will walk out of the slit quickly. This will generate an analogous stochastic oscillation. It implies that the system generates intermittent chaos. Of course, the area of k, which makes the system to be in intermittent chaos state, is not very broad, because the intermittent chaos needs a narrower slit between $y = x^*$ and $y = 1 + k\ \cos x^*$. If the slit becomes very wide, it does not generate intermittent chaos. Therefore, when k is close to 4.05, the system will evolve abruptly from period one state to intermittent chaos with the decline of k.

Figure 3(b) is the steady period one state solution diagram for a diagrammatized mode at $k = 5.2$. From Figure 3(b) we can find there are three points of intersection where x^* is close to -3.6, -2 and 1.4, respectively. It infers that there are three steady-state solutions. We can obtain the accurate values -3.6189, -2.2446 and 1.4789 by numerically solving the equation $x^* = 1 + 5.2\ \cos\ x^*$, and the slopes of $x^* = 1 + 5.2\ \cos\ x^*$ equals -2.3888, 4.0636 and -5.1781, respectively. Their absolute values are all more than 1, thus the three points of intersection are all unstable points. When $x^* = 6.0174$, $y = x^*$ and $y = 1 + 5.2\ \cos\ x^*$ are tangential, and the slope of $y = 1 + 5.2\ \cos\ x^*$ equals 1. It implies that this point is also a critical point (bifurcation point). When $k < 5.2$, there is no point of intersection. It infers that there is no steady-state solution, but at the side of $k > 5.2$, the slope's absolute value of $y = 1 + 5.2\ \cos\ x^*$ could be less than 1. That is to say, there is a steady-state solution. When $k < 5.2$, a slit will be made between $y = x^*$ and $y = 1 + k\ \cos\ x^*$, it can also generate intermittent chaos. Therefore, when k is close to 5.2, the system will evolve abruptly from period one state to intermittent chaos with the decline of k.

3.2. *Period two state*

When the system is in period two state, and let $x_{n+2} = x_n$, Equation (2) can be expressed as the following equation, x^* is the steady-state solution:

$$x^* = 1 + k\cos(1 + k\ \cos x^*). \tag{7}$$

If the stability of the electro-optical bistable system is analyzed with a diagrammatized mode, Equation (7) can be switched as the following equation:

$$\begin{aligned} y &= x^*, \\ y &= 1 + k\cos(1 + k\cos x^*). \end{aligned} \tag{8}$$

Figure 4 is the steady period two-state solution diagram for the diagrammatized mode. In this diagram, the points of intersection are between the beeline $y = x^*$ and the curves $y = 1 + k\ \cos(1 + k\ \cos x^*)$ are steady-state solutions. At the same time, when the absolute value of $dy/dx^* = k^2\ \sin\ x^*\ \sin(1 + k\ \cos x^*)$ equals 1, the point k is the critical point (bifurcation point).

From Figure 5(a) we find that there are three points of intersection where x^* are close to -1.2, 1.5 and 2.3. It implies that there are steady-state solutions whose accurate values are -1.188, 1.4408 and 2.2699. The slopes of $x^* = 1 + 3.4\ \cos(1 + 3.4\ \cos\ x^*)$ equal -8.2075, 11.36576 and -8.2079, their absolute values are all more than 1, and

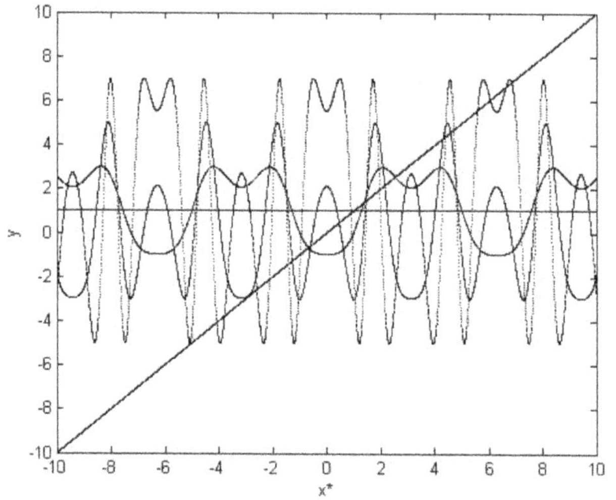

Figure 4. Steady period two-state solution diagram for the diagrammatized mode.

the three points of intersection are all unstable points. When x^* are close to 0 and 4.4, $y = x^*$ and $y = 1 + 3.4\ \cos(1 + 3.4\ \cos\ x^*)$ are tangential, the slope of $y = 1 + 3.4\ \cos(1 + 3.4\ \cos\ x^*)$ equals1. It infers that these points are also critical points (bifurcation points). When $k < 3.4$, there is no point of intersection. It implies there is no steady-state solution; but at the side of $k > 3.4$, the slope's absolute value of $y = 1 + 3.4 \cos(1 + 3.4 \cos x^*)$ could be less than 1. That is to say, there is a steady-state solution. When $k < 3.4$, a slit will be made between $y = x^*$ and $y = 1 + 3.4\ \cos(1 + 3.4\ \cos\ x^*)$, it also generates intermittent chaos. Thus, the system will evolve abruptly from period two state to intermittent chaos with the decline of k when k is close to 3.4. Figure 5(b) is the steady period two-state solution diagram for the diagrammatized mode at $k = 6.06$. The case of $k = 6.06$ is the same as $k = 3.4$. Therefore, when k is close to 6.06, the system will also evolve abruptly from period two state to intermittent chaos with the decline of k.

3.3. *Period three state*

When the system is in period three state, and let $x_{n+3} = x_n$, Equation (2) can be expressed as the following equation, x^* is the steady-state solution:

$$x^* = 1 + k \cos(1 + k \cos(1 + k \cos x^*)). \tag{9}$$

If the stability of the electro-optical bistable system is analyzed with the diagrammatized mode, Equation (9) can be switched as the following equation:

$$\begin{aligned} y &= x^*, \\ y &= 1 + k \cos(1 + k \cos(1 + k \cos x^*)). \end{aligned} \tag{10}$$

From Figure 6 one can find that the points of intersection are between the beeline $y = x^*$ and the curves $y = 1 + k\ \cos(1 + k\ \cos(1 + k\ \cos\ x^*))$ are steady-state solutions. At the same time, the absolute value of $dy/dx^* = -k^3\ \sin x^*\ \sin(1 + k\ \cos x^*)\ \sin(1 + k\ \cos(1 + k\ \cos\ x^*))$ equals 1; hence, the points of k are critical points (bifurcation points).

Figure 7(a)–(d) is the steady period three-state solution diagrams for the diagrammatized mode at k equalling 2.68, 3.18, 4.88 and 5.88. At the points of $k = 2.68$, 3.18, 4.88 and 5.88, the beeline $y = x^*$ and the curves $y = 1 + k\ \cos(1 + k\ \cos(1 + k\ \cos x^*))$ are tangential, the slope of $y = 1 + k\ \cos(1 + k\ \cos(1 + k\ \cos\ x^*))$ equals 1. It implies that these points are also critical points (bifurcation points). When $k < 2.68$, there is no point of intersection, i.e. there is no steady-state solution also; if $k > 2.68$, the slope's absolute value of $y = 1 + 2.68\ \cos(1 +$

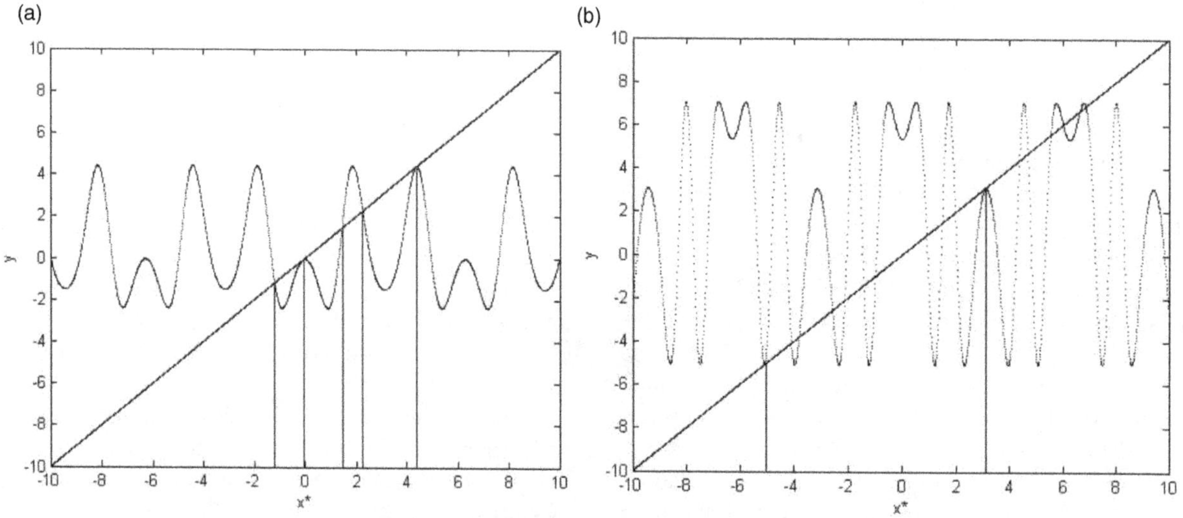

Figure 5. (a) Steady period two-state solution diagram for the diagrammatized mode at $k = 3.4$ and (b) steady period two-state solution diagram for the diagrammatized mode at $k = 6.06$.

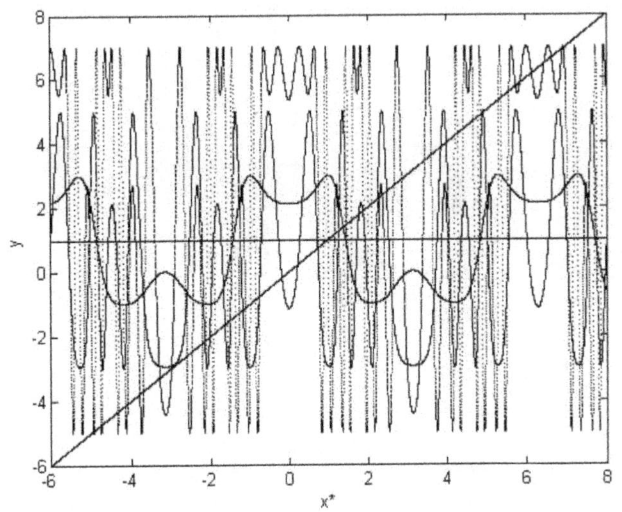

Figure 6. Steady period three-state solution diagram for the diagrammatized mode.

2.68 cos(1 + 2.68 cos x^*)) is less than 1, then there is a steady-state solution. When $k < 2.68$, a slit will be made between $y = x^*$ and $y = 1 + 2.68 \cos(1 + 2.68 \cos(1 + 2.68 \cos x^*))$, it can generate intermittent chaos. Therefore, when k is close to 2.68, the system will evolve abruptly from period three state to intermittent chaos with the decline of k. The cases where k equals 3.18, 4.88 and 5.88 are alike. The system also can generate intermittent chaos at these points.

3.4. Period m state

We can extend this method to a random period. When the system is in period m state (m is a random positive integer), and let $x_{n+m} = x_n$, Equation (2) can be expressed as the following equation, where x^* is the steady-state solution:

$$x^* = 1 + k \cos(1 + k \cos(1 + k \cos(\cdots))). \qquad (11)$$

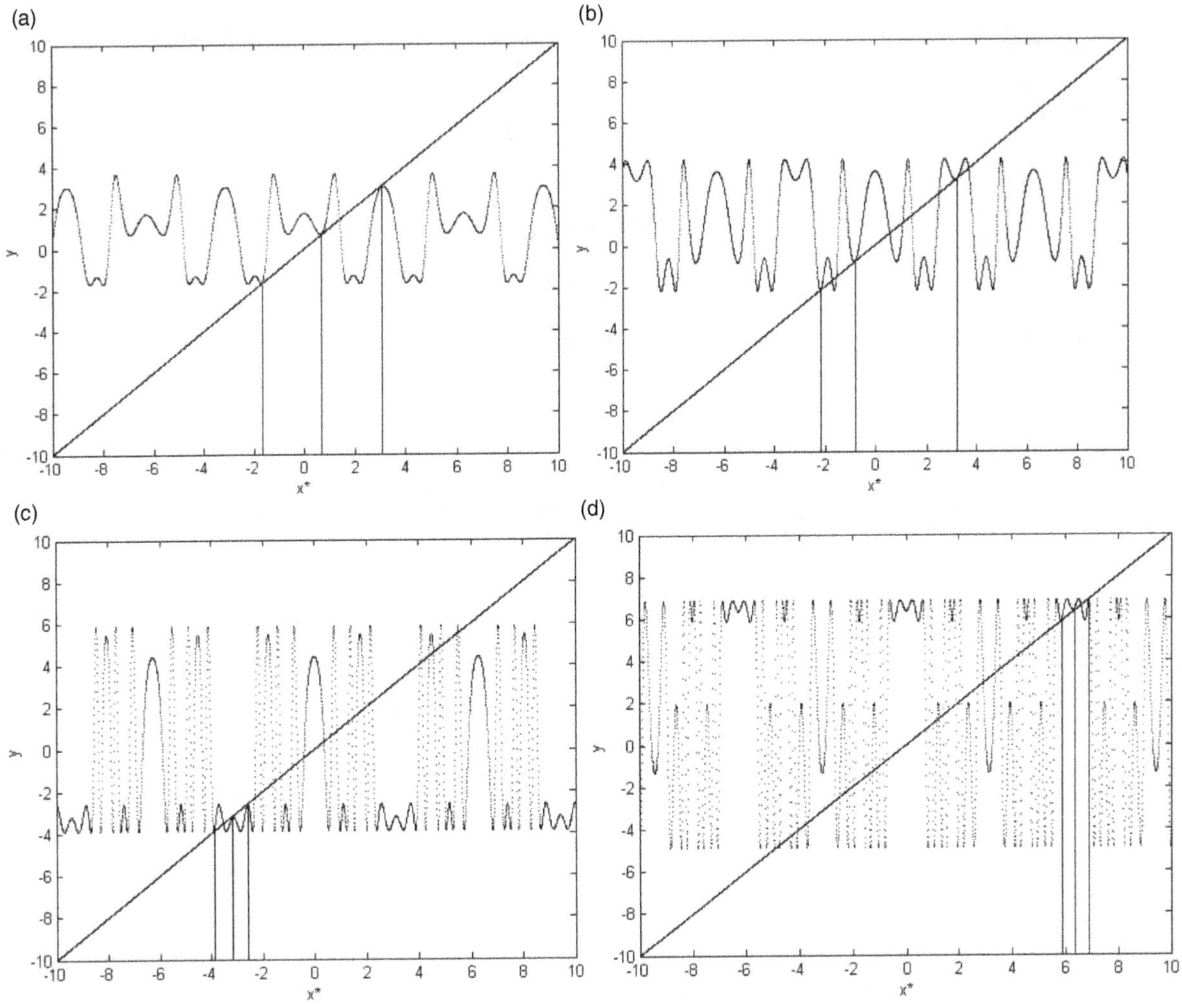

Figure 7. (a) Steady period three-state solution diagram for the diagrammatized mode at $k = 2.68$, (b) steady period three-state solution diagram for the diagrammatized mode at $k = 3.18$, (c) steady period three-state solution diagram for the diagrammatized mode at $k = 4.88$ and (d) steady period three-state solution diagram for the diagrammatized mode at $k = 5.88$.

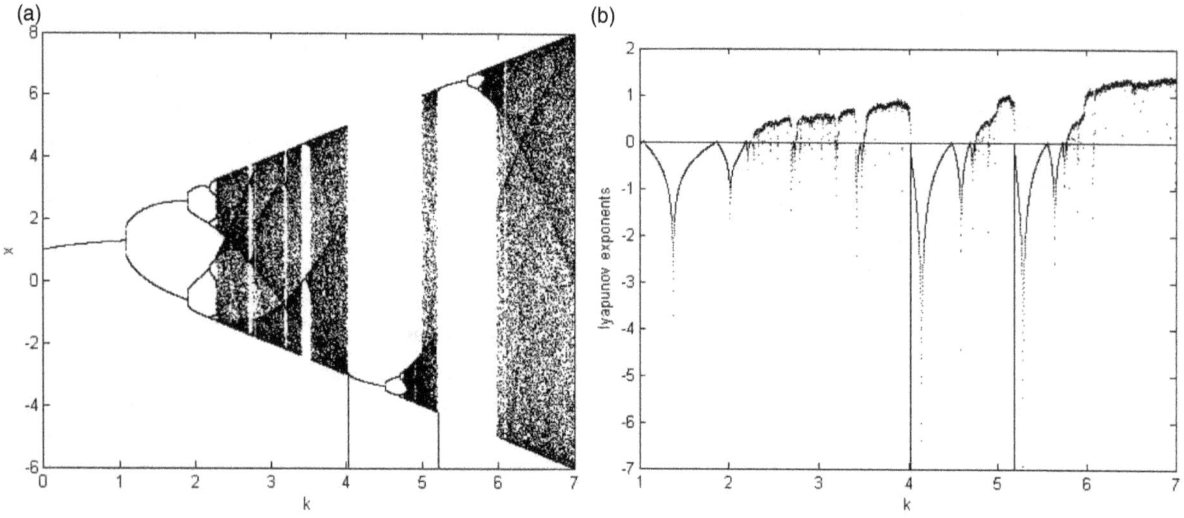

Figure 8. (a) Bifurcation diagram of the electro-optical bistable system and (b) the largest Lyapunov exponent diagram of the electro-optical bistable system.

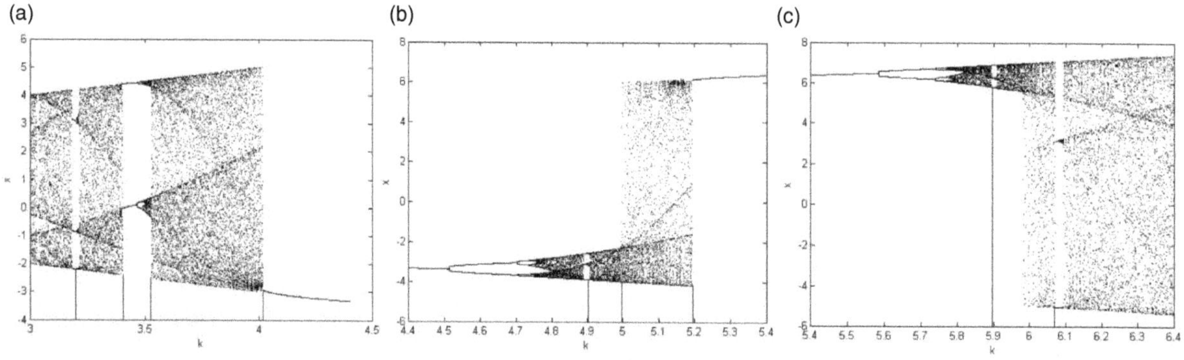

Figure 9. (a) Bifurcation diagram at $k \in (3, 4.4)$, (b) bifurcation diagram at $k \in (4.4, 5.4)$ and (c) bifurcation diagram at $k \in (5.4, 6.4)$.

If the stability of the electro-optical bistable system is analyzed with the diagrammatized mode, Equation (11) can be switched as the following equation:

$$\begin{aligned} y &= x^*, \\ y &= 1 + k\cos(1 + k\cos(1 + k\cos(\cdots))). \end{aligned} \tag{12}$$

4. Results of numerical calculation

The parameters I and θ are the same as mentioned in Section 3. Figure 8(a) shows the bifurcation diagram of the electro-optical bistable system with k changing. From Figure 8(a), one can find, as k changes the system goes into period one, period two, period four, period eight and chaotic state successively, on the other hand, there are many transparent windows (periodic windows) in the chaos area. For example, when k is close to 2.6, there is a window of period three, which agrees with the theory that period three implies chaos advanced by Li Tianyan and York in 1973 (Shen, 2000), and accords with the results analyzed by Figure 7(a).

Figure 8(b) is the largest Lyapunov exponent (λ_L) diagram corresponding to Figure 8(a). Figure 8(b) shows critical points (where $\lambda_L = 0$) are in accordance with the bifurcation points in Figure 8(a). By comparing Figure 8(a) and 8(b), one can find, for $\lambda_L > 0$, the system is in chaotic states including the intermittent chaos and the chaos being generated through doubling bifurcations.

Figure 8(a) shows that the system is in period one state for $k \in (0, 1)$, and is in period two state for $k \in (1, 1.8)$; obviously, $k = 1$ is the system bifurcate point from period one to period two state. Then, the system enters into chaos through period double bifurcates if $k > 2.2$. For $k \in (3.5, 4.5)$ there are a few fuscous lines in the chaotic region, and the boundary is orderly because the chaos is generated through period doubling bifurcations and intermittent chaos. The system changes from tangent bifurcation to period one state. As k increases the system goes again into period one, period two, period four, period eight and chaos state, successively. If $k \in (5, 5.2)$, there is a fuscous area in the hyponastic chaos. It is also caused by period doubling bifurcations and intermittent chaos, and others only caused by intermittent chaos. The system changes from tangent bifurcation to period one state again as k is close to 5.2. With k increasing the system goes again into period one, period two, period four, period eight and chaos state, successively. This phenomenon accords with the results which are analyzed by the diagrammatized mode.

Figure 9(a) is the bifurcation diagram at $k \in (3, 4.4)$. When k is close to 3.2, there is a window of period three, which is in accordance with the results analyzed in Figure 7(b). There is a window of period two when k is close to 3.4. It is in accordance with the results which are analyzed in Figure 5(a). The system brings tangent bifurcation at $k = 4.05$. The point with $k = 4.05$ and $x^* = 1.45$ is the end point of the branch in period doubling bifurcation, which is in accordance with the results that are analyzed in Figure 3(a). Figure 9(b) is the bifurcation diagram at $k \in (4.4, 5.4)$, when k is close to 4.9 there is a window of period three, which is in accordance with the results analyzed in Figure 7(c). The system brings tangent bifurcation at $k = 5.2$. The points that $k = 4.05$, $x^* = -3.6$, -2 and 1.4 are the end points of the branches in period doubling bifurcation, which is in accordance with the results analyzed in Figure 3(b). Figure 9(c) is the bifurcation diagram at $k \in (5.4, 6.4)$, when k is close to 5.9 there is a window of period three, which is in accordance with the results analyzed in Figure 7(d). When k is close to 6.06, there is a window of period two, which is in accordance with the results analyzed in Figure 5(b). These phenomena are all in accordance with the results that are analyzed by the diagrammatized mode.

5. Results and discussion

In this paper, the stability of the electro-optical bistable system is theoretically analyzed by the diagrammatized mode. It is analyzed for the stability of the electro-optical bistable system with change in the system parameter. It infers that the chaos can be generated through period doubling bifurcations and intermittent chaos in this system. The results of numerical calculation show that the analysis in theory is reasonable. We discover that periodic window implies intermittent chaos from Figure 3(a)–(b), Figure 5(a)–(b) and Figure 7(a)–(d). This work offers a useful method to research periodic window.

Acknowledgement

This project was supported by the National Natural Science Foundation of China (grant number 10975047).

References

Chen, X. F., Shen, X. M., Jiang, M. P., & Shi, D. F. (2007). Study on the bistability of nonlinear BRAGG microcavity. *Acta Photonica Sinica, 36*(4), 613–616.

Gao, J. Y., Narducci, L. M., Schulman, L. S., Squicciarini, M., & Yuan, J. M. (1983). Route to chaos in a hybrid bistable system with delay. *Physical Review A, 28*(5), 2910–2914.

Hopf, F. A., Kaplan, D. L., Rose, M. H., & Sanders, L. D. (1986). Characterization of chaos in a hybrid optically bistable device. *Physical Review Letters, 57*(12), 1394–1397.

Hua, C. C., & Lu, Q. S. (2000). The time-dependent bifurcation and dynamical behavior for absorptive optical bistability equation. *Acta Physica Sinica, 49*(4), 733–740.

Jin, Y., Chen, X. F., Huang, Z. Y., & Jiang, M. P. (2009). Optical bistability of a nonlinear microcavity. *Chinese Journal of Quantum Electronics, 26*(5), 591–595.

Li, F. L. (2006). *Advanced laser physics*. Beijing: Higher Education Press.

Liu, B. Z. (1994). *Nonlinear dynamics and chaos based*. Changchun: Northeast Normal University Press.

Liu, Y. H., Wang, S. Y., Gao, M. H., & Lu, Y. Q. (1999). Chaos modulation effect of an hcoustooptic bistable system. *Acta Physica Sinica, 48*(5), 795–801.

Lv, L., Li, G., & Cao, H. J. (2004). Chaos control of acousto-optical bistable system. *Chinese Journal of Lasers, 31*(2), 161–163.

Niu, Y. D., Ma, W. Q., & Wang, R. (2009). Chaos control and synchronization in electrical-optical bistable systems. *Acta Physica Sinica, 58*(5), 2934–2938.

Réal, V., & Claude, D. (1985). Route to chaos in an acousto-optic bistable device. *Physical Review A, 31*(4), 2390–2396.

Shen, K. (2000). *Chaos in optics*. Changchun: Northeast Normal University Press.

Szöke, A., Daneu, V., Goldhar, J., & Kurnit, N. A. (1969). Bistable optical element and its applications. *Applied Physics Letters, 15*(3), 376–380.

Wang, B., & Yan, S. P. (2009). Correlation function of an optical bistable system with cross-correlated additive white noise and multiplicative colored noise. *Chinese Optics Letters, 7*(9), 838–840.

Yang, Y., Dai, J. H., & Zhang, H. J. (1994). The dynamical behavior of a discrete model of optical bistable system. *Acta Physica Sinica, 43*(5), 699–706.

Zhang, H. J. (1997). *Optical chaos*. Shanghai: Shanghai Scientific and Technological Education Publishing House.

Zhang, T., Zhai, A. M., Zhang, X. Y., & Liu, P. T. (2001). Hyperchaos control with an external periodic stimulation signal to a two-order cascading acousto-optical bistale system. *Acta Photonica Sinica, 30*(7), 818–827.

Zhang, W., Pan, W., Luo, B., & Gao, Y. (2008). The chaos control by delayed feedback in Bragg acousto-optic bistable system. *Laser Journal, 29*(3), 38–40.

Zhang, Y., Li, J. B., Zheng, Z. R., Jiang, Y., & Gao, J. Y. (1998). Dynamic storage function by chaos control in a hybrid bistable system. *Physical Review E, 57*(2), 1611–1614.

6

Stability analysis of fractional-order systems with the Riemann–Liouville derivative

Zhiquan Qin*, Ranchao Wu and Yanfen Lu

School of Mathematics, Anhui University, Hefei 230601, People's Republic of China

In this paper, the stability of fractional-order systems with the Riemann–Liouville derivative is discussed. By applying the Mittag-Leffler function, generalized Gronwall inequality and comparison principle to fractional differential systems, some sufficient conditions ensuring stability and asymptotic stability are given.

Keywords: asymptotic stability; generalized Gronwall inequality; Riemann–Liouville derivative; fractional differential system

1. Introduction

Fraction calculus has more than 300 years history. With the development of science and engineering applications, fractional calculus has become one of the most hottest topics. Up to now, many fractional results have been presented which are very useful (Debnath, 2004; Miller & Ross, 1993; Podlubny, 1999; Samko, Kilbas, & Marichev, 1993; Zhang & Li, 2011).

Stability analysis is the most fundamental for studying fractional differential equations. Recently, many stability results of fractional-order systems are interesting in physical systems, so more and more stability results have been found, see, for instance, Ahn and Chen (2008), Ahmed, EI-Saka, and EI-Saka (2007), Deng, Li, and Liu, (2007), Li, Chen, and Podlubny (2009), Li and Zhang (2011), Miller and Ross (1993), Moze, Sabatier, and Oustaloup (2007), Odibat (2010), Qian, Li, Agarwal, and Wong (2010), Radwan, Soliman, Elwakil, and Sedeek (2009), Sabatier, Moze, and Farges (2010), Samko et al. (1993), Tavazoei and Haeri (2009), Wen, Wu, and Lu, (2008) and Zhang and Li (2011). These stability results are mainly concerned with the linear fractional differential system. For example, in Matignon (1996), a sufficient and necessary condition on asymptotic stability of linear fractional differential system with order $0 < \alpha < 1$ was first given. Then some other research on the stability of fractional-order systems appeared. Of course, there also exist fractional-order systems with order lying in (1, 2). In Zhang and Li (2011), authors dealt with the following fractional differential system:

$$D^{\alpha}_{t_0,t}x(t) = Ax(t) + B(t)x(t),$$

where $1 < \alpha < 2, D^{\alpha}_{t_0,t}$ denotes either the Caputo or the Riemann–Liouville fractional derivative operator. They analysed stability of the above fractional differential system by applying Gronwall's inequality (Corduneanu, 1971) and related results.

In this paper, three conditions about $B(t)$ are given as follows:

(I) $0 < \alpha < 1, \int_0^{\infty} \mathrm{P}B(t)\mathrm{P}dt$ is bounded;
(II) $1 < \alpha < 2, \|B(t)\|$ is bounded;
(III) $1 < \alpha < 2, B(t) = O(t - t_0)^{\theta}(\theta < -\alpha, t_0 > 0)$.

Under these conditions, the stability and asymptotic stability of nonautonomous linear fractional differential systems with the Riemann–Liouville derivative are analysed by using generalized Gronwall's inequality, some properties of the Mittag-Leffler function and relevant results. From the results derived in this paper, we can also analyse the stability of these nonlinear systems in the future.

This paper is organized as follows. In Section 2 some necessary definitions and lemmas are recalled, which will be used later. The main results are presented in Section 3. Finally, some conclusions are drawn in Section 4.

2. Preliminaries

In this section, the most commonly used definitions and results are stated, which will be used later.

DEFINITION 2.1 *The Riemann–Liouville fractional derivative with order α of function $x(t)$ is defined as*

$${}_{RL}D^{\alpha}_{a,t}x(t) = \frac{1}{\Gamma(m-\alpha)}\frac{\mathrm{d}^m}{\mathrm{d}t^m}\int_a^t (t-\tau)^{m-\alpha-1}x(\tau)\mathrm{d}\tau,$$

where $m - 1 \leq \alpha < m, \Gamma(\cdot)$ is the Gamma function.

*Corresponding author. Email: zquan2174@163.com

The Laplace transform of the Riemann–Liouville fractional derivative is

$$\int_0^{\infty} e^{-st}{}_{\mathrm{RL}}D^{\alpha}_{a,t}x(t)\mathrm{d}t = s^{\alpha}X(s) - \sum_{k=0}^{n-1} s^k[D^{\alpha-k-1}x(t)]_{t=a} \quad (n-1 \le \alpha < n).$$

DEFINITION 2.2 *The Mittag-Leffler function with two parameters is defined as*

$$E_{\alpha,\beta}(z) = \sum_{k=0}^{\infty} \frac{z^k}{\Gamma(k\alpha+\beta)},$$

where $\alpha > 0, \beta > 0, z \in C$. *When* $\beta = 1$, *one has* $E_{\alpha}(z) = E_{\alpha,1}(z)$, *furthermore,* $E_{1,1}(z) = e^z$.

The Laplace transform of the Mittag-Leffler function is

$$\int_0^{\infty} e^{-st}t^{k\alpha+\beta-1}E^{(k)}_{\alpha,\beta}(\pm at^{\alpha})\mathrm{d}t = \frac{k!s^{\alpha-\beta}}{(s^{\alpha} \mp a)^{k+1}} \quad (\Re(s) > |a|^{1/n}).$$

DEFINITION 2.3 *The zero solution of*

$${}_{RL}D^{\alpha}_{t_0,t}x(t) = f(t, x(t)),$$

with order $0 < \alpha \le 1 (1 < \alpha < 2)$ *is said to be stable if, for any initial values* $x_k(k = 0)(x_k(k = 0, 1))$, *there exists* $\varepsilon > 0$ *such that* $\|(t)\| < \varepsilon$ *for all* $t > t_0$. *The zero solution is said to be asymptotically stable if, in addition to being stable,* $\|x(t)\| \to 0$ *as* $t \to +\infty$.

LEMMA 1 *If* $A \in C^{n\times n}$ *and* $0 < \alpha < 2, \beta$ *is an arbitrary real number,* μ *satisfies* $\alpha\pi/2 < \mu < \min\{\pi, \alpha\pi\}$, *and* $C > 0$ *is a real constant, then*

$$\left\|E_{\alpha,\beta}(A)\right\| \le \frac{C}{1+\|A\|},$$

where $\mu \le |\arg(spec(A))| \le \pi$, *spec(A) denotes the eigenvalues of matrix A and* $\|\cdot\|$ *denotes the* l_2 *norm.*

LEMMA 2 *If* $A \in C^{n\times n}$ *and* $0 < \alpha < 2, \beta$ *is an arbitrary complex number and* μ *satisfies* $\alpha\pi/2 < \mu < \min\{\pi, \alpha\pi\}$, *then for an arbitrary integer* $p \ge 1$, *the following expansions hold:*

$$E_{\alpha,\beta}(z) = \frac{1}{\alpha}z^{1-\beta/\alpha}\exp(z^{1/\alpha}) - \sum_{k=1}^{p} \frac{z^{-k}}{\Gamma(\beta-k\alpha)} + O(|z|^{-p-1}),$$

with $|z| \to \infty, |\arg(z)| \le \mu$ *and*

$$E_{\alpha,\beta}(z) = -\sum_{k=1}^{p} \frac{z^{-k}}{\Gamma(\beta-k\alpha)} + O(|z|^{-p-1}),$$

with $|z| \to \infty$ *and* $\mu < |\arg(z)| \le \pi$.

Especially, in Zhang and Li (2011) it has been obtained that the matrix $(t-t_0)^{\alpha-k-1}E_{\alpha,\alpha-k}(A(t-t_0)^{\alpha})$ is bounded, i.e.

$$\|(t-t_0)^{\alpha-k-1}E_{\alpha,\alpha-k}(A(t-t_0)^{\alpha})\| \le M_k,$$

for some $M_k > 0$.

LEMMA 3 *Suppose* $\alpha > 0, a(t)$ *is a nonnegative locally integrable function on* $0 \le t < T$ *(some* $T \le \infty$*) and* $g(t)$ *is a nonnegative and nondecreasing continuous function defined on* $0 \le t < T, g(t) \le M$ *(constant), and suppose* $u(t)$ *is nonnegative and locally integrable on* $0 \le t < T$ *with*

$$u(t) \le a(t) + g(t)\int_0^t (t-s)^{\alpha-1}u(s)\mathrm{d}s,$$

on this interval, then

$$u(t) \le a(t) + \int_0^t \left[\sum_{n=1}^{\infty} \frac{(g(t)\Gamma(\alpha))^n}{\Gamma(n\alpha)}(t-s)^{n\alpha-1}a(s)\right]\mathrm{d}s.$$

Moreover, if $a(t)$ is a nondecreasing function on $[0, T)$, then

$$u(t) \le a(t)E_{\alpha}(g(t)\Gamma(\alpha)t^{\alpha}).$$

LEMMA 4 *Suppose that* $g(t)$ *and* $u(t)$ *are continuous on* $[t, t_0], g(t) \ge 0, \lambda \ge 0$ *and* $r \ge 0$ *are two constants, if*

$$u(t) \le \lambda + \int_{t0}^t [g(\tau)u(\tau) + r]d\tau,$$

then

$$u(t) \le (\lambda + r(t_1 - t_0))\exp\int_{t0}^t g(\tau)d\tau, \quad t_0 \le t \le t_1.$$

3. Stability of nonautonomous linear fractional differential systems

3.1. *Fractional-order* $\alpha : 0 < \alpha < 1$

Consider the nonautonomous fractional system

$${}_{\mathrm{RL}}D^{\alpha}_{0,t}x(t) = Ax(t) + B(t)x(t) \quad (0 < \alpha < 1), \tag{1}$$

with the initial condition

$${}_{\mathrm{RL}}D^{\alpha-1}_{0,t}x(t)|_{t=0} = x_0, \tag{2}$$

where $x \in R^n$, matrix $A \in R^{n\times n}, B(t) : [0, \infty] \to R^{n\times n}$ is a continuous t matrix.

THEOREM 1 *Suppose* $\|t^{\alpha-1}E_{\alpha,\alpha}(At^{\alpha})\| \le Me^{-\gamma t}, 0 \le t < \infty, \gamma > 0$ *and* $\int_0^{\infty}\|B(t)\|\mathrm{d}t$ *is bounded, i.e.* $\int_0^{\infty}\|B(t)\|\mathrm{d}t \le N$, *where* $M, N > 0$, *then the solution of Equation (1) is asymptotically stable.*

Proof By the Laplace transform and the inverse Laplace transform, the solution of Equations (1) with (2) can be written as

$$x(t) = t^{\alpha-1}E_{\alpha\alpha}(At^{\alpha})x_0 + \int_0^t (t-\tau)^{\alpha-1}E_{\alpha\alpha}(A(t-\tau)^{\alpha})B(\tau)x(\tau)\mathrm{d}\tau,$$

then we can obtain

$$\|x(t)\| \leq \|t^{\alpha-1}E_{\alpha\alpha}(At^{\alpha})\|\,\|x_0\| + \int_0^t \|(t-\tau)^{\alpha-1}E_{\alpha\alpha}(A(t-\tau)^{\alpha})\|\,\|B(\tau)\|\,\|x(\tau)\|\mathrm{d}\tau.$$

From the boundedness, we can obtain

$$\|x(t)\| \leq Me^{-\gamma t}\|x_0\| + \int_0^t Me^{-\gamma(t-\tau)}\|B(\tau)\|\,\|x(\tau)\|\mathrm{d}\tau. \quad (3)$$

Multiplying by $e^{\gamma t}$ both sides of Equation (3), we have

$$e^{\gamma t}\|x(t)\| \leq M\|x_0\| + \int_0^t Me^{\gamma\tau}\|B(\tau)\|x(\tau)\|\mathrm{d}\tau.$$

Let $e^{\gamma t}\|x(t)\| = u(t)$, then according to Lemma 4, one has

$$e^{\gamma t}\|x(t)\| \leq (M\|x_0\|)\exp\left(M\int_0^t \|B(t)\|\mathrm{d}t\right). \quad (4)$$

Multiplying by $e^{-\gamma t}$ both sides of Equation (4), we can obtain

$$\|x(t)\| \leq (M\|x_0\|)\exp\left(M\int_0^t \|B(t)\|\mathrm{d}t\right)e^{-\gamma t},$$

then $\|x(t)\| \leq (M\|x_0\|)e^{MN-\gamma t}$, so $\|x(t)\| \to 0, t \to \infty$.

That is, the solution of Equation (1) is asymptotically stable. ■

3.2. *Fractional-order* α : $1 < \alpha < 2$

Consider the following fractional-order system:

$${}_{\mathrm{RL}}D^{\alpha}_{t_0,t}x(t) = Ax(t) + B(t)x(t) \quad (1 < \alpha < 2), \quad (5)$$

with the initial conditions

$${}_{\mathrm{RL}}D^{\alpha-k}_{t_0,t}x(t)|_{t=t0} = x_{k-1} \quad (k = 1, 2), \quad (6)$$

where $x \in R^n$, matrix $A \in R^{n\times n}, B(t) : [t_0, \infty) \to R^{n\times n}$ is a continuous matrix.

THEOREM 2 *If the eigenvalues of matrix A satisfy* $|\arg(\lambda(A))| > \alpha\pi/2$ *and* $\|B(t)\|$ *is bounded, i.e.* $\|B(t)\| \leq M$ *for some* $M > 0$, *then the zero solution of Equation (5) is asymptotically stable.*

Proof By the Laplace transform and the inverse Laplace transform, the solution of Equations (5) with (6) can be written as

$$x(t) = (t-t_0)^{\alpha-1}E_{\alpha\alpha}(A(t-t_0)^{\alpha})x_0 + (t-t_0)^{\alpha-2}E_{\alpha\alpha-1}(A(t-t_0)^{\alpha})x_1 + \int_{t_0}^t (t-\tau)^{\alpha-1}E_{\alpha\alpha}(A(t-\tau)^{\alpha})B(\tau)x(\tau)\mathrm{d}\tau,$$

then we can obtain

$$\begin{aligned}\|x(t)\| &\leq \|(t-t_0)^{\alpha-1}E_{\alpha\alpha}(A(t-t_0)^{\alpha})\|\,\|x_0\| + \|(t-t_0)^{\alpha-2}E_{\alpha\alpha-1}(A(t-t_0)^{\alpha})\|\,\|x_1\| \\ &\quad + \int_{t0}^t (t-\tau)^{\alpha-1}\|E_{\alpha\alpha}(A(t-\tau)^{\alpha})\| \times \|B(\tau)\|\,\|x(\tau)\|\mathrm{d}\tau \\ &\leq M_0\|x_0\| + M_1\|x_1\| + LM\int_{t_0}^t (t-\tau)^{\alpha-1} \times \|x(\tau)\|\mathrm{d}\tau,\end{aligned}$$

where $L, M, M_0, M_1 > 0$ such that

$$\|(t-t_0)^{\alpha-k-1}E_{\alpha,\alpha-k}(A(t-t_0)^{\alpha})\| \leq M_k \quad (k = 0, 1),$$
$$\|E_{\alpha\alpha}(A(t-t_0)^{\alpha})\| \leq L.$$

Based on Lemmas 2 and 3, we can obtain

$$\begin{aligned}\|x(t)\| &\leq (M_0\|x_0\| + M_1\|x_1\|)E_{\alpha}(LM\Gamma(\alpha)(t-t_0)^{\alpha}) \\ &= (M_0\|x_0\| + M_1\|x_1\|) \times \left[-\sum_{k=1}^{p}\frac{(LM\Gamma(\alpha)(t-t_0)^{\alpha})^{-k}}{\Gamma(1-k\alpha)} + O(|LM\Gamma(\alpha)t^{\alpha}|)^{-1-p}\right].\end{aligned}$$

When $t \to \infty$, $\|x(t)\| \to 0$. That is, the solution of Equation (5) is asymptotically stable. ■

Remark 1 Suppose the Caputo derivative takes the place of the Riemann–Liouville derivative in Equation (1) and all other assumed conditions remain the same, then the conclusions of Theorem 2 still hold.

THEOREM 3 *If all eigenvalues of matrix A satisfy* $|\arg(\lambda(A))| > \alpha\pi/2$, $\|B(t)\|$ *is nondecreasing and* $B(t) = O(t-t_0)^{\theta}, (\theta < -\alpha, t_0 > 0)$, *then the zero solution is asymptotically stable.*

Proof By the Laplace transform and the inverse Laplace transform, the solution of Equations (5) with (6) can be written as

$$x(t) = (t - t_0)^{\alpha-1} E_{\alpha\alpha}(A(t - t_0)^{\alpha}) x_0 + (t - t_0)^{\alpha-2} E_{\alpha\alpha-1}(A(t - t_0)^{\alpha}) x_1 + \int_{t0}^{t} (t - \tau)^{\alpha-1} E_{\alpha\alpha}(A(t - \tau)^{\alpha}) B(\tau) x(\tau) \mathrm{d}\tau,$$

then one can obtain

$$\|x(t)\| \leq \|(t - t_0)^{\alpha-1} E_{\alpha\alpha}(A(t - t_0)^{\alpha})\| \|x_0\| + \|(t - t_0)^{\alpha-2} E_{\alpha\alpha-1}(A(t - t_0)^{\alpha})\| \|x_1\| + \int_{t_0}^{t} (t - \tau)^{\alpha-1} \|E_{\alpha\alpha}(A(t - \tau)^{\alpha})\| \|B(\tau)\| \|x(\tau)\| \mathrm{d}\tau. \quad (7)$$

Then

$$\|x(t)\| \leq M_0 \|x_0\| + M_1 \|x_1\| + L \int_{t_0}^{t} (t - \tau)^{\alpha-1} \|B(\tau)\| \|x(\tau)\| \,\mathrm{d}\tau,$$

where $L, M_0, M_1 > 0$ such that

$$\|(t - t_0)^{\alpha-k-1} E_{\alpha,\alpha-k}(A(t - t_0)^{\alpha})\| \leq M_k \quad (k = 0, 1),$$
$$\|E_{\alpha,\alpha}(A(t - t_0)^{\alpha})\| \leq L.$$

Multiplying by $\|B(t)\|$ on both sides of Equation (7), one obtains

$$\|B(t)\| \|x(t)\| \leq \|B(t)\| (M_0 \|x_0\| + M_1 \|x_1\|) + L \|B(t)\| \times \int_{t_0}^{t} (t - \tau)^{\alpha-1} \|B(\tau)\| \|x(\tau)\| \,\mathrm{d}\tau.$$

Applying Lemma 3 leads to

$$\|B(t)\| \|x(t)\| \leq \|B(t)\| (M_0 \|x_0\| + M_1 \|x_1\|) \times E_{\alpha}(L \|B(t)\| \Gamma(\alpha) t^{\alpha}).$$

Then

$$\|x(t)\| \leq (M_0 \|x_0\| + M_1 \|x_1\|) E_{\alpha}(L \|B(t)\| \Gamma(\alpha)(t - t_0)^{\alpha}) \leq (M_0 \|x_0\| + M_1 \|x_1\|) \sum_{k=0}^{\infty} \frac{(L\Gamma(\alpha) \|B(t)\| (t - t_0)^{\alpha})^k}{\Gamma(k\alpha + 1)}.$$

Since $B(t) = O(t - t_0)^{\theta}$, $(\theta < -\alpha, t_0 > 0)$, then $\|B(t)\| (t - t_0)^{\alpha} \to 0$ as $t \to \infty$, so $\|x(t)\|$ is bounded, i.e. $\exists N$, such that $\|x(t)\| \leq N$.

We also can obtain the following expression from the solution:

$$\|x(t)\| \leq (t - t_0)^{\alpha-2} L_1 \|x_0\| + (t - t_0)^{\alpha-2} L_2 \|x_1\| + L_1 \int_{t_0}^{t} (t - \tau)^{\alpha-2} \|B(\tau)\| \|x(\tau)\| \,\mathrm{d}\tau,$$

where

$$\|(t - t_0) E_{\alpha,\alpha}(A(t - t_0)^{\alpha})\| < L_1,$$
$$\|E_{\alpha,\alpha-1}(A(t - t_0)^{\alpha})\| < L_2.$$

Since $\|x(t)\| \leq N$, then

$$\|x(t)\| \leq (t - t_0)^{\alpha-2} (L_1 \|x_0\| + L_2 \|x_1\|) + L_1 N \int_{t_0}^{t} (t - \tau)^{\alpha-2} \|B(\tau)\| \,\mathrm{d}\tau \leq (t - t_0)^{\alpha-2} (L_1 \|x_0\| + L_2 \|x_1\|) + L_1 N \frac{\Gamma(\alpha - 1)\Gamma(1 + \theta)}{\Gamma(\alpha + \theta)} O(t - t_0)^{\alpha+\theta-1}.$$

When $t \to \infty$, $\|x(t)\| \to 0$. So the solution of Equation (5) is asymptotically stable. ■

Remark 2 Suppose the Caputo derivative takes the place of the Riemann–Liouville derivative in Equation (5) and all other assumed conditions remain the same, then the conclusion is stable.

4. Conclusions

In this paper, we have studied the stability and asymptotic stability of the nonautonomous linear differential system with the Riemann–Liouville fractional derivative and established the corresponding stability results of its zero solution. By using the Laplace transform, Mittag-Leffler function, the generalized Gronwall inequality, some sufficient conditions ensuring the stability and asymptotic stability of the perturbed linear fractional differential system with the Riemann–Liouville fractional derivative were given.

Acknowledgements

We would like to thank Ranchao Wu and Yanfen Lu for discussions, and the reviewers and the associate editor for their useful comments on our paper. Ranchao Wu is supported by the Specialized Research Fund for the Doctoral Program of Higher Education of China under grant 20093401120001, the Natural Science Foundation of Anhui Province under grant 11040606M12 and the 211 project of Anhui University under grant KJJQ1102.

References

Ahmed, E., El-Saka, A. M. A., & El-Saka, H. A. A. (2007). Equilibrium points, stability and numerical solutions of fractional-order predator–prey and rabies models. *Journal of Mathematical Analysis and Applications*, *325*(1), 542–553.

Ahn, H. S., & Chen, Y. Q. (2008). Necessary and sufficient stability condition of fractional-order interval linear systems. *Automatica*, *44*(11), 2985–2988.

Corduneanu, C. (1971). *Principles of differential and integral equations*. Bosyon, MA: Allyn & Bacon.

Debnath, L. (2004). A brief historical introduction to fractional calculus. *International Journal of Mathematical Education in Science and Technology*, *35*(4), 487–501.

Deng, W., Li, C., & Liu, J. (2007). Stability analysis of linear fractional differential system with multiple time delays. *Nonlinear Dynamics*, *48*(4), 409–416.

Li, C. P., & Zhang, F. R. (2011). A survey on the stability of fractional differential equations. *The European Physical Journal Special Topics*, *193*, 27–47.

Li, Y., Chen, Y. Q., & Podlubny, I. (2009). Mittag-Leffler stability of fractional order nonlinear dynamic systems. *Automatica*, *45*(8), 1965–1969.

Matignon, D. (1996). *Stability results for fractional differential equations with applications to control processing*. Proceedings of the IMACS-SMC, Lille, France, Vol. 2, pp. 963–968.

Miller, K. S., & Ross, B. (1993). *An introduction to the fractional calculus and fractional differential equation*. New York: John Wiley & Sons.

Moze, M., Sabatier, J., & Oustaloup, A. (2007). LMI characterization of fractional systems stability. In J. Sabatier, O. P. Agrawal, & J. A. Tenreiro Machado (Eds.), *Advances in Fractional Calculus* (pp. 419–434). Dordrecht: Springer.

Odibat, Z. M. (2010). Analytic study on linear systems of fractional differential equations. *Computers & Mathematics with Applications*, *59*(3), 1171–1183.

Podlubny, I. (1999). *Fractional differential equations*. San Diego, CA: Academic.

Qian, D., Li, C., Agarwal, R. P., & Wong, P. J. Y. (2010). Stability analysis of fractional differential system with Riemann–Liouville derivative. *Mathematical and Computer Modelling*, *52*(5–6), 862–874.

Radwan, A. G., Soliman, A. M., Elwakil, A. S., & Sedeek, A. (2009). On the stability of linear systems with fractional-order elements. *Chaos, Solitons and Fractals*, *40*(5), 2317–2328.

Sabatier, J., Moze, M., & Farges, C. (2010). LMI stability conditions for fractional order systems. *Computers & Mathematics with Applications*, *59*(5), 1594–1609.

Samko, S. G., Kilbas, A. A., & Marichev, O. I. (1993). *Fractional integrals and derivatives*. Yverdon, Switzerland: Gordon and Breach Science Publishers.

Tavazoei, M. S., & Haeri, M. (2009). A note on the stability of fractional order systems. *Mathematics and Computers in Simulation*, *79*(5), 1566–1576.

Wen, X. J., Wu, Z. M., & Lu, J. G. (2008). Stability analysis of a class of nonlinear fractional-order systems. *IEEE Transactions on Circuits and Systems II*, *55*(11), 1178–1182.

Zhang, F. R., & Li, C. P. (2011). Stability analysis of fractional differential systems with order lying (1,2). *Advances in Difference Equations*, Vol. 2011, Article ID 213485, doi:10.1155/2011/213485

7

Analytical solutions to LQG homing problems in one dimension

Mario Lefebvre[a]* and Foued Zitouni[b]

[a]*Département de mathématiques et de génie industriel, École Polytechnique, C.P. 6079, Succursale Centre-ville, Montréal, Québec, Canada H3C 3A7;* [b]*Département de mathématiques et de statistique, Université de Montréal, C.P. 6128, Succursale Centre-ville, Montréal, Québec, Canada H3C 3J7*

The problem of optimally controlling one-dimensional diffusion processes until they leave a given interval is considered. By linearizing the Riccati differential equation satisfied by the derivative of the value function in the so-called linear quadratic Gaussian homing problem, we are able to obtain an exact expression for the solution to the general problem. Particular problems are solved explicitly.

Keywords: optimal stochastic control; diffusion processes; first-passage time; survival time optimization

AMS Subject Classification: 93E20

1. Introduction

Consider the following problem in one dimension: let $\{X(t), t \geq 0\}$ be a controlled diffusion process that satisfies the stochastic differential equation

$$\begin{aligned} \mathrm{d}X(t) = & f[X(t)]\,\mathrm{d}t + b[X(t)]u[X(t)]\,\mathrm{d}t \\ & + v^{1/2}[X(t)]\,\mathrm{d}B(t) \end{aligned} \tag{1}$$

in which $u(\cdot)$ is the control variable, $b(\cdot), f(\cdot)$ and $v(\cdot) > 0$ are Borel measurable functions, and $\{B(t), t \geq 0\}$ is a standard Brownian motion. The set of admissible controls consists of Borel measurable functions. We assume that the solution of this equation exists for all $t \in [0, \infty)$ and is weakly unique.

We look for the control that minimizes the mathematical expectation of the cost function

$$J(x) = \int_0^{T(x)} \left(\frac{1}{2} q[X(t)]u^2[X(t)] + \lambda \right) \mathrm{d}t, \tag{2}$$

where $q(\cdot)$ is a positive Borel measurable function, λ is a real parameter and $T(x)$ is the first-passage time defined by

$$T(x) = \inf\{t > 0 : X(t) = d_1 \text{ or } d_2 \mid X(0) = x \in (d_1, d_2)\}. \tag{3}$$

Whittle (1982, p. 289) has termed this type of problem linear quadratic Gaussian (*LQG*) *homing*. Actually, Whittle considered the case of n-dimensional processes. In the general formulation, $T(x_1, \ldots, x_n)$ is the first time $(X_1(t), \ldots, X_n(t), t)$ enters a stopping set $D \subset \mathbb{R}^n \times (0, \infty)$. Moreover, there can be a termination cost $K[X_1(T(x)), \ldots, X_n(T(x)), T(x)]$.

Lefebvre has written a series of papers on LQG homing problems; see, for instance, Lefebvre (2011) and the references therein. Kuhn (1985) and Makasu (2009) solved homing problems with a risk-sensitive cost criterion; see also Whittle (1990, p. 222). Recent papers written on homing problems include the ones published by Makasu Lefebvre (2012a, 2012b).

A practical application of LQG homing problems is an optimal landing problem: assume that $X(t)$ denotes the height of an aircraft at time t. The optimizer controls the aircraft until the time $T(x)$ it touches the runway. Because of the *noise* in the system, $T(x)$ is a random variable. This problem was considered by Lefebvre (1998).

Another possible application would be to find the control that enables a dam manager to release water in an optimal way when there is a risk of flooding. Suppose that $X(t)$ is the flow of a certain river at time t, and let $T(x)$ be defined as in Equation (3). The constant d_2 would be the value of the flow from which flooding takes place, while d_1 would be a flow value that is considered to be safe. In this application, we would give a very large termination cost if $X(T(x)) = d_2$. Then, the optimal control would be such that the flow will never reach d_2. In practice, the dam manager does not want to release too much water, because of the economic losses due to the decrease in electricity production it entails.

Now, to obtain the optimal control, we can try to find the value function $F(x)$ defined by

$$F(x) = \inf_{u[X(t)],\ 0 \leq t \leq T(x)} E[J(x)]. \tag{4}$$

*Corresponding author. Emails: mario.lefebvre@polymtl.ca, mlefebvre@polymtl.ca

We assume that this function exists and is twice differentiable. It then satisfies the dynamic programming equation

$$\inf_u H(u) = 0, \tag{5}$$

where $u := u(x)$ and

$$H(u) := \frac{1}{2}q(x)u^2 + \lambda + [f(x) + b(x)u]F'(x) + \frac{1}{2}v(x)F''(x).$$

The optimal control can be expressed as

$$u^* = -\frac{b(x)}{q(x)}F'(x). \tag{6}$$

Hence, we have

$$H(u^*) = \lambda + f(x)F'(x) - \frac{b^2(x)}{2q(x)}[F'(x)]^2 + \frac{1}{2}v(x)F''(x) = 0. \tag{7}$$

That is,

$$\lambda + f(x)G(x) - \frac{b^2(x)}{2q(x)}G^2(x) + \frac{1}{2}v(x)G'(x) = 0, \tag{8}$$

with $G(x) := F'(x)$. Notice that this last equation is a particular Riccati equation.

Next, if the relation

$$\alpha v[X(t)] = \frac{b^2[X(t)]}{q[X(t)]} \tag{9}$$

holds for some positive constant α, then, setting

$$\phi(x) = e^{-\alpha F(x)}, \tag{10}$$

Whittle (1982) has shown that the differential equation (7) satisfied by the value function is transformed into the linear equation

$$\frac{1}{2}v(x)\phi''(x) + f(x)\phi'(x) = \alpha\lambda\phi(x). \tag{11}$$

Since $F(x) = 0$ if $x = d_1$ or d_2, the boundary conditions are

$$\phi(x) = 1 \quad \text{if } x = d_1 \text{ or } d_2. \tag{12}$$

Now, not only is the differential equation (11) linear, it is actually the Kolmogorov backward equation satisfied by the mathematical expectation (that is, the moment-generating function)

$$L(x;\theta) := E[e^{-\theta\tau(x)}],$$

where $\theta := \alpha\lambda$ and $\tau(x)$ is the same as the first-passage time $T(x)$, but for the uncontrolled process $\{\eta(t), t \geq 0\}$ obtained by setting $u[X(t)] \equiv 0$ in Equation (1). That is,

$$d\eta(t) = f[\eta(t)]\,dt + v^{1/2}[\eta(t)]\,dB(t). \tag{13}$$

Moreover, the boundary conditions (12) are the appropriate ones.

Thus, if $b^2[X(t)]/(q[X(t)]v[X(t)])$ is a constant, it is possible to transform the optimal stochastic control problem into a purely probabilistic problem. Notice, however, that obtaining an explicit expression for the function $\phi(x)$ defined in Equation (11) is itself often a difficult problem.

In Section 2, we will show that even if $b^2[X(t)]/(q[X(t)]v[X(t)])$ is not a constant, we can obtain an analytical solution to the LQG problem set up above, as long as $b[X(t)] \neq 0$ in the interval $[d_1, d_2]$. Particular problems will be solved explicitly in Section 3. We will end this paper with a few concluding remarks in Section 4.

2. Analytical solutions to LQG homing problems in one dimension

The transformation $\phi(x)$ defined in Equation (10) enables us to linearize the differential equation satisfied by the function $F(x)$, if the relation (9) is satisfied. Actually, as is well known, it is always possible to linearize the Riccati equation (8). Indeed, we can transform this first-order nonlinear ordinary differential equation into a second-order linear differential equation. However, we then need two boundary conditions to determine the value of the two arbitrary constants that will appear in the general solution to the second-order equation.

It is also sometimes possible to directly obtain an explicit solution to the Riccati equation. The problem is that, in general, we do not have a boundary condition for the function $G(x)$. Therefore, we cannot easily determine the value of the arbitrary constant in the expression obtained for $G(x)$. We must then try to integrate $G(x)$ to obtain $F(x)$ and make use of the boundary conditions $F(d_1) = F(d_2) = 0$ to find out the two arbitrary constants. Unfortunately, this integral is often very difficult to perform. Thus, we cannot find the optimal control explicitly.

The Lefebvre and Zitouni, in a paper published in 2012, showed that it is sometimes possible to use the symmetry present in the problem to determine the value of x_0 for which the function $F(x)$ should have either a maximum or a minimum, so that $G(x_0)$ is equal to zero. Then, if we are indeed able to solve the Riccati equation explicitly, we can obtain an exact expression for the optimal control u^*. However, in the general case, finding the exact value of x_0 that corresponds to an extremal point of the value function is not an easy problem.

Let

$$z(x) = \exp\left\{-\int \frac{b^2(x)}{q(x)v(x)}G(x)\,dx\right\}. \tag{14}$$

If $b^2(x)/[q(x)v(x)]$ is a constant, then the function $z(x)$ is equivalent to $\phi(x)$. Assuming that $b(x) \neq 0$ in the interval $[d_1, d_2]$, we can write that

$$G(x) = -\frac{z'(x)}{z(x)}\frac{q(x)v(x)}{b^2(x)}. \tag{15}$$

We find that the function $z(x)$ satisfies the following differential equation:

$$\lambda - \frac{f(x)q(x)v(x)}{b^2(x)}\frac{z'(x)}{z(x)} - \frac{1}{2}\frac{q(x)v^2(x)}{b^2(x)}\frac{z''(x)}{z(x)} - \frac{1}{2}\frac{v(x)[q'(x)v(x)+q(x)v'(x)]}{b^2(x)}\frac{z'(x)}{z(x)} + \frac{b'(x)q(x)v^2(x)}{b^3(x)}\frac{z'(x)}{z(x)} = 0. \quad (16)$$

Simplifying, we obtain the second-order linear differential equation

$$z''(x)+\left\{2\frac{f(x)}{v(x)}+\frac{[q'(x)v(x)+q(x)v'(x)]}{q(x)v(x)}-2\frac{b'(x)}{b(x)}\right\}z'(x) - 2\lambda\frac{b^2(x)}{q(x)v^2(x)}z(x)=0. \quad (17)$$

Let

$$z(x) = c_1 z_1(x) + c_2 z_2(x)$$

be the general solution of Equation (17), where c_1 and c_2 are arbitrary constants. We have

$$z'(x) = -z(x)\frac{b^2(x)}{q(x)v(x)}G(x). \quad (18)$$

Since $F(d_1) = F(d_2) = 0$, we can state that there exists a point $x_0 \in (d_1, d_2)$ for which $G(x_0) = 0$. We then deduce from the previous equation that $z'(x_0) = 0$ as well. Hence, we may write that

$$c_2 = -c_1\frac{z_1'(x_0)}{z_2'(x_0)}.$$

We assume that $z_i'(x_0) \neq 0$ for $i = 1, 2$. It follows that both c_1 and c_2 must be different from zero. Thus, from Equation (15), we obtain that

$$G(x) = -\frac{z_2'(x_0)z_1'(x) - z_1'(x_0)z_2'(x)}{z_2'(x_0)z_1(x) - z_1'(x_0)z_2(x)}\frac{q(x)v(x)}{b^2(x)}.$$

Next, we set

$$F(x) = \int_{d_1}^{x} G(y)\,\mathrm{d}y.$$

As we mentioned above, in general it is very difficult to obtain an exact expression for the point x_0. However, we can estimate x_0 by making use of the condition

$$0 = F(d_2) = -\int_{d_1}^{d_2}\frac{z_2'(x_0)z_1'(y) - z_1'(x_0)z_2'(y)}{z_2'(x_0)z_1(y) - z_1'(x_0)z_2(y)}\frac{q(y)v(y)}{b^2(y)}\,\mathrm{d}y. \quad (19)$$

Indeed, if we denote the integral in the previous equation by $I(x_0)$, then we can use a mathematical software to compute $I(x_0)$ for $x_0 \in (d_1, d_2)$. It is not difficult to estimate x_0 quite precisely.

Summing up, we can state the following proposition.

PROPOSITION 2.1 *The control $u^*(x)$ that minimizes the expected value of the cost function $J(x)$ defined in Equation (2) is given by*

$$u^*(x) = \frac{z_2'(x_0)z_1'(x) - z_1'(x_0)z_2'(x)}{z_2'(x_0)z_1(x) - z_1'(x_0)z_2(x)}\frac{v(x)}{b(x)},$$

in which $z_1(x)$ and $z_2(x)$ are two linearly independent solutions of Equation (17). Furthermore, x_0 is such that $G(x_0) = 0$ and can be obtained from the condition (19).

Remarks

(i) If we can determine the exact value of x_0 with the help of a symmetry argument, for instance, then of course we do not have to make use of the condition (19).

(ii) The proposition provides an expression for the optimal control $u^*(x)$, which only depends on the derivative of the value function $F(x)$. If one needs $F(x)$ as well, then one must be able to integrate the function $G(x)$. In general, as we mentioned above, this is not an easy task.

(iii) We have assumed above that $b(x) \neq 0$ in the interval $[d_1, d_2]$. If there exists a point x_1 in this interval for which $b(x_1) = 0$, then we deduce from Equation (18) that $z'(x_1) = 0$, which yields the following expression for $G(x)$:

$$G(x) = -\frac{z_2'(x_1)z_1'(x) - z_1(x_1)z_2'(x)}{z_2(x_1)z_1(x) - z_1(x_1)z_2(x)}\frac{q(x)v(x)}{b^2(x)},$$

in which there is no unknown. Hence, we cannot satisfy the boundary condition $F(d_2) = 0$ (respectively, $F(d_1) = 0$) by setting

$$F(x) = \int_{d_1}^{x} G(y)\,\mathrm{d}y \times \left(\text{respectively, } F(x) = \int_{x}^{d_2} G(y)\,\mathrm{d}y\right).$$

Actually, there could be some points in $[d_1, d_2]$ at which $b(x)$ is equal to zero, but we should then allow the functions v and q to be non-negative (rather than strictly positive) and the ratio $b^2(x)/[q(x)v(x)]$ should always be positive in the interval $[d_1, d_2]$. For instance, with $d_1 = 0$ and $d_2 = 1$, we could have: $b(x) = x$, $q(x) = q_0 > 0$ and $v(x) = x^2$, in which case we can use the result in Whittle (1982).

In the next section, two particular problems will be solved explicitly.

3. Particular examples

3.1. *Example 1*

In the first example that we present, we assume that $f[X(t)] = X(t)$ and $b[X(t)] = v[X(t)] = X^2(t) + 1$, so that the stochastic differential equation (1) becomes

$$\mathrm{d}X(t) = X(t)\,\mathrm{d}t + [X^2(t) + 1]u[X(t)]\,\mathrm{d}t + \{X^2(t) + 1\}^{1/2}\,\mathrm{d}B(t).$$

Moreover, we set $d_1 = -1$ and $d_2 = 1$, and we choose the cost function

$$J(x) = \int_0^{T(x)} \left(\frac{1}{2}u^2[X(t)] + \lambda\right) \mathrm{d}t. \tag{20}$$

That is, we take $q[X(t)] \equiv 1$. We assume that the parameter λ is positive. Therefore, the aim is to minimize the (expected) time spent by the process in the interval $(-1, 1)$, taking the quadratic control costs into account.

Remark We can choose the parameter λ as large as we want. However, when λ is negative, there is a minimum value that it can take. Otherwise, the value function will become infinite.

With the choices above, Equation (17) becomes

$$z''(x) - 2\lambda z(x) = 0. \tag{21}$$

The general solution of this equation can be written as

$$z(x) = c_1\,\mathrm{e}^{-\sqrt{2\lambda}x} + c_2\,\mathrm{e}^{\sqrt{2\lambda}x},$$

where c_1 and c_2 are arbitrary constants.

Now, by symmetry, it is clear that the value function $F(x)$ takes on its maximum value at $x = 0$. Hence (see Equation (18)),

$$z'(0) = -z(0)b(0)G(0) = 0.$$

It follows that

$$z(x) = 2c_1 \cosh(\sqrt{2\lambda}x)$$

and

$$G(x) = -\sqrt{2\lambda}\frac{\sinh(\sqrt{2\lambda}x)}{\cosh(\sqrt{2\lambda}x)}\frac{1}{x^2 + 1}. \tag{22}$$

The functions $G(x)$ and $F(x)$ and the optimal control $u^*(x) = -b(x)G(x)$ are shown in Figures 1–3, respectively, in the case when $\lambda = 1$.

Remarks

(i) If one did not deduce from the symmetry in the problem that the value of x_0 for which $G(x_0) = 0$ is $x_0 = 0$, then one can plot (with $\lambda = 1$)

$$F(1) = \sqrt{2}\int_{-1}^{1} \frac{\mathrm{e}^{\sqrt{2}x_0}\,\mathrm{e}^{-\sqrt{2}x} - \mathrm{e}^{-\sqrt{2}x_0}\,\mathrm{e}^{\sqrt{2}x}}{\mathrm{e}^{\sqrt{2}x_0}\,\mathrm{e}^{-\sqrt{2}x} + \mathrm{e}^{-\sqrt{2}x_0}\,\mathrm{e}^{\sqrt{2}x}}\frac{1}{x^2 + 1}\,\mathrm{d}x$$

for $x_0 \in (-1, 1)$. We easily deduce from Figure 4 that x_0 is indeed equal to 0, since $F(1) = 0$ for this value.

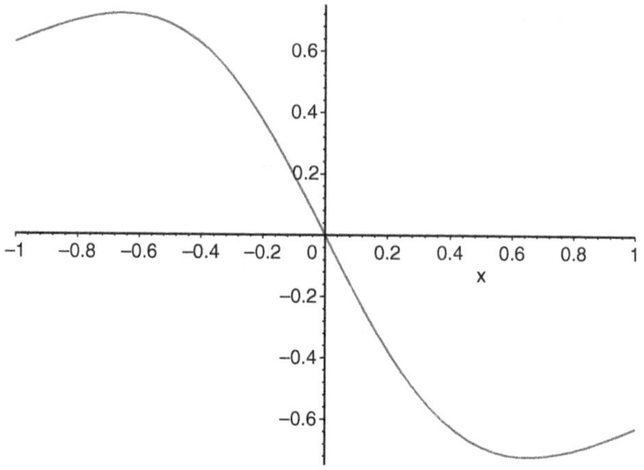

Figure 1. Derivative $G(x)$ of the value function when $d_1 = -1$, $d_2 = 1, f(x) = x, b(x) = v(x) = x^2 + 1, q(x) \equiv 1$ and $\lambda = 1$.

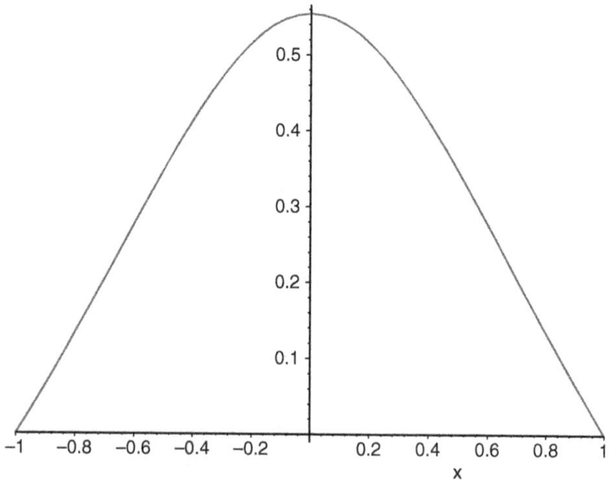

Figure 2. Value function $F(x)$ when $d_1 = -1, d_2 = 1, f(x) = x$, $b(x) = v(x) = x^2 + 1, q(x) \equiv 1$ and $\lambda = 1$.

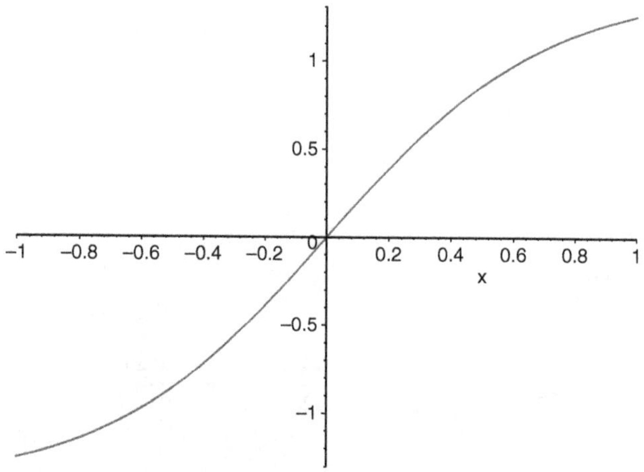

Figure 3. Optimal control $u^*(x)$ when $d_1 = -1$, $d_2 = 1$, $f(x) = x, b(x) = v(x) = x^2 + 1, q(x) \equiv 1$ and $\lambda = 1$.

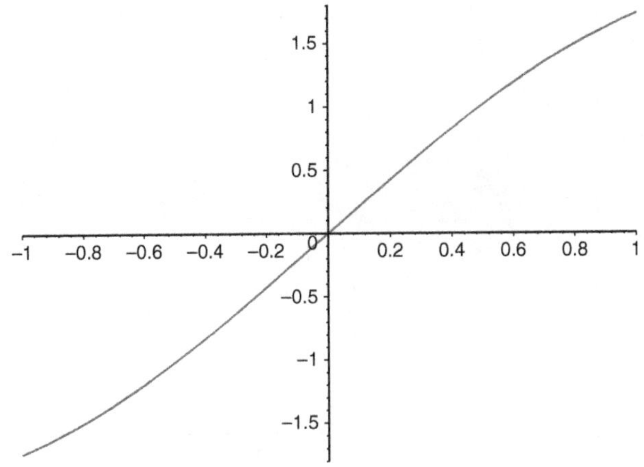

Figure 4. The value of $F(1)$ as a function of $x_0 \in (-1, 1)$ in Example 1.

Notice that $G(x)$ is an odd function when $x_0 = 0$, which implies that x_0 is exactly equal to 0.

(ii) It is interesting to compare the value of $F(x)$, which is obtained by using the optimal control above, and the expected value of the cost function $J(x)$ if the optimizer chooses $u[X(t)] \equiv 0$. We then have

$$E[J(x)] = E\left[\int_0^{T(x)} \lambda \, \mathrm{d}t\right] = \lambda E[T(x)].$$

Let $m(x)$ denote $E[T(x)]$. It is well known that this function satisfies (here) the ordinary differential equation

$$\tfrac{1}{2}(x^2+1)m''(x) + xm'(x) = -1,$$

subject to the boundary conditions $m(-1) = m(1) = 0$; see, for instance, Lefebvre (2007). We find that

$$m(x) = \ln\left(\frac{2}{x^2+1}\right).$$

We show in Figure 5 the difference $D(x) := F(x) - m(x)$.

(iii) If we consider the Riccati differential equation satisfied by $G(x)$, namely

$$\lambda + xG(x) - \frac{(x^2+1)^2}{2}G^2(x) + \frac{1}{2}(x^2+1)G'(x) = 0,$$

we find that

$$G(x) = \frac{\sqrt{2\lambda i}\tan(\sqrt{2\lambda i}x + c)}{x^2+1}.$$

It is not obvious to determine the constant c, which (contrary to x_0) can take any value. If we apply the condition $G(0) = 0$, we retrieve the solution given in Equation (22).

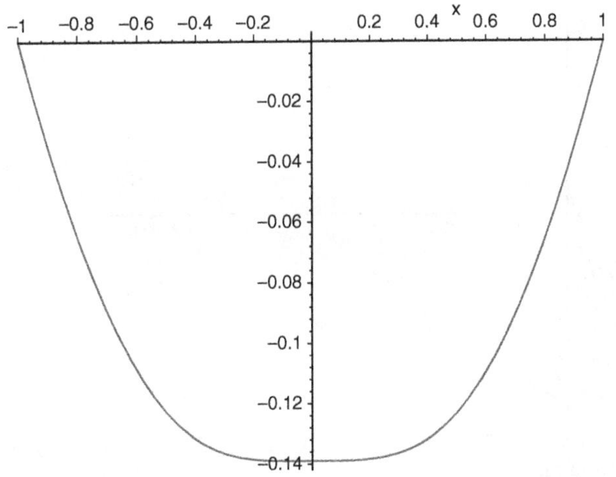

Figure 5. Difference between the value function $F(x)$ and the expected value of the cost function $J(x)$ when $u[X(t)] \equiv 0$ in Example 1.

3.2. *Example 2*

We now consider the case when $f[X(t)] \equiv 0$, $b[X(t)] = X(t)$ and $q[X(t)] = v[X(t)] \equiv 1$. The stochastic differential equation (1) is thus

$$\mathrm{d}X(t) = X(t)u[X(t)]\,\mathrm{d}t + \mathrm{d}B(t)$$

and the cost function is the same as in the previous example. Again, we assume that the parameter λ is positive. Finally, we take $d_1 = 1$ and $d_2 = 2$. Notice that the uncontrolled process in this example is a standard Brownian motion.

The function $z(x)$ satisfies the ordinary differential equation

$$z''(x) - \frac{2}{x}z'(x) - 2\lambda x^2 z(x) = 0,$$

whose general solution can be written as follows:

$$z(x) = x^{3/2}[c_1 I_{3/4}(\sqrt{\lambda/2}x^2) + c_2\sqrt{x}K_{3/4}(\sqrt{\lambda/2}x^2)],$$

where $I_{3/4}$ and $K_{3/4}$ are modified Bessel functions; see Abramowitz and Stegun (1965, p. 374).

Making use of the function $z(x)$ and choosing $\lambda = 1$, we obtain that

$$G(x) = -\frac{\sqrt{2}}{x} \frac{K_{1/4}(x_0^2/\sqrt{2})I_{-1/4}(x^2/\sqrt{2}) - I_{-1/4}(x_0^2/\sqrt{2})K_{1/4}(x^2/\sqrt{2})}{K_{1/4}(x_0^2/\sqrt{2})I_{3/4}(x^2/\sqrt{2}) + I_{-1/4}(x_0^2/\sqrt{2})K_{3/4}(x^2/\sqrt{2})}.$$

Contrary to Example 1, we cannot easily determine the value of x_0 for which $G(x_0) = 0$. Intuitively, this value should be near 1.5, namely near the middle of the interval $[1, 2]$.

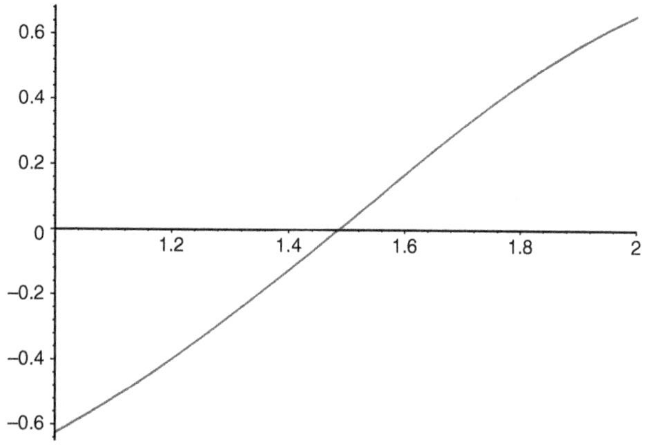

Figure 6. The value of $F(2)$ as a function of $x_0 \in (1, 2)$ in Example 2.

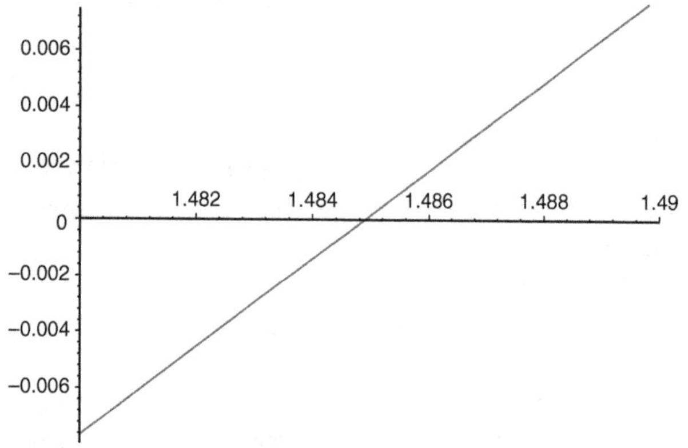

Figure 7. The value of $F(2)$ as a function of $x_0 \in [1.48, 1.49]$ in Example 2.

To obtain an approximate value for x_0, we plot

$$F(2) = \int_1^2 G(x)\,\mathrm{d}x$$

as a function of $x_0 \in (1, 2)$. We obtain the curve shown in Figure 6. We see that x_0 is slightly larger than 1.48. Therefore, we then plot $F(2)$ for $x_0 \in [1.48, 1.49]$; see Figure 7. We can now state that $x_0 \simeq 1.485$, which should be precise enough.

Next, we present the value function $F(x)$ and the optimal control $u^*(x)$ in Figures 8 and 9, respectively.

Finally, we compute the difference between $F(x)$ and the expected value of $J(x)$ when $u[X(t)] \equiv 0$. Proceeding as in Example 1, we easily find that this expected value is given by

$$E[J(x)] = E[T(x)] = -x^2 + 3x - 2.$$

See Figure 10.

Remark In this example, $G(x)$ satisfies the Riccati differential equation

$$1 - \frac{x^2}{2} G^2(x) + \frac{1}{2} G'(x) = 0.$$

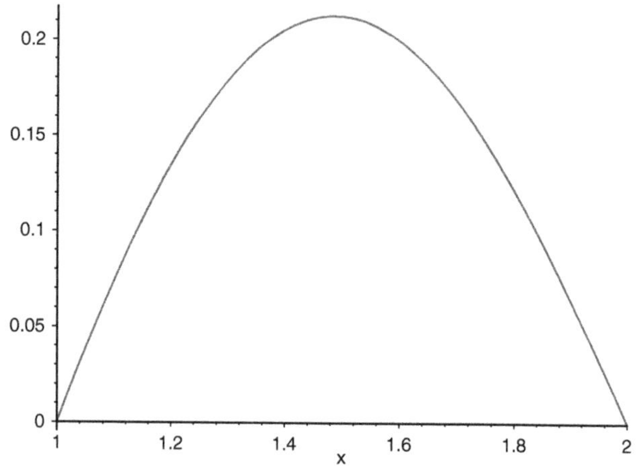

Figure 8. Value function $F(x)$ when $d_1 = 1$, $d_2 = 2$, $f(x) \equiv 0$, $b(x) = x$, $v(x) = q(x) \equiv 1$ and $\lambda = 1$.

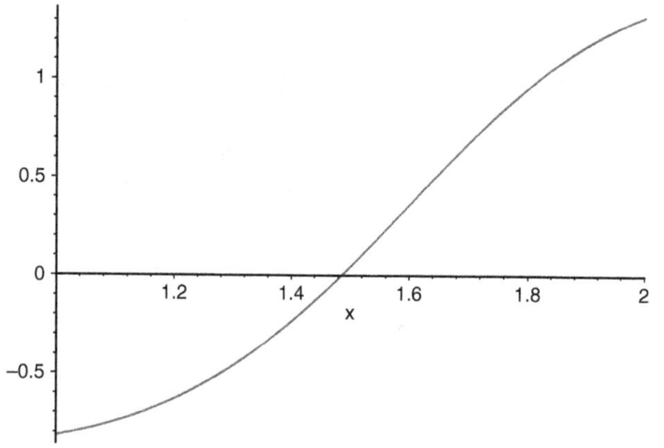

Figure 9. Optimal control $u^*(x)$ when $d_1 = 1$, $d_2 = 2$, $f(x) \equiv 0$, $b(x) = x$, $v(x) = q(x) \equiv 1$ and $\lambda = 1$.

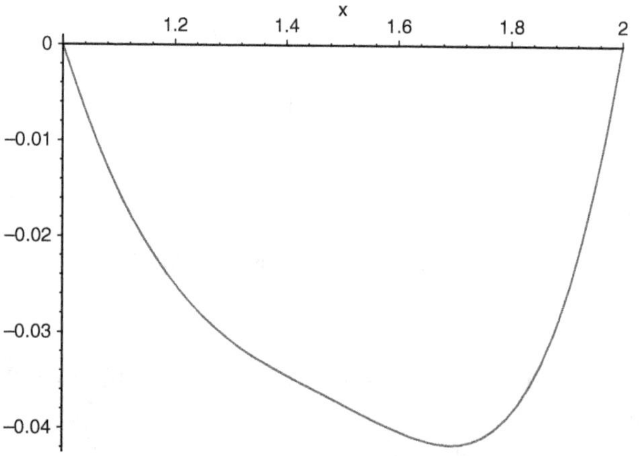

Figure 10. Difference between the value function $F(x)$ and the expected value of the cost function $J(x)$ when $u[X(t)] \equiv 0$ in Example 2.

Its solution can be written as

$$G(x) = -\frac{\sqrt{2}}{x} \frac{I_{-1/4}(x^2/\sqrt{2}) - cK_{1/4}(x^2/\sqrt{2})}{I_{3/4}(x^2/\sqrt{2}) + cK_{3/4}(x^2/\sqrt{2})}.$$

As in the previous example, it is more difficult to find the value of the constant c than to determine x_0.

4. Conclusion

By defining the function $z(x)$ in Equation (14) in terms of the derivative $G(x)$ of the value function, we were able to linearize the Riccati differential equation satisfied by $G(x)$. Moreover, making use of the fact that $z'(x_0) = 0$ for a certain $x_0 \in (d_1, d_2)$, we obtained analytical solutions to much more general LQG homing problems than the ones that can be solved when the relation in Equation (9) holds.

In Section 3, we presented two particular problems that we were able to solve explicitly, even though Equation (9) does not hold. In the first problem, we deduced from symmetry that $x_0 = 0$, while in the second one we showed that we can easily obtain a very good approximation for x_0.

Next, we could try to extend our results to the n-dimensional case. We could at least find problems in two or more dimensions for which symmetry arguments can be used to obtain explicit (and exact) expressions for the optimal control.

Finally, LQG problems can be modified in various ways: we can assume that the noise term is *uniform white noise* rather than *Gaussian white noise*, the formulation of the problem with linear state dynamics and quadratic control costs can be modified, a parameter that takes the risk-sensitivity of the optimizer into account can be introduced, etc.

Acknowledgements

The authors are grateful to the anonymous reviewers whose constructive comments helped to improve their paper.

References

Abramowitz, M., & Stegun, I. A. (1965). *Handbook of mathematical functions with formulas, graphs, and mathematical tables*. New York, NY: Dover.

Kuhn, J. (1985). The risk-sensitive homing problem. *Journal of Applied Probability, 22*, 796–803.

Lefebvre, M. (1998). A bidimensional optimal landing problem. *Automatica, 34*, 655–657.

Lefebvre, M. (2007). *Applied stochastic processes*. New York, NY: Springer.

Lefebvre, M. (2011). Stochastic optimal control in a danger zone. *International Journal of Systems Science, 42*, 653–659.

Lefebvre, M., & Zitouni, F. (2012). General LQG homing problems in one dimension. *International Journal of Stochastic Analysis, 2012*, Article ID 803724, 20 pages. doi:10.1155/2012/803724

Makasu, C. (2009). Risk-sensitive control for a class of homing problems. *Automatica, 45*, 2454–2455.

Makasu, C., & Lefebvre, M. (2012a). LQG homing problem with a maximin cost. *Journal of the Franklin Institute, 349*, 1944–1950. doi:10.1016/j.jfranklin.2012.01.014

Makasu, C., & Lefebvre, M. (2012b). Probabilistic solution to a two-dimensional LQG homing problem. *Control and Cybernetics, 41*, 5–12.

Whittle, P. (1982). *Optimization over time*, Vol. I. Chichester: Wiley.

Whittle, P. (1990). *Risk-sensitive optimal control*. Chichester: Wiley.

8

Robust and resilient state-dependent control of discrete-time nonlinear systems with general performance criteria

Xin Wang[a]*, Edwin E. Yaz[b] and James Long[c]

[a]*Department of Electrical and Computer Energy Engineering, Southern Illinois University Edwardsville, Edwardsville, IL 62026, USA;* [b]*Department of Electrical and Computer Engineering, Marquette University, Milwaukee, WI 53201, USA;* [c]*Computer Systems Engineering, Oregon Institute of Technology, Klamath Falls, OR 97601, USA*

A novel state-dependent control approach for discrete-time nonlinear systems with general performance criteria is presented. This controller is robust for unstructured model uncertainties, resilient against bounded feedback control gain perturbations in achieving optimality for general performance criteria to secure quadratic optimality with inherent asymptotic stability property together with quadratic dissipative type of disturbance reduction. For the system model, unstructured uncertainty description is assumed, which incorporates commonly used types of uncertainties, such as norm-bounded and positive real uncertainties as special cases. By solving a state-dependent linear matrix inequality (LMI) at each time step, sufficient condition for the control solution can be found which satisfies the general performance criteria. The results of this paper unify existing results on nonlinear quadratic regulator, H_∞ and positive real control to provide a novel robust control design. The effectiveness of the proposed technique is demonstrated by simulation of the control of inverted pendulum.

Keywords: nonlinear control; robust control; linear matrix inequality

1. Introduction

Optimal control of nonlinear systems is traditionally characterized in terms of Hamilton Jacobi Equations (HJEs). The solution of the HJEs provides the necessary and sufficient optimal control condition for nonlinear systems. Furthermore, when the controlled system is linear time invariant and the performance index is linear quadratic regulator (LQR), the HJEs are reduced to Algebraic Riccati Equations (AREs). As for H_∞ nonlinear control problem, the optimal control solution is equivalent to solving the corresponding Hamilton Jacobi Inequalities (HJIs). However, HJEs and HJIs, which are first-order partial differential equations and inequalities, cannot be solved for more than a few state variables. In the past few years, it has been shown that the problems of quadratic regulation and H_∞ nonlinear control can be effectively solved by state-dependent Riccati equation (SDRE) and nonlinear matrix inequality (NLMI) techniques (Huang & Lu, 1996). The state-dependent LMI control of nonlinear systems, as pointed out in Wang and Yaz (2009), Wang, Yaz, and Jeong (2010), and Wang, Yaz, and Yaz (2010, 2011), synthesizes a controller to achieve mixed nonlinear quadratic regulator (NLQR) and H_∞ control.

Dissipative control for linear systems has also received considerable attention over the past two decades. The concept of dissipative system was first introduced by Willems (1972a, 1972b), and further generalized by Hill and Moylan (1976, 1980), playing an important role in systems, circuits and controls. The theory of dissipative systems generalizes the basic tools including the passivity theorem, bounded real lemma, Kalman–Yakubovich lemma and circle criterion. Dissipativity performance includes H_∞ performance, passivity, positive realness and sector-bounded constraint as special cases. Research addressing the problems of H_∞ and positive real control systems can be found in Safonov, Jonckheere, Verma, and Limebeer (1987), Doyle, Glover, Khargonekar, and Francis (1989), Haddad and Bernstein (1991), Sun, Khargonekar, and Shim (1994), and Shim (1996). Control of uncertain linear systems with l_2-bounded structured uncertainty satisfying H_∞ and passivity criteria has been tackled in Petersen (1987) and Khargonekar, Petersen, and Zhou (1990). More recent development involving the quadratic dissipative control for linear systems problem has been tackled in Xie, Xie, and De Souza (1998) and Tan, Soh, and Xie (2000).

In this paper, we further consider the problem of optimal, robust and resilient LMI control of discrete-time nonlinear systems with general performance criteria. The controller is robust for model uncertainties and resilient for gain perturbations. As for the uncertain nonlinear systems, we consider a general form of l_2-bounded uncertainty description, without any standard structure, incorporating commonly used types of uncertainty, such as norm-bounded

*Corresponding author. Email: xwang@siue.edu

and positive real uncertainties as special cases. The purpose behind this novel approach is to convert a nonlinear system control problem into a convex optimization problem which is solved by state-dependent LMI. The recent development in convex optimization provides very efficient algorithms for solving LMIs. If a solution can be expressed in an LMI form, then there exist optimization means providing efficient global numerical solutions (Boyd, Ghaoui, Feron, & Balakrishnan, 1994). Therefore if the LMI is feasible, then the LMI control technique provides asymptotically stable solutions satisfying the general performance criteria. We further propose to employ general performance criteria to design the controller guaranteeing the quadratic sub-optimality with inherent stability property in combination with dissipativity type of disturbance attenuation. The general performance criteria is a generalization of the NLQR, H_∞, positive realness and sector-bounded constraint. The results of the paper unify existing results on NLQR, H_∞ and positive real control and provide a novel robust control design. The paper is organized as follows: Section 2 covers the general performance criteria including the performance of NLQR, H_∞, positive realness and sector-bounded constraint. Section 3 presents state-dependent LMI-based control for nonlinear systems achieving general performance criteria. In the final section, an inverted pendulum on a cart system is used to demonstrate the effectiveness and robustness of the new approach.

2. System model and general performance criteria

The following notation is used in this work: $\Re_+$ stands for the set of non-negative real numbers, $\Re^n$ stands for the n-dimensional Euclidean space. $x_k \in \Re^n$ denotes n-dimensional real vector with norm $\|x_k\| = (x_k^T x_k)^{1/2}$, where $(\cdot)^T$ indicates transpose. $\Re^{n\times m}$ is the set of $n \times m$ real matrices. I_n is the $n \times n$ identity matrix. $A \geq 0$ for a symmetric matrix denotes a positive semi-definite matrix. l_2 is the space of finite-dimensional vectors with finite energy: $\sum_{k=0}^{\infty} \|x_k\|^2 < \infty$. The inner product on $\Re^n$ is defined by $\langle u, v\rangle = \sum_{i=1}^{n} u_i v_i$.

Consider the nonlinear dynamical system and performance output equation as following:

$$\begin{aligned} x_{k+1} &= f(x_k, u_k, w_k) \\ &= (A(x_k) + \Delta_A(x_k))\, x_k + (B(x_k) + \Delta_B(x_k))\, u_k \\ &\quad + (E(x_k) + \Delta_E(x_k))\, w_k \\ &= (A + \Delta_A)\, x_k + (B + \Delta_B)\, u_k + (E + \Delta_E)\, w_k, \quad (1) \\ z_k &= g(x_k, u_k) = C_k \cdot x_k + D_k \cdot w_k, \quad (2) \end{aligned}$$

where $x_k \in \Re^n$ is the state of the dynamical system; $u_k \in \Re^m$ the applied input; $w_k \in \Re^p$ the l_2 type of disturbance; $z_k \in \Re^r$ the performance output; f, g the nonlinear vector functions; $A_k \in \Re^{n\times n}, B_k \in \Re^{n\times m}, E_k \in \Re^{n\times p}, C_k \in \Re^{r\times n}$ and $D_k \in \Re^{r\times p}$ the state-dependent coefficient (SDC) matrices and $\Delta_A \in \Re^{n\times n}, \Delta_B \in \Re^{n\times m}$ and $\Delta_E \in \Re^{n\times p}$ the state-dependent uncertainty matrices.

Note that the discrete-time, state feedback, input affine and autonomous nonlinear system must be fully controllable and state observable. The way of finding the nominal system parameter matrices A_k, B_k, E_k, C_k and D_k is a process of factorizing the nonlinear system into a linear-like structure which contains SDC matrices, so-called *mathematical factorization*. The l_2-bounded perturbation matrices Δ_A, Δ_B and Δ_E are the unstructured uncertainty matrices, which can also be time-varying state-dependent matrices. Without any standard structures, the uncertainty matrices provide us a general framework to compensate for effect of the unmodeled system dynamics, external disturbances, perturbation and noise. The commonly used types of uncertainty, such as norm-bounded, structured uncertainties and positive real uncertainties are special cases of the uncertainty matrices description in this work.

It is assumed that the full state is available for feedback and the state feedback control input is given by

$$u_k = (K(x_k) + \Delta_k(x_k))\, x_k = (K_k + \Delta_k)\, x_k, \quad (3)$$

where there is additive (possibly state dependent) perturbation on the feedback gain.

Introducing the quadratic energy supply function E associated with the system equations, defined by Hill and Moylan (1976, 1980) as

$$E(z_k, w_k) = \langle z_k, Qz_k\rangle + 2\langle z_k, Sw_k\rangle + \langle w_k, Rw_k\rangle, \quad (4)$$

where $Q \in \Re^{r\times r}, S \in \Re^{r\times p}, R \in \Re^{p\times p}$ are the chosen weighing matrices. Next, from the definition of dissipativity, we have

Definition 2.1 *Given matrices $Q \in \Re^{r\times r}, S \in \Re^{r\times p}, R \in \Re^{p\times p}$ with Q, R symmetric, the system* (1) *and* (2) *with energy function* (4) *is said to be (Q, S, R) dissipative if for some real function $\beta(\cdot)$ with $\beta(0) = 0$,*

$$E(z_k, w_k) + \beta(x_0) \geq 0, \forall w \in l_2, \forall k \geq 0. \quad (5)$$

Furthermore, if for some scalar $\alpha > 0$,

$$E(z_k, w_k) + \beta(x_0) \geq \alpha\langle w_k, w_k\rangle, \forall w \in l_2, \forall k \geq 0. \quad (6)$$

The system (1) *and* (2) *is said to be strictly (Q, S, R) dissipative.*

Theorem 1 *Consider the quadratic function $V_k = x_k^T P_k x_k > 0$, matrices $Q \in \Re^{r\times r}, S \in \Re^{r\times p}, R \in \Re^{p\times p}$ with Q, R symmetric, $M \in \Re^{n\times n}, M > 0, N \in \Re^{m\times m}, N > 0$ with M, N symmetric, the system (1) and (2) control will achieve mixed NLQR and dissipative performance if the following condition holds:*

$$\begin{aligned} &V_{k+1} - V_k + x_k^T M x_k + u_k^T N u_k \\ &\quad - \left(z_k^T Q z_k + 2 z_k^T S w_k + w_k^T R w_k\right) < 0, \forall k \geq 0. \quad (7) \end{aligned}$$

Proof Note that upon summation over k, we have

$$\sum_{i=0}^{N-1}\left[z_k^{\mathrm{T}}Qz_k+2z_k^{\mathrm{T}}Sw_k+w_k^{\mathrm{T}}Rw_k\right]$$
$$>\sum_{i=0}^{N-1}\left[x_k^{\mathrm{T}}Mx_k+u_k^{\mathrm{T}}Nu_k\right]+V_N-V_0. \quad (8)$$

Let $\beta(x_0)=V_0, V_k(x)=x_k^{\mathrm{T}}P_kx_k, V_N\geq 0$, Equation (8) implies

$$\sum_{i=0}^{N-1}\left(z_k^{\mathrm{T}}Qz_k+2z_k^{\mathrm{T}}Sw_k+w_k^{\mathrm{T}}Rw_k\right)+\beta(x_0)>0, \quad (9)$$

which is the condition for (Q,S,R) dissipativity. ■

Remark 1 By adding the terms $x_k^{\mathrm{T}}Mx_k+u_k^{\mathrm{T}}Nu_k$, we include the NLQR control performance into the original (Q,S,R) dissipative criteria.

Remark 2 Note that both H_∞ and passivity are special cases of (Q,S,R) dissipativity.

The special cases are summarized as follows:

Case 1 $Q=-I, S=0, R=\gamma^2 I$, the strict (Q,S,R) dissipativity reduces H_∞ design (Doyle et al., 1989). The overall control design satisfies mixed NLQR–H_∞ performance.

Case 2 $Q=0, S=I, R=0$, the strict (Q,S,R) dissipativity reduces to strict positive realness (Sun et al., 1994). The overall control design satisfies mixed NLQR–strict positive realness performance.

Case 3 $Q=-\theta I, S=(1-\theta)I, R=\theta\gamma^2 I$, the strict (Q, S,R) dissipativity reduces to mixed H_∞ and positive real performance design, when $\theta\in(0,1)$. The overall control design satisfies mixed NLQR–H_∞–positive real performance.

Case 4 $Q=-I, S=\frac{1}{2}(K_1+K_2)^{\mathrm{T}}, R=-\frac{1}{2}\left(K_1^{\mathrm{T}}K_2+K_2^{\mathrm{T}}K_1\right)^{\mathrm{T}}$, where K_1 and K_2 are constant matrices of appropriate dimensions, the strict (Q,S,R) dissipativity reduces to a sector-bounded constraint (Gupta & Joshi, 1994). The overall control design satisfies mixed NLQR–sector-bounded constraint performance.

Before introducing the main result of the paper, the following model of uncertainties is introduced.

Assumption 1 *The following general form of l_2-bounded unstructured uncertainties is considered:*

$$\Delta_A\Delta_A^{\mathrm{T}}\leq\gamma_A I,$$
$$\Delta_B\Delta_B^{\mathrm{T}}\leq\gamma_B I,$$
$$\Delta_E\Delta_E^{\mathrm{T}}\leq\gamma_E I,$$
$$\Delta_K\Delta_K^{\mathrm{T}}\leq\gamma_K I, \quad (10)$$

for $\forall x_k\in\Re^n$ and $k\geq 0$.

3. State-dependent LMI control

Lemma 1

$$AB^{\mathrm{T}}+BA^{\mathrm{T}}\leq\alpha AA^{\mathrm{T}}+\alpha^{-1}BB^{\mathrm{T}}. \quad (11)$$

This can be proven easily by considering

$$\left(\alpha^{1/2}A-\alpha^{-1/2}B\right)\left(\alpha^{1/2}A-\alpha^{-1/2}B\right)^{\mathrm{T}}\geq 0. \quad (12)$$

Also, by choosing A and B matrices as $A=\begin{bmatrix}a^{\mathrm{T}}\\0\end{bmatrix}$ *and* $B=\begin{bmatrix}0\\b^{\mathrm{T}}\end{bmatrix}$, *we have*

$$\begin{bmatrix}0 & a^{\mathrm{T}}b\\ b^{\mathrm{T}}a & 0\end{bmatrix}\leq\begin{bmatrix}\varsigma a^{\mathrm{T}}a & 0\\ 0 & \varsigma^{-1}b^{\mathrm{T}}b\end{bmatrix}. \quad (13)$$

The following theorem summarizes the main results of the paper:

Theorem 2 *Given the system equation (1), performance output (2) and control input (3), if there exist matrices $X_k=P_k^{-1}>0$ and Y_k for all $k>0$, such that the following state-dependent LMI holds:*

$$\text{If}\quad Q<0,\begin{bmatrix}X_k & \Upsilon_{12} & \Upsilon_{13} & Y_k^{\mathrm{T}} & \Upsilon_{15} & X_k\\ * & \Upsilon_{22} & E^{\mathrm{T}} & 0 & 0 & 0\\ * & * & \Upsilon_{33} & 0 & 0 & 0\\ * & * & * & \Upsilon_{44} & 0 & 0\\ * & * & * & * & \Upsilon_{55} & 0\\ * & * & * & * & * & \Upsilon_{66}\end{bmatrix}>0, \quad (14)$$

$$\text{If}\quad Q=0,\begin{bmatrix}X_k & \Upsilon_{12} & \Upsilon_{13} & Y_k^{\mathrm{T}} & X_k\\ * & \Upsilon_{22} & E^{\mathrm{T}} & 0 & 0\\ * & * & \Upsilon_{33} & 0 & 0\\ * & * & * & \Upsilon_{44} & 0\\ * & * & * & * & \Upsilon_{66}\end{bmatrix}>0, \quad (15)$$

$$\text{where}\quad \begin{aligned}\Upsilon_{12}&=X_kC_k^{\mathrm{T}}QD_k+X_kC_k^{\mathrm{T}}S,\\ \Upsilon_{13}&=X_kA_k^{\mathrm{T}}+Y_k^{\mathrm{T}}B_k^{\mathrm{T}},\\ \Upsilon_{15}&=X_kC_k^{\mathrm{T}},\\ \Upsilon_{22}&=D_k^{\mathrm{T}}S+S^{\mathrm{T}}D_k+D_k^{\mathrm{T}}QD_k+R+I,\\ \Upsilon_{33}&=X_k+(2\gamma_B+\gamma_E+1)I+B_kB_k^{\mathrm{T}},\\ \Upsilon_{44}&=N^{-1},\\ \Upsilon_{55}&=-Q^{-1},\\ \Upsilon_{66}&=M^{-1}-(\gamma_A+2\gamma_K)^{-1}I.\end{aligned} \quad (16)$$

Then the inequality (7) to guarantee mixed NLQR and dissipative performance is satisfied. The nonlinear feedback control gain is given by

$$K_k=Y_k\cdot P_k. \quad (17)$$

Proof In the proof below, the time and state argument will be dropped for notational simplicity. By applying system and performance output equations (1) and (2), and state feedback input equation (3), the performance index can be formed as follows:

$$\begin{aligned}&\{x_k^{\mathrm{T}}[A_k+\Delta_A+(B_k+\Delta_B)(K_k+\Delta_K)]^{\mathrm{T}}+w_k^{\mathrm{T}}[E_k+\Delta_E]^{\mathrm{T}}\}\\&\cdot P_{k+1}\cdot\{[A_k+\Delta_A+(B_k+\Delta_B)(K_k+\Delta_K)]x_k\\&+[E_k+\Delta_E]w_k\}-x_k^{\mathrm{T}}P_kx_k\\&+x_k^{\mathrm{T}}Mx_k+x_k^{\mathrm{T}}[K_k+\Delta_K]^{\mathrm{T}}N[K_k+\Delta_K]x_k\\&-[C_kx_k+Dw_k]^{\mathrm{T}}Q[C_kx_k+Dw_k]-2[C_kx_k+D_kw_k]^{\mathrm{T}}\\&Sw_k-w_k^{\mathrm{T}}Rw_k<0.\end{aligned} \tag{18}$$

By grouping the terms, we have

$$\begin{bmatrix}x_k^{\mathrm{T}} & w_k^{\mathrm{T}}\end{bmatrix}\Psi\begin{bmatrix}x_k & w_k\end{bmatrix}^{\mathrm{T}}=\begin{bmatrix}x_k^{\mathrm{T}} & w_k^{\mathrm{T}}\end{bmatrix}\begin{bmatrix}\Psi_{11} & \Psi_{12}\\ * & \Psi_{22}\end{bmatrix}\begin{bmatrix}x_k\\ w_k\end{bmatrix}<0, \tag{19}$$

where

$$\begin{aligned}\Psi_{11}&=\{(A_k+\Delta_A)+(B_k+\Delta_B)(K_k+\Delta_K)\}^{\mathrm{T}}\cdot P_{k+1}\\&\quad\cdot\{(A_k+\Delta_A)+(B_k+\Delta_B)(K_k+\Delta_K)\}\\&\quad+M-P_k+[K_k+\Delta_K]^{\mathrm{T}}N[K_k+\Delta_K]-C_k^{\mathrm{T}}QC_k,\\\Psi_{12}&=\{(A_k+\Delta_A)+(B_k+\Delta_B)(K_k+\Delta_K)\}^{\mathrm{T}}\\&\quad P_{k+1}[E_k+\Delta_E]-C_k^{\mathrm{T}}QD_k-C_k^{\mathrm{T}}S,\\\Psi_{22}&=[E_k+\Delta_E]^{\mathrm{T}}P_{k+1}[E_k+\Delta_E]-D_k^{\mathrm{T}}QD_k\\&\quad-\left(D_k^{\mathrm{T}}S+S^{\mathrm{T}}D_k\right)-R.\end{aligned} \tag{20}$$

Denote the following terms:

$$\begin{aligned}\mathrm{A}&=(A_k+\Delta_A)+(B_k+\Delta_B)(K_k+\Delta_K),\\\mathrm{K}&=K_k+\Delta_K,\\\mathrm{E}&=E_k+\Delta_E.\end{aligned} \tag{21}$$

Then Equation (19) is equivalent to

$$\begin{bmatrix}\mathrm{A}^{\mathrm{T}}P_{k+1}\mathrm{A}-P_k & \mathrm{A}^{\mathrm{T}}P_{k+1}\mathrm{E}\\ * & \mathrm{E}^{\mathrm{T}}P_{k+1}\mathrm{E}\end{bmatrix}+\begin{bmatrix}M+\mathrm{K}^{\mathrm{T}}N\mathrm{K}-C_k^{\mathrm{T}}QC_k & -C_k^{\mathrm{T}}QD_k-C_k^{\mathrm{T}}S\\ * & -D_k^{\mathrm{T}}S-S^{\mathrm{T}}D_k-D_k^{\mathrm{T}}QD_k-R\end{bmatrix}<0. \tag{22}$$

By adding and subtracting P_k term, we have

$$\begin{bmatrix}\mathrm{A}^{\mathrm{T}}\\ \mathrm{E}^{\mathrm{T}}\end{bmatrix}(P_{k+1}-P_k+P_k)\begin{bmatrix}\mathrm{A} & \mathrm{E}\end{bmatrix}-\begin{bmatrix}I\\0\end{bmatrix}P_k\begin{bmatrix}I & 0\end{bmatrix}+\begin{bmatrix}M+\mathrm{K}^{\mathrm{T}}N\mathrm{K}-C_k^{\mathrm{T}}QC_k & -C_k^{\mathrm{T}}QD_k-C_k^{\mathrm{T}}S\\ * & -D_k^{\mathrm{T}}S-S^{\mathrm{T}}D_k-D_k^{\mathrm{T}}QD_k-R\end{bmatrix}<0. \tag{23}$$

Imposing the property $P_{k+1}\le P_k$, the sufficient condition for Equation (23) is given as follows:

$$\begin{bmatrix}\mathrm{A}^{\mathrm{T}}\\ \mathrm{E}^{\mathrm{T}}\end{bmatrix}P_k\begin{bmatrix}\mathrm{A} & \mathrm{E}\end{bmatrix}-\begin{bmatrix}I\\0\end{bmatrix}P_k\begin{bmatrix}I & 0\end{bmatrix}+\begin{bmatrix}M+\mathrm{K}^{\mathrm{T}}N\mathrm{K}-C_k^{\mathrm{T}}QC_k & -C_k^{\mathrm{T}}QD_k-C_k^{\mathrm{T}}S\\ * & -D_k^{\mathrm{T}}S-S^{\mathrm{T}}D_k-D_k^{\mathrm{T}}QD_k-R\end{bmatrix}<0. \tag{24}$$

Equivalently, we obtain

$$\begin{bmatrix}P_k-M-\mathrm{K}^{\mathrm{T}}N\mathrm{K}+C_k^{\mathrm{T}}QC_k & C_k^{\mathrm{T}}QD_k+C_k^{\mathrm{T}}S\\ * & D_k^{\mathrm{T}}S+S^{\mathrm{T}}D_k+D_k^{\mathrm{T}}QD_k+R\end{bmatrix}-\begin{bmatrix}\mathrm{A}^{\mathrm{T}}\\ \mathrm{E}^{\mathrm{T}}\end{bmatrix}P_k\begin{bmatrix}\mathrm{A} & \mathrm{E}\end{bmatrix}>0. \tag{25}$$

Applying the Schur complement (Boyd et al., 1994), we have

$$\begin{bmatrix}\begin{pmatrix}P_k-M-\mathrm{K}^{\mathrm{T}}N\mathrm{K}\\+C_k^{\mathrm{T}}QC_k\end{pmatrix} & C_k^{\mathrm{T}}QD_k+C_k^{\mathrm{T}}S & \mathrm{A}^{\mathrm{T}}\\ * & \begin{pmatrix}D_k^{\mathrm{T}}S+S^{\mathrm{T}}D_k\\+D_k^{\mathrm{T}}QD_k+R\end{pmatrix} & \mathrm{E}^{\mathrm{T}}\\ * & * & P_k^{-1}\end{bmatrix}>0. \tag{26}$$

Taking $Q<0$ (the case where $Q=0$ will be considered later), we apply the Schur complement twice to Equation (26), then

$$\begin{bmatrix}P_k-M & C_k^{\mathrm{T}}QD_k+C_k^{\mathrm{T}}S & \mathrm{A}^{\mathrm{T}} & \mathrm{K}^{\mathrm{T}} & C_k^{\mathrm{T}}\\ * & D_k^{\mathrm{T}}S+S^{\mathrm{T}}D_k+D_k^{\mathrm{T}}QD_k+R & \mathrm{E}^{\mathrm{T}} & 0 & 0\\ * & * & P_k^{-1} & 0 & 0\\ * & * & * & N^{-1} & 0\\ * & * & * & * & -Q^{-1}\end{bmatrix}>0. \tag{27}$$

Let $X_k=P_k^{-1}$, by pre- and post-multiplying the above matrix inequality by $\operatorname{diag}\{X_k\quad I\quad I\quad I\quad I\}$, we have

$$\begin{bmatrix}X_k-X_kMX_k & X_kC^{\mathrm{T}}QD_k+X_kC^{\mathrm{T}}S & X_k\mathrm{A}^{\mathrm{T}} & X_k\mathrm{K}^{\mathrm{T}} & X_kC_k^{\mathrm{T}}\\ * & \begin{pmatrix}D_k^{\mathrm{T}}S+S^{\mathrm{T}}D_k+\\D_k^{\mathrm{T}}QD_k+R\end{pmatrix} & \mathrm{E}^{\mathrm{T}} & 0 & 0\\ * & * & X_k & 0 & 0\\ * & * & * & N^{-1} & 0\\ * & * & * & * & -Q^{-1}\end{bmatrix}>0 \tag{28}$$

By applying the Schur complement again, we have

$$\begin{bmatrix} X_k & X_kC_k^{\mathrm{T}}QD_k+X_kC_k^{\mathrm{T}}S & X_k\mathrm{A}^{\mathrm{T}} & X_k\mathrm{K}^{\mathrm{T}} & X_kC^{\mathrm{T}} & X_k \\ * & \begin{pmatrix} D_k^{\mathrm{T}}S+S^{\mathrm{T}}D_k+ \\ D_k^{\mathrm{T}}QD_k+R \end{pmatrix} & \mathrm{E}^{\mathrm{T}} & 0 & 0 & 0 \\ * & * & X_k & 0 & 0 & 0 \\ * & * & * & N^{-1} & 0 & 0 \\ * & * & * & * & -Q^{-1} & 0 \\ * & * & * & * & * & M^{-1} \end{bmatrix} > 0 \quad (29)$$

Denote

$$Y_k = K_kX_k \quad (30)$$

By replacing the variables with Equation (21) and applying Lemma 1 and Assumption 1, the sufficient condition for inequality (29) is given below

$$\begin{bmatrix} X_k & \begin{pmatrix} X_kC_k^{\mathrm{T}}QD_k+ \\ X_kC_k^{\mathrm{T}}S \end{pmatrix} & \begin{pmatrix} X_kA_k^{\mathrm{T}}+ \\ Y_k^{\mathrm{T}}B_k^{\mathrm{T}} \end{pmatrix} & X_kK_k^{\mathrm{T}} & X_kC_k^{\mathrm{T}} & X_k \\ * & \begin{pmatrix} D_k^{\mathrm{T}}S+S^{\mathrm{T}}D_k+ \\ D_k^{\mathrm{T}}QD_k+R \end{pmatrix} & E_k^{\mathrm{T}} & 0 & 0 & 0 \\ * & * & X_k & 0 & 0 & 0 \\ * & * & * & N^{-1} & 0 & 0 \\ * & * & * & * & -Q^{-1} & 0 \\ * & * & * & * & * & M^{-1} \end{bmatrix} + \begin{bmatrix} \Omega_{11} & 0 & 0 & 0 & 0 & 0 \\ * & \Omega_{22} & 0 & 0 & 0 & 0 \\ * & * & \Omega_{33} & 0 & 0 & 0 \\ * & * & * & \alpha_1^{-1}I & 0 & 0 \\ * & * & * & * & 0 & 0 \\ * & * & * & * & * & 0 \end{bmatrix} > 0 \quad (31)$$

where

$$\Omega_{11} = (\alpha_1\gamma_K+\alpha_1\gamma_A+\alpha_4\gamma_K)X_kX_k+\alpha_2Y_k^{\mathrm{T}}Y_k,$$
$$\Omega_{22} = \alpha_2I,$$
$$\Omega_{33} = \alpha_1^{-1}\gamma_BI+\alpha_2^{-1}(\gamma_B+\gamma_E)I+\alpha_3^{-1}I+\alpha_4^{-1}B_kB_k^{\mathrm{T}}. \quad (32)$$

Finally, by applying the Schur complement twice, we have

$$\begin{bmatrix} X_k & \Upsilon_{12} & \Upsilon_{13} & Y_k^{\mathrm{T}} & \Upsilon_{15} & X_k \\ * & \Upsilon_{22} & E^{\mathrm{T}} & 0 & 0 & 0 \\ * & * & \Upsilon_{33} & 0 & 0 & 0 \\ * & * & * & \Upsilon_{44} & 0 & 0 \\ * & * & * & * & \Upsilon_{55} & 0 \\ * & * & * & * & * & \Upsilon_{66} \end{bmatrix} > 0, \quad (33)$$

where

$$\Upsilon_{12} = X_kC_k^{\mathrm{T}}QD_k+X_kC_k^{\mathrm{T}}S,$$
$$\Upsilon_{13} = X_kA_k^{\mathrm{T}}+Y_k^{\mathrm{T}}B_k^{\mathrm{T}},$$
$$\Upsilon_{15} = X_kC_k^{\mathrm{T}},$$
$$\Upsilon_{22} = D_k^{\mathrm{T}}S+S^{\mathrm{T}}D_k+D_k^{\mathrm{T}}QD_k+R+\alpha_2I,$$
$$\Upsilon_{33} = X_k+\alpha_1^{-1}\gamma_BI+\alpha_2^{-1}(\gamma_B+\gamma_E)I+\alpha_3^{-1}I + \alpha_4^{-1}B_kB_k^{\mathrm{T}},$$
$$\Upsilon_{44} = N^{-1}+\left(\alpha_1^{-1}-\alpha_2^{-1}\right)I,$$
$$\Upsilon_{55} = -Q^{-1},$$
$$\Upsilon_{66} = M^{-1}-(\alpha_1\gamma_A+\alpha_2\gamma_K+\alpha_4\gamma_K)^{-1}I. \quad (34)$$

Note that Equation (33) is derived under the condition that $Q < 0$. However, when strict positive realness criteria are chosen for control design, then condition $Q = 0$ must be satisfied. In this case, matrix inequality (33) should be replaced by

$$\begin{bmatrix} X_k & \Upsilon_{12} & \Upsilon_{13} & Y_k^{\mathrm{T}} & X_k \\ * & \Upsilon_{22} & E^{\mathrm{T}} & 0 & 0 \\ * & * & \Upsilon_{33} & 0 & 0 \\ * & * & * & \Upsilon_{44} & 0 \\ * & * & * & * & \Upsilon_{66} \end{bmatrix} > 0. \quad (35)$$

Since positive constants $\alpha_1, \ldots, \alpha_5$ are arbitrary, choosing all of them as 1, we obtain Equations (14) and (15). Therefore, if LMI e (14) or (15) holds under different conditions on Q, the inequality (7) is satisfied. By solving the LMI at each step, the values of P_k, Y_k can be obtained. The nonlinear feedback control gain can be found by $K_k = Y_k \cdot P_k$. This concludes the proof. ■

Remark 3 At this point, it is to be noted that other choices of constants $\alpha_1, ..., \alpha_4$ are possible and can be tried if the value 1 for all these constants does not work.

4. Application to the inverted pendulum on a cart

We test the novel robust and resilient state-dependent LMI approach with the inverted pendulum on a cart (Wang et al. 2010) to compare the performance of different controllers. Using the Euler–Lagrange Equation technique, the complete equations of motion for the inverted pendulum on a cart are found to be

$$(M+m)\ddot{x}+b\dot{x}+mL\ddot{\theta}\cos(\theta)-mL\dot{\theta}\sin(\theta)=F,$$
$$(I+mL^2)\ddot{\theta}+mgL\sin(\theta)+mL\ddot{x}\cos(\theta)=0. \quad (36)$$

The following system parameters are assumed

$M = 0.5\,\mathrm{kg}, m = 0.5\,\mathrm{kg}, b = 0.1\,\mathrm{N\,s/m}, L = 0.3\,\mathrm{m},$

$I = 0.06\,\mathrm{kg\,m^2}.$

Sampling time : $T = 0.01\,s$.

Denote the following state variables:

$$x_{1,k} = x(kT), x_{2k} = \dot{x}(kT), x_{3k} = \theta(kT), x_{4k} = \dot{\theta}(kT).$$

The following initial conditions are assumed:

$$x_1 = 1, x_2 = 0, x_3 = \pi/4, x_4 = 0.$$

The following design parameters are chosen to satisfy different mixed criteria:

Mixed NLQR–H_∞design (predominant NLQR)

$$C = \begin{bmatrix} 0.01 & 0.01 & 0.01 & 0.01 \end{bmatrix}, D = [0.01], M = I_4,$$
$$N = 1, Q = -1, S = 0, R = 5.$$

Mixed NLQR–H_∞ design (predominant H_∞)

$$C = \begin{bmatrix} 1 & 1 & 1 & 1 \end{bmatrix}, D = [1], M = 0.01 \times I_4, N = 0.01,$$
$$Q = -1, S = 0, R = 5.$$

Mixed NLQR–H_∞–positive real design (NLQR passivity)

$$C = \begin{bmatrix} 1 & 1 & 1 & 1 \end{bmatrix}, D = [1], M = I_4, N = 1,$$
$$Q = -0.01,$$
$$S = 0.5, R = 0.01.$$

All of the above-mixed criteria control performance results are shown in the Figures 1–5, in comparison with the traditional LQR technique based on linearization. From these figures, we find that the novel state-dependent LMI control has better performance compared with the traditional LQR technique based on linearization. Especially, Figures 1 and 2 show that the traditional LQR technique loses control of the position and velocity of the cart, respectively, while the state-dependent LMI approach effectively stabilizes the position and the velocity of the cart. It should also be noted that predominant NLQR and predominant H_∞ control techniques lead to faster response times than the NLQR-passivity technique. We observe that predominant H_∞ control shows the fastest response. Figure 5 shows that the highest magnitude of control is needed by the

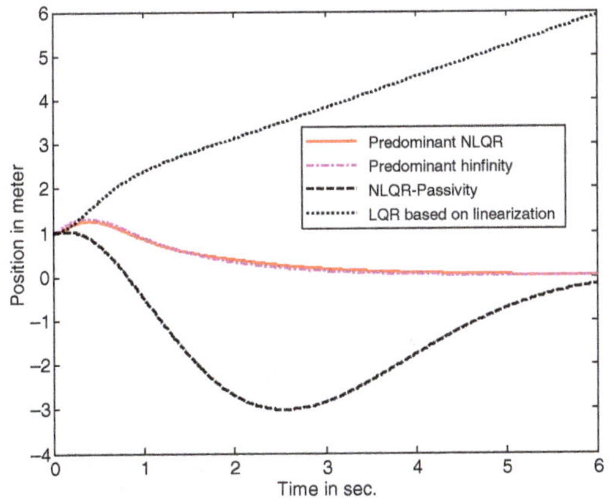

Figure 1. Position trajectory of the inverted pendulum.

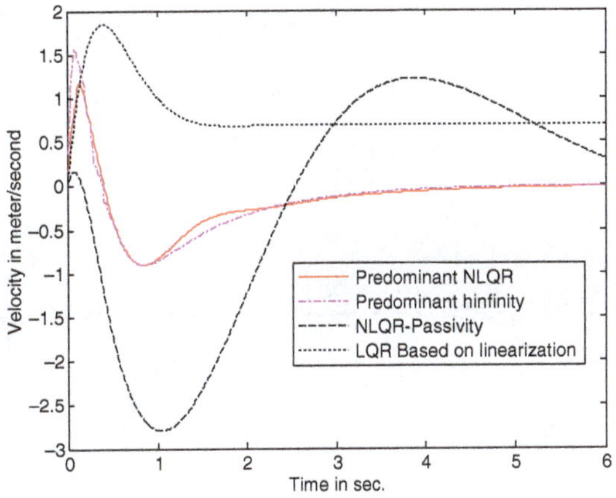

Figure 2. Velocity trajectory of the inverted pendulum.

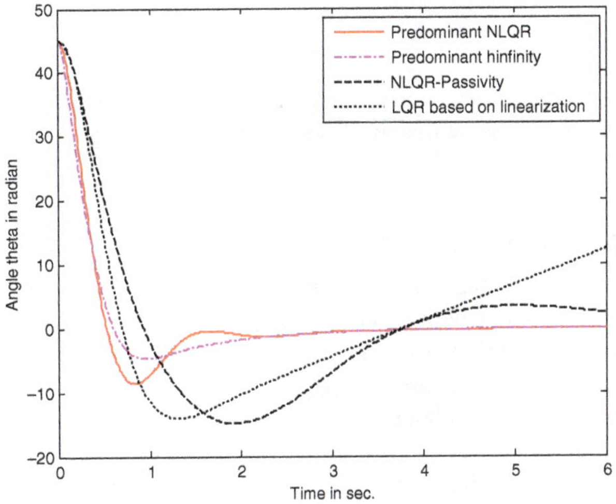

Figure 3. Angle "theta" trajectory of the inverted pendulum.

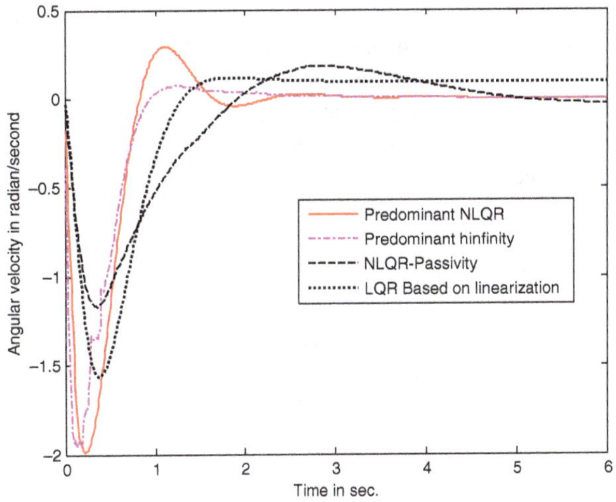

Figure 4. Angular velocity trajectory of the inverted pendulum.

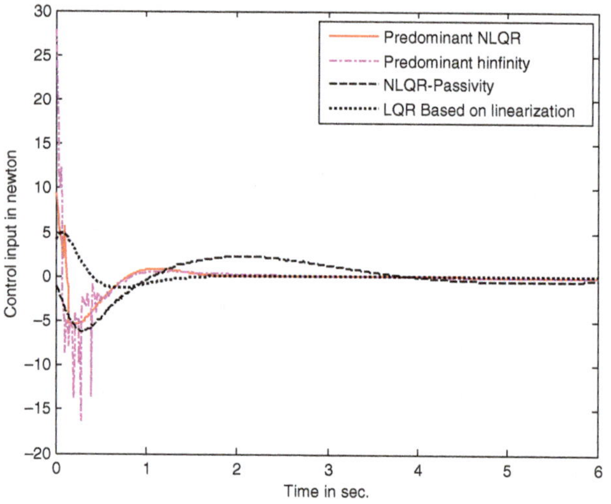

Figure 5. Control input.

predominant H_∞ control and the lowest control magnitude is needed by the linearization-based LQR technique.

5. Conclusions

This paper has addressed discrete-time nonlinear control system design with general NLQR and quadratic dissipative criteria to achieve asymptotic stability, quadratic optimality and strict quadratic dissipativeness. For systems with unstructured but bounded uncertainty, the LMI-based sufficient conditions are derived for the control solution. These results unify the existing results on SDRE control, robust H_∞ and positive real control. The relative weighting matrices of these criteria can be achieved by choosing different coefficient matrices. The optimal control can be obtained by solving LMI at each time step. The inverted pendulum on a cart is used as an example to demonstrate the effectiveness and robustness of the proposed method. The numerical simulation studies show that the proposed method provides a satisfactory alternative to the existing nonlinear control approaches.

Acknowledgements

We thank the reviewers for their remarks, which have improved the quality of the paper. This research has been financially supported by a grant from NITC (National Institute for Transportation and Communities), which is gratefully acknowledged.

References

Boyd, S., Ghaoui, L. E., Feron, E., & Balakrishnan, V. (1994). *Linear matrix inequalities in system and control theory*. Philadelphia, PA: SIAM Studies in Applied Mathematics.

Doyle, J. C., Glover, K., Khargonekar, P. P., & Francis, B. A. (1989). State space solution to the standard H2 and H-infinity control problems. *IEEE Transactions on Automatic Control, 34*, 831–847.

Gupta, S., & Joshi, S. M. (1994). *Some properties and stability results for sector bounded LTI systems*. Proceedings of the 33rd IEEE conference on decision and control, Orlando, FL, pp. 2973–2978.

Haddad, W. M., & Bernstein, D. S. (1991). Robust stabilization with positive real uncertainty, beyond the small gain theorem. *Systems and Control Letters, 17*, 191–208.

Hill, D. J., & Moylan, P. J. (1976). Stability of nonlinear dissipative systems. *IEEE Transactions on Automatic Control, 21*, 708–711.

Hill, D. J., & Moylan, P. J. (1980). Dissipative dynamical systems: Basic input-output and state properties. *Journal of Franklin Institute, 309*, 327–357.

Huang, Y., & Lu, W.-M. (1996). *Nonlinear optimal control: Alternatives to Hamilton-Jacobi equation*. Proceedings of 35th IEEE conference on decision and control, Kobe, Japan, pp. 3942–3947.

Khargonekar, P. P., Petersen, I. R., & Zhou, K. (1990). Robust stabilization of uncertain linear systems: Quadratic stabilizability and H-infinity control theory. *IEEE Transactions on Automatic Control, 35*, 356–361.

Petersen, I. R. (1987). A stabilization algorithm for a class of uncertain systems. *Systems and Control Letters, 8*, 181–188.

Safonov, M. G., Jonckheere, E. A., Verma, M., & Limebeer, D. J. N. (1987). Synthesis of positive real multivariable feedback systems. *International Journal of Control, 45*, 817–842.

Shim, D. (1996). Equivalence between positive real and norm-bounded uncertainty. *IEEE Transactions on Automatic Control, 41*, 1190–1193.

Sun, W., Khargonekar, P. P., & Shim, D. (1994). Solution to the positive real control problem for linear time-invariant systems. *IEEE Transactions on Automatic Control, 39*, 2034–2046.

Tan, Z., Soh, Y. C., & Xie, L. (2000). Dissipative control of linear discrete-time systems with dissipative uncertainty. *International Journal of Control, 73* (4), 317–328.

Wang, X., & Yaz, E. E. (2009). *The state dependent control of continuous time nonlinear systems with mixed performance criteria*. Proceedings of IASTED international conference on identification and control applications, Honolulu, HI, pp. 98–102.

Wang, X., Yaz, E. E., & Jeong, C. S. (2010). *Robust nonlinear feedback control of discrete-time systems with mixed performance criteria*. Proceedings of American control conference, Baltimore, MD, pp. 6357–6362.

Wang, X., Yaz, E. E., & Yaz, Y. I. (2010). *Robust and resilient state dependent control of continuous time nonlinear systems with general performance criteria*. Proceedings of the 49th IEEE conference on decision and control, Atlanta, GA, pp.603–608.

Wang, X., Yaz, E. E., & Yaz, Y. I. (2011). *Robust and resilient state dependent control of discrete time nonlinear systems with general performance criteria*. Proceedings of the18th IFAC congress, Milano, Italy.

Willems, J. C. (1972a). Dissipative dynamical systems-part 1: General theory. *Archive for Rational Mechanics and Analysis, 45*, 321–351.

Willems, J. C. (1972b). Dissipative dynamical systems-part 2: Linear systems with quadratic supply rates. *Archive for Rational Mechanics and Analysis, 45*, 352–393.

Xie, S., Xie, L., & De Souza, C. E. (1998). Robust dissipative control for linear systems with dissipative uncertainty. *International Journal of Control, 70* (2), 169–191.

9

Parameter modulation for secure communication via the synchronization of Chen hyperchaotic systems

Jianbin He* and Jianping Cai

College of Mathematics and Statistics, Minnan Normal University, Zhangzhou, People's Republic of China

The Chen hyperchaotic systems are synchronized via linear feedback control and the parameter is identified by using the adaptive control techniques even though the parameter is unknown. It is proved by the Lyapunov stability theory that the response system is able to track the driving system well and the parameter is estimated exactly. Based on the synchronization of Chen hyperchaotic systems, a scheme of secure communication using the parameter modulation method is presented and the transmitted plaintext message can be successfully recovered. Finally, white Gaussian noise in different kinds of signal-to-noise ratio is conducted to evaluate the performance of the proposed secure communication scheme. The return maps of the transmitted signals are provided to show the higher degree of security. Numerical simulation shows its feasibility.

Keywords: secure communication; parameter modulation; hyperchaos; adaptive control; return map

1. Introduction

Chaos is a very interesting nonlinear phenomenon because it is sensitively dependent on the initial conditions. In the past few decades, many chaotic systems have been proposed such as the Lorenz system, Chua's circuit, Chen system, and hyperchaotic system (Chen & Lü, 2002; Chen, Lu, Lü, & Yu, 2006; Lorenz, 1963; Ueta & Chen, 2000). Their complex behaviors have been widely studied, and several techniques have been used to control the chaotic system. The synchronization of chaotic system has been investigated since the pioneering work of Pecora and Carroll (1990), a series of synchronization schemes are proposed such as active–passive decomposition (Kocarev & Parlitz, 1995), linear feedback control (Tao, Xiong, & Hu, 2006), slide mode control (SMC) (Cai, Jing, & Zhang, 2010), and controllable probabilistic particle swarm optimization algorithm (Tang, Wang, & Fang, 2011). Wang, Han, Xie, & Zhang (2009) develop the chaos control problem for a general class of chaotic systems using a feedback controller to guarantee asymptotical stability of the chaotic system based on the SMC theory. The distributed synchronization of networks composed of agent systems with multiple randomly occurring nonlinearities, multiple randomly occurring controllers, and multiple randomly occurring updating laws has been achieved by Tang, Gao, Zou, and Kurths (2013) in mean square under certain criteria. The synchronization criteria and the observed phenomena are demonstrated by several numerical simulation examples and the advantage of distributed adaptive controllers over conventional adaptive controllers is illustrated.

Since 1990, the synchronization of chaotic systems has attracted much attention due to its potential applications in secure communication, analysis of chemical reactions, information processing, and so on (Arman, Kia, Naser, & Henry, 2009; Xiang, Cheng, & Liao, 2008; Xiao & Cao, 2009). The characteristics of broadband, noiselike, and unpredictability make chaotic signals ideal for information encryption or hiding in secure communication. Chaotic secure communications have been proposed, and there are many methods such as chaos masking (Cuomo, Oppenheim, & Strogatz, 1993), chaos parameter modulation (Yang & Chua, 1996), chaotic shift keying (Dedieu, Kennedy, & Hasler, 1993), and impulsive synchronization to secure communication (Yang, 2004). The key is to complete the synchronization of the slave–master systems by driving the slave with a signal derived from the master. It is easy to synchronize the master–slave systems when the parameters of the master system are known. But their parameters are usually unknown in advance, such an adaptive controller should be designed to synchronize the master–slave systems and identify the unknown parameters. Thus, the synchronization of chaotic systems in the presence of unknown parameters is more essential and useful in real-world applications (Chen, 2012). Tang represents the first attempt to include two measures of controllability into one unified framework, and the detection problem of controlling regions in cortical

*Corresponding author. Email: jbh2012yml@126.com

networks is converted into a constrained optimization problem, then the detection of controlling regions of a weighted and directed complex network is thoroughly investigated (Tang, Wang, Gao, Stephen, & Kurths, 2012). Song and Cao (2004) propose a secure communication scheme via Chua chaos using an adaptive learning mechanism that the parameters are applied to modulate the discrete message signals, but it does not make any security analysis of the proposed scheme. Based on the modified adaptive method, Yu proposes some secure communication schemes that the transmitted signal is masked by chaotic signal or modulated into the system, which effectively blurs the constructed return map and can resist this return map attack (Yu, Cao, Wong, & Lü, 2007). Li and Zhao (2011) proposed a scheme in which controllers not only realize the synchronization of the state vectors, but also synchronize the unknown response parameters to the given drive parameter as time goes to infinity; however, the update laws of the parameters are questionable. One can use the return map attack to break both chaotic masking and chaotic modulating systems. Li points out that the security of the modulation-based schemes proposed by Wu, Hu, and Zhang (2004) is not so satisfactory from a pure cryptographical point of view and the improved scheme is still insecure against a new attack (Li, Alvarez, & Chen, 2005). To overcome the security problems of most traditional chaos-based secure communication schemes, a number of new countermeasures have been proposed in recent years. One widely suggested measure is to use more complex chaotic systems rather than three-dimensional systems like the Lorenz and Chua systems (Li, Alvarez, Li, & Halang, 2007). Much different from the method above (Song & Cao, 2004; Yu et al., 2007), a fourth dimension hyperchaotic system is employed as the transmitter in this scheme. Using the high-dimension hyperchaotic systems that have multiple positive Lyapunov exponents may produce a more complex chaotic behavior to resist the return map attack. We explore the simple parameter modulation for secure communication by making use of hyperchaotic systems, and this method can resist the well-known return map attack, but note that the parameter embedded with message signals of driven system is unknown.

In the present paper, we propose an adaptive synchronization method for the Chen hyperchaotic systems with unknown parameter, and a simple parameter modulation scheme for secure communication is explored. Based on the Lyapunov stability theory, a linear feedback controller is used to synchronize the hyperchaotic systems. An adaptive update law is derived which enables the receiver to retrieve the message signals sent by the transmitter. Simulations show that synchronization is achieved asymptotically and the modulated message signals are recovered well. Finally, the robustness to noise and the security analysis of the proposed scheme are given. It is found out that the return maps generated from the chaotic carrier blur and diffuse with each other; thus, to distinguish them is not so easy.

2. Systems description and synchronization between the Chen hyperchaotic systems

2.1. *Systems description*

The Chen hyperchaotic system is given by

$$\begin{aligned}\frac{dx_1}{dt} &= a(x_2 - x_1) + x_4,\\ \frac{dx_2}{dt} &= dx_1 - x_1x_3 + cx_2,\\ \frac{dx_3}{dt} &= x_1x_2 - bx_3,\\ \frac{dx_4}{dt} &= x_2x_3 + rx_4,\end{aligned} \tag{1}$$

where $x = [x_1, x_2, x_3, x_4]$ are state variables and a, b, c, d, r are real constants. When $a = 35$, $b = 3$, $c = 12$, $d = 7$, $0 \le r \le 0.798$ system (1) is chaotic, when $a = 35$, $b = 3$, $c = 12$, $d = 7$, $0.0085 \le r \le 0.798$, system (1) is hyperchaotic, when $a = 35$, $b = 3$, $c = 12$, $d = 7$, $0.798 \le r \le 0.9$, system (1) is periodic (Li, Tang, & Chen, 2005; Park, 2005).

We found that hyperchaos does exist in the Chen system. In the numerical simulations, the parameters are always chosen as $a = 35$, $b = 3$, $c = 12$, $d = 7$, $r = 0.5$, then hyperchaotic attractors can be found.

2.2. *Simple parameter modulation for secure communication*

We assume that the Chen hyperchaotic system is the master system, and it can be presented in the form of

$$\begin{aligned}\frac{dx_1}{dt} &= a(x_2 - x_1) + x_4,\\ \frac{dx_2}{dt} &= dx_1 - x_1x_3 + cx_2,\\ \frac{dx_3}{dt} &= x_1x_2 - bx_3,\\ \frac{dx_4}{dt} &= x_2x_3 + rx_4,\\ \frac{dr}{dt} &= 0.\end{aligned} \tag{2}$$

The response system can be presented in the form of

$$\begin{aligned}\frac{dy_1}{dt} &= a(y_2 - y_1) + y_4 - k_1e_1,\\ \frac{dy_2}{dt} &= dy_1 - y_1y_3 + cy_2 - k_2e_2,\\ \frac{dy_3}{dt} &= y_1y_2 - by_3 - k_3e_3,\\ \frac{dy_4}{dt} &= y_2y_3 + \hat{r}y_4 - k_4e_4,\\ \frac{d\hat{r}}{dt} &= f_r,\end{aligned} \tag{3}$$

where $y = [y_1, y_2, y_3, y_4]$ are the response system state variables, k_1, k_2, k_3, k_4 are feedback gains, $e_1 = y_1 - x_1$, $e_2 = y_2 - x_2$, $e_3 = y_3 - x_3$, $e_4 = y_4 - x_4$, $e_r = \hat{r} - r$, and f_r is a function for learning to be determined. The parameter r is the unknown constant parameter, $\hat{r}$ is the estimated parameter in the receiver side. Subtracting Equation (2) from Equation (3) we would obtain

$$\begin{aligned}
\frac{de_1}{dt} &= ae_2 - ae_1 + e_4 - k_1e_1,\\
\frac{de_2}{dt} &= (d - e_3)e_1 + ce_2 - k_2e_2 - x_3e_1 - x_1e_3,\\
\frac{de_3}{dt} &= x_1e_2 - be_3 - k_3e_3 + x_2e_1 + e_1e_2,\\
\frac{de_4}{dt} &= x_3e_2 + \hat{r}e_4 + x_4e_r - k_4e_4 + e_2e_3 + x_2e_3,\\
\frac{d\hat{r}}{dt} &= f_r.
\end{aligned} \tag{4}$$

We can try to find the function f_r in Equation (4) so that the synchronization between Equations (2) and (3) is realized, then all the states of receiver will track the corresponding states in transmitter, of course, $\hat{r}$ will track r. Therefore, r can be taken as the message signal carrier, and the modulation of r will be done.

To this end, we take the Lyapunov function

$$V(e_1, e_2, e_3, e_4, e_r) = \frac{1}{2}\left(\frac{1}{\alpha}e_1^2 + e_2^2 + e_3^2 + e_4^2 + \frac{1}{\beta}e_r^2\right), \quad \alpha, \beta > 0. \tag{5}$$

Its derivative along the error dynamics (4) is

$$\begin{aligned}
\dot{V} &= \frac{1}{\alpha}e_1\dot{e}_1 + e_2\dot{e}_2 + e_3\dot{e}_3 + e_4\dot{e}_4 + \frac{1}{\beta}e_r\dot{e}_r\\
&= \frac{a}{\alpha}e_1e_2 - \frac{a}{\alpha}e_1^2 + \frac{1}{\alpha}e_1e_4 - \frac{k_1}{\alpha}e_1^2 + de_1e_2 + ce_2^2\\
&\quad - k_2e_2^2 - e_2e_1e_3 - x_3e_1e_2 - x_1e_2e_3 - be_3^2 - k_3e_3^2\\
&\quad + e_1e_2e_3 + x_1e_2e_3 + x_1e_1e_3 + x_2e_3e_4 + e_4e_2e_3\\
&\quad + \hat{r}e_4^2 + x_4e_4e_r + x_3e_2e_4 - k_4e_4^2 + \frac{1}{\beta}e_r\dot{e}_r\\
&= e_r\left(x_4e_4 + \frac{1}{\beta}\dot{e}_r\right) - \left(\frac{k_1}{\alpha} + \frac{a}{\alpha} - 3\right)e_1^2\\
&\quad - \left(k_2 - c - 2 - \frac{(a/\alpha + d - x_3)^2}{4}\right)e_2^2\\
&\quad - \left(k_3 + b - \frac{x_3^2}{4} - 1\right)e_3^2\\
&\quad - \left(k_4 - \frac{1}{4\alpha^2} - \frac{x_3^2}{4} - \frac{y_3^2}{4} - \hat{r}\right)e_4^2\\
&\quad - \left(e_1 - \frac{a/\alpha + d - x_3}{2}e_2\right)^2 - \left(e_1 - \frac{1}{2\alpha}e_4\right)^2\\
&\quad - \left(e_1 - \frac{x_1}{2}e_3\right)^2 - \left(e_2 - \frac{y_3}{2}e_4\right)^2 - \left(e_3 - \frac{x_2}{2}e_4\right)^2.
\end{aligned}$$

Obviously, if we let

$$\begin{aligned}
&\dot{e}_r = -\beta x_4e_4, \beta > 0, k_1 \geq \left(3 - \frac{a}{\alpha}\right)\alpha, k_2 \geq c + 2 + \frac{(a/\alpha + d - x_3)^2}{4},\\
&k_3 \geq \frac{x_2^2}{4} - b + 1, k_4 \geq \frac{1}{4\alpha^2} + \frac{y_3^2 + x_3^2}{4} + \hat{r} \quad \text{then}\\
&\dot{V} \leq -\left(e_1 - \frac{a/\alpha + d - x_3}{2}e_2\right)^2\\
&\quad - \left(e_1 - \frac{1}{2\alpha}e_4\right)^2 - \left(e_1 - \frac{x_1}{2}e_3\right)^2 - \left(e_2 - \frac{y_3}{2}e_4\right)^2\\
&\quad - \left(e_3 - \frac{x_2}{2}e_4\right)^2 < 0.
\end{aligned} \tag{6}$$

According to the Lyapunov theory, the inequality $\dot{V}(t) < 0$ indicates that $V(t)$ converges to zero and is bounded for all time, i.e. $V \in L_\infty$. The definition of $V(t)$ in Equation (5) indicates $e(t) \in L_\infty, \hat{r} \in L_\infty$. Inequalities of $\dot{V} < 0$ imply $\dot{e}(t) \in L_\infty$, it is noted that $e(t) \to 0$ as $t \to \infty$ by Babalat's lemma (Khalil, 1992). We have

$$e_1 \to 0, e_2 \to 0, e_3 \to 0, e_4 \to 0 \quad \text{as } t \to \infty.$$

That means asymptotical tracking of all states will be realized.

Meanwhile, from Equations (4) and (6), we have

$$\begin{aligned}
&\dot{e}_r = \dot{\hat{r}} - \dot{r} = -\beta x_4e_4 \quad \text{and}\\
&\dot{e}_4 = x_3e_2 + \hat{r}e_4 + x_4e_r - k_4e_4 = 0.
\end{aligned}$$

Hence $\dot{\hat{r}} = \dot{r} - \beta x_4e_4, x_4e_r = 0$, since $\dot{r} = 0$, and x_4 is not identically equal to zero. Therefore,

$$e_r \to 0.$$

Then the resulting receiver for modulating r is written as

$$\begin{aligned}
\frac{dy_1}{dt} &= a(y_2 - y_1) + y_4 - k_1e_1,\\
\frac{dy_2}{dt} &= dy_1 - y_1x_3 + cy_2 - k_2e_2,\\
\frac{dy_3}{dt} &= x_1y_2 - by_3 - k_3e_3,\\
\frac{dy_4}{dt} &= y_2x_3 + \hat{r}y_4 - k_4e_4,\\
\frac{d\hat{r}}{dt} &= -\beta x_4e_4.
\end{aligned} \tag{7}$$

3. Application to secure communication

In this section, the preceding adaptive law-based synchronization scheme is applied to chaotic secure communication. Figure 1 illustrates the proposed communication system consisting of a transmitter, modulation, and receiver at the receiving end of communication. At the transmitter side, the original message $m(t)$ is multiplied by a factor f and modulated by the parameter into the chaotic signals $x(t)$.

The scaling of factor f must be well chosen to reduce the message signal to a degree that it can be successfully modulated by the parameters and masked by $x(t)$.

We first use a "modulation rule" to modulate $s(t)$ in a parameter of the transmitter in Equation (2). Then an adaptive controller is used at the receiver to maintain synchronization by continuously tracking the changes in the modulation parameter. So that $s(t)$ can be recovered by this adaptive controller. In this section, we discuss the case when only a parameter of the transmitter is modulated while others remain constant.

3.1. *Parameter r-modulation*

In this case, the parameter r is used to modulate the signal $s(t)$, the modulation rule is given by

$$R(t) = f(s(t)) = \frac{s(t)}{d} + 0.5, \tilde{R}(t) = f^{-1}\tilde{s}(t), \qquad (8)$$

where $d = 10$ and $\tilde{s}(t)$ is the recovered message signal. We choose the transmitted message signal as follows:

$$s(t) = \frac{\sin(t)}{2} + \frac{\cos(2t)}{3}.$$

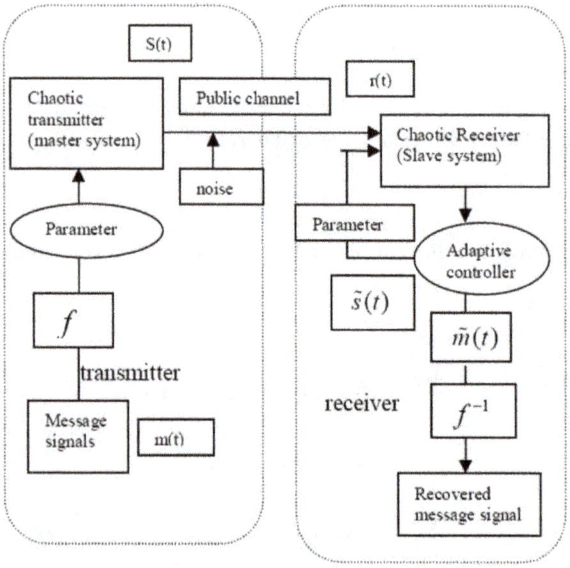

Figure 1. Secure communication systems based on parameters modulation and adaptive controller.

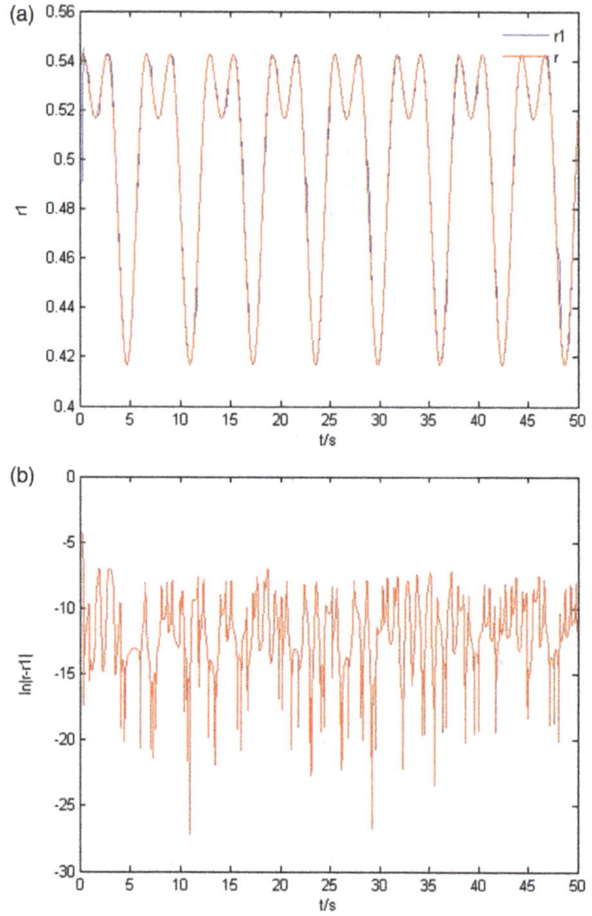

Figure 2. The estimation of unknown parameters r (a) and signal errors of $\ln|r(t) - \hat{r}(t)|$ (b).

The unknown parameters are chosen as $a = 35$, $b = 3$, $c = 12$, $d = 7$, $r = 0.5$. So the master system can exhibit a chaotic behavior. According to the rules (6), we assume $k_1 = 50$, $k_2 = 100$, $k_3 = 200$, $k_4 = 500$, $\alpha = 10$, $\beta = 100$. Simulation result is shown in Figure 2.

3.2. *Security analyses*

In the communication scheme, the transmitted message signals may be disturbed by random noise, attackers may retrieve the plaintext via some methods, such as return maps analysis, power-spectral (filtering) analysis, etc. So a good secure communication should resist all kinds of unknown attacks, some security analysis has been performed on the proposed secure communication scheme in this section.

3.2.1. *Noise analyses*

Channel noise is inevitable in secure communication. Not only is this transmission scheme accurate, but it is robust noise to some extent. Assume that the white Gaussian noise is considered in this scheme.

First, the signal-to-noise ratio (SNR) is defined as

$$dB = 10^* \log\left(\frac{S}{N}\right),$$

where S is the power of signal and N is the power of noise.

Then the continuous signal is taken as follows:

$$s(t) = \frac{\sin(2t)}{2} + \frac{\cos(t)}{3};$$

it is modulated by the rule in Equation (8). White Gaussian noise is added into the transmitter system as follows:

$$\frac{dx_1}{dt} = a(x_2 - x_1) + x_4 + n_1(t),$$
$$\frac{dx_2}{dt} = dx_1 - x_1x_3 + cx_2 + n_2(t),$$
$$\frac{dx_3}{dt} = x_1x_2 - bx_3 + n_3(t),$$
$$\frac{dx_4}{dt} = x_2x_3 + rx_4 + n_4(t),$$
$$\frac{dr}{dt} = 0,$$

where $n_i(t)(i = 1, 2, 3, 4)$ are random white Gaussian noise. The transmitted message is modulated into the parameter r, so we take $n_i(t) \equiv 0 \ \ (i = 1, 2, 3)$.

When the SNR is 103.7 dB, the transmission of a continuous signal can be recovered well as shown in Figures 3 and 4. The encrypted signals are almost chaotic and difficult to be identified by the attackers.

Similarly, when the SNR is 32.91 dB, the transmission of a continuous signal can be recovered as shown in Figure 5, as it is disturbed by white Gaussian noise, we can see that the recovered signal has been blurred. There are some experimental results in Table 1 with the different kinds of SNR. In Table 1, the $|r(t) - \hat{r}(t)|$ is the errors of the recovered signals and sum $|r(t) - \hat{r}(t)|/n$ is the average value of $|r(t) - \hat{r}(t)|$. As we can see, the high SNR of the transmitted signals can be recovered much better than the low one in Table 1.

In summary, the transmitted signals may be disturbed by the unknown channel noises. Compared with the results in Table 1, the original signal can be recovered well when the SNR is higher than 65.44 dB, and the average error of the $|r(t) - \hat{r}(t)|$ is lower than 0.0020 and bounds from 1.4789e–08 to 0.0560. On the other side, if the SNR is below to 32.91 dB, the original signal can be just identified roughly and even get poor results. So the proposed secure communication scheme is robust noise to some extent.

3.2.2. *Security of Chen hyperchaotic parameter modulation*

In this digital mode secure communication, as described by Perez and Cerdeira (1995), the key to extracting message from the chaotic mask is to recognize that a small change in

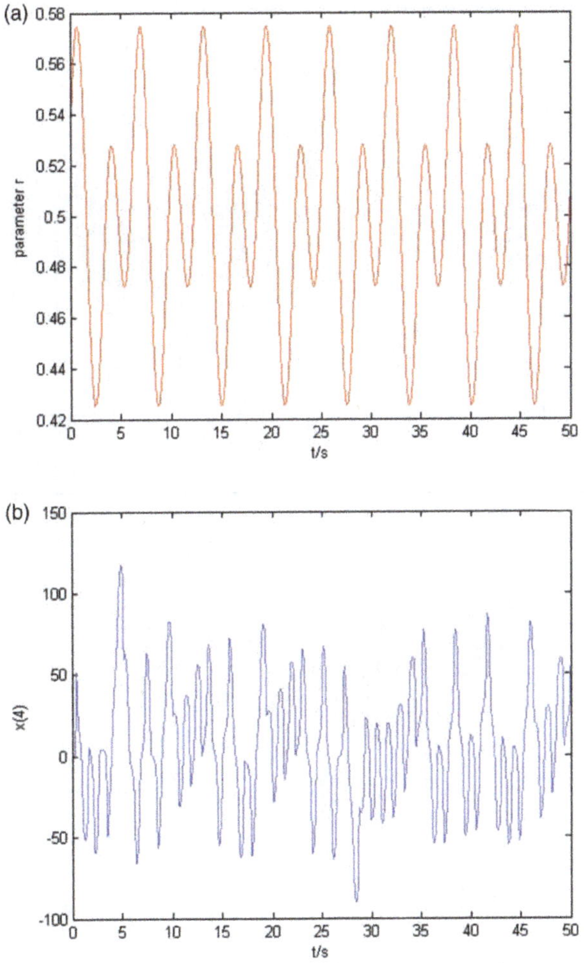

Figure 3. Original signal (a) and encrypted signal (b).

the parameters of the sender not only frustrates the synchronization but also affects the attractor obtained in the return map. Following Perez and Cerdeira, two modified return maps are defined by

$$A_n = \frac{(X_n + Y_m)}{2}, \quad B_n = X_n - Y_m \quad \text{and}$$
$$C_n = \frac{(X_{n+1} + Y_m)}{2}, \quad D_n = X_n - Y_{m+1},$$

where X_n, Y_m denote the nth (local) maximum and mth (local) minimum of the transmitted signal, respectively.

In this paper, we choose the following parameters as the standard parameters: $a = 35, b = 3, c = 12, d = 7, r = 0.5$, and the system is hyperchaotic. First, a comparison of the return maps with a small error in the standard parameters is explored. Figure 6 shows the return maps A_n vs. B_n and $-C_n$ vs. $-D_n$ of the Chen hyperchaotic system with the standard parameters. From Figure 6(a), choose parameter $r = 0.5$ and $r = 0.5001$, and we can see that the return maps are scattered and diffused. It is sensitive to the parameter r, so the transmitted signals are difficult to be identified exactly. Then based on the proposed modulation rules in

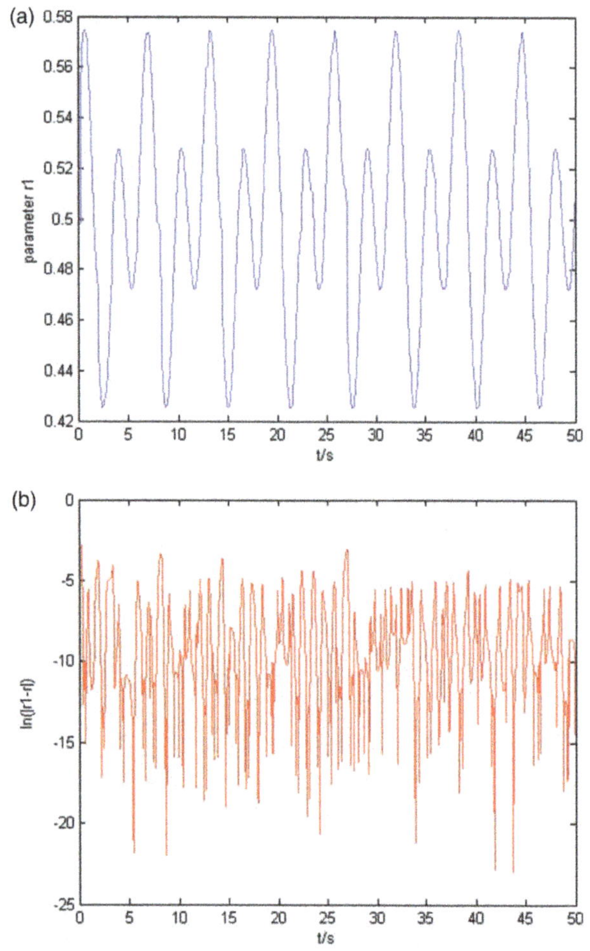

Figure 4. Recovered signal (a) and signal errors of $\ln|r(t) - \hat{r}(t)|$ (b).

Section 3.1, the return maps of the transmitted signal are shown in Figure 6(b), and it is difficult to distinguish the changes in the attractors because the two return maps are also scattered.

In many cases, only amplitude return maps are not enough to detect message signals. They also need some return maps which can reveal frequency information. Suppose that the state $X_4(t)$ is the transmitted condition. From $X_4(t)$ we can construct different kinds of return maps. Let $t_n^{\max}$ be the moment when $X_4(t)$ get its nth local maximum $X_{\max}(n)$, and $t_n^{\min}$ be the moment when $X_4(t)$ gets its nth local minimum $X_{\min}(n)$. Assume that Y_n is the value of $X_4(t)$ at that minimum moment, let $T_{\max}(n) = t_n^{\max} - t_{n-1}^{\max}$ and $T_{\min}(n) = t_n^{\min} - t_{n-1}^{\min}$ be two time intervals, then we define the following return maps (Yang, Yang, & Yang, 1998):

$$r_{\max}^A : X_{\max}(n) \mapsto X_{\max}(n+1),$$

$$r_{\min}^A : Y_{\min}(n) \mapsto Y_{\min}(n+1),$$

$$r_{\max}^T : T_{\max}(n) \mapsto X_{\max}(n),$$

$$r_{\min}^T : T_{\min}(n) \mapsto Y_{\min}(n).$$

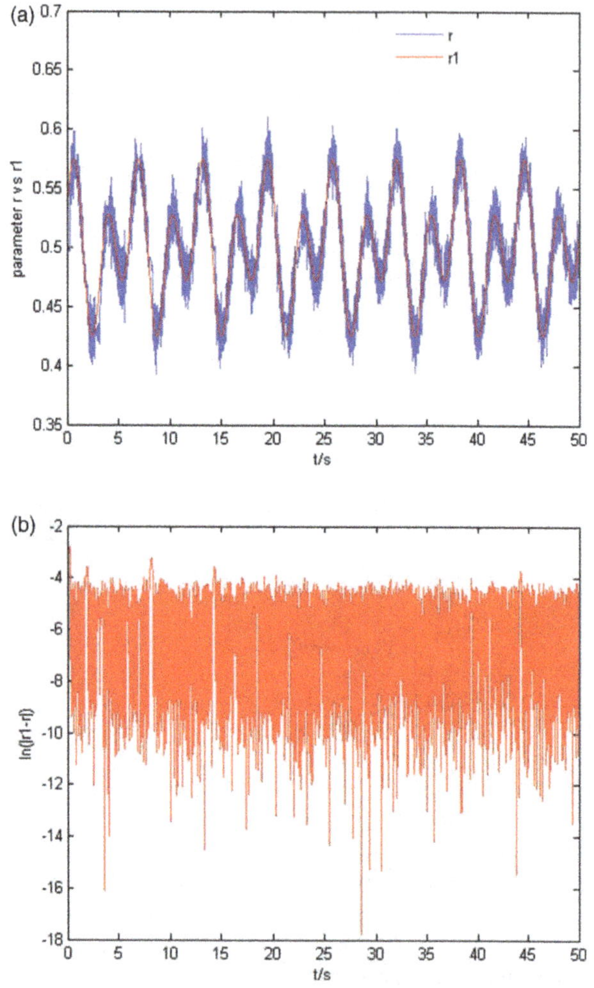

Figure 5. Recovered signal (a) and signal errors of $\ln|r(t) - \hat{r}(t)|$ (b).

Table 1. The errors of estimated parameters $\hat{r}$ with different SNR.

Case	SNR (db)	sum $\lvert r(t) - \hat{r}(t)\rvert/n$	$\lvert r(t) - \hat{r}(t)\rvert$
1	103.7	0.0017	1.1138e–09∼0.0560
2	89.88	0.0018	5.0526e–09∼0.0560
3	83.98	0.0018	7.6003e–09∼0.0560
4	70.05	0.0017	1.3541e–08∼0.0561
5	65.44	0.0020	1.4789e–08∼0.0560
6	44.57	0.0032	1.5475e–09∼0.0598
7	38.80	0.0048	1.3459e–07∼0.0596
8	32.91	0.0085	5.8395e–07∼0.0603

In Figure 7(a), the return maps with standard parameters and modulated parameters are shown. We were not able to find some shape deformations because they are scattered and the two return maps are mixed or overlapped.

Compared with others existing works, such as the works of Li, Chen, and Alvarez (2006), the return maps of the Lorenz system is given in Figure 8; note that there are three segments in the return map, and each segment is further

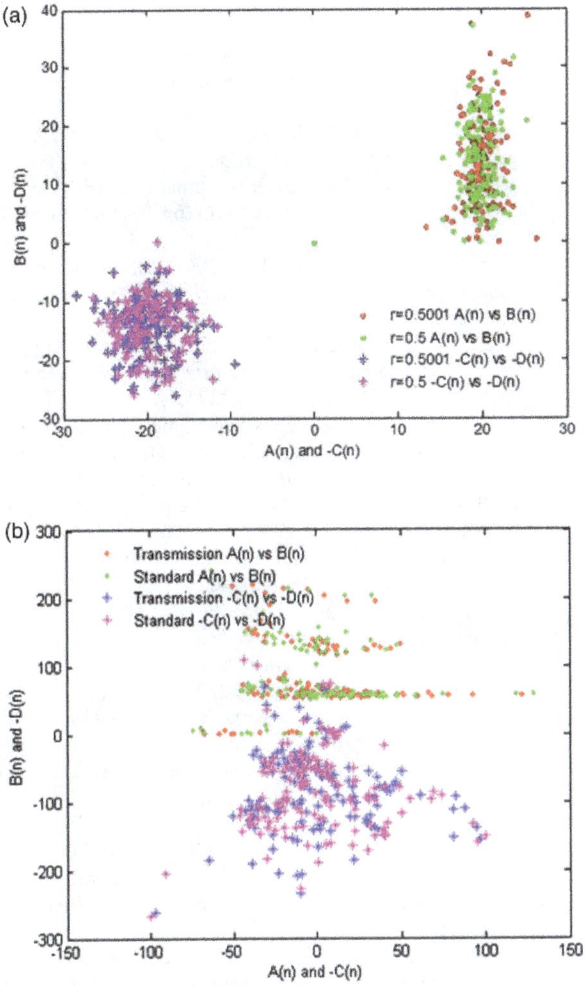

Figure 6. The return maps of the Chen hyperchaotic system with the parameter $r = 0.5$ and $r = 0.5001$ (a). The return maps ($A(n)$ and $-C(n)$ vs. $B(n)$ and $-D(n)$) of the transmitted signals with modulated parameters and standard parameters (b).

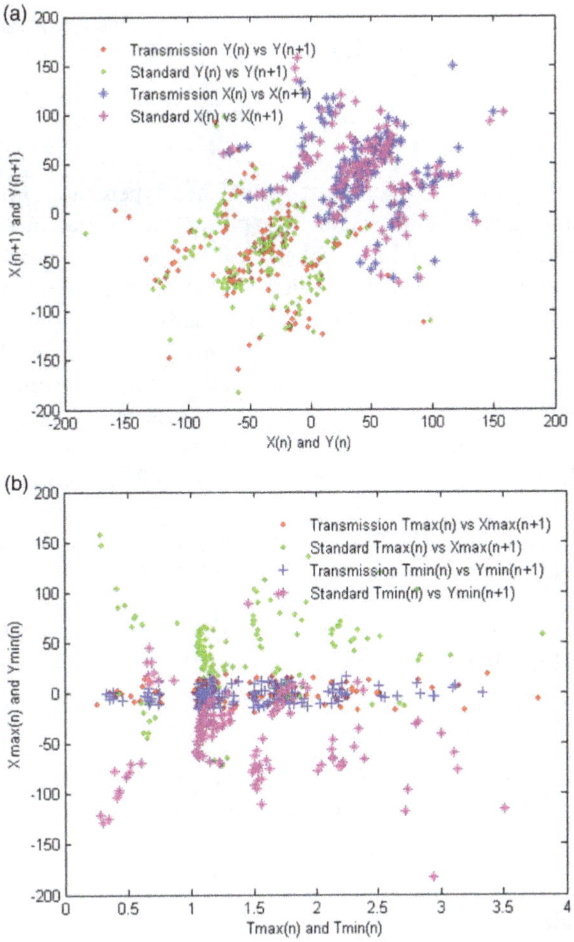

Figure 7. The return maps ($X(n)$ and $Y(n)$ vs. $X(n+1)$ and $Y(n+1)$) of the Chen hyperchaotic system with standard parameters and modulated parameters (a). The return maps ($T_{\max}(n)$ and $T_{\min}(n)$ vs. $X_{\max}(n)$ and $Y_{\min}(n)$) of $r^T_{\max}$ and $r^T_{\min}$ (b).

split into 10 strips. It is obvious that the split of the map is caused by the switching of the value of parameter b between $b0$ and $b1$, where $b0 = 3.1$ and $b1 = 4.0$. From Figure 6(a), however, it can be seen that the return maps of the Chen hyperchaotic system do not have a clear strip and it is scattered. In addition, it is much sensitive to the parameter r, so our scheme certainly can resist the return map attacks and the degree of security is high enough.

On the other hand, the changes of parameters not only change the sizes of the attractors but also their natural frequencies (Yang, 1995). From Figure 7(b), which shows $r^T_{\max}$ and $r^T_{\min}$, we obviously cannot find some vertical shifts between $r^T_{\max}$ and $r^T_{\min}$, which denote the changes in natural frequency. All the maps are blurred and diffused with each other; thus, distinguishing them is not easy. Therefore, by using the parameter modulation, this method can resist the well-known return map attack. This secure communication schemes whose return maps are complicated

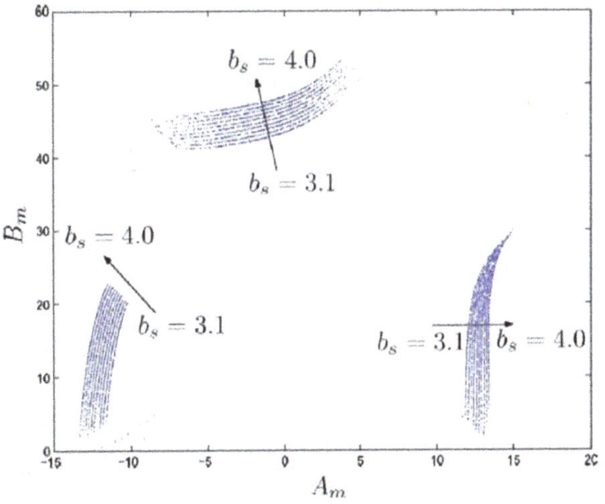

Figure 8. The return maps of the Lorenz system with different parameters.

and the changes of return maps are irregular, we know that the degree of security is high enough.

4. Conclusions

In this paper, the synchronization of Chen hyperchaotic systems with unknown parameter is presented. The parameter of transmitter can be identified exactly while the synchronization is completed. Theoretical analysis and numerical simulations are shown to verify the results. A new application with parameter modulation in the continuous signals transmission is given. We know that the continuous signals are recovered well from the simulations using the method of parameter modulation. The proposed scheme achieved robustness to noise to some extent, and different kinds of return maps show the higher degree of security. In the future, we will do more analysis and design a chaos-based system to secure communication.

Acknowledgements

The authors would like to thank the Editor and the referees for their valuable comments and suggestions that helped to improve the quality of this paper. Research is supported by the National Natural Science Foundation of China (Grant No: 61074012) and the Natural Science Foundation of Fujian Province, China (Grant No: 2011J01025).

References

Arman, K. B., Kia, F., Naser, P., & Henry, L. (2009). A chaotic secure communication scheme using fractional chaotic systems based on an extended fractional Kalman filter. *Communications in Nonlinear Science and Numerical Simulation, 14*, 863–879.

Cai, N., Jing, Y., & Zhang, S. (2010). Modified projective synchronization of chaotic systems with disturbances via active sliding mode control. *Communications in Nonlinear Science and Numerical Simulation, 15*, 1613–1620.

Chen, A. M., Lu, J. A., Lü, J. H., & Yu, S. M. (2006). Generating hyper chaotic Lü attractor via state feedback control. *Physica A: Statistical Mechanics and its Applications, 364*, 103–110.

Chen, C. J. (2012). Robust synchronization of uncertain unified chaotic systems subject to noise and its application to secure communication. *Applied Mathematics and Computation, 219*, 2698–2712.

Chen, S. H., & Lü, J. H. (2002). Synchronization of an uncertain unified chaotic system via adaptive control. *Chaos, Solitons & Fractals, 14*, 643–647.

Cuomo, K. M., Oppenheim, A. V., & Strogatz, S. H. (1993). Synchronization of Lorenz-based chaotic circuits with applications to communications. *IEEE Transactions on Circuits and Systems II, 40*, 626–633.

Dedieu, H., Kennedy, M. P., & Hasler, M. (1993). Chaos shift keying: Modulation and demodulation of a chaotic carrier using self-synchronizing Chua's circuits. *IEEE Transactions on Circuits and Systems II, 40*, 634–642.

Khalil, H. K. (1992). *Nonlinear systems*. New York: Macmillan Publishing Company.

Kocarev, L., & Parlitz, U. (1995). General approach for chaotic synchronization with applications to communication. *Physical Review Letters, 74*, 5028–5031.

Li, S. J., Alvarez, G., & Chen, G. R. (2005). Breaking a chaos-based secure communication scheme designed by an improved modulation method. *Chaos, Solitons & Fractals, 25*, 109–120.

Li, S. J., Alvarez, G., Li, Z., & Halang, W. A. (2007). *Analog chaos-based secure communications and cryptanalysis: A brief survey*. Proceedings of 3rd international IEEE scientific conference on physics and control (pp. 92–98), Potsdam, Germany.

Li, S. J., Chen, G. R., & Alvarez, G. (2006). Return-map cryptanalysis revisited. *International Journal of Bifurcation and Chaos, 16*, 1557–1568.

Li, Y. X., Tang, W. K. S., & Chen, G. R. (2005). Generating hyperchaos via state feedback control. *International Journal of Bifurcation and Chaos, 15*, 3367–3375.

Li, Z. B., & Zhao, X. S. (2011). The parametric synchronization scheme of chaotic system. *Communications in Nonlinear Science and Numerical Simulation, 16*, 2936–2944.

Lorenz, E. (1963). Deterministic nonperiodic flow. *Journal of the Atmospheric Sciences, 20*, 130–141.

Park, J. H. (2005). Adapive synchronization of hyperchaotic Chen system with uncertain parameters. *Chaos, Solitons & Fractals, 26*(3), 959–964.

Pecora, L., & Carroll, T. (1990). Synchronization in chaotic systems. *Physical Review Letters, 64*, 821–824.

Perez, G., & Cerdeira, H. A. (1995). Extracting message masked by chaos. *Physical Review Letters, 74*, 1970–1973.

Song, Y. X., & Cao, Z. W. (2004). *Secure communication via multiparameter modulation using Chua systems*. 5th Asian control conference (vol. 2, pp. 1107–1110), Melbourne, Victoria, Australia.

Tang, Y., Gao, H. J., Zou, W., & Kurths, J. (2013). Distributed synchronization in networks of agent systems with nonlinearities and random switchings. *IEEE Transactions on Cybernetics, 43*, 358–370.

Tang, Y., Wang, Z. D., & Fang, J. A. (2011). Controller design for synchronization of an array of delayed neural networks using a controllable probabilistic PSO. *Information Sciences, 181*, 4715–4732.

Tang, Y., Wang, Z. D., Gao, H. J., Stephen, S., & Kurths, J. (2012). A constrained evolutionary computation method for detecting controlling regions of cortical networks. *IEEE/ACM Transactions on Computational Biology and Bioinformatics, 9*, 1569–1581.

Tao, C. H., Xiong, H. X., & Hu, F. (2006). Two novel synchronization criterions for a unified chaos system. *Chaos, Solitons & Fractals, 27*, 115–120.

Ueta, T., &. Chen, G. R. (2000). Bifurcation analysis of Chen's equation. *International Journal of Bifurcation and Chaos, 10*, 1917–1931.

Wang, H., Han, Z. Z., Xie, Q. Y., & Zhang, W. (2009). Sliding mode control for chaotic systems based on LMI. *Communications in Nonlinear Science and Numerical Simulation, 14*, 1410–1417.

Wu, X., Hu, H., & Zhang, B. (2004). Analyzing and improving a chaotic encryption method. *Chaos, Solitons & Fractals, 22*, 367–373.

Xiang, T., Cheng, K. W., & Liao, X. (2008). An improved chaotic cryptosystem with external key. *Communications in Nonlinear Science and Numerical Simulation, 13*, 1879–1887.

Xiao, M., & Cao, J. (2009). Synchronization of a chaotic electronic system with cubic term via adaptive feedback control. *Communications in Nonlinear Science and Numerical Simulation, 14*, 3379–3388.

Yang, T. (1995). Recovery of digital signals from chaotic switching. *International Journal of Circuit Theory and Applications, 23*, 611–615.

Yang, T. (2004). A survey of chaotic secure communication systems. *International Journal of Computational Cognition, 2*, 81–130.

Yang, T., & Chua, L. O. (1996). Secure communication via chaotic parameter modulation. *IEEE Transactions on Circuits and Systems I, 43*, 817–819.

Yang, T., Yang, L. B., & Yang, C. M. (1998). Cryptanalyzing chaotic secure communication using return maps. *Physics Letters A, 245*, 495–510.

Yu, W. W., Cao, J. D., Wong, K. W., & Lü, J. H. (2007). New communication schemes based on adaptive synchronization. *Chaos, 17*, 033114–033127.

10

A novel approach in the finite-time controller design

T. Binazadeh and M.H. Shafiei*

Department of Electrical and Electronic Engineering, Shiraz University of Technology, Modares Blvd., Shiraz, Iran

This paper deals with the finite-time stabilization of a class of nonlinear systems. Based on the backstepping technique, a new recursive procedure is proposed which entwines the choice of the Lyapunov function with the design of the feedback control laws. The main efficiency of the proposed technique is due to adding a dummy state variable to the state vector. The dynamic equation of this state variable has a special structure which makes the design procedure of the finite-time controller more feasible. The designed controller guarantees the stabilization of the closed-loop system in a finite time. Computer simulations reveal the efficiency of the proposed technique and also verify the theoretical results.

Keywords: backstepping technique; finite-time stabilization; strict-feedback form; Lyapunov function

1. Introduction

The classical stability concepts, such as Lyapunov stability, asymptotical stability and bounded input–bounded output (BIBO) stability, study the stability of systems in an infinite time-interval. The concept of finite-time stabilization naturally arises from finite-time optimal control problems (Bhat & Bernstein, 2000) and deals with the dynamical systems whose operation time is limited to a fixed finite time-interval. From practical considerations, for such systems which should operate only over a finite time-interval, finite-time stability is the only meaningful description of stability. Additionally, when the classical concepts of stability require that system states be bounded, the bound values are not prescribed while the finite-time stability requires prescribed bounds on system states (Dorato, 2006). It also should be noted that the term finite-time stability has been used with different meanings in the literature. In this paper, this definition is used to describe the dynamical systems whose states approach to zero in a finite time (Hong, Wang, & Xi, 2005).

Recently, finite-time control of nonlinear systems has received increasing attentions (Amato, Ariola, & Dorato, 2001; Guo & Vincent, 2010; Honga, Xub, & Huangb, 2002; Zhu, Shen, & Li, 2009). Finite-time stabilization of higher-order systems (Hong, 2002), lower-triangular systems (Hong, Wang, & Cheng, 2006; Huang, Lin, & Yang, 2005; Pongvuthithum, 2009; Zhanga, Fengb, & Sunb, 2012), switched systems (Orlov, 2005), non-autonomous systems (Moulay & Perruquetti, 2008), time-delay system (Moulay, Dambrine, Yeganefar, & Perruquetti, 2008) and finite-time stabilization using output feedback (Hong, Huang, & Xu, 2001), dynamic gain (Praly & Jiang, 2004), backstepping (Reichhartinger & Horn, 2011) and control vector Lyapunov function (Nersesov, Haddad, & Hui, 2008) have been developed in the literature. Also, finite-time control via the terminal sliding mode (Chen, Wu, & Cui, 2013; Chuan-Kai, 2006; Feng, Yu, & Man, 2002; Lu, Chiu, & Chen, 2010) and fast terminal sliding mode (Hao, Lihua, & Zhong, 2013) has been studied, extensively. In all of these methods, in addition to structural limitations in each approach, the design procedure of the finite-time controller is almost complicated.

This paper presents a simple design method for the finite-time stabilization of a class of nonlinear systems. In the proposed method, first a dummy state variable is augmented to the state vector. The dynamic equation of this state variable has a special structure which makes the design procedure of the finite-time control law more feasible. Then, based on the backstepping technique, a recursive procedure that entwines the choice of a Lyapunov function with the design of the feedback control law is proposed. In this procedure, the controller design for the whole system breaks into a sequence of design problems for some lower order systems. In order to show the great positive effects of adding the dummy state variable, a design example is considered. The finite-time stabilizing controller is designed with and without adding the dummy state variable to the equations of the design example. Finally, simulation results of the closed-loop system verify the theoretical result and also reveal the great improvements, due to adding the dummy

*Corresponding author. Email: shafiei@sutech.ac.ir

state variable, on the stability of the closed-loop system and also the transient responses of the state variables and the control input.

The remainder of the paper is arranged as follows. First, the preliminaries about the finite-time stability are given in Section 2. In Section 3, the design procedure of the finite-time stabilizing control law is explained in detail. Next, in Section 4, the proposed approach is applied to a design example. Finally, conclusions are made in Section 5.

2. Preliminaries

Consider the following nonlinear system:

$$\dot{x} = f(x), \quad x(t_0) = x_0, \tag{1}$$

where $x \in U \subset R^n (0 \in U)$ is the state vector, $f : U \to R^n$ is a continuous vector function and $f(0) = 0$.

DEFINITION 1 *The equilibrium point ($x = 0$) of system (1) is finite-time stable, if it is the Lyapunov stable and finite-time convergent. In other words, for every initial condition $x(0) \in U/\{0\}$, there is a settling time $T > 0$ such that $\lim_{t\to T} x(t) = 0$ and $x(t) = 0$ for all $t \geq T$ (Hong et al., 2005).*

LEMMA 1 *Suppose that there exists a continuously differentiable function $V(x) : U \to R$, and real numbers $c > 0$ and $0 < \alpha < 1$, such that,*

$$V(0) = 0,$$
$$V(x) \text{ is positive definite on } U,$$
$$\dot{V}(x) \leq -cV^{\alpha}(x), \quad \forall x \in U. \tag{2}$$

Then the settling time T (refer to Definition 1) exists and satisfies,

$$T(x_0) \leq \frac{V(x_0)^{1-\alpha}}{c(1-\alpha)}, \tag{3}$$

where $V(x_0)$ is the initial value of $V(x)$.

Proof See (Bhat & Bernstein, 2000) ■

LEMMA 2 *For real numbers $l_i, i = 1, 2, \ldots, n$ and every $\alpha \in (0, 1)$, the following inequality holds:*

$$(|l_1| + \cdots + |l_n|)^{\alpha} \leq |l_1|^{\alpha} + \cdots + |l_n|^{\alpha}. \tag{4}$$

Proof See (Yu, Yu, Shirinzadeh, & Man, 2005) ■

3. Design of the finite-time controller

Consider the following nonlinear system which is given in the strict-feedback form:

$$\begin{aligned}
\dot{x}_1 &= f_1(x_1) + g_1(x_1)x_2,\\
&\vdots\\
\dot{x}_{n-1} &= f_{n-1}(x_1, x_2, \ldots, x_{n-1}) + g_{n-1}(x_1, x_2, \ldots, x_{n-1})x_n,\\
\dot{x}_n &= f_n(x_1, x_2, \ldots, x_n) + g_n(x_1, x_2, \ldots, x_n)u,
\end{aligned} \tag{5}$$

where $x_i \in R (i = 1, \ldots, n), u \in R$ and f_1 to f_n are continuous functions which vanish at origin, and also over the domain of interest, $g_i \neq 0$ for $i = 1, 2, \ldots, n$.

Moreover, consider a dummy state variable with the following dynamical equation:

$$\dot{x}_0 = -x_0^{\beta} + g_0(x_0)x_1, \tag{6}$$

where $g_0 \neq 0$ and $\beta = (2q - p)/p$. (where p and q are positive odd integers and $q < p < 2q$).

The approach of considering a new state variable x_0 increases the dimension of state vector; however, the structure of Equation (6) is such that x_0 is finite-time stable (in the absence of x_1) and adding this equation to state-space equations (5) makes the design procedure of the finite-time controller more feasible. Therefore, the augmented state-space equations are as follows:

$$\begin{aligned}
\dot{x}_0 &= -x_0^{\beta} + g_0(x_0)x_1,\\
\dot{x}_1 &= f_1(x_1) + g_1(x_1)x_2,\\
&\vdots\\
\dot{x}_{n-1} &= f_{n-1}(x_1, x_2, \ldots, x_{n-1}) + g_{n-1}(x_1, x_2, \ldots, x_{n-1})x_n,\\
\dot{x}_n &= f_n(x_1, x_2, \ldots, x_n) + g_n(x_1, x_2, \ldots, x_n)u.
\end{aligned} \tag{7}$$

The goal is to design a finite-time stabilizing control law u for system (7). In order to show the design procedure, let us start with the following special case of Equations (7) and then gradually complete it:

$$\dot{x}_0 = -x_0^{\beta} + g_0(x_0)x_1, \tag{8a}$$
$$\dot{x}_1 = u. \tag{8b}$$

Equation (8a) is finite-time stable for $x_1 = \varphi_1(x_0) = 0$. To show this point, consider $V_0(x_0) = 0.5x_0^2$ as a Lyapunov function candidate for equation $\dot{x}_0 = -x_0^{\beta}$. Then,

$$\begin{aligned}
\dot{V}_0 &= \frac{\partial V_0}{\partial x_0}\dot{x}_0\\
&= -x_0^{\beta+1} = -2^{(\beta+1)/2}(\frac{1}{2}x_0^2)^{(\beta+1)/2}\\
&= -2^{(\beta+1)/2}V_0^{(\beta+1)/2} = -cV_0^{\alpha},
\end{aligned} \tag{9}$$

where $\alpha = (\beta + 1)/2 = q/p \in (0, 1)$ and $c = 2^{(\beta+1)/2} = 2^{\alpha} > 0$. Therefore, according to Lemma 1, Equation (8a) is finite-time stable with $x_1 = 0$. Now, the finite-time stabilizing control law u will be designed for system (8). For this purpose, consider the following Lyapunov function:

$$V_1(x_0, x_1) = V_0(x_0) + |x_1|. \tag{10}$$

Therefore,

$$\begin{aligned} \dot{V}_1 &= \frac{\partial V_0}{\partial x_0}(-x_0^\beta + g_0 x_1) + \operatorname{sgn}(x_1)u \\ &= -cV_0^\alpha + \frac{\partial V_0}{\partial x_0} g_0 x_1 + \operatorname{sgn}(x_1)u \\ &= -cV_0^\alpha + g_0 x_0 x_1 + \operatorname{sgn}(x_1)u. \end{aligned} \tag{11}$$

Now, choosing,

$$u = -g_0|x_1|x_0 - cx_1^\alpha \tag{12}$$

and considering that $\operatorname{sgn}(x_1)x_1^\alpha = |x_1|^\alpha$ (because $\alpha = q/p$ and q and p are odd integers), thus,

$$\dot{V}_1 = -cV_0^\alpha - c|x_1|^\alpha. \tag{13}$$

According to Lemma 2, $V_1^\alpha = (V_0 + |x_1|)^\alpha \le V_0^\alpha + |x_1|^\alpha$, thus,

$$\dot{V}_1 = -cV_0^\alpha - c|x_1|^\alpha \le -cV_1^\alpha. \tag{14}$$

Therefore, according to Lemma 1, the closed-loop system (8) is finite-time stable with the proposed control law (12). Now, if the following more general structure is considered:

$$\begin{aligned} \dot{x}_0 &= -x_0^\beta + g_0(x_0)x_1 \\ \dot{x}_1 &= f_1(x_1) + g_1(x_1)u \end{aligned} \tag{15}$$

Then the input transformation $u = (w_1 - f_1)/g_1$ will reduce Equation (15) to the following equations:

$$\begin{aligned} \dot{x}_0 &= -x_0^\beta + g_0(x_0)x_1, \\ \dot{x}_1 &= w_1, \end{aligned} \tag{16}$$

where according to above discussion, Equations (16) can be stabilized in a finite time by $w_1 = -g_0|x_1|x_0 - cx_1^\alpha$. Therefore, the finite-time stabilizing controller u for system (15) is as follows:

$$\begin{aligned} u &= \varphi_2(x_0, x_1) \\ &= \frac{1}{g_1}(\underbrace{-g_0|x_1|x_0 - cx_1^\alpha}_{w_1} - f_1). \end{aligned} \tag{17}$$

The above discussions are summarized in the following Lemma.

LEMMA 3 *Considering system (15), the state feedback control law (17) stabilizes the origin of Equation (15) in a finite time and the corresponding Lyapunov function is* $V_1(x_0, x_1) = 0.5x_0^2 + |x_1|$.

Now, consider the following third-order system:

$$\begin{aligned} \dot{x}_0 &= -x_0^\beta + g_0(x_0)x_1, \\ \dot{x}_1 &= f_1(x_1) + g_1(x_1)x_2, \\ \dot{x}_2 &= f_2(x_1, x_2) + g_2(x_1, x_2)u. \end{aligned} \tag{18}$$

After one step of backstepping, the first two equations in Equation (18), with x_2 as the control input, can be globally stabilized in a finite time by $x_2 = \varphi_2(x_0, x_1)$ (where $\varphi_2(x_0, x_1)$ is given in Equation (17)) and $V_1(x_0, x_1) = 0.5x_0^2 + |x_1|$ is the corresponding Lyapunov function. To backstep, apply the following change of variables $z_2 = x_2 - \varphi_2$. Thus, state-space equations (18) are transformed to the following equations:

$$\begin{aligned} \dot{x}_0 &= -x_0^\beta + g_0 x_1, \\ \dot{x}_1 &= f_1 + g_1\varphi_2 + g_1 z_2, \\ \dot{z}_2 &= f_2 + g_2 u \\ &\quad \underbrace{- \frac{\partial \varphi_2}{\partial x_0}(-x_0^\beta + g_0 x_1) - \frac{\partial \varphi_2}{\partial x_1}(f_1 + g_1\varphi_2 + g_1 z_2)}_{-\dot{\varphi}_2}. \end{aligned} \tag{19}$$

By the input transformation $u = (w_2 - f_2 + \dot{\varphi}_2)/g_2$, one has

$$\begin{aligned} \dot{x}_0 &= -x_0^\beta + g_0 x_1, \\ \dot{x}_1 &= f_1 + g_1\varphi_2 + g_1 z_2, \\ \dot{z}_2 &= w_2. \end{aligned} \tag{20}$$

Consider the following Lyapunov function for system (20),

$$V_2(x_0, x_1, z_2) = V_1(x_0, x_1) + |z_2|. \tag{21}$$

Thus, $\dot{V}_2$ can be easily calculated as follows:

$$\begin{aligned} \dot{V}_2 = {} & \frac{\partial V_1}{\partial x_0}(-x_0^\beta + g_0 x_1) + \frac{\partial V_1}{\partial x_1}(f_1 + g_1\varphi_2) \\ & + \frac{\partial V_1}{\partial x_1} g_1 z_2 + \operatorname{sgn}(z_2)w_2. \end{aligned} \tag{22}$$

Since φ_2 is previously designed such that $((\partial V_1/\partial x_0)(-x_0^\beta + g_0 x_1) + (\partial V_1/\partial x_1)(f_1 + g_1\varphi_2) \le -cV_1^\alpha)$, thus

$$\dot{V}_2 \le -cV_1^\alpha + \frac{\partial V_1}{\partial x_1} g_1 z_2 + \operatorname{sgn}(z_2)w_2, \tag{23}$$

Choosing,

$$\begin{aligned} w_2 &= -|z_2|\frac{\partial V_1}{\partial x_1} g_1 - cz_2^\alpha \\ &= -|z_2|\operatorname{sgn}(x_1)g_1 - cz_2^\alpha, \end{aligned} \tag{24}$$

results in,

$$\dot{V}_2 \le -cV_1^\alpha - c|z_2|^\alpha. \tag{25}$$

Using Lemma 2 and the inequality (25) (similar to the case of V_1), it can be obtained that $\dot{V}_2 \leq -cV_2^{\alpha}$. Thus, the following control law guarantees finite-time stabilizing of system (18):

$$u = \varphi_3(x_0, x_1, x_2) = \frac{1}{g_2}(\underbrace{-|z_2|\text{sgn}(x_1)g_1 - cz_2^{\alpha} + \dot{\varphi}_2 - f_2}_{w_2}), \quad (26)$$

with the following Lyapunov function:

$$V_2(x_0, x_1, x_2) = \frac{1}{2}x_0^2 + |x_1| + |x_2 - \varphi_2|. \quad (27)$$

Now, consider the following forth-order system:

$$\begin{aligned} \dot{x}_0 &= -x_0^{\beta} + g_0(x_0)x_1, \\ \dot{x}_1 &= f_1(x_1) + g_1(x_1)x_2, \\ \dot{x}_2 &= f_2(x_1, x_2) + g_2(x_1, x_2)x_3, \\ \dot{x}_3 &= f_3(x_1, x_2, x_3) + g_3(x_1, x_2, x_3)u. \end{aligned} \quad (28)$$

After two step of backstepping, the first three equations in Equation (28) can be globally stabilized in a finite time by $x_3 = \varphi_3(x_0, x_1, x_2)$ (where $\varphi_3(x_0, x_1, x_2)$ is given in Equation (26)) and $V_2(x_0, x_1, x_2) = \frac{1}{2}x_0^2 + |x_1| + |x_2 - \varphi_2|$ is the corresponding Lyapunov function. To backstep, apply the change of variables $z_3 = x_3 - \varphi_3$. Similar to the previous case, the process can be repeated to obtain the finite-time stabilizing control law for the system (28).

Ultimately, the backstepping method may be applied in a systematic way to design the finite-time stabilizing control law, $u = \varphi_n(x_0, x_1, x_2, \ldots, x_n)$ for state-space equations (7). If the proposed process be repeated n times, then the corresponding Lyapunov function $V_n(x_0, x_1, x_2, \ldots, x_n)$ for system (7) will be obtained as

$$V_n(x_0, x_1, x_2, \ldots, x_n) = \frac{1}{2}x_0^2 + |x_1 - \varphi_1| + |x_2 - \varphi_2| + \cdots |x_n - \varphi_n|,$$

where φ_1 is zero function and the scalar functions φ_is $(i = 2, 3, \ldots, n)$ should be computed in a backstepping procedure (similar to the proposed procedure in the design of φ_2 and φ_3). Finally, the finite-time stabilizing control law $u = \varphi_n(x_0, x_1, x_2, \ldots, x_n)$ will be achieved.

4. Design example

In this section, a design example is considered to show the efficiency of the proposed method in the design of a finite-time stabilizing controller and also to show the positive effects of adding a dummy state variable to the system equations. For this purpose, the design of a finite-time stabilizing control law is done by the proposed method once without adding a dummy state variable and another time by adding this state variable.

Consider the following state-space equations:

$$\begin{aligned} \dot{x}_1 &= x_1^2 + x_2, \\ \dot{x}_2 &= -x_1 + u. \end{aligned} \quad (29)$$

4.1. *First approach (without adding the dummy state variable x_0)*

Consider the first state equation $\dot{x}_1 = x_1^2 + x_2$. Choosing $x_2 = \varphi(x_1) = -x_1^2 - x_1^{\beta}$, leads to the finite-time stabilization of this equation with the Lyapunov function $V_1(x_1) = 0.5x_1^2$, where $\dot{V}_1(x_1) = -x_1^{\beta+1} = -cV_1^{\alpha}$. Set $z = x_2 - \varphi$, then,

$$\begin{aligned} \dot{x}_1 &= x_1^2 + \varphi + z, \\ \dot{z} &= -x_1 + u - \dot{\varphi}. \end{aligned} \quad (30)$$

By input transformation $u = w + \dot{\varphi} + x_1$, one has

$$\begin{aligned} \dot{x}_1 &= x_1^2 + \varphi + z, \\ \dot{z} &= w. \end{aligned} \quad (31)$$

Choosing $V_2(x_1, z) = V_1(x_1) + |z|$ as the Lyapunov function for Equation (31), then,

$$\dot{V}_2 = \frac{\partial V_1}{\partial x_1}(x_1^2 + \varphi) + \frac{\partial V_1}{\partial x_1}z + \text{sgn}(z)w. \quad (32)$$

Putting

$$\begin{aligned} w &= -\frac{\partial V_1}{\partial x_1}|z| - cz^{\alpha} \\ &= -x_1|z| - cz^{\alpha}, \end{aligned} \quad (33)$$

then $\dot{V}_2 \leq -cV_2^{\alpha}$. Therefore, the following control law guarantees the finite-time stabilization of system (29):

$$\begin{aligned} u &= w + \dot{\varphi} + x_1 \\ &= x_1 - x_1|z| - cz^{\alpha} + \frac{\partial \varphi}{\partial x_1}(x_1^2 + x_2), \end{aligned} \quad (34)$$

where $\varphi = -x_1^2 - x_1^{\beta}$ and $z = x_2 - \varphi = x_2 + x_1^2 + x_1^{\beta}$.

4.2. *Second approach (adding the dummy state variable x_0)*

After adding a dummy state variable with state equation (6) to the equations (29), one has

$$\begin{aligned} \dot{x}_0 &= -x_0^{\beta} + x_1, \\ \dot{x}_1 &= x_1^2 + x_2, \\ \dot{x}_2 &= -x_1 + u, \end{aligned} \quad (35)$$

which has the same structure as state-space equations (18). By substituting

$$g_0 = 1, \quad f_1 = x_1^2, \quad g_1 = 1$$
$$f_2 = -x_1, \quad g_2 = 1. \tag{36}$$

And according to the proposed procedure in the previous section and considering Equations (26) and (36), the finite-time controller is as follows:

$$u = (-|z_2|\text{sgn}(x_1) - cz_2^\alpha + \frac{\partial \varphi_2}{\partial x_0}(-x_0^\beta + x_1) + \frac{\partial \varphi_2}{\partial x_1}(x_1^2 + x_2) + x_1), \tag{37}$$

where $\varphi_2 = -|x_1|x_0 - cx_1^\alpha - x_1^2$ (refer to Equation (17)) and $z_2 = x_2 - \varphi_2$.

4.3. *Computer simulations*

The time responses of the state variables related to designed control laws (34) and (37) (in the first and second approaches) are shown in Figures 1 and 2. As it can be seen, adding the dummy state variable has great improvements on the transient response and settling time of the closed-loop system. The time responses of control signals (34) and (37), which are related to the first and second approaches, respectively, are shown in Figure 3. As it is seen in Figure 3, the time response of control law (34) is highly oscillating. In fact, computer simulations show that the proposed idea in adding the dummy state variable leads to a better transient response, less settling time and less control effort.

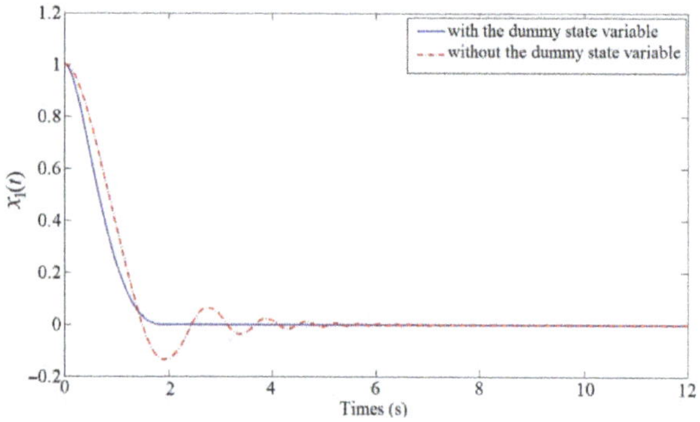

Figure 1. Time responses of $x_1(t)$ in the first and second approaches.

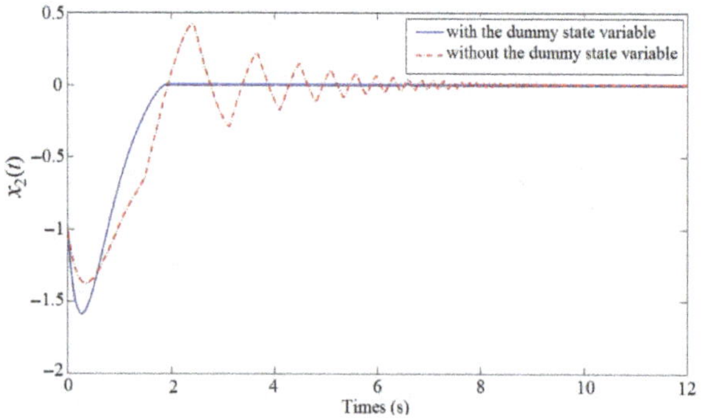

Figure 2. Time responses of $x_2(t)$ in the first and second approaches.

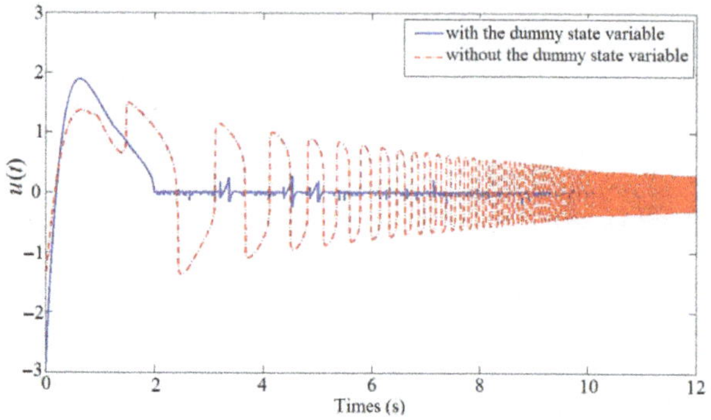

Figure 3. Time responses of $u(t)$ in the first and second approaches.

5. Conclusions

In this paper, a new procedure was presented to design a finite-time stabilizing controller. In the proposed method, a dummy state variable with a finite-time stable dynamic equation was added to the state vector. Then, based on backstepping idea, a recursive procedure that interlaced the choice of a Lyapunov function with the design of the finite-time control law was proposed. Finally, by means of computer simulations, the theoretical results were verified and the great improvements due to adding the dummy state variable on the stability of the closed-loop system and transient responses of state variables and the control input were shown.

References

Amato, F., Ariola, M., & Dorato, P., (2001). Finite-time control of linear systems subject to parametric uncertainties and disturbances. *Automatica, 37* (9), 1459–1463.

Bhat, S. P., & Bernstein, D. S. (2000). Finite-time stability of continuous autonomous systems. *SIAM Journal on Control and Optimization, 38* (3), 751–766.

Chen, M., Wu, Q. X., & Cui, R. X. (2013). Terminal sliding mode tracking control for a class of SISO uncertain nonlinear systems. *ISA Transactions, 52* (2), 198–206.

Chuan-Kai, L. (2006). Nonsingular terminal sliding mode control of robot manipulators using fuzzy wavelet networks. *IEEE Transactions on Fuzzy Systems, 14* (6), 849–859.

Dorato, P., (2006). *An overview of finite-time stability in current trends in nonlinear systems and control: In Honor of Petar Kokotovic and Turi Nicosia* (pp. 185–194). Boston: Birkhäuser.

Feng, Y., Yu, X., & Man, Z. (2002). Non-singular terminal sliding mode control of rigid manipulators. *Automatica, 38*(12), 2159–2167.

Guo, R. & Vincent, U.E., (2010). Finite time stabilization of chaotic systems via single input. *Physics Letters A, 375*, 119–124.

Hao, L., Lihua, D., & Zhong, S. (2013). Adaptive nonsingular fast terminal sliding mode control for electromechanical actuator. *International Journal of Systems Science, 44*(3), 401–415.

Hong, Y. (2002). Finite-time stabilization and stabilizability of a class of controllable systems. *System & Control Letters, 46*, 231–236.

Hong, Y., Huang, J., & Xu, Y. (2001). On an output feedback finite-time stabilization problem. *IEEE Transaction on Automatic Control, 46*, 305–309.

Hong, Y., Wang, J., & Cheng, D., (2006). Adaptive finite-time control of nonlinear systems with parametric uncertainty. *IEEE Transactions on Automatic Control, 51*(5), 858–862.

Hong, Y., Wang, J., & Xi, Z. (2005). Stabilization of uncertain chained form systems within finite settling time. *IEEE Transactions on Automatic Control, 50*(9), 1379–1384.

Honga, Y., Xub, Y., & Huangb, J. (2002). Finite-time control for robot manipulators. *Systems & Control Letters, 46*, 243–253.

Huang, X., Lin, W., & Yang, B., (2005). Global finite-time stabilization of a class of uncertain nonlinear systems. *Automatica, 41*, 881–888.

Lu, Y. S., Chiu, C. W., & Chen, J. S. (2010). Time-varying sliding-mode control for finite-time convergence. *Electrical Engineering, 92*, 257–268.

Moulay, E., Dambrine, M., Yeganefar, N., & Perruquetti, W. (2008). Finite time stability and stabilization of time-delay systems. *Systems & Control Letters, 57*(7), 561–566.

Moulay, E., & Perruquetti, W. (2008). Finite time stability conditions for non-autonomous continuous systems. *International Journal of Control, 81*(5), 797–803.

Nersesov, S. G., Haddad, W. M., & Hui, Q. (2008). Finite time stabilization of nonlinear dynamical systems via control vector Lyapunov functions. *Journal of the Franklin Institute, 345*(7), 819–837.

Orlov, Y. (2005). Finite time stability and robust control synthesis of uncertain switched systems. *SIAM Journal on Control and Optimization, 43*(4), 1253–1271.

Pongvuthithum, R., (2009). A time-varying feedback approach for triangular systems with nonlinear parameterization. *SIAM Journal on Control and Optimization, 48*(3), 1660–1674.

Praly, L., & Jiang, Z. (2004). Linear output feedback with dynamic high gain for nonlinear systems. *Systems & Control Letters, 53*(2), 107–116.

Reichhartinger, M., & Horn, M., (2011). *Finite-time stabilization by robust backstepping for a class of mechanical systems*. IEEE international conference on control applications (CCA). Denver, CO.

Yu, S., Yu, X., Shirinzadeh, B., & Man, Z. (2005). Continuous finite-time control for robotic manipulators with terminal sliding mode. *Automatica, 41*, 1957–1964.

Zhanga, X., Fengb, G., & Sunb, Y., (2012). Finite-time stabilization by state feedback control for a class of time-varying nonlinear systems, *Automatica, 48*, 499–504.

Zhu, L., Shen, Y., & Li, C. (2009). Finite-time control of discrete-time systems with time-varying exogenous disturbance. *Communications in Nonlinear Science and Numerical Simulation, 14*, 361–370.

Research on a chaotic circuit based on an active TiO_2 memristor

Wei Wang*, Guangyi Wang and Xiaoyuan Wang

School of Electronics Information, Hangzhou Dianzi University, Hangzhou, People's Republic of China

The memristor is the fourth fundamental circuit element besides the resistor, inductor and capacitor. As a two-terminal nonlinear resistor, the memristor has a broad application prospect. In this paper, a negative memconductance expression of a flux-controlled memristor is derived from the relationship between voltage and current for the Hewlett-Packard memristor. By replacing Chua's diode with the active flux-controlled TiO_2 memristor, a chaotic circuit is obtained. By means of the conventional dynamic analysis method, dynamic behaviors of the chaotic circuit are investigated. Software simulation and theoretical analysis all indicate that this active memristor-based chaotic circuit has more complex behaviors. Furthermore, the integrated circuit experiment on the digital signal processor chip of this circuit was also realized.

Keywords: TiO_2 memristor; Chua's oscillator; chaotic system; dynamic analysis

1. Introduction

In 1971, Leon O Chua of UC Berkeley first proposed the memristor as the fourth fundamental circuit clement besides the resistor, inductor and capacitor (Chua, 1971; Tour & He, 2008). Almost four decades from then on, the actual memristor had not been manufactured, which prevents the development of memristor research and application. Fortunately, in 2008, a research team in the Hewlett-Packard (HP) Company fabricated a nanometer-sized TiO_2 memristor (Strukov et al., 2008), which not only confirms the existence of the memristor, but also spirits the research upsurge of the memristor like the i–v characteristics studying and modeling of the memristor (Hu et al., 2011), using the memristor model to structure a memcapacitor circuit (Wang, Fitch, Iu, Sreeram, & Qi, 2012), chaotic circuit building based on memristor (Bao, Liu, & Xu, 2010; Itoh & Chua, 2008; Muthuswamy, 2009; Qi, Bian, & Li, 2011; Wang, Qi, & Wang, 2012; Wang, Wang, Chen, & Tan, 2011; Wang, Wang, & Tan, 2011), memristor-based analog circuits (Batas & Fiedler, 2011; Mutlu & Karakulak, 2010) and so on. The arrangement of this paper is as follows: in the next section, we review the proposal of the memristor and focus on the structure and formulas of the HP memristor, and as well as the theoretical foundation of the memristor model building help to derive the negative memristor model in this paper. In Section 3 by replacing Chua's diode with the active memristor, a chaotic system was built and theoretical simulation with Matlab was shown to confirm the chaotic behavior of this system. In Section 4, the digital signal processor (DSP) realization is done.

2. Memristor and TiO_2 memristor

As one of the four fundamental circuit element, the memristor has different characteristics from the traditional circuit devices, especially in memory character, which enables it to be used in neural network (Kim, Sah, Yang, & Chua, 2012; Jo et al., 2010) and non-volatile random access memory (Duan, Hu, Wang, Li, & Mazumder, 2012). The memristor is an element proposed to indicate the direct relations between magnetic flux ϕ and charge q, as is shown in Figure 1. The symbol of the memristor and the simplified structure of TiO_2 memristor are shown in Figure 2.

The relationship between voltage across the memristor v and current i through it in both charge-controlled memristor and flux-controlled memristor can be described by the following equation (Wang, Fitch et al. 2012)

$$\begin{aligned} v &= M(q)i, \\ i &= W(\phi)v, \end{aligned} \tag{1}$$

where $M(q)$ is the memristance of a memristor and $W(\phi)$ is memconductance in each kind of definitions. As shown in Figure 2, by analyzing the moving process inside the solid-state HP memristor, the following i–v relationship can be obtained (Kim et al., 2012; Strukov et al., 2008):

$$v(t) = \left(R_{\mathrm{ON}}\frac{w(t)}{D} + R_{\mathrm{OFF}}\left(1 - \frac{w(t)}{D}\right)\right)i(t), \tag{2}$$

where R_{ON} and R_{OFF} represent the resistances of the doped region and undoped region, $w(t)$ is assigned as the width

*Corresponding author. Email: zouwei2924@126.com

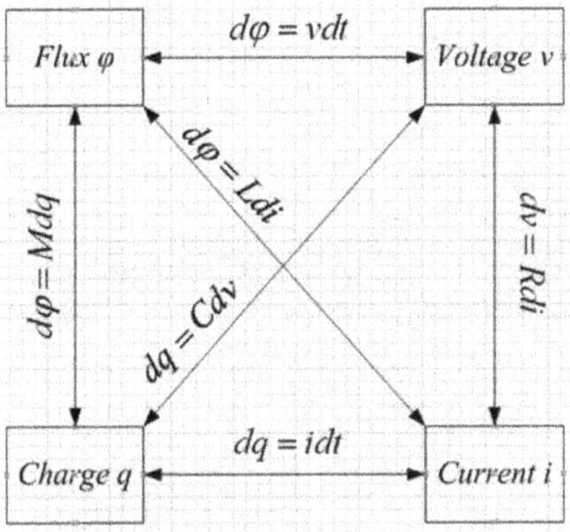

Figure 1. Relationship between fundamental circuit variables and the memristor.

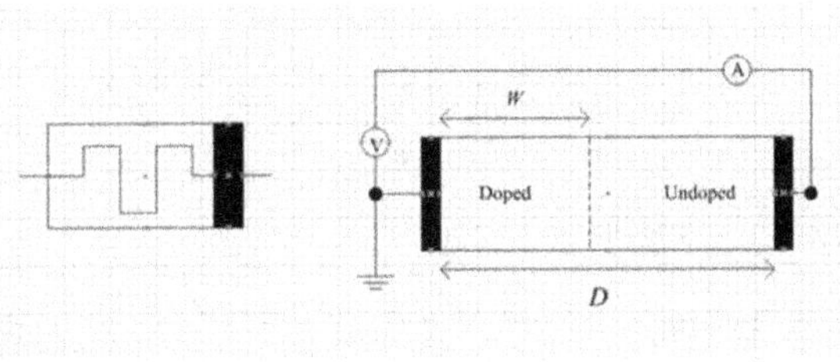

Figure 2. Symbol of the memristor and structure of the TiO_2 memristor.

of the doped region and D is the total width of the titanium dioxide film. In the TiO_2 memristor, $w(t)$ ranges from 0 to D and is determined by the following equation:

$$\frac{\mathrm{d}w(t)}{\mathrm{d}t} = \mu_V \frac{R_{ON}}{D} i(t), \tag{3}$$

where μ_V is the dopant mobility.

By resolving Equation (3), we can obtain

$$w(t) = \mu_V \frac{R_{ON}}{D} \int_{-\infty}^{t} i(\tau)\,\mathrm{d}\tau = \mu_V \frac{R_{ON}}{D} q(t) + w_0, \tag{4}$$

where w_0 is the initial width of w (Kim et al., 2012). Applying Equation (4) to Equation (2), we can rewrite Equation (2) as the following relation:

$$v(t) = \left(\frac{R_{ON}}{D} \left(\mu_V \frac{R_{ON}}{D} q(t) + w_0 \right) + R_{OFF} - \frac{R_{OFF}}{D} \left(\frac{\mu_V R_{ON}}{D} q(t) + w_0 \right) \right) i(t). \tag{5}$$

Integrating both sides of Equation (5), we can obtain

$$\phi(t) = R_{OFF} \left\{ q(t) \left[1 + \frac{w_0}{D} \left(\frac{R_{ON}}{R_{OFF}} - 1 \right) \right] - \frac{\mu_V R_{ON}}{2D^2} \left(1 - \frac{R_{ON}}{R_{OFF}} \right) q(t)^2 \right\} + \phi_0. \tag{6}$$

Now, let $a = 1 + (w_0/D)(R_{ON}/R_{OFF} - 1)$, $b = (\mu_V R_{ON}/2D^2)(1 - R_{ON}/R_{OFF})$, Equation (6) will be simplified as

$$\phi(t) = R_{OFF}[aq(t) - bq(t)^2] + \phi_0. \tag{7}$$

Combining Equation (7) with the definition of memductance, $W(\phi)$ can be derived as

$$W(\phi) = \frac{\mathrm{d}q(\phi)}{\mathrm{d}\phi} = \pm \frac{1}{R_{OFF}} \left[a^2 - \frac{4b}{R_{OFF}} (\phi - \phi_0) \right]^{-1/2}. \tag{8}$$

In terms of the characteristics of the TiO_2 memristor, and under the assumption of $R_{ON} \ll R_{OFF}$ and $w_0/D \approx 0$, we can obtain

$$a \approx 1, \quad b \approx \frac{\mu_V R_{ON}}{2D^2}. \tag{9}$$

So Equation (8) can be simplified as

$$W(\phi) = \pm \frac{1}{R_{OFF}} \left[1 - \frac{2\mu_V R_{ON}}{R_{OFF} D^2} (\phi - \phi_0) \right]^{-1/2}. \tag{10}$$

Let $a_1 = 1/R_{OFF}$, $a_2 = 2\mu_V R_{ON}/R_{OFF}D^2$, $a_3 = \phi_0$, then Equation (10) can be described as

$$W(\phi) = \pm a_1[1 - a_2(\phi - a_3)]^{-1/2}. \tag{11}$$

As the memconductance has been obtained, and in this paper we only discuss when using the negative value of Equation (11) as the mathematic model of the memristor in Chua's circuit instead of the nonlinear diode, to see what the dynamic characteristics will appear.

3. Construction of the memristor-based chaotic circuit

In 2008, Chua derived several oscillators based on the memristor whose characteristics were described by a monotone-increasing piecewise-linear curve (Wang, Fitch et al., 2012). In Muthuswamy (2009), the author used a monotone-increasing cubic model of the memristor built a chaotic ciruit. And in Wang, Qi et al. (2012), a novel memristor-based chaotic system was proposed, and the mathematic model used as the memristor is based on the memristive system definition. All of these chaotic circuits based on the memristor are by means of replacing Chua's diode with a memristor. So in the same way, we replace Chua's diode with an active flux-controlled TiO_2 memristor model which is shown in Equation (11), and a five-order chaotic oscillator based on an active memristor is derived as shown in Figure 3.

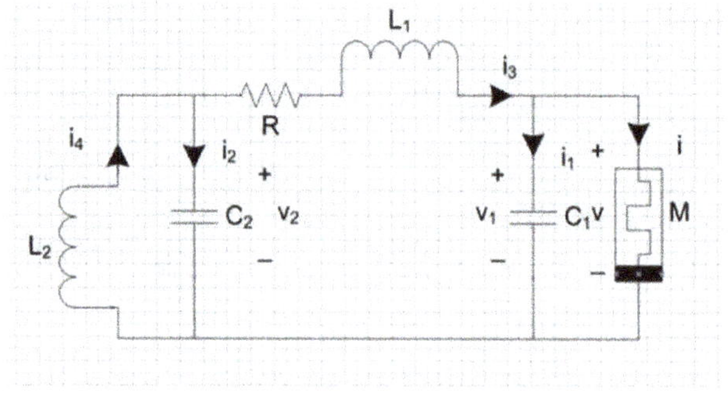

Figure 3. A five-order chaotic oscillator based on the TiO_2 memristor.

Applying Kirchoff's laws to the system in Figure 3, we can obtain

$$\begin{aligned}
\frac{\mathrm{d}v_1}{\mathrm{d}t} &= \frac{1}{C_1}[i_3 - W(\phi)v_1],\\
\frac{\mathrm{d}v_2}{\mathrm{d}t} &= \frac{1}{C_2}(-i_3 + i_4),\\
\frac{\mathrm{d}i_3}{\mathrm{d}t} &= \frac{1}{L_1}(v_2 - v_1 - Ri_3),\\
\frac{\mathrm{d}i_4}{\mathrm{d}t} &= -\frac{1}{L_2}v_2,\\
\frac{\mathrm{d}\phi}{\mathrm{d}t} &= v_1.
\end{aligned} \tag{12}$$

If $x = v_1, y = v_2, z = i_3, w = i_4, u = \phi$, then Equation (12) can be transformed as the following:

$$\begin{aligned}
\dot{x} &= p(z - W(u)x),\\
\dot{y} &= -z + w,\\
\dot{z} &= q(y - x - z),\\
\dot{w} &= -ry,\\
\dot{u} &= x.
\end{aligned} \tag{13}$$

Let $R = 1$, $C_2 = 1$, $p = 1/C_1 = 8.985$, $q = 1/L_1 = 15.125, r = 1/L_2 = 24.625$, and $a_1 = 1/1.2, a_2 = 0.98$ and

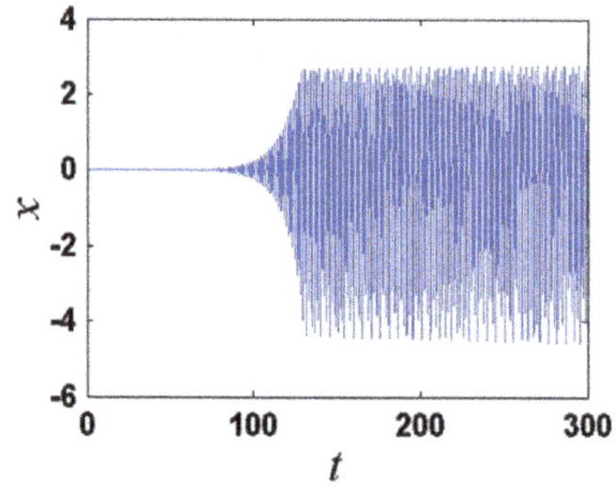

Figure 5. x time domain waveform.

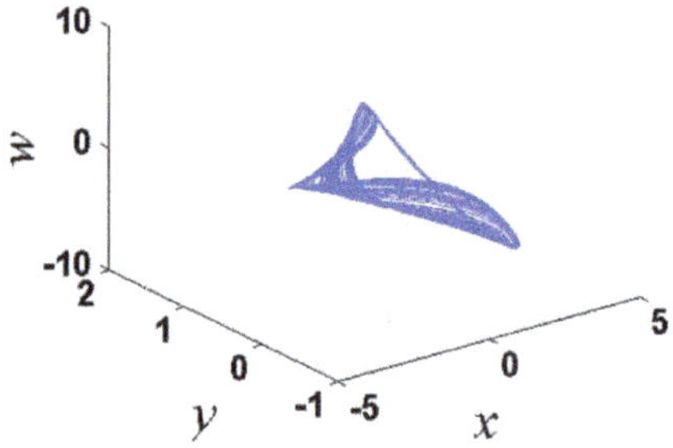

Figure 6. x–y–w.

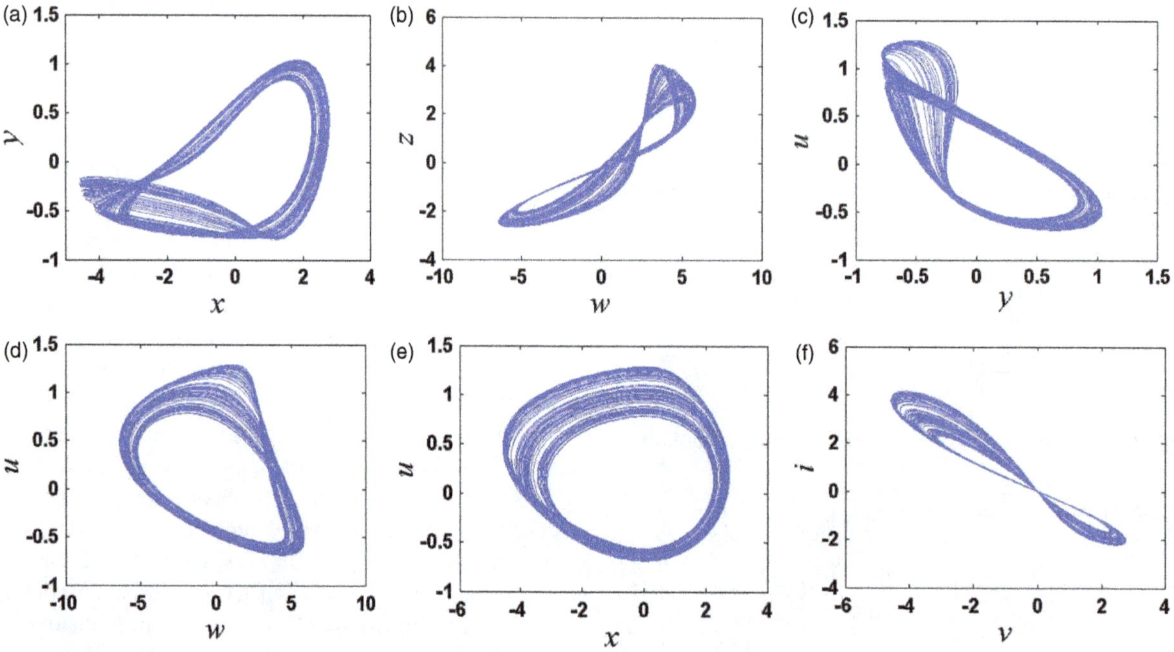

Figure 4. Chaotic attractors of the memristor-based circuit.

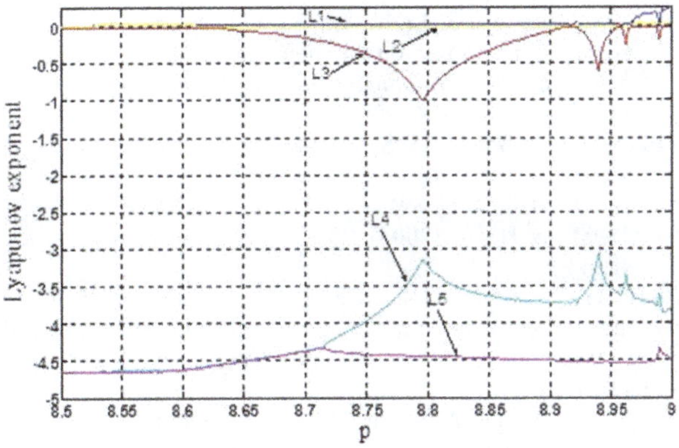

Figure 7. Lyapunov exponents with a changing p.

$a_3 = 0.5$, the system shown in Figure 3 has a chaotic attractor as shown in Figure 4. Also, the x time domain waveform and the 3-D phase trajectory are shown in Figures 5 and 6. By means of the Jacobi method, the Lyapunov exponents are obtained, they are $L_1 = 0.2315$, $L_2 = 0.0043$, $L_3 = -0.0014$, $L_4 = -3.4515$ and $L_5 = -4.5058$, which confirms that the system is a chaotic dynamic one. The Lyapunov exponents diagram is shown in Figure 7, and it is obvious that when p changes there are two exponents more than zero, three are negative, and the summary of these five exponents is negative, which satisfies the feature of the chaotic system.

4. DSP realization of the chaotic system

To realize the above circuit with DSP, we use Euler's formula method as the discretization method. With the help of the core processor, a 16-bit fixed-point digital signal processor TMS320C5509 and a two channels analog oscilloscope, the chaotic attractors shown in Figure 8 were observed in the oscilloscope, which coincides with the results in Figure 4.

5. Conclusions

In this paper, an active memristor model based on the HP memristor was formed, and by replacing Chua's diode with this active memristor, a five-order chaotic circuit is derived. By means of the conventional dynamic analysis, the dynamic characteristics of this system are discussed in detail. Simulation results as well as theoretical analysis indicate that such an active memristor-based chaotic system easily generates complex chaotic dynamic behaviors. Finally, the DSP experiment results were shown.

Acknowledgements

This work was supported by the National Nature Science Foundation of China (Grant No: 61271064) and the National Natural Science Foundation of Zhejiang province (Grant No: LZ12F01001).

References

Bao, B.-C., Liu, Z., & Xu, J.-P. (2010). Dynamical analysis of memristor chaotic oscillator. *Acta Physica Sinica*, *59*, 3785–3793.

Batas, D., & Fiedler, H. (2011). A memristor SPICE implementation and a new approach for magnetic flux-controlled memristor modeling. *IEEE Transactions on Nanotechnology*, *10*, 250–255.

Chua, L. O. (1971). Memristor – The missing circuit element. *IEEE Transactions on Circuit Theory*, *CT-18*, 507–519.

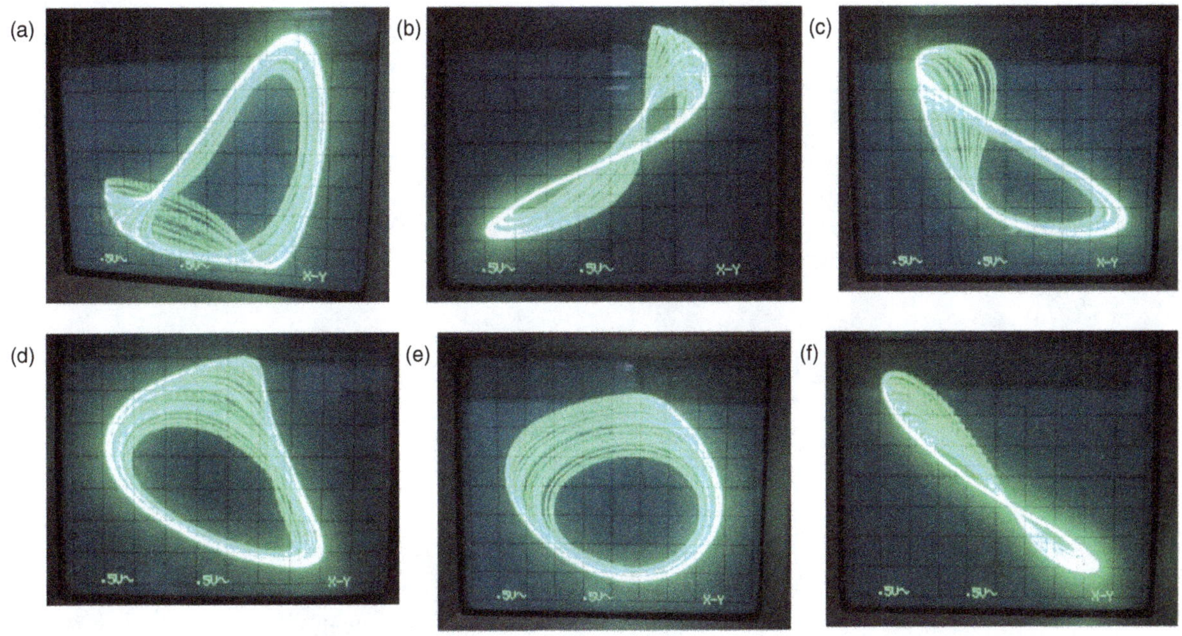

Figure 8. Chaotic attractors observed in the oscilloscope.

Duan, S.-K., Hu, X.-F., Wang, L.-D., Li, C.-D., & Mazumder, P. (2012). Memristor-based RRAM with applications. *Science China Press, 42*, 754–769.

Hu, B.-L., Wang, L.-D., Huang, Y.-W., Hu, X.-F., Zhang, Y.-Y., & Duan, S.-K. (2011). Simulink modeling and graphical user interface design for a memristor. *Journal of Southwest University (Natural Science Edition), 33*, 50–56.

Itoh, M., & Chua, L. O. (2008). Memristor oscillators. *International Journal of Bifurcation and Chaos, 18*, 3183–3206.

Jo, S. H., Chang, T., Ebong, I., Bhadviya, B. B., Mazumder, P., & Lu, W. (2010). Nanoscale memristor device as synapse in neuromorphic systems. *Nano Letters, 10*, 1297–1301. doi:10.102/nl904092h

Kim, H., Sah, M. Pd., Yang, C., & Chua, L. O. (2012). Neural synaptic weighting with a pulse-based memristor circuit. *IEEE Transactions on Circuits and Systems, 59*, 148–158.

Muthuswamy, B. (2009). *Memristor-based chaotic circuits.* Retrieved from http://www.eecs.berkeley.edu/Pubs/TechRpts/2009/EECS-2009-6.html

Mutlu, R., & Karakulak, E. (2010). *Emulator circuit of TiO_2 memristor with linear dopant drift made using analog multiplier.* Electrical, Electronics and Computer Engineering (ELECO), 2010 National Conference on, 380–384.

Qi, A.-X., Bian, L., & Li, W.-B. (2011). Realization of Chua's chaotic circuit based on the memristor and dynamic performance analysis. *Periodical of Ocean University of China, 41*, 119–124.

Strukov, D. B., Snider, G. S., Stewart, D. R., & Williams, R. S. (2008). The missing memristor found. *Nature, 453*, 80–83. doi:10.1038/nature0693

Tour, J. M., & He, T. (2008). The fourth element. *Nature, 453*, 42–43.

Wang, G.-Y., Wang, C.-L., Chen, W., & Tan, D. (2011). A magnetic-controlled memristor based chaotic circuit and its FPGA realization. *Journal of Circuits and Systems, 16*, 114–119.

Wang, X. Y., Fitch, A. L., Iu, H. H. C., Sreeram, V., & Qi, W. G. (2012). Design of a memcapacitor emulator based on a memristor. *Physics Letters A, 376*, 394–399.

Wang, X. Y., Qi, W. G., & Wang, X. Y. (2012). Novel circuit of memristor and its chaotic dynamics characteristics. *Journal of Beijing University of Aeronautics and Astronautics, 38*, 1–5.

Wang, W., Wang, G., & Tan, D. (2011). *A new memristor based chaotic circuit.* The Fourth International Workshop on Chaos-Fractals Theories and Applications, Hangzhou.

12

Stability analysis for a class of switched nonlinear time-delay systems

M. Kermani* and A. Sakly

Research Unit of Industrial Systems Study and Renewable Energy (ESIER), National Engineering School of Monastir (ENIM), Ibn El Jazzar, Skaness, 5019, Monastir, Tunisia

This paper investigates the stability analysis for a class of discrete-time switched nonlinear time-delay systems. These systems are modelled by a set of delay difference equations, which are represented in the state form. Then, another transformation is made towards an arrow form. Therefore, by applying the Kotelyanski conditions combined to the M − matrix properties, new delay-independent sufficient stability conditions under arbitrary switching which correspond to a Lyapunov function vector are established. The obtained results are explicit and easy to use. A numerical example is provided to show the effectiveness of the developed results.

Keywords: discrete-time switched nonlinear time-delay systems; arbitrary switching; global asymptotic stability; M-matrix; Kotelyanski lemma; arrow matrix

1. Introduction

A switched system is a type of hybrid dynamical system that consists of a number of subsystems which are described by differential or difference equations and a switching signal selecting a subsystem to be active during a certain time interval. As a special class of hybrid systems, many dynamical systems can be modelled as switched systems (Branicky, 1998; Darouach & Chadli, 2013; Dong, Liu, Mei, & Li, 2011; Liberzon, 2003; Liberzon & Morse, 1999; Mahmoud, 2010; Phat & Ratchagit, 2011; Sun & Ge, 2011; Zhang, Abate, Hu, & Vitus, 2009; Zhang & Yu, 2009).

Recently, switched systems have strong engineering background in various areas and are often used as a unified modelling tool for a great number of real-world systems such as mechanical systems, automotive industry, power electronics, chemical engineering processing, communication networks, aircraft and air traffic control, chemical and electrical engineering and models for epidemiology (Pellanda, Apkarian, & Tuan, 2002; Tse & Bernardo, 2002; Wu, Shi, Su, & Chu, 2011; Zhang, Liu, & Huang, 2010).

The last decade has witnessed increasing research activities in the study of switched systems. Among of those research topics, stability analysis has attracted most of attention. Hence, several methods have been proposed for this matter (Arunkumar, Sakthivel, Mathiyalagan, & Marshal Anthoni, 2012; Branicky, 1998; Hespanha & Morse, 1999; Ishii & Francis, 2002; Jianyin & Kai, 2010; Liberzon, 2003; Lien, Yu, Chang, Chung, & Chen, 2012; Lien et al., 2011; Shim, Noh, & Seo, 2001; Sun, Wang, Liu, & Zhao, 2008; Vu & Liberzon, 2005; Zhai, Hu, Yasuda, & Michel, 2000; Zhang, Shi, & Basin, 2008; Zhang & Yu, 2009). In fact, stability under arbitrary switching is a fundamental and challenging research issue in the design and analysis of these systems (Mathiyalagan, Sakthivel, Marshal, & Anthoni, 2012; Zhang et al., 2008; Zhang & Yu, 2009).

Within this context, it is well known that a switched system is asymptotically stable if all the individual systems are stable and the switching is sufficiently slow, so as to allow the transient effects to dissipate after each switch. In this case, the existence of a common Lyapunov function for all the subsystems is a sufficient condition to guarantee the stability of the switched system under arbitrary switching law (Liberzon, 2003). Therefore, this method is usually very difficult to apply even for discrete-time switched linear systems; however, it becomes more complicated for the nonlinear case. However, some attempts are presented to construct a common Lyapunov function for nonlinear exponentially stable systems in Shim et al. (2001), and for general nonlinear asymptotically stable systems in Vu & Liberzon (2005).

To avoid the problem of existence of the common Lyapunov function, frequently, we are required to seek conditions to guarantee the stability of the switched systems for any admissible switched law. For this, many methodologies' efficient approaches have been developed, such as the multiple Lyapunov function approach (Branicky, 1998) and the average dwell time method (Hespanha & Morse, 1999; Ishii & Francis, 2002; Zhai et al., 2000; Zhang & Gao, 2010). Whereas, stability analysis under arbitrary switching

*Corresponding author. Email: kermanimarwen@gmail.com

remains the most essential issue in practical switched systems.

Furthermore, time-delay phenomenon also cannot be avoided in practical systems, for example, chemical processes, long-distance transportation systems, hydraulic pressure systems, hybrid procedure, electron network, network control systems and so on. The problem of time-delay may cause instability and poor performance of practical systems (Gao & Chen, 2007; Gao, Lam, & Wang, 2007). Therefore, the stability analysis of switched time-delay systems is very worthy to be researched.

Basically, current efforts to achieve stability in time-delay systems can be divided into two categories, namely, delay-independent criteria and delay-dependent criteria. In this paper, in view of delay-independent analysis, we expect to aid in studying stability analysis under arbitrary switching law.

Up to date, most of the previous results on stability analysis for discrete-time switched time-delay systems were interested in the linear case (Arunkumar et al., 2013; Branicky, 1998; Gao & Chen, 2007; Gao et al., 2007; Liberzon & Morse, 1999; Phat & Ratchagit, 2011; Sun & Ge, 2011; Wu et al., 2011; Zhang & Yu, 2009). Therefore, few results are developed for discrete-time switched nonlinear time-delay systems (Xu, 2002).

Motivated by these gaps, the objective of this paper is to present a novel approach for asymptotic stability of a class of discrete-time switched nonlinear time-delay systems. In fact, based on transforming the representation of the system under consideration into the arrow form matrix representation (Benrejeb & Borne, 1978; Benrejeb, Borne, & Laurent, 1982; Benrejeb & Gasmi, 2001; Benrejeb, Gasmi, & Borne, 2005; Benrejeb, Soudani, Sakly, & Borne, 2006; Elmadssia, Saadaoui, & Benrejeb, 2011, 2013; Filali, Hammami, Benrejeb, & Borne, 2012; Jabbali, Kermani, & Sakly, 2013; Kermani, Sakly, & M'sahli, 2012; Sfaihi & Benrejeb, 2013), the use of an appropriate Lyapunov function associated with the Kotelyanski conditions (Benrejeb et al., 2006; Borne, 1987; Borne, Gentina, & Laurent, 1972; Borne, Vanheeghe, & Duflos, 2007; Kotelyanski, 1952) and the $M-$ matrix proprieties (Gantmacher, 1966; Robert, 1966), new sufficient stability conditions are derived under arbitrary switching. The obtained stability conditions which correspond to a Lyapunov function vector are simple to employ, explicit and allow us to avoid the search for a common Lyapunov function which appears a difficult matter in this case.

Within the frame of studying the stability analysis, the said approach has already been introduced in Elmadssia et al. (2011, 2013) for continuous-time delay systems and in our previous work (Jabbali et al., 2013) for discrete-time switched linear time-delay systems, in a field related to the study of convergence. This proposed approach could be further used as a constructive solution to the problems of state feedback stabilization.

The rest of this paper is organized as follows. The problem is formulated and some basic notations and definitions are given in Section 2. The main results of this paper are presented in Section 3. Section 4 is devoted to derive new delay-independent conditions for asymptotic stability of switched nonlinear systems defined by difference equations. Remarks and numerical example are presented in Section 5 to illustrate the theoretical results, and the conclusions are drawn in Section 6.

2. Problem statements and preliminaries

2.1. Problem statements

Consider the following discrete-time switched nonlinear time-delay system formed by N subsystems given in the state form:

$$x(k+1) = \sum_{i=1}^{N} \zeta_i(k)(A_i(\cdot)x(k) + D_i(\cdot)x(k-d)), \quad x(l) = \phi(l), \quad l = -d, \dots, -1, 0, \tag{1}$$

where $x(k) \in \Re^n$ is the system state, $d > 0$ is the time delay, $\phi(l) : \{-d, -d+1, \dots, 0\} \rightarrow \Re^n$: is a vector-valued initial function, $A_i(\cdot)$ $(i = 1, \dots, N)$ and $D_i(\cdot)$ $(i = 1, \dots, N)$ are matrices that have nonlinear elements of appropriate dimensions denoting the subsystems and $N \geq 1$ denotes the number of subsystems.

The switching sequence is defined through a switching vector: $\zeta(k) = [\zeta_1(k), \dots, \zeta_N(k)]^{\mathrm{T}}$ whose components $\zeta_i(k)$ $\Re_+ \rightarrow M = \{0, 1\}$ are exogenous functions that depend only on the time and not on the state, they are defined through:

$$\zeta_i(k) = \begin{cases} 1 & \text{when } A_i(\cdot) \text{ and } D_i(\cdot) \text{ are active,} \\ 0 & \text{otherwise,} \end{cases} \quad i \in N, \tag{2}$$

it is obvious that $\sum_{i=1}^{N} \zeta_i(k) = 1$.

2.2. Notations and definitions

Throughout this paper, if not explicitly stated, matrices are assumed to have compatible dimensions. I_n is an identity matrix with appropriate dimension. Let $\Re^n$ denote an n dimensional linear vector space over the reals $\|\cdot\|$ which stands for the Euclidean norm of vectors. For any $u = (u_i)_{1 \leq i \leq n}$, $v = (v_i)_{1 \leq i \leq n} \in \Re^n$, we define the scalar product of the vectors u and v as: $\langle u, v \rangle = \sum_{i=1}^{n} u_i v_i$.

We are here interested to establish delay-independent stability conditions for system (1) by using the $M-$ matrix properties combined to the Kotelyanski conditions. In order to prepare for a precise formulation of our results, we introduce the following definitions and lemma that will play a key role in deriving our main results.

DEFINITION 1 *The hybrid system* (1) *is said to be uniformly asymptotically stable if for any* $\varepsilon > 0$, *there is a* $\delta(\varepsilon) > 0$

such that $\sup_{-d \le l \le 0} \| \phi(l) \| < \delta$ *implies* $\| x(k,\phi) \| \le \varepsilon$, $k \ge 0$. *For arbitrary switching signal* $\zeta(k)$ *and there is also a* $\delta' > 0$ *such that* $\sup_{-d \le l \le 0} \| \phi(k) \| < \delta'$ *implies* $\| x(k,\phi) \| \to 0$ *as* $k \to \infty$ *for arbitrary switching signal* (2).

Next, we introduce several useful tools, including Kotelyanski lemma and definition of an $M-$ matrix.

KOTELYANSKI LEMMA (Borne, 1987) *The real parts of the eigenvalues of matrix A, with non-negative off-diagonal elements, are less than a real number* μ *if and only if all those of matrix* M, $M = \mu I_n - A$, *are positive, with* I_n *the n identity matrix.*

When the principal minors of matrix $(-A)$ are positive, the Kotelyanski lemma permits to conclude on stability property of the system characterized by A.

DEFINITION 2 (Borne, 1987) *The matrix* $A \in \Re^{n\times n}$ *is called a Z-matrix if it has null or negative off-diagonal elements.*

THEOREM 1 (Borne, 1987) *In order that a Z-matrix* $A(\cdot)$ *is said an M-matrix if the following properties are verified:*

- *All the eigenvalues of* $A(\cdot)$ *have a positive real part.*
- *The real eigenvalues are positive.*
- *The principal minors of* $A(\cdot)$ *are positive*

$$(A(\cdot))\begin{pmatrix}1 & 2 & \cdots & j\\ 1 & 2 & \cdots & j\end{pmatrix} > 0 \quad \forall j \in 1,\ldots,n. \quad (3)$$

- *For any positive vector* $x = (x_1,\ldots,x_n)^{\mathrm{T}}$, *the algebraic equations* $A(\cdot)x$ *have a positive solution* $w = (w_1,\ldots,w_n)^{\mathrm{T}}$.

Remark 1 A discrete-time system characterized by a matrix $A(\cdot)$ is stable if the matrix $(I_n - A(\cdot))$ verifies the Kotelyanski conditions, in this case $(I_n - A(\cdot))$ is an M-matrix.

3. Main results

In this part, after previous formulation, we can now state the main result of this paper summarized in the following theorem which gives stability conditions for the discrete-time nonlinear switched time-delay systems (1).

THEOREM 2 *The discrete-time switched nonlinear time-delay system* (1) *is asymptotically stable under arbitrary switching rule* (2) *if the matrix* $(I_n - T_c(\cdot))$ *is an M-matrix where*

$$T_c(\cdot) = \max_{1\le i\le N}(T_{\zeta(k)}(\cdot)) \quad (4)$$

and

$$T_{\zeta(k)}(\cdot) = (|A_{\zeta(k)}(\cdot)| + |D_{\zeta(k)}(\cdot)|). \quad (5)$$

$$A_{\zeta(k)}(\cdot) = \begin{bmatrix} \sum_{i=1}^N \zeta_i(k)(a_i^{11}(\cdot)) & \cdots & \cdots & \sum_{i=1}^N \zeta_i(k)(a_i^{1n}(\cdot)) \\ \vdots & \vdots & \vdots & \vdots \\ \vdots & \vdots & \vdots & \vdots \\ \sum_{i=1}^N \zeta_i(k)(a_i^{n1}(\cdot)) & \cdots & \cdots & \sum_{i=1}^N \zeta_i(k)(a_i^{nn}(\cdot)) \end{bmatrix}, \quad (6)$$

$$D_{\zeta(k)}(\cdot) = \begin{bmatrix} \sum_{i=1}^N \zeta_i(k)(d_i^{11}(\cdot)) & \cdots & \cdots & \sum_{i=1}^N \zeta_i(k)(d_i^{1n}(\cdot)) \\ \vdots & \vdots & \vdots & \vdots \\ \vdots & \vdots & \vdots & \vdots \\ \sum_{i=1}^N \zeta_i(k)(d_i^{n1}(\cdot)) & \cdots & \cdots & \sum_{i=1}^N \zeta_i(k)(d_i^{nn}(\cdot)) \end{bmatrix}. \quad (7)$$

So, we obtain sufficient conditions for asymptotic stability of system (1).

Proof Let us consider the system (1) of any switching signal (2), let $(w_l > 0, \forall l = 1,\ldots,n)$ next, we choose the following Lyapunov function candidate:

$$v(k) = v_0(k) + \sum_{j=1}^{r} v_j(k), \quad (8)$$

where

$$\begin{aligned} v_0(k) &= \langle |x(k)|, w\rangle, \\ v_j(k) &= \langle |D_{\zeta(k)}(\cdot)||x(k-j)|, w\rangle, \quad (j = 1,\ldots,r). \end{aligned} \quad (9)$$

It suffices to show that

$$\Delta v(k) = v(k+1) - v(k) < \langle (T_c(\cdot))|x(k)|, w\rangle, \quad r > 0, \quad (10)$$

where

$$\Delta v(k) = \Delta v_0(k) + \sum_{j=1}^{r} \Delta v_j(k) \quad (11)$$

and

$$\Delta v_0 = \langle |x(k+1)|, w\rangle - \langle |x(k)|, w\rangle. \quad (12)$$

For any $r > 0$, we obtain from Equation (9) that

$$\begin{aligned} \Delta v_j =& \langle |D_{\zeta(k)}(\cdot)||x(k-j+1)|, w\rangle \\ &- \langle |D_{\zeta(k)}(\cdot)||x(k-j)|, w\rangle, \quad j = 1,\ldots,r. \end{aligned} \quad (13)$$

Knowing that

$$\begin{aligned}&\langle||x(k+1)|,w\rangle\\&\quad=\langle|A_{\zeta(k)}(\cdot)x(k)+D_{\zeta(k)}(\cdot)x(k-r)|,w\rangle\\&\quad<\langle|A_{\zeta(k)}(\cdot)||x(k)|+|D_{\zeta(k)}(\cdot)||x(k-r)|,w\rangle\\&\quad=\langle|A_{\zeta(k)}(\cdot)||x(k)|,w\rangle+\langle|D_{\zeta(k)}(\cdot)||x(k-r)|,w\rangle.\end{aligned}\tag{14}$$

On the other hand, we have

$$\begin{aligned}&\sum_{j=1}^{r}\Delta v_j(k)\\&\quad=\Delta v_1(k)+\Delta v_2(k)+\cdots+\Delta v_{r-1}(k)+\Delta v_r(k).\end{aligned}\tag{15}$$

Therefore, by Equations (9) and (15), we have that

$$\begin{aligned}&\sum_{j=1}^{r}\Delta v_j(k)\\&\quad=(\langle|D_{\zeta(k)}(\cdot)||x(k)|,w\rangle-\langle|D_{\zeta(k)}(\cdot)||x(k-1)|,w\rangle)\\&\qquad+(\langle|D_{\zeta(k)}(\cdot)||x(k-1)|,w\rangle\\&\qquad-\langle|D_{\zeta(k)}(\cdot)||x(k-2)|,w\rangle)\\&\qquad+(\langle|D_{\zeta(k)}(\cdot)||x(k-r+1)|,w\rangle\\&\qquad-\langle|D_{\zeta(k)}(\cdot)||x(k-r)|,w\rangle)\\&\quad=(\langle|D_{\zeta(k)}(\cdot)||x(k)|,w\rangle-\langle|D_{\zeta(k)}(\cdot)||x(k-r)|,w\rangle).\end{aligned}\tag{16}$$

Thus, we can eventually obtain that

$$\begin{aligned}\Delta v(k)&=\Delta v_0(k)+(\langle|D_{\zeta(k)}(\cdot)||x(k)|,w\rangle\\&\quad-\langle|D_{\zeta(k)}(\cdot)||x(k-r)|,w\rangle).\end{aligned}\tag{17}$$

Moreover, it is not difficult to remark that

$$\begin{aligned}\langle|x(k+1)|,w\rangle&<\langle|A_{\zeta(k)}(\cdot)||x(k)|,w\rangle\\&\quad+\langle|D_{\zeta(k)}(\cdot)||x(k-r)|,w\rangle.\end{aligned}\tag{18}$$

By Equations (11)–(14), we have that

$$\begin{aligned}\Delta v(k)&<\langle|A_{\zeta(k)}(\cdot)||x(k)|,w\rangle+\langle|D_{\zeta(k)}(\cdot)||x(k-r)|,w\rangle\\&\quad-\langle|x(k)|,w\rangle\\&\quad+\langle|D_{\zeta(k)}(\cdot)||x(k)|,w\rangle-\langle|D_{\zeta(k)}(\cdot)||x(k-r)|,w\rangle.\end{aligned}\tag{19}$$

We have

$$\begin{aligned}\Delta v(k)&<\langle|A_{\zeta(k)}(\cdot)||x(k)|,w\rangle-\langle|x(k)|,w\rangle\\&\quad+\langle|D_{\zeta(k)}(\cdot)||x(k)|,w\rangle,\end{aligned}\tag{20}$$

and finally we obtain

$$\Delta v(k)<\langle(|A_{\zeta(k)}(\cdot)|+|D_{\zeta(k)}(\cdot)|-I_n)|x(k)|,w\rangle,\tag{21}$$

it follows that

$$\Delta v(k)\le\langle(T_c(\cdot)-I_n)|x(k)|,w\rangle,\tag{22}$$

where the matrix $T_c(\cdot)$ is defined in Equation (4).

Now, suppose that $I_n-T_c(\cdot)$ is an $M-$ matrix, according to the proprieties of the $M-$ matrix, we can find a vector $\rho\in\Re_+^{*n}$ ($\rho_l\in\Re_+^*$, $l=1,\ldots,n$) satisfying the relation $(I_n-T_c(\cdot))^{\mathrm{T}}w=\rho$, $\forall w\in\Re_+^{*n}$, so we can write

$$\begin{aligned}\langle(I_n-T_c(\cdot))|x(k)|,w\rangle&=\langle(I_n-T_c(\cdot))^{\mathrm{T}}w,|x(k)|\rangle\\&=\langle\rho,|x(k)|\rangle.\end{aligned}\tag{23}$$

Then, we have

$$\langle(T_c(\cdot)-I_n)|x(k)|,w\rangle=\langle-\rho,|x(k)|\rangle.\tag{24}$$

Finally, we obtain

$$\Delta v(k)\le\langle(T_c(\cdot)-I_n)|x(k)|,w\rangle\le-\sum_{l=1}^{n}\rho_l|x_l(k)|<0.\tag{25}$$

This completes the proof of Theorem 2. ■

Then, the discrete-time switched nonlinear time-delay system given in Equation (1) is asymptotically stable.

Remark 2 By Theorem 2, the stability conditions of the switched time-delay system (1) are independent of time-delay.

4. Stability conditions for switched systems defined by difference equations

The purpose of this section consists in applying the established result to discrete-time switched nonlinear time-delay systems modelled by the following switched nonlinear difference equation (Jabbali et al., 2013):

$$\begin{aligned}&y(k+n)+\sum_{i=1}^{N}\zeta_i(k)\left[\sum_{p=0}^{n-1}a_i^{n-p}(\cdot)y(k+p)\right.\\&\qquad\left.+\sum_{p=0}^{n-1}d_i^{n-p}(\cdot)y(k+p-\tau)\right]=0,\end{aligned}\tag{26}$$

where $\zeta_i(k)$ are the components of the switching function $\zeta(k)$, $i=1,\ldots,N$, given in Equation (2). Then, the presence of both delay-time terms and nonlinearities of the coefficients makes the stability analysis of the problem (26) difficult. To solve this matter as a solution, we will adopt the following change of variable:

$$x_{p+1}(k) = y(k+p), \quad p = 0, \ldots, n-1, \tag{27}$$

in fact, Equation (26) becomes

$$\begin{aligned} x_p(k+1) &= x_{p+1}(k), \quad p = 1, \ldots, n-1, \\ x_n(k+1) &= \sum_{i=1}^{N} \zeta_i(k) \left[-\sum_{p=0}^{n-1} a_i^{n-p}(\cdot) x_{p+1}(k) \right. \\ &\quad \left. - \sum_{p=0}^{n-1} d_i^{n-p}(\cdot) x_{p+1}(k-\tau) \right], \end{aligned} \tag{28}$$

or under matrix representation, we obtain the following state form:

$$\begin{aligned} x(k+1) &= \sum_{i=1}^{N} \zeta_i (A_i(\cdot)x(k) + D_i(\cdot)x(k-\tau)), \\ x(l) &= \phi(l), \quad l = -\tau, \ldots, -1, 0, \end{aligned} \tag{29}$$

where $x(k)$ is the state vector of components $x_p(k)$, $p = 0, \ldots, n-1$.

$\zeta(k)$ is the switched function defined in Equation (2).

The matrices $A_i(\cdot)$ and $D_i(\cdot)$ are given as follows:

$$A_i(\cdot) = \begin{bmatrix} 0 & 1 & \cdots & 0 \\ 0 & 0 & \ddots & \vdots \\ \vdots & \vdots & \ddots & 1 \\ -a_i^n(\cdot) & -a_i^{n-1}(\cdot) & \cdots & -a_i^1(\cdot) \end{bmatrix}, \tag{30}$$

$$D_i(\cdot) = \begin{bmatrix} 0 & 0 & \cdots & 0 \\ 0 & 0 & \ddots & \vdots \\ \vdots & \vdots & \ddots & 0 \\ -d_i^n(\cdot) & -d_i^{n-1}(\cdot) & \cdots & -d_i^1(\cdot) \end{bmatrix}, \tag{31}$$

where $a_i^j(\cdot)$ is a coefficient of the instantaneous characteristic polynomial $P_i(\lambda)$ of the matrix $A_i(\cdot)$ given by

$$P_i(\lambda) = \lambda^n + \sum_{q=0}^{n-1} a_i^{n-q}(\cdot)\lambda^q, \tag{32}$$

and $d_i^j(\cdot)$ is a coefficient of the instantaneous characteristic polynomial $Q_i(\lambda)$ of the matrix $D_i(\cdot)$ defined such as

$$Q_i(\lambda) = \sum_{q=0}^{n-1} d_i^{n-q}(\cdot)\lambda^q. \tag{33}$$

In Jabbali et al. (2013), a change of coordinate for the system given in Equation (29) under the arrow form allows the synthesis of sufficient stability conditions easy to test.

This leads to the following state-space description:

$$z(k+1) = \sum_{i=1}^{N} \zeta_i(k)(M_i(\cdot)z(k) + N_i(\cdot)z(k-\tau)), \tag{34}$$

where $z(k)$ is the new state vector so as $z(k) = Px(k)$, P is the corresponding passage matrix and $M_i(\cdot)$ and $N_i(\cdot)$ are matrices in the arrow form represented as following:

$$M_i(\cdot) = P^{-1}A_i(\cdot)P = \begin{bmatrix} \alpha_1 & 0 & \cdots & 0 & \beta_1 \\ 0 & \ddots & \ddots & \vdots & \vdots \\ \vdots & \ddots & \ddots & 0 & \vdots \\ 0 & \cdots & 0 & \alpha_{n-1} & \beta_{n-1} \\ \gamma_i^1(\cdot) & \cdots & \cdots & \gamma_i^{n-1}(\cdot) & \gamma_i^n(\cdot) \end{bmatrix}, \tag{35}$$

$$N_i(\cdot) = P^{-1}D_i(\cdot)P = \begin{bmatrix} 0_{n-1,n-1} \cdots & 0_{n-1,1} \\ \delta_i^1(\cdot) \cdots \delta_i^{n-1}(\cdot) & \delta_i^n(\cdot) \end{bmatrix}, \tag{36}$$

and P is the corresponding passage matrix defined as follows:

$$P = \begin{bmatrix} 1 & 1 & \cdots & 1 & 0 \\ \alpha_1 & \alpha_2 & \cdots & \alpha_{n-1} & 0 \\ (\alpha_1)^2 & (\alpha_2)^2 & \cdots & (\alpha_{n-1})^2 & \vdots \\ \vdots & \vdots & \cdots & \vdots & 0 \\ (\alpha_1)^{n-1} & (\alpha_2)^{n-1} & \cdots & (\alpha_{n-1})^{n-1} & 1 \end{bmatrix}. \tag{37}$$

For defined distinct arbitrary constant parameters α_j ($j = 1, \ldots, n-1$) and $\alpha_j \neq \alpha_q\ \forall j \neq q$, $q = 1, \ldots, n-1$.

Let us introduce the notation elements of the matrix $M_i(\cdot)$ ($i = 1, \ldots, N$) which are defined as follows:

$$\begin{aligned} \beta_j &= \prod_{\substack{q=1 \\ q\neq j}}^{n-1} (\alpha_j - \alpha_q)^{-1} \quad \forall j = 1, \ldots, n-1, \\ \gamma_i^j(\cdot) &= -P_i(\alpha_j) \quad \forall j = 1, \ldots, n-1, \\ \gamma_i^n(\cdot) &= -a_i^1(\cdot) - \sum_{j=1}^{n-1} \alpha_j, \end{aligned} \tag{38}$$

and the elements of the matrices $N_i(\cdot)$ ($i = 1, \ldots, N$) are as follows:

$$\begin{aligned} \delta_i^j(\cdot) &= -Q_i(\alpha_j) \quad \forall j = 1, \ldots, n-1, \\ \delta_i^n(\cdot) &= -d_i^1(\cdot). \end{aligned} \tag{39}$$

Taking into account the previous relations, the matrices $T_i(\cdot)$ ($i = 1, \ldots, N$) will be defined as follows:

$$T_i(\cdot) = \begin{bmatrix} |\alpha_1| & 0 & \cdots & 0 & |\beta_1| \\ 0 & \ddots & \ddots & \vdots & \vdots \\ \vdots & \ddots & \ddots & 0 & \vdots \\ 0 & \cdots & 0 & |\alpha_{n-1}| & |\beta_{n-1}| \\ t_i^1(\cdot) & \cdots & \cdots & t_i^{n-1}(\cdot) & t_i^n(\cdot) \end{bmatrix}, \tag{40}$$

and the matrix $T_{\zeta(k)}(\cdot)$ is given as follows:

$$T_{\zeta(k)}(\cdot) = \begin{bmatrix} |\alpha_1| & 0 & \cdots & 0 & |\beta_1| \\ 0 & \ddots & \ddots & \vdots & \vdots \\ \vdots & \ddots & \ddots & 0 & \vdots \\ 0 & \cdots & 0 & |\alpha_{n-1}| & |\beta_{n-1}| \\ \sum_{i=1}^{N} \zeta_i(k) t_i^1(\cdot) & \cdots & \cdots & \sum_{i=1}^{N} \zeta_i(k) t_i^{n-1}(\cdot) & \sum_{i=1}^{N} \zeta_i(k) t_i^n(\cdot) \end{bmatrix}, \tag{41}$$

where

$$\begin{aligned} t_i^j(\cdot) &= |\gamma_i^j(\cdot)| + |\delta_i^j(\cdot)|, \quad j = 1, \ldots, n-1, \\ t_i^n(\cdot) &= |\gamma_i^n(\cdot)| + |\delta_i^n(\cdot)|. \end{aligned} \tag{42}$$

After this formulation, we can deduce the following theorem for the stability for the discrete-time switched time-delay systems (29).

THEOREM 3 *The discrete-time switched nonlinear time-delay system* (29) *is globally asymptotically stable under arbitrary switching rule* (2) *if there exist* α_j $(j = 1, \ldots, n-1)$, $\alpha_j \neq \alpha_q$, $\forall j \neq q$, *such as follows:*

$$\text{(i)} \quad 1 - |\alpha_j| > 0 \quad \forall j = 1, \ldots, n-1 \tag{43}$$

$$\text{(ii)} \quad 1 - (\bar{t}^n(\cdot)) - \sum_{j=1}^{n-1} (\bar{t}^j(\cdot)) |\beta_j| (1 - |\alpha_j|)^{-1} > 0, \tag{44}$$

where

$$\begin{aligned} \bar{t}^n(\cdot) &= \max_{1 \le i \le N} (t_i^n(\cdot)), \\ \bar{t}^j(\cdot) &= \max_{1 \le i \le N} (t_i^j(\cdot)), \quad j = 1, \ldots, n-1. \end{aligned} \tag{45}$$

Proof It suffices to verify that the matrix $(I_n - T_c(\cdot))$ is an $M -$ matrix:

$$T_c(\cdot) = \begin{bmatrix} |\alpha_1| & 0 & \cdots & 0 & |\beta_1| \\ 0 & \ddots & \ddots & \vdots & \vdots \\ \vdots & \ddots & \ddots & 0 & \vdots \\ 0 & \cdots & 0 & |\alpha_{n-1}| & |\beta_{n-1}| \\ \bar{t}^1(\cdot) & \cdots & \cdots & \bar{t}^{n-1}(\cdot) & \bar{t}^n(\cdot) \end{bmatrix}. \tag{46}$$

Since the elements α_j $(j = 1, \ldots, n-1)$ can be arbitrarily selected, the choice $|\alpha_j| \in]0, 1[$ with $\alpha_j \neq \alpha_q$, $\forall j \neq q$ then, the matrix $T_c(\cdot)$ with all elements positive.

Thus, the conditions of Theorem 3 can be deduced from the Kotelyanski conditions in the discrete case applied to the matrix $(I_n - T_c(\cdot))$.

In these conditions $(I_n - T_c(\cdot))$ is an $M -$ matrix, we determine sufficient stability conditions for the system (29). The $n - 1$ first conditions are checked because $|\alpha_j| \in]0, 1[$ $j = 1, \ldots, n-1$, however, the last condition yields to

$$\det(I_n - T_c(\cdot)) = \begin{vmatrix} 1 - |\alpha_1| & 0 & \cdots & 0 & -|\beta_1| \\ 0 & \ddots & \ddots & \vdots & \vdots \\ \vdots & \ddots & \ddots & 0 & \vdots \\ 0 & \cdots & 0 & 1 - |\alpha_{n-1}| & -|\beta_{n-1}| \\ -\bar{t}^1(\cdot) & \cdots & \cdots & -\bar{t}^{n-1}(\cdot) & (1 - \bar{t}^n(\cdot)) \end{vmatrix} > 0. \tag{47}$$

Then, we obtain: $1 - (\bar{t}^n(\cdot)) - \sum_{j=1}^{n-1} (\bar{t}^j(\cdot)) |\beta_j| (1 - |\alpha_j|)^{-1} > 0$. ■

To simplify the use of the stability conditions, Theorem 3 can be reduced to the following corollary.

COROLLARY 1 *If the system* (29) *is asymptotically stable under arbitrary switching rule* (2), *the following conditions are satisfied* $\forall \alpha_j \in]0\,1[$ $(j = 1, \ldots, n-1)$, $\alpha_j \neq \alpha_q$ $\forall j \neq q$ $\forall i = 1, \ldots, N$:

$$\text{(i)} \quad \beta_j (P_i(\cdot, \alpha_j) + Q_i(\cdot, \alpha_j)) < 0, \tag{48}$$

$$\text{(ii)} \quad (P_i(\cdot, \lambda = 1) + Q_i(\cdot, \lambda = 1)) > 0 \tag{49}$$

$$\text{(iii)} \quad (\gamma_i^n(\cdot) + \delta_i^n(\cdot)) > 0. \tag{50}$$

Proof It is easy to see that relation (44) gives us (Sfaihi and Benrejeb, 2013)

$$\max_{1 \le i \le N} (|(\gamma_i^n(\cdot) + \delta_i^n(\cdot))|) + \sum_{j=1}^{n-1} \max_{1 \le i \le N} (|(\gamma_i^j(\cdot) + \delta_i^j(\cdot))|) |\beta_j| (1 - |\alpha_j|)^{-1} < 1. \tag{51}$$

It implies that

$$1 - \max_{1 \le i \le N} (|(\gamma_i^n(\cdot) + \delta_i^n(\cdot))|) - \sum_{j=1}^{n-1} \max_{1 \le i \le N} (|(\gamma_i^j(\cdot) + \delta_i^j(\cdot))|) |\beta_j| (1 - |\alpha_j|)^{-1} > 0. \tag{52}$$

Moreover, it is clear to see that $\forall 1 \le i \le N$ and $\forall 1 \le j \le n-1$:

$$1 - (|(\gamma_i^n(\cdot) + \delta_i^n(\cdot))|) - \sum_{j=1}^{n-1} (|(\gamma_i^j(\cdot) + \delta_i^j(\cdot))|) |\beta_j| (1 - |\alpha_j|)^{-1} > 0 \tag{53}$$

is more restrictive than relation (52).

On the other hand, to prove the Corollary 1, it is simple to consider conditions (48) and (50), in order to find Equation (49). For this, substitute relations (38) and (39) in Equation (53).

Then, relation (53) becomes

$$1-(\gamma_i^n(\cdot)+\delta_i^n(\cdot)) - \sum_{j=1}^{n-1}(\gamma_i^j(\cdot)+\delta_i^j(\cdot))\beta_j(1-|\alpha_j|)^{-1}>0. \tag{54}$$

By Equations (38) and (39), we have that

$$1+(a_i^1(\cdot)+d_i^1(\cdot))+\sum_{j=1}^{n-1}\alpha_j + \sum_{j=1}^{n-1}\left(\frac{1}{(1-\alpha_j)}\left(\frac{(\lambda-\alpha_j)(P_i+Q_i)}{F(\lambda)}\right)\right)_{\lambda=\alpha_j}>0, \tag{55}$$

where

$$F(\lambda)=\prod_{j=1}^{n-1}(\lambda-\alpha_j). \tag{56}$$

Therefore, to deduce the stability conditions of the switched system given in Equation (29), let us first observe that

$$\frac{(P_i(\cdot,\lambda)+Q_i(\cdot,\lambda))}{F(\lambda)} = \lambda+(a_i^1(\cdot)+d_i^1(\cdot))+\sum_{j=1}^{n-1}\alpha_j + \sum_{j=1}^{n-1}\left(\left(\frac{(\lambda-\alpha_j)(P_i(\cdot,\lambda)+Q_i(\cdot,\lambda))}{(1-\alpha_j)F(\lambda)}\right)\right)_{\lambda=\alpha_j}. \tag{57}$$

Then, following that the developed stability condition (54) is equivalent to

$$\left(\frac{(P_i(\cdot,\lambda)+Q_i(\cdot,\lambda))}{F(\lambda)}\right)_{\lambda=1}>0, \tag{58}$$

where

$$F(\lambda=1)=\prod_{j=1}^{n-1}(1-\alpha_j)>0 \quad \forall\alpha_j\in]0\quad 1[,$$

that is: $(P_i(\cdot,\lambda=1)+Q_i(\cdot,\lambda=1))>0$.

Then, asymptotical stability of the system (29) is under switching law (2).

This completes the proof. ■

5. Illustrative example

In this section, we give a numerical example to illustrate the performance of the proposed approach.

Consider the discrete-time switched nonlinear time-delay system described by the following switched difference equation:

$$y(k+2)+\sum_{i=1}^{2}\zeta_i(k)\left(\sum_{j=0}^{1}a(\cdot)_i^{2-j}y(k+j)+\sum_{j=0}^{1}d(\cdot)_i^{2-j}y(k+j-\tau)\right)=0.$$

For the time-delay τ, according to Equations (27)–(31), we can obtain the following state representation:

$$x(k+1)=\sum_{i=1}^{2}\zeta_i\left(\begin{pmatrix}0 & 1\\ -a(\cdot)_i^2 & -a(\cdot)_i^1\end{pmatrix}x(k) + \begin{pmatrix}0 & 0\\ -d(\cdot)_i^2 & -d(\cdot)_i^1\end{pmatrix}x(k-\tau)\right),$$

$$x(k+1)=\sum_{i=1}^{2}\zeta_i(k)(A_i(\cdot)x(k)+D_i(\cdot)x(k-\tau)),$$

$$x(l)=\phi(l),\quad l=-\tau,\ldots,-1,0,$$

where the matrices are defined as follows:

$$A_1(\cdot)=\begin{bmatrix}0 & 1\\ -0.5+0.8f(\cdot) & 1-0.4f(\cdot)\end{bmatrix} \text{ and } A_2(\cdot)=\begin{bmatrix}0 & 1\\ -0.6+0.7f(\cdot) & 1.1-0.8f(\cdot)\end{bmatrix},$$

$$D_1(\cdot)=\begin{bmatrix}0 & 0\\ -0.4+0.25\Phi(\cdot) & 0.8-0.6\Phi(\cdot)\end{bmatrix} \text{ and } D_2(\cdot)=\begin{bmatrix}0 & 0\\ -0.2+0.4\Phi(\cdot) & 0.8-0.5\Phi(\cdot)\end{bmatrix},$$

where $f(\cdot)$ and $\Phi(\cdot)$ are unknown nonlinear functions.

Then, according to Equations (34)–(39), a change of base under the arrow form gives us the following matrices:

$$M_1(\cdot)=\begin{bmatrix}\alpha & 1\\ \gamma_1^1(\cdot) & \gamma_1^2(\cdot)\end{bmatrix} \text{ and } M_2(\cdot)=\begin{bmatrix}\alpha & 1\\ \gamma_2^1(\cdot) & \gamma_2^2(\cdot)\end{bmatrix},$$

$$N_1=\begin{bmatrix}0 & 0\\ \delta_1^1(\cdot) & \delta_1^2(\cdot)\end{bmatrix} \text{ and } N_2=\begin{bmatrix}0 & 0\\ \delta_2^1(\cdot) & \delta_2^2(\cdot)\end{bmatrix},$$

where

$$\gamma_1^1(\cdot) = -P_1(\alpha) = -[\alpha^2 + (1 - 0.4f(\cdot))\alpha - 0.5 + 0.8f(\cdot)],$$
$$\gamma_1^2(\cdot) = -(-1 + 0.4f(\cdot) + \alpha),$$
$$\gamma_2^1(\cdot) = -P_2(\alpha) = -[\alpha^2 + (1.1 - 0.8f(\cdot))\alpha - 0.6 + 0.7f(\cdot)],$$
$$\gamma_2^2(\cdot) = -(-1.1 + 0.8f(\cdot) + \alpha),$$

and

$$\delta_1^1(\cdot) = -Q_1(\alpha) = -[(0.8 - 0.6\Phi(\cdot))\alpha - 0.4 + 0.25\Phi(\cdot)],$$
$$\delta_1^2(\cdot) = -(-0.8 + 0.6\Phi(\cdot)),$$
$$\delta_2^1(\cdot) = -Q_{12}(\alpha) = -[(0.8 - 0.5\Phi(\cdot))\alpha - 0.2 + 0.4\Phi(\cdot)],$$
$$\delta_2^2(\cdot) = -(0.8 - 0.5\Phi(\cdot)).$$

Then, the stability conditions for the example given by the Corollary 1 are the following:

(i) $0 < \alpha < 1$,
(ii) $(P_1(\alpha) + Q_1(\alpha)) < 0$,
(iii) $(P_2(\alpha) + Q_2(\alpha)) < 0$,
(iv) $(P_1(1) + Q_1(1)) > 0$,
(v) $(P_2(1) + Q_2(1)) > 0$,
(vi) $(\gamma_1^2(\cdot) + \delta_1^2) > 0$, and
(vii) $(\gamma_2^2(\cdot) + \delta_2^2) > 0$.

Therefore, in case $\alpha = 0.2$, $\beta = 1$, conditions (ii), (iii), (iv), (v), (vi) and (vii) allow for deducing the following stability conditions:

(i) $f(\cdot) > 0.8 - 0.18\Phi(\cdot)$,
(ii) $f(\cdot) > 0.85 - 0.55\Phi(\cdot)$,
(iii) $f(\cdot) < 0.25 + 0.875\Phi(\cdot)$,
(iv) $f(\cdot) > 1 - \Phi(\cdot)$,
(v) $f(\cdot) < 4 - 1.5\Phi(\cdot)$, and
(vi) $f(\cdot) < 2.125 - 0.625\Phi(\cdot)$.

In the following, we determine the stability domain for the chosen α. Figure 1 illustrates the stability domain given by the nonlinear $f(\cdot)$ relative to the nonlinear $\Phi(\cdot)$.

Then, according to the stability domain shown in Figure 1, for particular values of the nonlinearities $f(\cdot) = 1$ and $\Phi(\cdot) = 1.5$, by choosing the sampling time $T_e = 0.2\,s$, $t_f = kT_e = 10\,s$ the switched time $t_1 = k_1 T_e = 5\,s$, the simulation results are shown in Figures 2 and 3, where $\phi(l) = [-1 \quad 2]^T$ for all $l = -5, -4, -3, -2, -1, 0$. Figure 2 shows the state responses; the state trajectories are depicted in Figure 3, which shows the stability of the system given in the example. Then, for the same values

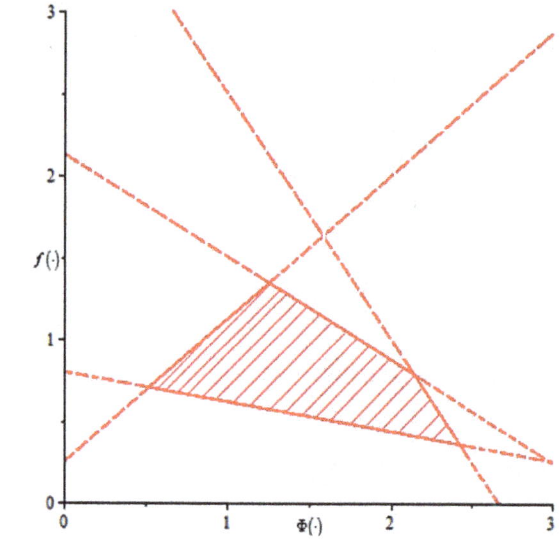

Figure 1. Stability domain given of the system illustrate in example obtained from Corollary 1.

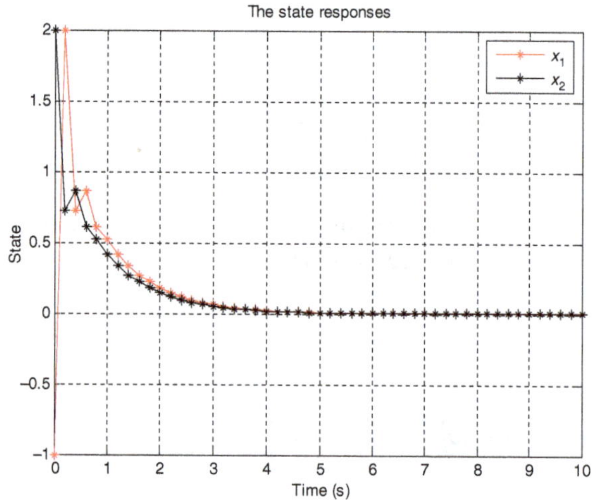

Figure 2. The state responses of the system given in example for $t_1 = k_1 T_e = 5\,s$.

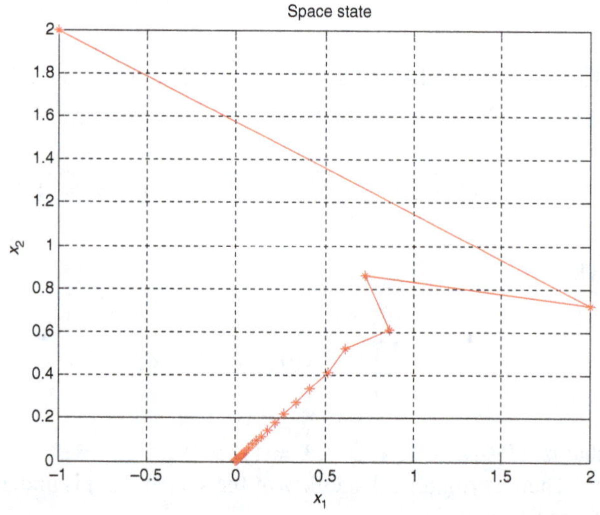

Figure 3. Trajectory response of the system given in example for $t_1 = k_1 T_e$ =5 s.

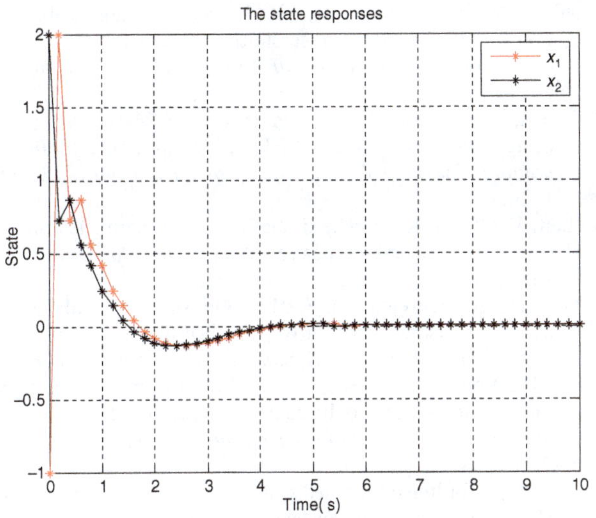

Figure 4. The state responses of the system given in example for $t_1 = k_1 T_e = 0.4\,s$.

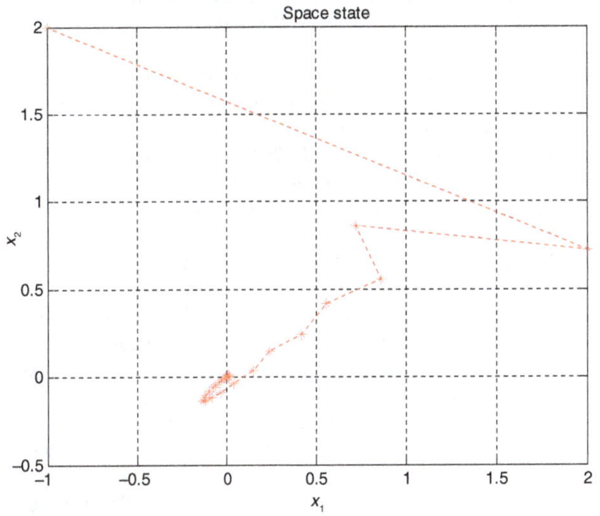

Figure 5. Trajectory response of the system given in example for $t_1 = k_1 T_e = 0.4\,s$.

of the nonlinearities $f(\cdot) = 1$ and $\Phi(\cdot) = 1.5$ and the sampling time $T_e = 0.2\ s$. The switched time $t_1 = k_1 T_e = 0.4\,s$ and $\phi(l) = [-1 \quad 2]^{\mathrm{T}}$ for all $l = -1, 0$ the evolution of the states and the state space are shown in Figures 4 and 5, respectively.

This example has been treated to show that these stability conditions are sufficient and very close to be necessary. On the other hand, it is uncertain to treat an example for which there exists a common Lyapunov function, in order to do a comparison with our approach.

6. Conclusions

In this paper, we have developed a new method for the stability analysis under arbitrary switching for a class of discrete-time switched nonlinear time-delay systems. These stability conditions were derived from an appropriate Lyapunov function associated with the application of the Kotelyanski stability conditions. The main benefit of this technique is that it is easy to use and it gives us an explicitly stability condition that guaranteed the stability under arbitrary switching law. Moreover, it can avoid the problem of existence of Lyapunov functions which are usually very difficult to apply, or even not possible. Simulation results have been presented to illustrate the effectiveness of the developed method.

The limits of this paper are that it has been confined to the boundaries of numerical examples. It would be beneficial to extend the research further so as to include real systems.

References

Arunkumar, A., Sakthivel, R., Mathiyalagan, K., & Marshal Anthoni, S. (2012). Robust stability criteria for discrete-time switched neural networks with various activation functions. *Applied Mathematics and Computation, 218*, 10803–10816.

Arunkumar, A., Sakthivel, R., Mathiyalagan, K., & Marshal Anthoni, S. (2013). State estimation for switched discrete-time stochastic BAM neural networks with time varying delay. *Nonlinear Dynamics, 73*, 1565–1585.

Benrejeb, M., & Borne, P. (1978). *On an algebraic stability criterion for non-linear processes. Interpretation in the frequency domain*. Proceedings of the measurement and control international symposium MECO'78, Eugenidis Foundation, Athens, Greece, pp. 678–682.

Benrejeb, M., Borne, P., & Laurent, F. (1982). Sur une application de la représentation en flèche à l'analyse des processus. *RAIRO Automatique – Systems analysis and control, 16*, 133–146.

Benrejeb, M., & Gasmi, M. (2001). On the use of an arrow form matrix for modelling and stability analysis of singularly perturbed nonlinear systems. *System Analysis Modelling Simulation, 40*, 509–525.

Benrejeb, M., Gasmi, M., & Borne, P. (2005). New stability conditions for TS fuzzy continuous nonlinear models. *Nonlinear Dynamics System Theory, 5*, 369–379.

Benrejeb, M., Soudani, D., Sakly, A., & Borne, P. (2006). New discrete Tanaka domain Sugeno Kang fuzzy systems characterization and stability domain. *International Journal of Computers, Communications and Control, I*, 9–19.

Borne, P. (1987). Nonlinear system stability: Vector norm approach. *System and Control Encyclopedia, t.5*, 3402–3406.

Borne, P., Gentina, J. C., & Laurent, F. (1972). Sur la stabilité des systèmes Échantillonnés non linéaires. *Revue française d'automatique, informatique et de recherche opérationnelle, 2*, 96–105.

Borne, P., Vanheeghe, P., & Duflos, E. (2007). *Automatisation des processus dans l'espace d'état*. Paris: Ed. Technip.

Branicky, M. S. (1998). Multiple Lyapunov functions and other analysis tools for switched and hybrid systems. *IEEE Transactions on Automatic Control, 43*(4), 475–482.

Darouach, M., & Chadli, M. (2013). Admissibility and control of switched discrete-time singular systems. *Systems Science and Control Engineering: An Open Access Journal, 1*(1), 43–51.

Dong, Y., Liu, J., Mei, S., & Li, M. (2011). Stabilization for switched nonlinear time-delay systems. *Nonlinear Analysis: Hybrid Systems, 5*, 78–88.

Elmadssia, S., Saadaoui, K., & Benrejeb, M. (2011). *New delay-dependent stability conditions for linear systems with delay*.

International conference on communications, computing and control applications (CCCA), Hammamet, Tunisia, pp. 1–6.

Elmadssia, S., Saadaoui, K., & Benrejeb, M. (2013). New delay-dependent stability conditions for linear systems with delay. *Systems Science & Control Engineering, 1*, 37–41.

Filali, R. L., Hammami, S., Benrejeb, M., & Borne, P. (2012). On synchronization, anti-synchronization and hybrid synchronization of 3D discrete generalized H12 non map. *Nonlinear Dynamics System Theory, 12*, 81–96.

Gantmacher, F. R. (1966). *Théorie des matrices*. T.1 et 2. Paris: Dunod.

Gao, H., & Chen, T. (2007). New results on stability of discrete-time systems with time-varying state delay. *IEEE Transactions on Automatic Control, 52*(2), 328–334.

Gao, H., Lam, J., & Wang, Z. (2007). Discrete bilinear stochastic systems with time-varying delay: Stability analysis and control synthesis. *Chaos, Solitons Fractals, 34*(2), 394–404.

Hespanha, J. P., & Morse, A. S. (1999). *Stability of switched systems with average dwell-time*. 38th IEEE conference decision control, Arizona, USA, pp. 2655–2660.

Ishii, H., & Francis, B. A. (2002). Stabilizing a linear system by switching control with dwell time. *IEEE Transactions on Automatic Control, 47*, 1962–1973.

Jabbali, A., Kermani, M., & Sakly, A. (2013). A new stability analysis for discrete-time switched time-delay systems. *Special Issue-international conference on control, engineering & information technology (CEIT'13), proceedings engineering & technology (PET), 1*, 200–204.

Jianyin, F., & Kai, L. (2010). *Stabilization of nonlinear systems under arbitrary switching law*. Proceedings of the 30th Chinese control conference, Yantai, China, pp. 641–643.

Kermani, M., Sakly, A., & M'sahli, F. (2012). A new stability analysis and stabilization of discrete-time switched linear systems using vector norms approach. *World Academy of Science, Engineering and Technology, 71*, 1302–1307.

Kotelyanski, D. M. (1952). Some properties of matrices with positive elements. *Matematicheskii Sbornik [in Russian], 31*, 497–505.

Liberzon, D. (2003). *Switching in systems and control*. Boston, MA: Birkhäuser.

Liberzon, D., & Morse, A. S. (1999). Basic problems in stability and design of switched systems. *IEEE Control Systems Magazine, 19*, 57–70.

Lien, C. H., Yu, K. W., Chang, H. C., Chung, L. Y., & Chen, J. D. (2012). Switching signal design for exponential stability of discrete switched systems with interval time varying delay. *Journal of the Franklin Institute, 349*, 2182–2192.

Lien, C. H., Yu, K. W., Chung, Y. J., Chang, H. C., Chung, L. Y., & Chen, J. D. (2011). Switching signal design for global exponential stability of uncertain switched nonlinear systems with time-varying delay. *Nonlinear Analysis: Hybrid Systems, 5*, 10–19.

Mahmoud, M. S. (2010). *Switched time-delay systems*. Boston, MA: Springer-Verlag.

Mathiyalagan, K., Sakthivel, R., Marshal, & Anthoni, S. (2012). New robust exponential stability results for discrete-time switched fuzzy neural networks with time delays. *Computers and Mathematics with Applications, 64*, 2926–2938.

Phat, V. N., & Ratchagit, K. (2011). Stability and stabilization of switched linear discrete-time systems with interval time-varying delay. *Nonlinear Analysis: Hybrid Systems, 5*, 605–612.

Pellanda, P., Apkarian, P., & Tuan, H. (2002). Missile autopilot design via a multi-channel LFT/LPV control method. *International Journal of Robust and Nonlinear Control, 12*(1), 1–20.

Robert, F. (1966). Recherche d'une M-matrice parmi les minorantes d'un opérateur liéaire. *Numerische Mathematik, 9*, 188–199.

Sfaihi, B., & Benrejeb, M. (2013). On stability analysis of nonlinear discrete singularly perturbed T-S fuzzy models. *International Journal of Dynamics and Control, 1*, 20–31.

Shim, H., Noh, D. J., & Seo, J. H. (2001). Common Lyapunov function for exponentially stable nonlinear systems. *Journal of Korean Institute of Electrical Engineers, 11*, 108–111.

Sun, X. M., Wang, W., Liu, G. P., & Zhao, J. (2008). Stability analysis for linear switched systems with time-varying delay. *IEEE Transactions on Systems, Man, and Cybernetics, Part B, 38*, 528–533.

Sun, Z., & Ge, S. S. (2011). *Stability theory of switched dynamical systems*. London: Springer.

Tse, C. K., & Bernardo, M. Di., (2002). Complex behavior in switching power converters. *Proceedings of the IEEE, 90*(5), 768–781.

Vu, L., & Liberzon, D. (2005). Common Lyapunov functions for families of commuting nonlinear systems. *Systems and Control Letters, 54*, 405–416.

Wu, Z., Shi, P., Su, H., & Chu, J. (2011). Delay-dependent stability analysis for switched neural networks with time-varying delay. *IEEE Transactions on Systems, Man, and Cybernetics, Part B, 41*(6), 1522–1530.

Xu, S. (2002). Robust H1 filtering for a class of discrete-time uncertain nonlinear systems with state delay. *IEEE Transactions on Circuits and Systems* (I), *49*, 1853–1859.

Zhai, G., Hu, B., Yasuda, K., & Michel, A. N. (2000). *Stability analysis of switched systems with stable and unstable subsystems: An average dwell time approach*. American control conference, Chicago, USA, vol. 1, pp. 200–204.

Zhang, H., Liu, Z., & Huang, G. (2010). Novel delay-dependent robust stability analysis for switched neutral-type neural networks with time-varying delays via sc technique. *IEEE Transactions on Systems, Man, and Cybernetics, Part B, 40*(6), 1480–1491.

Zhang, L., Shi, P., & Basin, M. (2008). Robust stability and stabilization of uncertain switched linear discrete time-delay systems. *IET Proceedings Control Theory and Applications, 2*, 606–614.

Zhang, L. X., & Gao, H. J. (2010). Asynchronously switched control of switched linear systems with average dwell time. *Automatica, 46*(5), 953–958.

Zhang, W., Abate, A., Hu, J., & Vitus, M. P. (2009). Exponential stabilization of discrete-time switched linear systems. *Automatica, 45*, 2526–2536.

Zhang, W. A., & Yu, L. (2009). Stability analysis for discrete-time switched time-delay systems. *Automatica, 45*, 2265–2271.

13

Analysis and circuit design of a fractional-order Lorenz system with different fractional orders

H.Y. Jia[a]*, Q. Tao[a] and Z.Q. Chen[b]

[a]*Department of Automation, Tianjin University of Science and Technology, Tianjin 300222, People's Republic of China;* [b]*Department of Automation, Nankai University, Tianjin 300071, People's Republic of China*

The paper first discusses the recently reported fractional-order Lorenz system, analyzes it by using the frequency-domain approximation method and the time-domain approximation method, and finds its chaotic dynamics when the order of the fractional-order system varies from 2.8 to 2.9 in steps of 0.1. Especially for the fractional-order Lorenz system of the order as low as 2.9, the results obtained by the frequency-domain method are consistent with those obtained by the time-domain method. Some Lyapunov exponent diagrams, bifurcation diagrams, and phase orbits diagrams have also been shown to verify the chaotic dynamics of the fractional-order Lorenz system. Then, an analog circuit for the fractional-order Lorenz system of the order as low as 2.9 is designed to confirm its chaotic dynamic, the results from circuit experiment show that it is chaotic.

Keywords: fractional-order; bifurcation; Lyapunov exponents; analog circuit

1. Introduction

The research for the theory of the fractional integrals and derivatives has begun to attract more and more attention since 1960. Generally, chaotic cannot be found in continuous autonomous systems whose order is less than three. However, with the development of fractional calculus, some fractional-order systems whose dimensions are less than three are reported to display chaotic motion recently, such as the fractional Chua's circuit (Hartley, Lorenzo, & Qammer, 1995), the fractional Rössler system (Li & Chen, 2004), the fractional Lorenz system (Grigorenko & Grigorenko, 2003), the fractional Chen system (Lu & Chen, 2006), the fractional-modified Duffing system (Ge & Qu, 2007), and so on (Ahmad & Sprott, 2003; Chen, Liu, Wang, & Li, 2008; Huang, Zhao, Wang, & Li, 2012; Li, Liao, & Lou, 2012; Lu, 2006; Pan, Zhou, Zhou, & Sun, 2011; Ahmad, 2005; Wang, Huang, & Zhao, 2012). All these works may bring new motivation for the research on the theory and application of chaos.

However, most of the above researches only display chaotic dynamics of some fractional-order systems by phase portraits, Lyapunov exponents for fixing system parameters, bifurcations, and Poincaré section. Compared with the integer-order chaotic systems, these seem insufficient for showing chaotic characteristics of the fractional-order systems. Therefore, the paper studies the fractional-order Lorenz system, gives its Lyapunov exponents diagram which changes with system parameter, and designs an analog circuit which displays its chaotic behavior from physical implementation. Our work may help to display the chaotic characteristic of the fractional-order Lorenz system from different points of view.

Grigorenko and Grigorenko (2003) reported the fractional-order Lorenz system, and gave a conclusion that chaotic dynamic cannot be found when system dimension of the fractional-order Lorenz is less than or equal to 2.91. Subsequently, Li and Chen (2004) reported some mistakes. In 2009, Yu discussed Hopf bifurcation behavior and showed the phase orbits when system dimension is 2.96 (Yu, Li, Wang, & Yu, 2009). Recently, Jia, Chen, and Xue (2013) found chaos existed in the lower dimensional fractional-order Lorenz system. However, they only discuss its chaotic dynamics when the fractional orders α, β, and γ are same. Therefore, the paper will further discuss its chaotic dynamics when the fractional orders α, β, and γ are different.

2. Fractional-order Lorenz system

Grigorenko and Grigorenko (2003) reported the fractional-order Lorenz system

$$\begin{aligned} \frac{d^{\alpha}x}{dt^{\alpha}} &= a(y-x), \\ \frac{d^{\beta}y}{dt^{\beta}} &= cx - xz + y, \\ \frac{d^{\gamma}z}{dt^{\gamma}} &= xy - bz, \end{aligned} \qquad (1)$$

*Corresponding author. Email: jiahy@tust.edu.cn

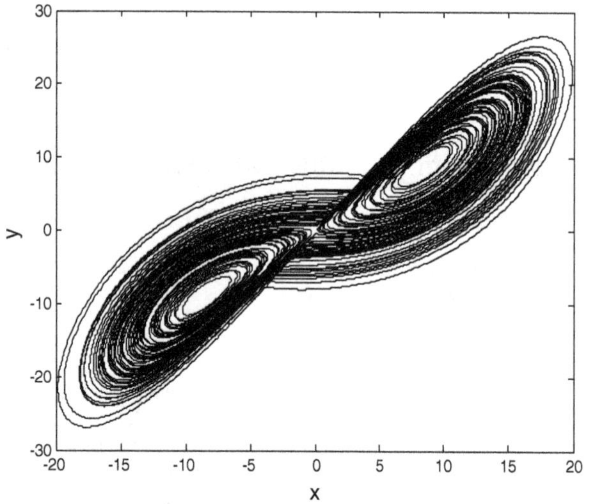

Figure 1. Chaotic attractor of the fractional-order Lorenz system based on the frequency-domain approximation method.

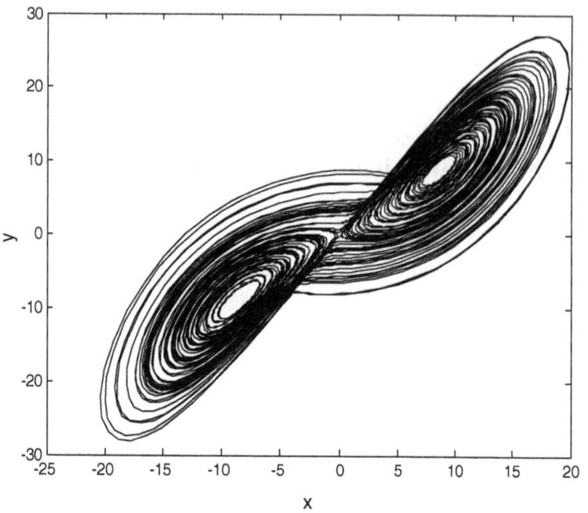

Figure 2. Chaotic attractor of the fractional-order Lorenz system based on the time-domain approximation method.

where a, b, and c is system parameters, α, β, and γ is the fractional order.

Generally, computing a fractional differential equation is complex and difficult work, therefore, approximation methods are often adopted to analyze some fractional-order systems numerically such as the time-domain approximation methods and the frequency-domain The former are more reliable to show chaos, the latter is more convenient for circuit implementation. In order to display its chaotic characteristics from different point of view, the two approximation methods will be used to display the chaotic dynamic of the fractional-order Lorenz in this paper.

Let $a = 10, b = \frac{8}{3}, c = 28, \alpha = 0.9, \beta = 1$, and $\gamma = 1$, based on the frequency-domain approximation method, using approximation function (Ahmad & Sprott, 2003)

$$\frac{1}{s^{0.9}} = \frac{2.2675(s+1.292)(s+215.4)}{(s+0.01292)(s+2.154)(s+359.4)}, \quad (2)$$

A chaotic attractor can be observed, as shown in Figure 1.

Under the same condition, based on the time-domain approximation method, using the advised Adams–Bashforth–Moulton method, a same chaotic attractor can also be found, as shown in Figure 2. That is, the chaotic dynamic indeed exists in the fractional-order Lorenz system of the order as low as 2.9.

2.1. *Fractional-order Lorenz system of the order as low as 2.9*

In order to show the chaotic behaviors of the fractional-order Lorenz system of the order as low as 2.9, when fixing $b = \frac{8}{3}$, $c = 28$, $\alpha = 0.9$, $\beta = 1$, $\gamma = 1$ and varying a, by using the both frequency-domain method and time-domain method, the paper computes Lyapunov exponents diagrams and bifurcation diagrams, as shown in Figure 3. Furthermore, the paper analyses Lyapunov exponents diagram and bifurcation diagrams for the fractional-order Lorenz system, finds the results obtained by the frequency-domain method are consistent with those obtained by the time-domain method. That is, using the frequency-domain method, the paper gives Lyapunov exponents diagram and bifurcation diagram, as shown in Figure 3(a) and 3(b), respectively. Generally speaking, when the system displays chaotic dynamics, its biggest Lyapunov exponent is larger than zero, the points in the corresponding bifurcation diagram are dense. When the system does not display chaotic dynamics, its biggest Lyapunov exponent is equal to or less than zero, the points in the corresponding bifurcation diagram are sparse or few. That is, Figure 3(a) and 3(b) is consistent. Furthermore, when using the time-domain method to analyze the system, we only need to give bifurcation diagram. That is, it is enough for showing the consistence between the frequency-domain method and time-domain method to analyze bifurcation diagram, therefore, the paper only computes bifurcation diagram by using the time-domain method, as shown in Figure 3(c), which is same as Figure 3(b). By comparing Figure 3(b) and 3(c), a conclusion can be obtained that the chaotic behaviors exist in the fractional-order Lorenz system of the order as low as 2.9, and the results are consistent when using the frequency-domain method and time-domain method.

By analyzing Lyapunov exponents diagram and bifurcation diagrams in Figure 3, the following results can be obtained for the fractional-order Lorenz system, when $b = \frac{8}{3}$, $c = 28$, $\alpha = 0.9$, $\beta = 1$, $\gamma = 1$ and varying a, the fractional system (1) can display the periodic attractors, and asymptotically stable orbits besides the chaotic attractors. Now let $a = 3$, using the time-domain method and frequency-domain method, the fractional-order Lorenz system will show periodic attractors, as shown in Figure 4(a) and 4(b), respectively. In addition, let $a = 1$, using the

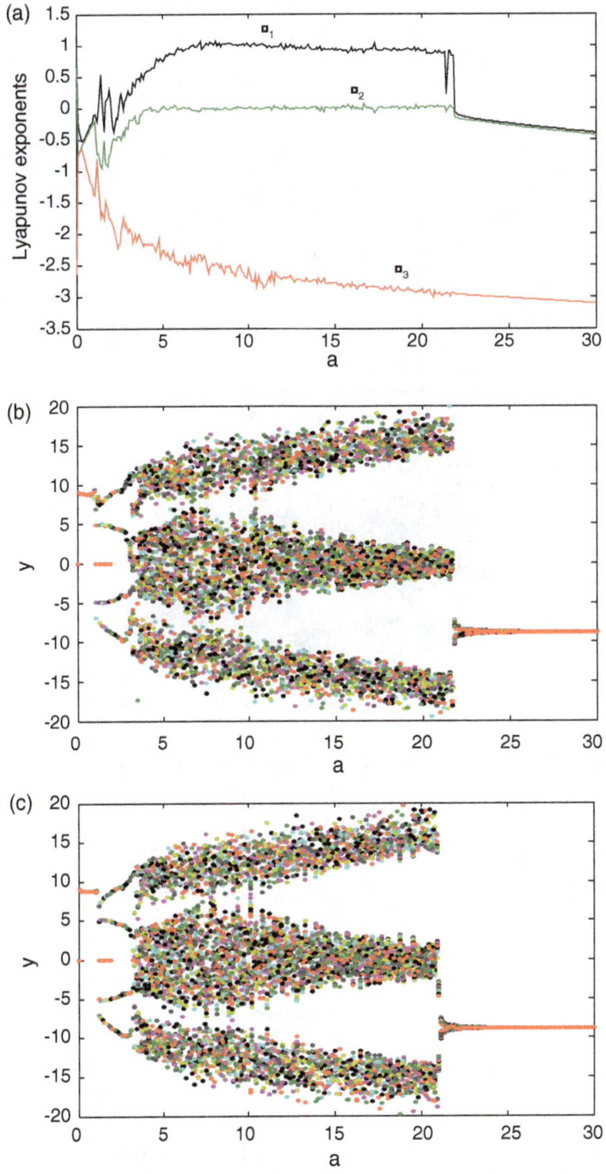

Figure 3. Lyapunov exponents diagram and bifurcations for the 2.9-order Lorenz system. (a) Lyapunov exponents diagram based on the frequency-domain method. (b) Bifurcation diagram based on the frequency-domain method. (c) Bifurcation diagram based on the time-domain method.

time-domain method and frequency-domain method, the fractional-order Lorenz system will show asymptotically stable orbits, as shown in Figure 5(a) and 5(b), respectively. That is, phase portraits are consistent when using two different approximation methods, and verify the analysis from Lyapunov exponents diagram and bifurcation diagram.

2.2. *Fractional-order Lorenz system of the order as low as 2.8*

When fixing $b = \frac{8}{3}$, $c = 28$, $\alpha = 0.9$, $\beta = 0.9$, $\gamma = 1$ and varying a, using the frequency-domain method, the paper

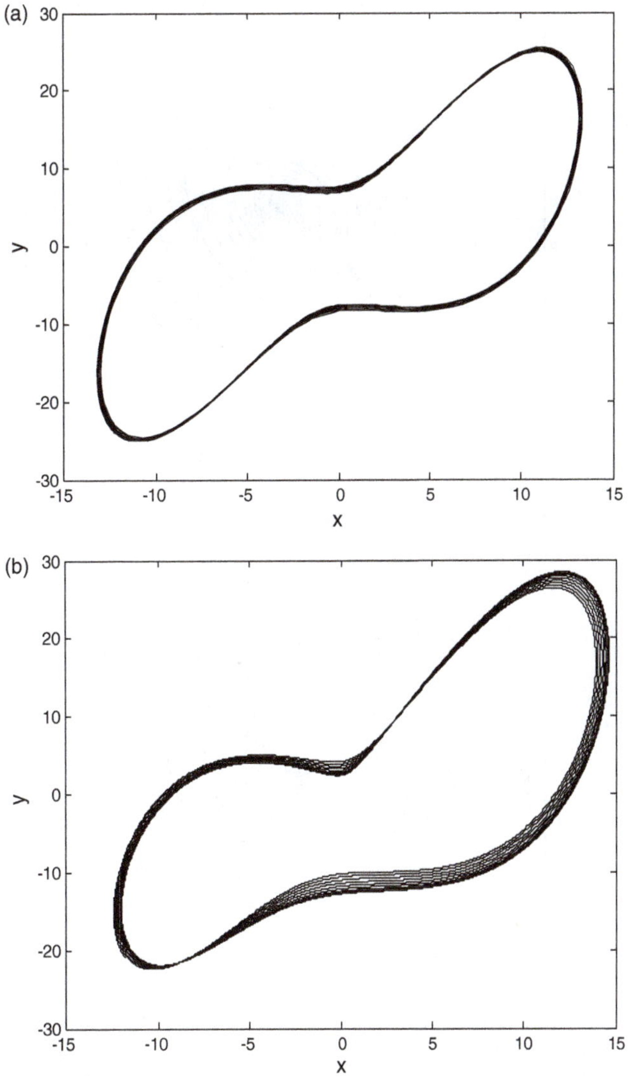

Figure 4. Periodic attractors of the fractional-order Lorenz system. (a) Periodic attractor based on the time-domain method. (b) Periodic attractor based on the frequency-domain method.

also gives Lyapunov exponents diagram and bifurcation diagram, as shown in Figure 6(a) and 6(b), respectively, and the Lyapunov exponents diagram is consistent with bifurcation diagram. By analyzing Figure 6(a) and 6(b), a conclusion can be obtained that the chaotic dynamics exit in the fractional-order Lorenz system of the order as low as 2.8 when using the frequency-domain method.

By comparing Lyapunov exponents diagram in Figure 3(a) with that in Figure 6(a), the following results can be obtained, when fixing $b = \frac{8}{3}$, $c = 28$ and $a \in [2.5, 21.8]$, the 2.9-order Lorenz system displays chaotic dynamics; when fixing $b = \frac{8}{3}$, $c = 28$, and $a \in [2.5, 17.3]$, the 2.8-order Lorenz system also displays chaotic dynamics. That is, under the same condition, the chaotic range of the 2.9-order Lorenz system may be larger.

In addition, under the same condition, using the time-domain approximation methods, the paper also obtains

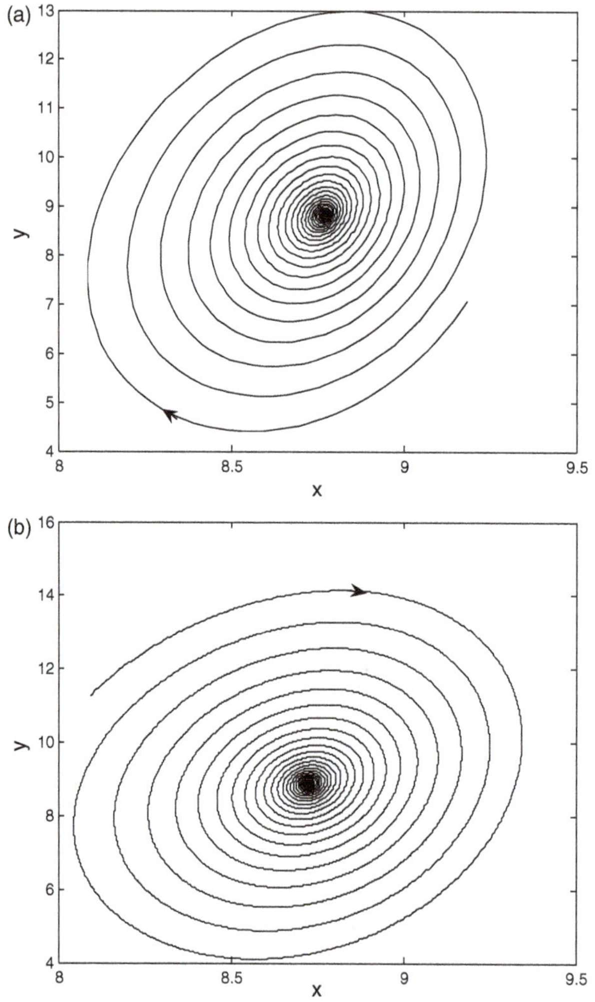

Figure 5. Asymptotically stable orbits of the fractional-order Lorenz system. (a) Asymptotically stable orbit based on the time-domain method. (b) Asymptotically stable orbit based on the frequency-domain method.

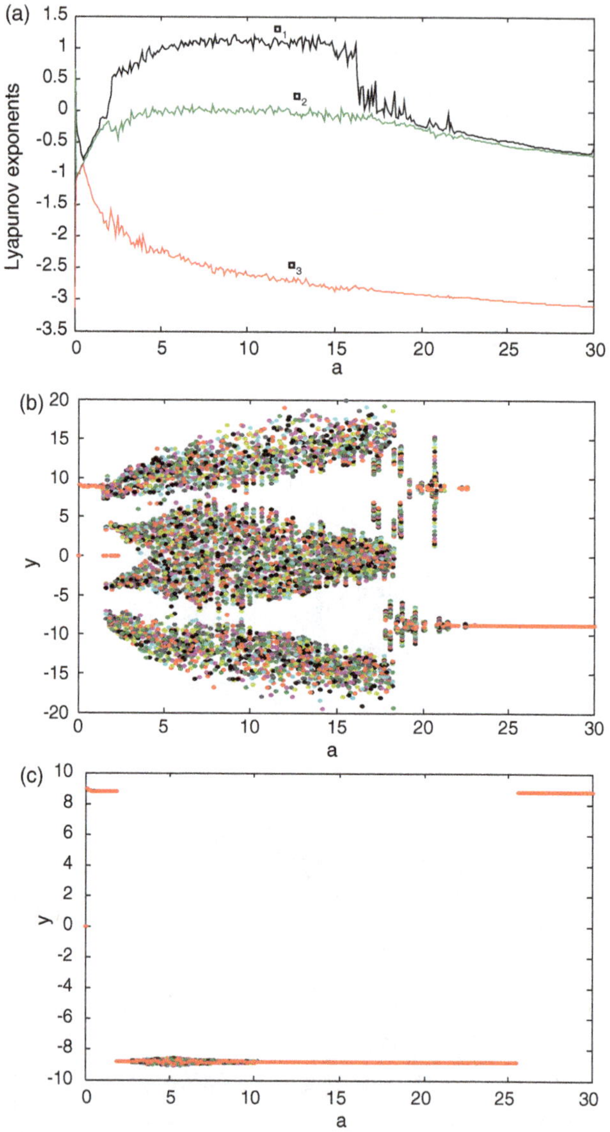

Figure 6. Lyapunov exponents diagram and bifurcation for 2.8-order Lorenz system. (a) Lyapunov exponents diagram based on the frequency-domain method. (b) Bifurcation diagram based on the frequency-domain method. (c) Bifurcation diagram based on the time-domain method.

another bifurcation diagram, as shown in Figure 6(c). By analyzing Figure 6(c), the chaos is not found for the fractional-order Lorenz system of the order as low as 2.8. That is, a conclusion can be obtained that the chaotic dynamics do not exist in the fractional-order Lorenz system of the order as low as 2.8 when using the time-domain method. Furthermore, when using the frequency-domain method and time-domain method, the 2.8-order Lorenz system displays different dynamics. The former shows that the system is chaotic, and the latter shows that the system is not chaotic. Which result is on earth true? This will need further study in the future.

3. Circuit design for the fractional-order Lorenz system

Although the fractional-order Lorenz system has been proved by the Lyapunov exponents, the numerical simulations and bifurcation diagrams, these seem insufficient to show complex dynamics of the fractional-order Lorenz system. Therefore, the paper also designs an analog circuit to implement the system by using resistors, capacitors, analog multipliers AD633, and analog operational amplifiers LF347N, as shown in Figure 7. That is, the dynamics of the system are confirmed by physical implementation.

In the circuit, three state variables x, y, and z are implemented by three channels, respectively, the operations of addition, subtraction, and integration in every channel are performed by the operational amplifiers LF347Ns, and the nonlinear terms of system (1) are implemented by the analog multiplier AD633. In addition, based on fractional frequency-domain approximation (Ahmad & Sprott, 2003)

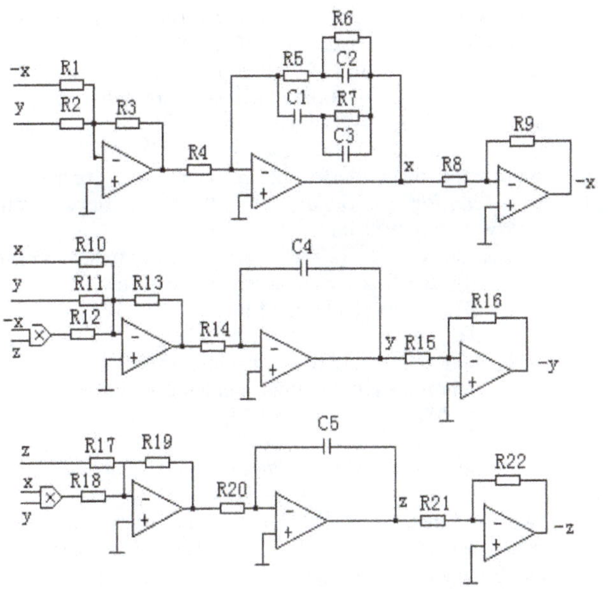

Figure 7. Analog circuit of the fractional-order Lorenz system.

and tree circuit approximation (Chen et al., 2008), an integration circuit for the fractional order $\alpha = 0.9$ is designed and used in the first channel of Figure 7. and thus an analog circuit for the fractional-order Lorenz system of the order as low as 2.9 is finished. The resistors shown in Figure 7 are that $R_1 = R_2 = R_8 = R_9 = R_{12} = R_{13} = R_{15} = R_{16} = R_{18} = R_{19} = R_{21} = R_{22} = 10\,\text{k}\Omega$, $R_4 = 1\,\text{k}\Omega$, $R_5 = 1.55\,\text{M}\Omega$, $R_6 = 62\,\text{M}\Omega$, $R_7 = 2.5\,\text{k}\Omega$, $R_{10} = 3.57\,\text{k}\Omega$, $R_{11} = R_{14} = R_{20} = 100\,\text{k}\Omega, R_{17} = 37.5\,\text{k}\Omega$, R_3 is adjusted, and capacitors are that $C_1 = 0.73\,\text{uF}$, $C_2 = 0.52\,\text{uF}$, $C_3 = 1.1\,\text{uF}, C_4 = C_5 = 1\,\text{nF}$. When adjusting R_3, phase portraits of some attractors can be observed by oscilloscope. Furthermore, when $R_3 = 20\,\text{k}\Omega$, we can observe that periodic attractor in $x - z$ plane is showed in Figure 8(a) and periodic attractor in $x - y$ plane is showed in Figure 8(b); when $R_3 = 50\,\text{k}\Omega$, we can observe chaotic attractor in $x - z$ plane is showed in Figure 8(c) and chaotic attractor in $x - y$ plane is showed in Figure 8(d); All these attractors observed by oscilloscope not only show the chaotic dynamics of the fractional-order Lorenz system of the order as low as 2.9, but also prove that analog circuit for the fractional-order system is well coincident with numerical simulations.

4. Conclusions

Using the frequency-domain approximation method, time-domain approximation method and numerical simulation analysis, the paper finds chaotic behaviors of the fractional-Lorenz system of the order as low as 2.8 and 2.9. That is, chaos exists in the fractional-order Lorenz system. Especially, chaotic dynamics are same for the fractional-order Lorenz system of the order as low as 2.9 when using the two different approximations. In addition, we also design an analog circuit to implement the fractional-order system,

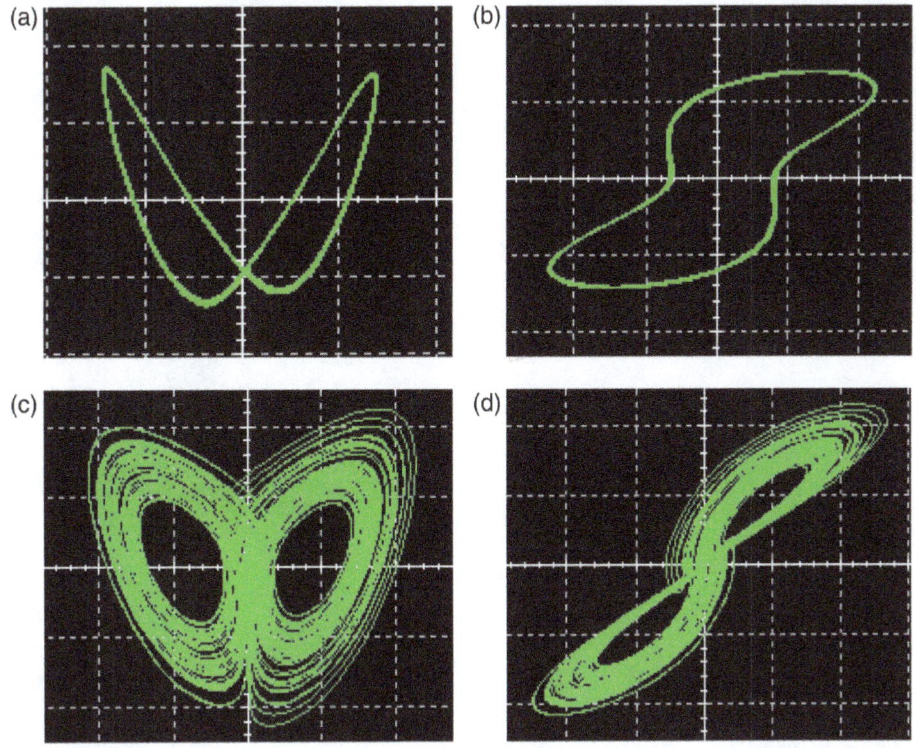

Figure 8. Attractors of the fractional-order Lorenz system observed by oscilloscope (1 V/Div). (a) periodic attractor in $x - z$ plane, (b) periodic attractor in $x - y$ plane, (c) chaotic attractor in $x - z$ plane, (d) chaotic attractor in $x - y$ plane.

and the results from circuit experiment are well consistent with numerical simulation.

Acknowledgements

This work was supported in part by the Young Scientists Fund of the National Natural Science Foundation of China (Grant No. 11202148), the Natural Science Foundation of China (Grant No. 61174094), the Specialized Research Fund for the Doctoral Program of Higher Education of China (Grant No. 20090031110029), and the Research Fund of Tianjin University of Science and Technology (Grant No. 20110124).

References

Ahmad, W. M. (2005). Hyperchaos in fractional order nonlinear systems. *Chaos Solitons & Fractals, 27*, 1459–1465. doi:10.1016/j.chaos.2005.03.031

Ahmad, W. M., & Sprott, J. C. (2003). Chaos in fractional-order autonomous nonlinear systems. *Chaos, Solitons & Fractals, 16*, 339–351. doi:10.1016/S0960-0779(02)00438-1

Chen, X. R., Liu, C. X., Wang, F. Q., & Li, Y. X. (2008). Study on the fractional-order Liu chaotic system with circuit experiment and its control. *Acta Physica Sinica, 57*(3), 1416–1422. Retrieved form http://wulixb.iphy.ac.cn

Ge, Z. M., & Qu, C. Y. (2007). Chaos in a fractional-order modified duffing system. *Chaos Solitons & Fractals, 34*, 262–291. doi:10.1016/j.chaos.2005.11.059

Grigorenko, I., & Grigorenko, E. (2003). Chaotic dynamics of the fractional Lorenz system. *Physical Review Letters, 91*(3), 034101-1-4. doi:10.1103/PhysRevLett.96.199902

Hartley, T. T., Lorenzo, C. F., & Qammer, H. K. (1995). Chaos in a fractional order Chua's system. *IEEE Transactions on Circuits and Systems-I: Fundamental Theory and Applications, 42*(8), 485–490. doi:10.1109/81.404062

Huang, X., Zhao, Z., Wang, Z., & Li, Y. X. (2012). Chaos and hyperchaos in fractional-order cellular neural networks. *Neurocomputing, 94*, 13–21. doi:10.1016/j.neucom.2012.01.011

Jia, H. Y., Chen, Z. Q., & Xue, W. (2013). Analysis and circuit implementation for the fractional-order Lorenz system. *Acta Physica Sinica, 14*, 140503. Retrieved from http://wulixb.iphy.ac.cn

Li, C. G., & Chen, G. R. (2004). Chaos and hyperchaos in the fractional-order Rossler equations. *Physica A, 341*, 55–61. doi:10.1016/j.physa.2004.04.113

Li, H. Q., Liao, X. F., & Lou, M. W. (2012). A novel non-equilibrium fractional-order chaotic system and its complete synchronization by circuit implementation. *Nonlinear Dynamics, 68*, 137–149. doi:10.1007/s11071-011-0210-4

Lu, J. G. (2006). Chaotic dynamics of the fractional-order Lü system and its synchronization. *Physics Letter A, 354*, 305–311. doi:10.1016/j.physleta.2006.01.068

Lu, J. G., & Chen, G. R. (2006). A note on the fractional-order Chen system. *Chaos Solitons & Fractals, 27*, 685–688. doi:10.1016/j.chaos.2005.04.037

Pan, L., Zhou, W. N., Zhou, L., & Sun, K. H. (2011). Chaos synchronization between two different fractional-order hyperchaotic systems. *Communications in Nonlinear Science and Numerical Simulation, 16*, 2628–2640. doi:10.1016/j.cnsns.2010.09.016

Wang, Z., Huang, X., & Zhao, Z. (2012). Synchronization of nonidentical chaotic fractional-order systems with different orders of fractional derivatives. *Nonlinear Dynamics, 69*, 999–1007. doi:10.1007/s11071-011-0322-x

Yu, Y. G., Li, H. X., Wang, S., & Yu, J. Z. (2009). Dynamic analysis of a fractional-order Lorenz chaotic system. *Chaos Solitons & Fractals, 42*, 1181–1189. doi:10.1016/j.chaos.2009.03.016

An interesting method for the exponentials for some special matrices

Jing Chen[a*] and Hongfen Zou[b]

[a]School of IoT Engineering, Jiangnan University, Wuxi 214122, People's Republic of China; [b]Wuxi Professional College of Science and Technology, Wuxi 214028, People's Republic of China

The matrix exponential e^{At} plays a central role in linear system and control theory. This paper develops a method to compute the accurate solution for the matrix exponential e^{At} with the assumption that the matrix A has an eigenvalue $s_1 = 0$. The examples show the effectiveness of the proposed method.

Keywords: matrix exponential; the eigenvalue; matrix theory; matrix equation

1. Introduction

It is well known that matrix is widely used in many areas (Dehghan & Hajarian, 2010, 2012; Hagiwara, 2011). For example, Al Zhour and Kilicman discussed some different matrix products for partitioned and non-partitioned matrices and some useful connections of the matrix products (Zhour & Kilicman, 2007). Ding and Chen defined a new operation – the block-matrix inner product – and presented a least square-based and a gradient-based iterative solutions of coupled matrix equations (Ding & Chen, 2005, 2006). Ding studied the transformations and relationships between some special matrices (Ding, 2010).

The solution $e^{At}x(0)$ of the differential equation $\dot{x}(t) = Ax(t)$ plays an important role in linear system and control theory. It is well known that e^{At} can be defined by a convergent power series $e^{At} = \sum_{i=0}^{\infty}((At)^i/i!)$. The infinite series $\sum_{i=0}^{\infty}((At)^i/i!)$ makes researchers design accurate controllers difficultly in theory and application, so it is important to develop a frame work to get the accurate solution of e^{At}.

In recent years, there exist many methods for computing e^{At} (Ben Taher & Rachidi, 2002; Bernstein & So, 1993; Cheng & Yau, 1997; Moler & Loan, 2003; Skaflestad & Wright, 2009; Wu, 2011; Zafer, 2008). Among these methods, the explicit formulas can overcome the truncation errors which are widely used (Ben Taher & Rachidi, 2002; Bernstein & So, 1993; Cheng & Yau, 1997; Wu, 2011). Based on the work in Bernstein and So (1993) and Wu (2011), the objective of this paper is to propose a method to compute the accurate solution of e^{At}, where the matrix A satisfies $A^n = \rho_1 A^{n-1} + \rho_2 A^{n-2}$. If the parameter $\rho_1 = 0$ or $\rho_2 = 0$, the matrix is the same as the matrix in Wu (2011), so our work is more widely used.

Briefly, this paper is organised as follows. Section 2 describes the main results. Section 3 provides two illustrative examples. Finally, concluding remarks are given in Section 4.

2. The main results

Let us introduce some notations first. The symbol $\boldsymbol{I}$ stands for an identity matrix of appropriate sizes, $\mathbb{C}$ denotes the set of complex number and $\mathbb{C}^{n\times n}$ denotes the set of $n \times n$ complex matrix.

As is well known, $e^{At}, A \in \mathbb{C}^{n\times n}$, can be written as the following convergent power series

$$e^{At} = \boldsymbol{I} + \frac{t}{1!}A + \frac{t^2}{2!}A^2 + \frac{t^3}{3!}A^3 + \cdots + \frac{t^{n-2}}{(n-2)!}A^{n-2} + \frac{t^{n-1}}{(n-1)!}A^{n-1} + \cdots. \quad (1)$$

Bernstein gave explicit formulas for $A^2 = A$, $A^2 = \rho \boldsymbol{I}_n$ and $A^3 = \rho A$ in Bernstein and So (1993). Wu gave explicit formulas for $A^{k+1} = \rho \mathrm{A}^k$, $A^{k+2} = \rho^2 A^k$ and $A^{k+3} = \rho^3 \mathrm{A}^k$ in Wu (2011). In this paper, we will propose a method for $A^n = \rho_1 A^{n-1} + \rho_2 A^{n-2} + \cdots + \rho_k A^{n-k}$, $k < n$.

First, let $A \in \mathbb{C}^{n\times n}$ and $A^n = A^{n-1} + A^{n-2}$, then $A^{n+1} = 2A^{n-1} + A^{n-2}$, $A^{n+2} = 3A^{n-1} + 2A^{n-2}, \ldots, A^{n+k} = \beta_1 A^{n-1} + \beta_2 A^{n-2}$, and we conclude

$$e^{At} = \boldsymbol{I} + \frac{t}{1!}A + \frac{t^2}{2!}A^2 + \frac{t^3}{3!}A^3 + \cdots + a_1(t)A^{n-2} + a_2(t)A^{n-1}, \quad (2)$$

*Corresponding author. Email: chenjing1981929@126.com

where the parameters $a_1(t)$ and $a_2(t)$ be computed as

$$a_1(t) = \frac{t^{n-2}}{(n-2)!} + \frac{t^n}{n!} + \frac{t^{n+1}}{(n+1)!} + \frac{2t^{n+2}}{(n+2)!} + \frac{3t^{n+3}}{(n+3)!} + \frac{5t^{n+4}}{(n+4)!} + \cdots, \quad (3)$$

$$a_2(t) = \frac{t^{n-1}}{(n-1)!} + \frac{t^n}{n!} + \frac{2t^{n+1}}{(n+1)!} + \frac{3t^{n+2}}{(n+2)!} + \frac{5t^{n+3}}{(n+3)!} + \frac{8t^{n+4}}{(n+4)!} + \cdots. \quad (4)$$

Equations (3) and (4) are infinite series, so it is difficult to obtain the exact figures of $a_1(t)$ and $a_2(t)$. In this paper, the solution is using the matrix theory to overcome the difficulty.

In order to compute the parameters, some mathematical preliminaries are required.

LEMMA 1 *A matrix*

$$\begin{bmatrix} 1 & s_1 & s_1^2 & \cdots & s_1^{n-1} \\ 1 & s_2 & s_2^2 & \cdots & s_2^{n-1} \\ 1 & s_3 & s_3^2 & \cdots & s_3^{n-1} \\ \vdots & \vdots & \vdots & \ddots & \vdots \\ 1 & s_n & s_n^2 & \cdots & s_n^{n-1} \end{bmatrix} \quad (5)$$

is called Vandermonde matrix, and

$$\begin{vmatrix} 1 & s_1 & s_1^2 & \cdots & s_1^{n-1} \\ 1 & s_2 & s_2^2 & \cdots & s_2^{n-1} \\ 1 & s_3 & s_3^2 & \cdots & s_3^{n-1} \\ \vdots & \vdots & \vdots & \ddots & \vdots \\ 1 & s_n & s_n^2 & \cdots & s_n^{n-1} \end{vmatrix} = \prod_{1 \leqslant j < i \leqslant n} (s_i - s_j). \quad (6)$$

LEMMA 2 *The matrix equation $AX = 0, X \in \mathbb{R}^{n\times 1}$, has only one solution $X = \mathbf{0}$, where $\mathbf{0}$ being a column vector whose entries are all 0 and the matrix $\boldsymbol{A}$ satisfies*

$$|\boldsymbol{A}| = \begin{vmatrix} a_{11} & a_{12} & \cdots & a_{1n} \\ a_{21} & a_{22} & \cdots & a_{2n} \\ \vdots & \vdots & \ddots & \vdots \\ a_{n1} & a_{n2} & \cdots & a_{nn} \end{vmatrix} \neq 0. \quad (7)$$

LEMMA 3 *The matrix $\boldsymbol{A} \in \mathbb{C}^{n\times n}$, the characteristic polynomial of $\boldsymbol{A}$ is $f(\lambda)$, then*

$$f(\boldsymbol{A}) = \boldsymbol{A}^n - \alpha_{n-1}\boldsymbol{A}^{n-1} - \alpha_{n-2}\boldsymbol{A}^{n-2} - \cdots - \alpha_1 \boldsymbol{I} = 0.$$

Using Lemma 3, Equation (1) can also be simplified as

$$e^{\boldsymbol{A}t} = b_0(t)\boldsymbol{I} + b_1(t)\boldsymbol{A} + b_2(t)\boldsymbol{A}^2 + \cdots + b_{n-2}(t)\boldsymbol{A}^{n-2} + b_{n-1}(t)\boldsymbol{A}^{n-1}. \quad (8)$$

Because the matrix $\boldsymbol{A}$ satisfies $\boldsymbol{A}^n = \boldsymbol{A}^{n-1} + \boldsymbol{A}^{n-2}$, we conclude that the matrix $\boldsymbol{A}$ has three different eigenvalues: $s_1 = 0, s_2$ and s_3, and the eigenvalue $s_1 = 0$ of the matrix $\boldsymbol{A}$ is an $n-2$ eigenvalue. Comparing Equations (2) and (8), we get

$$(b_0(t) - 1)1 + \left(b_1(t) - \frac{t}{1!}\right)s_i + \left(b_2(t) - \frac{t^2}{2!}\right)s_i^2 + \cdots + (b_{n-2}(t) - a_1(t))s_i^{n-2} + (b_{n-1}(t) - a_2(t))s_i^{n-1} = 0. \quad (9)$$

The matrix $\boldsymbol{A}$ has three eigenvalues, we built the matrix equation as

$$\begin{bmatrix} 1 & s_1 & s_1^2 & \cdots & s_1^{n-1} \\ 1 & s_2 & s_2^2 & \cdots & s_2^{n-1} \\ 1 & s_3 & s_3^2 & \cdots & s_3^{n-1} \end{bmatrix} \begin{bmatrix} b_0(t) - 1 \\ b_1(t) - \frac{t}{1!} \\ \vdots \\ b_{n-2}(t) - a_1(t) \\ b_{n-1}(t) - a_2(t) \end{bmatrix} = 0, \quad (10)$$

where $3 < n$, the above matrix equation has two or more solutions, so we cannot obtain the exact figures of $a_1(t)$ and $a_2(t)$ by $b_i(t), i = 0, 1, \ldots, n-1$.

Substituting $s_1 = 0$ into Equation (9) gets $b_0(t) = 1$, and after the derivation of s_1 to Equation (9), we get

$$\left(b_1(t) - \frac{t}{1!}\right) + 2\left(b_2(t) - \frac{t^2}{2!}\right)s_i + \cdots + (n-2)(b_{n-2}(t) - a_1(t))s_i^{n-3} + (n-1)(b_{n-1}(t) - a_2(t))s_i^{n-2} = 0 \quad (11)$$

taking $s_1 = 0$ into Equation (11) gets $b_1(t) = t/1!$ After the $2, 3, \ldots, n-3$ derivation of s_i to Equation (9), we can get $n-4$ equations. Taking $s_1 = 0$ into these equations gets $b_2(t) = t^2/2!$, $b_3(t) = t^3/3!, \ldots, b_{n-3}(t) = (t^{n-3}/(n-3)!)$. Finally, we get

$$(b_{n-2}(t) - a_1(t))s_i^{n-2} + (b_{n-1}(t) - a_2(t))s_i^{n-1} = 0. \quad (12)$$

Taking s_2 and s_3 into Equation (12) gets

$$\begin{bmatrix} s_2^{n-2} & s_2^{n-1} \\ s_3^{n-2} & s_3^{n-1} \end{bmatrix} \begin{bmatrix} cb_{n-2}(t) - a_1(t) \\ b_{n-1}(t) - a_2(t) \end{bmatrix} = 0. \quad (13)$$

Because

$$\begin{vmatrix} s_2^{n-2} & s_2^{n-1} \\ s_3^{n-2} & s_3^{n-1} \end{vmatrix} \neq 0, \quad (14)$$

and based on Lemma 2, we can conclude that $b_{n-2}(t) = a_1(t)$ and $b_{n-1}(t) = a_2(t)$.

THEOREM 1 *Let $A \in \mathbb{C}^{n\times n}$ and $A^n = A^{n-1} + A^{n-2}$, then we have*

$$e^{At} = I + \frac{t}{1!}A + \frac{t^2}{2!}A^2 + \cdots + \frac{t^{n-3}}{(n-3)!}A^{n-3} + b_{n-2}(t)A^{n-2} + b_{n-1}(t)A^{n-1}, \quad (15)$$

where $b_{n-2}(t)$ and $b_{n-1}(t)$ are computed by

$$\begin{bmatrix} b_{n-2}(t) \\ b_{n-1}(t) \end{bmatrix} = \begin{bmatrix} s_2^{n-2} & s_2^{n-1} \\ s_3^{n-2} & s_3^{n-1} \end{bmatrix}^{-1} \begin{bmatrix} e^{s_2 t} - \sum_{i=0}^{n-3} \frac{t^i s_2^i}{i!} \\ e^{s_3 t} - \sum_{i=0}^{n-3} \frac{t^i s_3^i}{i!} \end{bmatrix}, \quad (16)$$

and $s_2 = (1+\sqrt{5})/2$, $s_3 = (1-\sqrt{5})/2$.

Proof Rewritten Equations (2) and (8)

$$e^{At} = I + \frac{t}{1!}A + \frac{t^2}{2!}A^2 + \frac{t^3}{3!}A^3 + \cdots + a_1(t)A^{n-2} + a_2(t)A^{n-1},$$

$$e^{At} = b_0(t)I + b_1(t)A + b_2(t)A^2 + \cdots + b_{n-2}(t)A^{n-2} + b_{n-1}(t)A^{n-1}.$$

Because the parameters $b_0(t) = 1, b_1(t) = t/1!, \ldots, b_{n-3}(t) = (t^{n-3}/(n-3)!)$, $a_1(t) = b_{n-2}(t)$ and $a_2(t) = b_{n-1}(t)$, we conclude

$$e^{At} = I + \frac{t}{1!}A + \frac{t^2}{2!}A^2 + \frac{t^3}{3!}A^3 + \cdots + b_{n-2}(t)A^{n-2} + b_{n-1}(t)A^{n-1}.$$

The matrix A has three eigenvalues $s_1 = 0$, $s_2 = (1+\sqrt{5})/2$ and $s_3 = (1-\sqrt{5})/2$. Replacing A by $s_i, i = 2, 3$, we get

$$e^{s_i t} = 1 + \frac{t}{1!}s_i + \frac{t^2}{2!}s_i^2 + \frac{t^3}{3!}s_i^3 + \cdots + b_{n-2}(t)s_i^{n-2} + b_{n-1}(t)s_i^{n-1}. \quad (17)$$

In order to compute the parameters $b_{n-2}(t)$ and $b_{n-1}(t)$, we simplify Equation (17) as

$$e^{s_i t} - \sum_{i=0}^{n-3} \frac{t^i s_i^i}{i!} = b_{n-2}(t)s_i^{n-2} + b_{n-1}(t)s_i^{n-1}. \quad (18)$$

Taking s_2 and s_3 in Equation (18) gets

$$\begin{bmatrix} b_{n-2}(t) \\ b_{n-1}(t) \end{bmatrix} = \begin{bmatrix} s_2^{n-2} & s_2^{n-1} \\ s_3^{n-2} & s_3^{n-1} \end{bmatrix}^{-1} \begin{bmatrix} e^{s_2 t} - \sum_{i=0}^{n-3} \frac{t^i s_2^i}{i!} \\ e^{s_3 t} - \sum_{i=0}^{n-3} \frac{t^i s_3^i}{i!} \end{bmatrix}. \quad (19)$$

■

THEOREM 2 *Let $A \in \mathbb{C}^{n\times n}$ and $A^n = \rho_1 A^{n-1} + \rho_2 A^{n-2}$, then we have*

$$e^{At} = I + \frac{t}{1!}A + \frac{t^2}{2!}A^2 + \cdots + \frac{t^{n-3}}{(n-3)!}A^{n-3} + b_{n-2}(t)A^{n-2} + b_{n-1}(t)A^{n-1}, \quad (20)$$

where $b_{n-2}(t)$ and $b_{n-1}(t)$ are computed by

$$\begin{bmatrix} b_{n-2}(t) \\ b_{n-1}(t) \end{bmatrix} = \begin{bmatrix} s_2^{n-2} & s_2^{n-1} \\ s_3^{n-2} & s_3^{n-1} \end{bmatrix}^{-1} \begin{bmatrix} e^{s_2 t} - \sum_{i=0}^{n-3} \frac{t^i s_2^i}{i!} \\ e^{s_3 t} - \sum_{i=0}^{n-3} \frac{t^i s_3^i}{i!} \end{bmatrix}, \quad (21)$$

ρ_1 and ρ_2 satisfy $\rho_1^2 + 4\rho_2 > 0$, $s_2 = (\rho_1 + \sqrt{\rho_1^2 + 4\rho_2})/2$ and $s_3 = (\rho_1 - \sqrt{\rho_1^2 + 4\rho_2})/2$, when ρ_1 and ρ_2 satisfy $\rho_1^2 + 4\rho_2 < 0$, $s_2 = (\rho_1 + \sqrt{-\rho_1^2 - 4\rho_2} i)/2$ and $s_3 = (\rho_1 - \sqrt{-\rho_1^2 - 4\rho_2} i)/2$.

Remark 1 If ρ_1 and ρ_2 satisfy $\rho_1^2 + 4\rho_2 = 0$, then $s_2 = s_3$. Taking s_2 in Equation (9) gives

$$e^{s_2 t} - \sum_{i=0}^{n-3} \frac{t^i s_2^i}{i!} = b_{n-2}(t)s_2^{n-2} + b_{n-1}(t)s_2^{n-1}, \quad (22)$$

after the derivation of s_2 to Equation (9), we get

$$te^{s_2 t} - \sum_{i=1}^{n-3} \frac{it^i s_2^{i-1}}{i!} = (n-2)b_{n-2}(t)s_2^{n-3} + (n-1)b_{n-1}(t)s_2^{n-2},$$

then the parameters $b_{n-2}(t)$ and $b_{n-1}(t)$ are computed by

$$\begin{bmatrix} b_{n-2}(t) \\ b_{n-1}(t) \end{bmatrix} = \begin{bmatrix} s_2^{n-2} & s_2^{n-1} \\ (n-2)s_2^{n-3} & (n-1)s_2^{n-2} \end{bmatrix}^{-1} \times \begin{bmatrix} e^{s_2 t} - \sum_{i=0}^{n-3} \frac{t^i s_2^i}{i!} \\ te^{s_2 t} - \sum_{i=1}^{n-3} \frac{it^i s_2^{i-1}}{i!} \end{bmatrix}. \quad (23)$$

THEOREM 3 *Let $A \in \mathbb{C}^{n\times n}$ and $A^n = A^{n-1} + A^{n-2} + A^{n-3}$, then we have*

$$e^{At} = I + \frac{t}{1!}A + \frac{t^2}{2!}A^2 + \cdots + \frac{t^{n-4}}{(n-4)!}A^{n-4} + b_{n-3}(t)A^{n-3} + b_{n-2}(t)A^{n-2} + b_{n-1}(t)A^{n-1}, \quad (24)$$

where $b_{n-3}(t)$, $b_{n-2}(t)$ *and* $b_{n-1}(t)$ *are computed by*

$$\begin{bmatrix} b_{n-3}(t) \\ b_{n-2}(t) \\ b_{n-1}(t) \end{bmatrix} = \begin{bmatrix} s_2^{n-3} & s_2^{n-2} & s_2^{n-1} \\ s_3^{n-3} & s_3^{n-2} & s_3^{n-1} \\ s_4^{n-3} & s_4^{n-2} & s_4^{n-1} \end{bmatrix}^{-1} \begin{bmatrix} e^{s_2 t} - \sum_{i=0}^{n-4} \frac{t^i s_2^i}{i!} \\ e^{s_3 t} - \sum_{i=0}^{n-4} \frac{t^i s_3^i}{i!} \\ e^{s_4 t} - \sum_{i=0}^{n-4} \frac{t^i s_4^i}{i!} \end{bmatrix}, \quad (25)$$

s_2, s_3 *and* s_4 *are the roots of* $s^3 - s^2 - s - 1 = 0$.

THEOREM 4 *Let* $A \in \mathbb{C}^{n\times n}$ *and* $A^n = \rho_1 A^{n-1} + \rho_2 A^{n-2} + \rho_3 A^{n-3}$, *then we have*

$$e^{At} = I + \frac{t}{1!}A + \frac{t^2}{2!}A^2 + \cdots + \frac{t^{n-4}}{(n-4)!}A^{n-4} + b_{n-3}(t)A^{n-3} + b_{n-2}(t)A^{n-2} + b_{n-1}(t)A^{n-1}, \quad (26)$$

where $b_{n-3}(t)$, $b_{n-2}(t)$ *and* $b_{n-1}(t)$ *are computed by*

$$\begin{bmatrix} b_{n-3}(t) \\ b_{n-2}(t) \\ b_{n-1}(t) \end{bmatrix} = \begin{bmatrix} s_2^{n-3} & s_2^{n-2} & s_2^{n-1} \\ s_3^{n-3} & s_3^{n-2} & s_3^{n-1} \\ s_4^{n-3} & s_4^{n-2} & s_4^{n-1} \end{bmatrix}^{-1} \begin{bmatrix} e^{s_2 t} - \sum_{i=0}^{n-4} \frac{t^i s_2^i}{i!} \\ e^{s_3 t} - \sum_{i=0}^{n-4} \frac{t^i s_3^i}{i!} \\ e^{s_4 t} - \sum_{i=0}^{n-4} \frac{t^i s_4^i}{i!} \end{bmatrix}, \quad (27)$$

s_2, s_3 *and* s_4 *are the different roots of* $s^3 - \rho_1 s^2 - \rho_2 s - \rho_3 = 0$.

Remark 2 If s_2, s_3 and s_4 are not the different roots, we can also compute the parameters $b_{n-3}(t)$, $b_{n-2}(t)$ and $b_{n-1}(t)$ by using the method as in Remark 1.

THEOREM 5 *Let* $A \in \mathbb{C}^{n\times n}$ *and* $\mathbf{A}^n = \rho_1 A^{n-1} + \rho_2 A^{n-2} + \cdots + \rho_k \mathbf{A}^{n-k}, k < n$, *then we have*

$$e^{At} = I + \frac{t}{1!}A + \frac{t^2}{2!}A^2 + \cdots + \frac{t^{n-k-1}}{(n-k-1)!}A^{n-k-1} + b_{n-k}(t)A^{n-k} + b_{n-k+1}(t)A^{n-k+1} + \cdots + b_{n-1}(t)A^{n-1}, \quad (28)$$

where $b_{n-k}(t), b_{n-k+1}(t), \ldots, b_{n-1}(t)$ *are computed by*

$$\begin{bmatrix} cb_{n-k}(t) \\ b_{n-k+1}(t) \\ \vdots \\ b_{n-1}(t) \end{bmatrix} = \begin{bmatrix} s_2^{n-k} & s_2^{n-k+1} & \cdots & s_2^{n-1} \\ s_3^{n-3} & s_3^{n-2} & \cdots & s_3^{n-1} \\ \vdots & \vdots & \ddots & \vdots \\ s_{k+1}^{n-k} & s_{k+1}^{n-k+1} & \cdots & s_4^{n-1} \end{bmatrix}^{-1} \times \begin{bmatrix} e^{s_2 t} - \sum_{i=0}^{n-k-1} \frac{t^i s_2^i}{i!} \\ e^{s_3 t} - \sum_{i=0}^{n-k-1} \frac{t^i s_3^i}{i!} \vdots \\ e^{s_{k+1} t} - \sum_{i=0}^{n-k-1} \frac{t^i s_{k+1}^i}{i!} \end{bmatrix}, \quad (29)$$

$s_2, s_3, \ldots, s_{k+1}$ *are the different roots of* $s^k - \rho_1 s^{k-1} - \rho_2 s^{k-2} - \cdots - \rho_{k-1}s - \rho_k = 0$.

3. Examples

In this section, we will use some matrices to show the effectiveness of the proposed method.

Example 1 Let

$$A = \begin{bmatrix} 0 & 2 & 0 & 0 \\ 0 & 0 & 0 & 0 \\ 0 & 0 & 2 & 0 \\ 1 & 0 & 0 & -1 \end{bmatrix}.$$

It satisfies that $A^4 = A^3 + 2A^2$. For $\rho_1 = 1$, $\rho_2 = 2$ in Theorem 2, we have

$$\begin{aligned} e^{At} &= I_4 + tA + b_2(t)A^2 + b_3(t)A^3 \\ &= I_4 + tA + \left(\frac{1}{12}e^{2t} + \frac{2}{3}e^{-t} + 0.5t - 0.75\right)A^2 \\ &\quad + \left(\frac{1}{12}e^{2t} - \frac{1}{3}e^{-t} - 0.5t + 0.25\right)A^3 \\ &= \begin{bmatrix} 1 & 2t & 0 & 0 \\ 0 & 1 & 0 & 0 \\ 0 & 0 & e^{2t} & 0 \\ -e^{-t}+1 & 2e^{-t}+2t-2 & 0 & e^{-t} \end{bmatrix}. \end{aligned} \quad (30)$$

Example 2 Let

$$A = \begin{bmatrix} 2 & 0 & 0 & 0 \\ 0 & 1 & 1 & 0 \\ 0 & 0 & 0 & 2 \\ 0 & 0 & 0 & -1 \end{bmatrix}.$$

It satisfies that $A^4 = 2A^3 + A^2 - 2A$. For $\rho_1 = 2$, $\rho_2 = 1$ and $\rho_3 = -2$ in Theorem 4, we have

$$\begin{aligned} e^{At} &= I_4 + b_1(t)A + b_2(t)A^2 + b_3(t)A^3 \qquad (31) \\ &= I_4 + \left(-\frac{1}{6}e^{2t} + e^t - \frac{1}{3}e^{-t} - 0.5\right)A \\ &\quad + (0.5e^t + 0.5e^{-t} - 1)A^2 \\ &\quad + \left(\frac{1}{6}e^{2t} - 0.5e^t - \frac{1}{6}e^{-t} + 0.5\right)A^3 \\ &= \begin{bmatrix} e^{2t} & 0 & 0 & 0 \\ 0 & e^t & e^t - 1 & e^t + e^{-t} - 2 \\ 0 & 0 & 1 & -2e^{-t} + 2 \\ 0 & 0 & 0 & e^{-t} \end{bmatrix}. \qquad (32) \end{aligned}$$

4. Conclusions

One method to compute the accurate solution of e^{At} is presented in this letter. The basic idea of this method is using the matrix theory, the matrices satisfy the special case $A^n = \rho_1 A^{n-1} + \rho_2 A^{n-2} + \cdots + \rho_k A^{n-k}, k < n$. Furthermore, this method can be extended to the more general case $A^k = \rho_1 A^{k-1} + \rho_2 \mathbf{A}^{k-2} + \cdots + \rho_m A^{k-m}, k < n$, $m < k$.

Acknowledgements

This work was supported by the National Natural Science Foundation of China, the 111 Project (B12018) the Jiangsu Province Ordinary College Graduate Student Research Innovative Project (CXZZ11_0462) and the natural science foundation of Jiangsu Province (BK20131109).

References

Ben Taher, R., & Rachidi, M. (2002). Some explicit formulas for the polynomial decomposition of the matrix exponential and applications. *Linear Algebra and Its Applications, 350*(1–3), 171–184.

Bernstein, D. S., & So, W. (1993). Some explicit formulas for the matrix exponential. *IEEE Transactions on Automatic Control, 38*(8), 1228–1232.

Cheng, H. W., & Yau, S. S.-T. (1997). More explicit formulas for the matrix exponential. *Linear Algebra and Its Applications, 262*(1), 131–163.

Dehghan, M., & Hajarian, M. (2010). On the reflexive and anti-reflexive solutions of the generalised coupled Sylvester matrix equations. *International Journal of Systems Science, 41*(6), 607–625.

Dehghan, M., & Hajarian, M. (2012). The generalised Sylvester matrix equations over the generalised bisymmetric and skew-symmetric matrices. *International Journal of Systems Science, 43*(8), 1580–1590.

Ding, F. (2010). Transformations between some special matrices. *Computers and Mathematics with Applications, 59*(8), 2676–2695.

Ding, F., & Chen, T. (2005). Iterative least squares solutions of coupled Sylvester matrix equations. *Systems & Control Letters, 54*(2), 95–107.

Ding, F., & Chen, T. (2006). On iterative solutions of general coupled matrix equations. *SIAM Journal on Control and Optimization, 44*(6), 2269–2284.

Hagiwara, T. (2011). Block checker/diagonal transformation matrices, their properties, and the interplay with fast-lifting. *International Journal of Systems Science, 42*(8), 1293–1303.

Moler, C., & Loan, C. V. (2003). Nineteen dubious ways to compute the exponential of a matrix, twenty-five years later. *SIAM Review, 45*(1), 3–49.

Skaflestad, B., & Wright, W. M. (2009). The scaling and modified squaring method for matrix functions related to the exponential. *Applied Numerical Mathematics, 59*(3–4), 783–799.

Wu, B. B. (2011). Explicit formulas for the exponentials of some special matrices. *Applied Mathematics Letters, 24*(5), 642–647.

Zafer, A. (2008). Calculating the matrix exponential of a constant matrix on time scales. *Applied Mathematics Letters, 21*(6), 612–616.

Zhour, Z. A., & Kilicman, A. (2007). Some new connections between matrix products for partitioned and non-partitioned matrices. *Computers and Mathematics with Applications, 54*(6), 763–784.

15

Robust and resilient state-dependent control of continuous-time nonlinear systems with general performance criteria

Xin Wang[a]*, Edwin E. Yaz[b] and James Long[c]

[a]Department of Electrical and Computer Energy Engineering, Southern Illinois University Edwardsville, Edwardsville, IL 62026, USA; [b]Department of Electrical and Computer Engineering, Marquette University, Milwaukee, WI 53201, USA; [c]Computer Systems Engineering, Oregon Institute of Technology, Klamath Falls, OR 97601, USA

A novel state-dependent control approach for continuous-time nonlinear systems with general performance criteria is presented in this paper. This controller is optimally robust for model uncertainties and resilient against control feedback gain perturbations in achieving general performance criteria to secure quadratic optimality with inherent asymptotic stability property together with quadratic dissipative type of disturbance reduction. For the system model, unstructured uncertainty description is assumed, which incorporates commonly used types of uncertainties, such as norm-bounded and positive real uncertainties as special cases. By solving a state-dependent linear matrix inequality at each time, sufficient condition for the control solution can be found which satisfies the general performance criteria. The results of this paper unify existing results on nonlinear quadratic regulator, H_∞ and positive real control. The efficacy of the proposed technique is demonstrated by numerical simulations of the nonlinear control of the inverted pendulum on a cart system.

Keywords: robust control; nonlinear systems; optimal control

1. Introduction

Optimal control of nonlinear systems is traditionally characterized in terms of Hamilton Jacobi Equations (HJEs). The solution of the HJEs provides the necessary and sufficient optimal control condition for nonlinear systems. Furthermore, when the controlled system is linear time-invariant and the performance index is linear quadratic regulator (LQR), the HJEs reduced to Algebraic Riccati Equations (AREs). As for H_∞ nonlinear control problem, the optimal control solution is equivalent to solving the corresponding Hamilton Jacobi Inequalities (HJIs) Basar and Bernhard (1995). However, HJEs and HJIs, which are first-order partial differential equations and inequalities, cannot be solved for more than a few state variables. In the past few years, it has been shown that the problems of quadratic regulation and H_∞ nonlinear control can be approached by the state-dependent Riccati equation (SDRE) and nonlinear matrix inequality (NLMI) techniques (Cloutier, 1997; Cloutier, D'Souza, & Mracek, 1996; Huang & Lu, 1996). The state-dependent linear matrix inequality (LMI) control of nonlinear systems, as pointed out in Wang and Yaz (2009), Wang, Yaz, and Jeong (2010), and Wang, Yaz, and Yaz (2010, 2011) synthesizes a controller to achieve mixed nonlinear quadratic regulator (NLQR) and H_∞ control objectives.

Dissipative control for linear systems has also received considerable attention over the past two decades. The concept of dissipative systems was first introduced by Willems (1972a, 1972b), and further generalized by Hill and Moylan (1975, 1976, 1980), playing an important role in systems, circuits and controls. The theory of dissipative systems generalizes the basic tools including the passivity theorem, bounded real lemma, Kalman–Yakubovich lemma and circle criterion. Dissipativity performance includes H_∞ performance, passivity, positive realness and sector-bounded constraint as special cases. Research addressing the problems of H_∞ and positive real control systems can be found in Zhou and Khargonekar (1988) Doyle, Glover, Khargonekar, and Francis (1989), Haddad and Bernstein (1991), Sun, Khargonekar, and Shim (1994), Safonov, Jonckheere, Verma, and Limebeer (1987) and Shim (1996). Control of uncertain linear systems with L_2-bounded structured uncertainty satisfying H_∞ and passivity criteria has been tackled in Khargonekar, Petersen, and Zhou (1990) and Petersen (1987). More recent development involving the quadratic dissipative control for linear systems problem has been tackled in Tan, Soh, and Xie (2000) and Xie, Xie, and De Souza (1998).

In this paper, we further consider the problem of optimal, robust and resilient LMI control of continuous-time

*Corresponding author. Email: xwang@siue.edu

nonlinear systems with general performance criteria. The controller is robust for model uncertainties and resilient for control gain perturbations. As for uncertain nonlinear systems, we consider a general form of L_2-bounded uncertainty description, without any standard structure, incorporating commonly used types of uncertainty, such as norm-bounded and positive real uncertainties as special cases. The purpose behind this novel approach is to convert a nonlinear system control problem into a convex optimization problem which is solved by state-dependent LMI. The recent development in convex optimization provides very efficient means for solving LMIs. If a solution can be expressed in a LMI form, then there exist optimization algorithms providing efficient global numerical solutions (Boyd, Ghaoui, Feron, & Balakrishnan, 1994). Therefore, if the LMI is feasible, the LMI control technique provides asymptotically stable solutions satisfying various general performance criteria. We further propose to employ general performance criteria to design the controller guaranteeing the quadratic sub-optimality with inherent stability property in combination with dissipativity type of disturbance attenuation. The general performance criteria are a generalization of the NLQR, H_∞, positive realness and sector-bounded constraint; therefore, the results of the paper unify existing control results and provide a more general control design framework.

The paper is organized as follows. In Section 2, we present the general performance criteria including the performance of NLQR, H_∞, positive realness and sector-bounded constraint. Section 3 presents the state-dependent LMI-based control for nonlinear systems achieving general performance criteria. Finally, the inverted pendulum on a cart is used for applying the algorithm to an under-actuated robot with nonlinear dynamics to examine the effectiveness and robustness of the new approach in Section 4.

2. System model and general performance criteria analysis

The following notation is used in this work: $\Re_+$ stands for the set of non-negative real numbers and $\Re^n$ stands for the n-dimensional Euclidean space. $x \in \Re^n$ denotes n-dimensional real vector with norm $\|x\| = (x^{\mathrm{T}}x)^{1/2}$, where $(\cdot)^{\mathrm{T}}$ indicates transpose. $\Re^{n\times m}$ is the set of $n \times m$ real matrices. I_n is the $n \times n$ identity matrix. $A \geq 0$ for a symmetric matrix denotes a positive semi-definite matrix. L_2 is the space of finite dimensional vectors with finite energy: $\int_0^\infty |x(t)|^2 \mathrm{d}t < \infty$. Let L_{2e}^n be the extended space of L_2 space defined by

$$L_{2e}^n = \{f : f \text{ is a measureable function} : \Re_+ \to \Re^n, \text{ with} \times \text{ property that } F_T f \in L_2 \text{ for all finite } T \in \Re_+\}$$

where $F_T f(t) = \begin{cases} f(t), & 0 \leq t \leq T \\ 0, & T < t \end{cases}$ is called the truncation function on $\Re_+$ with values in $\Re^n$. The inner product in this space is defined as $\langle u(t), v(t)\rangle_T = \int_0^T u(t)v(t)\mathrm{d}t$, for u, $v \in L_{2e}^n$.

Consider the following nonlinear dynamical system equation and performance output equation

$$\begin{aligned} \dot{x} &= f(x(t), u(t), w(t)), \\ &= (A(x,t) + \Delta_A(x,t))x(t) + (B(x,t) + \Delta_B(x,t))u(t) \\ &\quad + (E(x,t) + \Delta_E(x,t))w(t), \\ &= (A + \Delta_A)x + (B + \Delta_B)u + (E + \Delta_E)w, \qquad (1) \\ z(t) &= g(x(t), w(t)) = C \cdot x + D \cdot w, \qquad (2) \end{aligned}$$

where $x(t) \in \Re^n$ is the state variable of the dynamical system, $u(t) \in \Re^m$ the applied input, $w(t) \in \Re^p$ the L_2 type of disturbance, $z(t) \in \Re^r$ the performance output function, f, g the smooth real vector functions, $A \in \Re^{n\times n}$, $B \in \Re^{n\times m}$, $E \in \Re^{n\times p}$, $C \in \Re^{r\times n}$ and $D \in \Re^{r\times p}$ the state-dependent coefficient matrices, and $\Delta_A \in \Re^{n\times n}$, $\Delta_B \in \Re^{n\times m}$ and $\Delta_E \in \Re^{n\times p}$ the time-varying uncertainty matrices.

It is assumed that the state feedback is available and the state feedback control input is given by

$$u(t) = (K(x,t) + \Delta_K(x,t))x(t) = (K + \Delta_K)x. \qquad (3)$$

Introducing the following quadratic energy supply function E: $L_{2e}^r \times L_{2e}^p \times \Re^+ \to \Re$ associated with the system equations, defined in Hill and Moylan (1975, 1976, 1980):

$$E(z, w, T) = \langle z, Qz\rangle_{\mathrm{T}} + 2\langle z, Sw\rangle_{\mathrm{T}} + \langle w, Rw\rangle_{\mathrm{T}}, \qquad (4)$$

where $Q \in \Re^{r\times r}$, $S \in \Re^{r\times p}$, $R \in \Re^{p\times p}$ are the chosen weighing matrices. Next, from the definition of dissipativity (Hill & Moylan 1975, 1976, 1980), we have

DEFINITION 1 *Given matrices $Q \in \Re^{r\times r}$, $S \in \Re^{r\times p}$, $R \in \Re^{p\times p}$ with Q, R symmetric, the system* (1) *and* (2) *with energy function* (4) *is said to be (Q, S, R) dissipative if for some real function $\beta(\cdot)$ with $\beta(0) = 0$. The physical meaning for $\beta(\cdot)$ is the stored energy at the initial time, given by the initial condition x_0.*

$$E(z, w, T) + \beta(x_0) \geq 0, \forall w \in L_{2e}, \forall T \geq 0. \qquad (5)$$

Furthermore, if for some scalar $\alpha > 0$,

$$E(z, w, T) + \beta(x_0) \geq \alpha\langle w, w\rangle_{\mathrm{T}}, \forall w \in L_{2e}, \forall T \geq 0. \qquad (6)$$

The system (1) *and* (2) *is said to be strictly (Q, S, R) dissipative.*

THEOREM 1 *Consider the quadratic function $V = x^{\mathrm{T}}Px > 0$, matrices $Q \in \Re^{r\times r}$, $S \in \Re^{r\times p}$, $R \in \Re^{p\times p}$ with Q, R symmetric, $M \in \Re^{n\times n}$, $M > 0$, $N \in \Re^{m\times m}$, $N > 0$ with M, N symmetric, the control of nonlinear system* (1) *and* (2) *will achieve mixed NLQR and dissipative performance if the following condition holds:*

$$\begin{aligned} &\dot{V} + x^{\mathrm{T}}Mx + u^{\mathrm{T}}Nu \\ &\quad - (z^{\mathrm{T}}Qz + 2z^{\mathrm{T}}Sw + w^{\mathrm{T}}Rw) < 0, \forall T \geq 0. \qquad (7) \end{aligned}$$

Proof By integrating Equation (7) from 0 to T, we have

$$\int_0^T (z^\mathrm{T}Qz + 2z^\mathrm{T}Sw + w^\mathrm{T}Rw)\,\mathrm{d}t >,$$

$$\int_0^T x^\mathrm{T}Mx\mathrm{d}t + \int_0^T u^\mathrm{T}Nu\mathrm{d}t + V(x(T)) - V(x(0)), \forall T \geq 0. \tag{8}$$

Let $\beta(x_0) = V(x(0)), V(x) = x^\mathrm{T}Px, V(x(T)) \geq 0$, Equation (8) implies

$$\int_0^T (z^\mathrm{T}Qz + 2z^\mathrm{T}Sw + w^\mathrm{T}Rw)\mathrm{d}t + \beta(x_0) > 0, \forall T \geq 0, \tag{9}$$

which is the condition of (Q, S, R) dissipative. By adding the terms $x^\mathrm{T}Mx + u^\mathrm{T}Nu$, we include the NLQR control performance into the original (Q, S, R)-dissipative criteria. ■

Remark 1 Notice that both H_∞ and passivity are special cases of (Q, S, R) dissipativity. The special cases are summarized as follows:

Case 1 $Q = -I$, $S = 0$, $R = \gamma^2 I$, the strict (Q, S, R) dissipativity reduces to H_∞ design (Doyle et al., 1989). The overall control design satisfies mixed NLQR–H_∞ performance.

Case 2 $Q = 0$, $S = I$, $R = 0$, the strict (Q, S, R) dissipativity reduces to strict positive realness (Sun et al., 1994). The overall control design satisfies mixed NLQR–strict positive realness performance.

Case 3 $Q = -\theta I$, $S = (1-\theta)I, R = \theta\gamma^2 I$, the strict ($Q$, S, R) dissipativity reduces to mixed H_∞ and positive real performance design, when $\theta \in (0, 1)$. The overall control design satisfies mixed NLQR–H_∞–positive real performance.

Case 4 $Q = -I$, $S = \frac{1}{2}(K_1 + K_2)^\mathrm{T}$, $R = -\frac{1}{2}(K_1^\mathrm{T}K_2 + K_2^\mathrm{T}K_1)$, where K_1 and K_2 are constant matrices of appropriate dimensions, the strict (Q, S, R) dissipativity reduces to a sector-bounded constraint (Gupta & Joshi, 1994). The overall control design satisfies mixed NLQR–sector-bounded constraint performance.

Before introducing the main result of the paper, the following model of unstructured uncertainties is introduced.

Assumption 1 *The following general form of L_2-bounded unstructured uncertainties is considered:*

$$\begin{aligned}\Delta_A\Delta_A^\mathrm{T} &\leq \gamma_A I,\\ \Delta_B\Delta_B^\mathrm{T} &\leq \gamma_B I,\\ \Delta_E\Delta_E^\mathrm{T} &\leq \gamma_E I,\\ \Delta_K\Delta_K^\mathrm{T} &\leq \gamma_K I,\end{aligned} \tag{10}$$

for $\forall x \in \Re^n$ *and* $t \geq 0$.

3. State-dependent LMI control

Lemma 1

$$AB^\mathrm{T} + BA^\mathrm{T} \leq \alpha AA^\mathrm{T} + \alpha^{-1}BB^\mathrm{T}. \tag{11}$$

This can be proven easily by considering

$$(\alpha^{1/2}A - \alpha^{-1/2}B)(\alpha^{1/2}A - \alpha^{-1/2}B)^\mathrm{T} \geq 0. \tag{12}$$

Also, by choosing A and B matrices as $A = \begin{bmatrix} a^\mathrm{T} \\ 0 \end{bmatrix}$ *and* $B = \begin{bmatrix} 0 \\ b^\mathrm{T} \end{bmatrix}$, *we have*

$$\begin{bmatrix} 0 & a^\mathrm{T}b \\ b^\mathrm{T}a & 0 \end{bmatrix} \leq \begin{bmatrix} \zeta a^\mathrm{T}a & 0 \\ 0 & \zeta^{-1}b^\mathrm{T}b \end{bmatrix}. \tag{13}$$

Lemma 2 *Denote $X = P^{-1}$ for positive definite matrix $P > 0$, then the following equality always holds:*

$$X\dot{P}X = -\dot{X}. \tag{14}$$

This can be proven easily by considering

$$0 = \frac{\mathrm{d}}{\mathrm{d}t}(I) = \frac{\mathrm{d}}{\mathrm{d}t}(PP^{-1}) = \frac{\mathrm{d}}{\mathrm{d}t}(P)\cdot P^{-1} + P\cdot\frac{\mathrm{d}}{\mathrm{d}t}(P^{-1}). \tag{15}$$

Remark 2 Since in the presentation below, P will be used to describe the energy content, which needs to decrease due to the asymptotic stability requirement, $X = P^{-1}$ matrix will be increasing in time. Therefore, we have $X\dot{P}X = -\dot{X} < 0$.

The following theorem summarizes the main results of the paper:

Theorem 2 *Given the system Equation* (1), *performance output Equation* (2) *and control Equation* (3), *if there exist matrices $X = P^{-1} > 0$ and Y for all $t > 0$, such that the following state-dependent LMI holds:*

$$\begin{bmatrix} \Upsilon_{11} & \Upsilon_{12} & X & Y^\mathrm{T} \\ * & \Upsilon_{22} & 0 & 0 \\ * & * & \Upsilon_{33} & 0 \\ * & * & * & \Upsilon_{44} \end{bmatrix} < 0, \tag{16}$$

where

$$\begin{aligned}\Upsilon_{11} &= XA^\mathrm{T} + AX + Y^\mathrm{T}B^\mathrm{T} + BY + BB^\mathrm{T} + [\gamma_A + 2\gamma_B + \gamma_E]I,\\ \Upsilon_{12} &= E - XC^\mathrm{T}QD - XC^\mathrm{T}S,\\ \Upsilon_{22} &= -D^\mathrm{T}QD - 2D^\mathrm{T}S - R + I,\\ \Upsilon_{33} &= -\{I + [3 + \lambda_{\max}(N)]\gamma_K I + M + C^\mathrm{T}QC\}^{-1},\\ \Upsilon_{44} &= -\{I + N^2 + N\}^{-1}.\end{aligned} \tag{17}$$

Then the inequality (7), *which guarantees mixed NLQR and dissipative performance, is satisfied. The nonlinear feedback control gain is given by*

$$K = Y \cdot P. \tag{18}$$

Proof In the proof below, the time and state argument will be dropped for notational simplicity. By applying system and performance output Equations (1) and (2), and state feedback input Equation (3), the performance index can be formed as follows:

$$\begin{aligned} &x^{\mathrm{T}}\{A+\Delta_A+(B+\Delta_B)(K+\Delta_K)\}^{\mathrm{T}}Px+w^{\mathrm{T}}[E+\Delta_E]^{\mathrm{T}}Px \\ &+x^{\mathrm{T}}P\{A+\Delta_A+(B+\Delta_B)(K+\Delta_K)\}x \\ &+x^{\mathrm{T}}P[E+\Delta_E]w \\ &+x^{\mathrm{T}}\dot{P}x+x^{\mathrm{T}}Mx+x^{\mathrm{T}}[K+\Delta_K]^{\mathrm{T}}N[K+\Delta_K]x \\ &-[Cx+Dw]^{\mathrm{T}}Q[Cx+Dw]-2[Cx+Dw]^{\mathrm{T}}Sw \\ &-w^{\mathrm{T}}Rw<0. \end{aligned} \tag{19}$$

Equivalently,

$$\begin{bmatrix}x^{\mathrm{T}} & w^{\mathrm{T}}\end{bmatrix}\Psi\begin{bmatrix}x & w\end{bmatrix}^{\mathrm{T}}=\begin{bmatrix}x^{\mathrm{T}} & w^{\mathrm{T}}\end{bmatrix}\begin{bmatrix}\Psi_{11} & \Psi_{12}\\ * & \Psi_{22}\end{bmatrix}\begin{bmatrix}x\\ w\end{bmatrix}<0, \tag{20}$$

where

$$\begin{aligned} \Psi_{11}&=\dot{P}+[A+\Delta_A+(B+\Delta_B)(K+\Delta_K)]^{\mathrm{T}}P \\ &\quad+P[A+\Delta_A+(B+\Delta_B)(K+\Delta_K)]+M \\ &\quad+[K+\Delta_K]^{\mathrm{T}}N[K+\Delta_K]+C^{\mathrm{T}}QC, \\ \Psi_{12}&=P[E+\Delta_E]-C^{\mathrm{T}}QD-C^{\mathrm{T}}S, \\ \Psi_{22}&=-D^{\mathrm{T}}QD-2D^{\mathrm{T}}S-R. \end{aligned} \tag{21}$$

Pre-multiplying and post-multiplying the matrix Ψ with the block $diag\{X,I\}$, where $X=P^{-1}$, $Y=K\cdot P^{-1}=KX$. Then the following matrix inequality holds:

$$\begin{bmatrix}\Lambda_{11} & \Lambda_{12}\\ * & \Lambda_{22}\end{bmatrix}<0, \tag{22}$$

where

$$\begin{aligned} \Lambda_{11}&=X[A+\Delta_A+(B+\Delta_B)(K+\Delta_K)]^{\mathrm{T}} \\ &\quad+[A+\Delta_A+(B+\Delta_B)(K+\Delta_K)]X+X\dot{P}X \\ &\quad+XMX+X[K+\Delta_K]^{\mathrm{T}}N[K+\Delta_K]X+XC^{\mathrm{T}}QCX, \\ \Lambda_{12}&=[E+\Delta_E]-XC^{\mathrm{T}}QD-XC^{\mathrm{T}}S, \\ \Lambda_{22}&=-D^{\mathrm{T}}QD-2D^{\mathrm{T}}S-R. \end{aligned} \tag{23}$$

Applying Lemma 2 and Remark 2, we have $X\dot{P}X<0$.

$$\text{Denote}\quad W=XA^{\mathrm{T}}+AX+Y^{\mathrm{T}}B^{\mathrm{T}}+BY. \tag{24}$$

The sufficient condition for matrix inequality (22) to be held is to change term Λ_{11} as follows:

$$\begin{aligned} \Lambda_{11}&=XA^{\mathrm{T}}+AX+Y^{\mathrm{T}}B^{\mathrm{T}}+BY+X\{\Delta_A+\Delta_BK+B\Delta_K \\ &\quad+\Delta_B\Delta_K\}^{\mathrm{T}}+\{\Delta_A+\Delta_BK+B\Delta_K+\Delta_B\Delta_K\}X \\ &\quad+XMX+Y^{\mathrm{T}}NY+X\Delta_K^{\mathrm{T}}NY+Y^{\mathrm{T}}N\Delta_KX \\ &\quad+X\Delta_K^{\mathrm{T}}N\Delta_KX+XC^{\mathrm{T}}QCX \\ &=W+\{\Delta_A+\Delta_BK\}X+X\{\Delta_A+\Delta_BK\}^{\mathrm{T}} \\ &\quad+X\{B\Delta_K+\Delta_B\Delta_K\}^{\mathrm{T}}+\{B\Delta_K+\Delta_B\Delta_K\}X \\ &\quad+\{X\Delta_K^{\mathrm{T}}NY+Y^{\mathrm{T}}N\Delta_KX\}+\{XMX+Y^{\mathrm{T}}NY \\ &\quad+X\Delta_K^{\mathrm{T}}N\Delta_KX+XC^{\mathrm{T}}QCX\}, \end{aligned} \tag{25}$$

By applying Lemma 1 to Equation (25) and using Assumption 1, we obtain

$$\begin{aligned} &\{\Delta_A+\Delta_BK\}X+X\{\Delta_A+\Delta_BK\}^{\mathrm{T}} \\ &=X\begin{bmatrix}I & K^{\mathrm{T}}\end{bmatrix}\begin{bmatrix}\Delta_A^{\mathrm{T}}\\ \Delta_B^{\mathrm{T}}\end{bmatrix}+\begin{bmatrix}\Delta_A & \Delta_B\end{bmatrix}\begin{bmatrix}I\\ K\end{bmatrix}X \\ &\le\alpha_1\begin{bmatrix}\Delta_A & \Delta_B\end{bmatrix}\begin{bmatrix}\Delta_A^{\mathrm{T}}\\ \Delta_B^{\mathrm{T}}\end{bmatrix}+\alpha_1^{-1}X\begin{bmatrix}I & K^{\mathrm{T}}\end{bmatrix}\begin{bmatrix}I\\ K\end{bmatrix}X \\ &\le\alpha_1(\gamma_A+\gamma_B)I+\alpha_1^{-1}X\begin{bmatrix}I & K^{\mathrm{T}}\end{bmatrix}\begin{bmatrix}I\\ K\end{bmatrix}X, \\ &X\Delta_K^{\mathrm{T}}B^{\mathrm{T}}+B\Delta_KX\le\alpha_2X\Delta_K^{\mathrm{T}}\Delta_KX \\ &\quad+\alpha_2^{-1}BB^{\mathrm{T}}\le\alpha_2\gamma_KX^2+\alpha_2^{-1}BB^{\mathrm{T}}, \\ &X\Delta_K^{\mathrm{T}}\Delta_B^{\mathrm{T}}+\Delta_B\Delta_KX\le\alpha_3X\Delta_K\Delta_K^{\mathrm{T}}X \\ &\quad+\alpha_3^{-1}\Delta_B\Delta_B^{\mathrm{T}}\le\alpha_3\gamma_KX^2+\alpha_3^{-1}\gamma_BI, \\ &X\Delta_K^{\mathrm{T}}NY+Y^{\mathrm{T}}N\Delta_KX\le\alpha_4X\Delta_K^{\mathrm{T}}\Delta_KX \\ &\quad+\alpha_4^{-1}Y^{\mathrm{T}}N^2Y\le\alpha_4\gamma_KX^2+\alpha_4^{-1}Y^{\mathrm{T}}N^2Y, \\ &X\Delta_K^{\mathrm{T}}N\Delta_KX\le X\Delta_K^{\mathrm{T}}\Delta_KX\cdot\lambda_{\max}(N)\le\lambda_{\max}(N)\gamma_KX^2. \end{aligned} \tag{26}$$

Therefore, we have

$$\begin{aligned} \Lambda_{11}&\le W+\alpha_1(\gamma_A+\gamma_B)I+\alpha_1^{-1}X\begin{bmatrix}I & K^{\mathrm{T}}\end{bmatrix}\begin{bmatrix}I\\ K\end{bmatrix}X \\ &\quad+\alpha_2\gamma_KX^2+\alpha_2^{-1}BB^{\mathrm{T}}+\alpha_3\gamma_KX^2+\alpha_3^{-1}\gamma_BI \\ &\quad+\alpha_4\gamma_KX^2+\alpha_4^{-1}Y^{\mathrm{T}}N^2Y+\{XMX+Y^{\mathrm{T}}NY \\ &\quad+\lambda_{\max}(N)\gamma_KX^2+XC^{\mathrm{T}}QCX\}, \end{aligned} \tag{27}$$

$$\begin{aligned} \Lambda_{12}&=[E+\Delta_E]-XC^{\mathrm{T}}QD-XC^{\mathrm{T}}S, \\ \Lambda_{22}&=-D^{\mathrm{T}}QD-2D^{\mathrm{T}}S-R. \end{aligned} \tag{28}$$

Using Lemma 1 and Assumption 1, we have

$$\begin{bmatrix}0 & \Delta_E\\ \Delta_E^{\mathrm{T}} & 0\end{bmatrix}\le\begin{bmatrix}\alpha_5\Delta_E\Delta_E^{\mathrm{T}} & 0\\ 0 & \alpha_5^{-1}I\end{bmatrix}\le\begin{bmatrix}\alpha_5\gamma_EI & 0\\ 0 & \alpha_5^{-1}I\end{bmatrix}. \tag{29}$$

Therefore, by applying the results of Equations (24)–(29) to Equation (22), we find:

$$\begin{bmatrix} \Theta_{11} & \Theta_{12} \\ * & \Theta_{22} \end{bmatrix} < 0, \quad (30)$$

where

$$\begin{aligned}
\Theta_{11} &= W + \left[\alpha_1\gamma_A + \alpha_1\gamma_B + \alpha_5\gamma_E + \alpha_3^{-1}\gamma_B\right] I \\
&\quad + \alpha_1^{-1} X \begin{bmatrix} I & K^{\mathrm{T}} \end{bmatrix} \begin{bmatrix} I \\ K \end{bmatrix} X + \alpha_4^{-1} Y^{\mathrm{T}} N^2 Y \\
&\quad + [\alpha_2 + \alpha_3 + \alpha_4 + \lambda_{\max}(N)]\, \gamma_K X^2 \\
&\quad + \alpha_2^{-1} B B^{\mathrm{T}} + \left\{ XMX + Y^{\mathrm{T}} N Y + X C^{\mathrm{T}} Q C X \right\} \\
&= W + [\alpha_1\gamma_A + \alpha_1\gamma_B + \alpha_5\gamma_E + \alpha_3^{-1}\gamma_B] I \\
&\quad + X\{\alpha_1^{-1} I + [\alpha_2 + \alpha_3 + \alpha_4 + \lambda_{\max}(N)]\gamma_K I \\
&\quad + M + C^{\mathrm{T}} Q C\} X \\
&\quad + Y^{\mathrm{T}}\{\alpha_1^{-1} + \alpha_4^{-1} N^2 + N\} Y + \alpha_2^{-1} B B^{\mathrm{T}}, \\
\Theta_{12} &= E - X C^{\mathrm{T}} Q D - X C^{\mathrm{T}} S, \\
\Theta_{22} &= -D^{\mathrm{T}} Q D - 2 D^{\mathrm{T}} S - R + \alpha_5^{-1} I. \qquad (31)
\end{aligned}$$

By applying the Schur complement, we conclude the final LMI solution as

$$\begin{bmatrix} \Upsilon_{11} & \Upsilon_{12} & X & Y^{\mathrm{T}} \\ * & \Upsilon_{22} & 0 & 0 \\ * & * & \Upsilon_{33} & 0 \\ * & * & * & \Upsilon_{44} \end{bmatrix} < 0, \quad (32)$$

where

$$\begin{aligned}
\Upsilon_{11} &= W + [\alpha_1\gamma_A + \alpha_1\gamma_B + \alpha_5\gamma_E + \alpha_3^{-1}\gamma_B] I + \alpha_2^{-1} B B^{\mathrm{T}} \\
&= X A^{\mathrm{T}} + A X + Y^{\mathrm{T}} B^{\mathrm{T}} + B Y + \alpha_2^{-1} B B^{\mathrm{T}} \\
&\quad + [\alpha_1\gamma_A + \alpha_1\gamma_B + \alpha_5\gamma_E + \alpha_3^{-1}\gamma_B] I, \\
\Upsilon_{12} &= E - X C^{\mathrm{T}} Q D - X C^{\mathrm{T}} S, \\
\Upsilon_{22} &= -D^{\mathrm{T}} Q D - 2 D^{\mathrm{T}} S - R + \alpha_5^{-1} I, \\
\Upsilon_{33} &= -\{\alpha_1^{-1} I + [\alpha_2 + \alpha_3 + \alpha_4 + \lambda_{\max}(N)]\gamma_K I \\
&\quad + M + C^{\mathrm{T}} Q C\}^{-1}, \\
\Upsilon_{44} &= -\{\alpha_1^{-1} + \alpha_4^{-1} N^2 + N\}^{-1}. \qquad (33)
\end{aligned}$$

Since positive constants $\alpha_1, \ldots, \alpha_5$ are arbitrary, choosing all of them as 1, we obtain Equation (16). Therefore, if LMI (16) holds, the inequality (7) is satisfied. This concludes the proof of the theorem. ■

Remark 3 At this point, it is to be noted that other choices of positive constants $\alpha_1, \ldots, \alpha_5$ are possible and can be tried if the value 1 for all these constants does not work.

4. Application to the inverted pendulum on a cart

The inverted pendulum on a cart problem (Wang & Yaz, 2010) is used for testing the novel robust and resilient state-dependent LMI approach to compare the performance. Using the Euler–Lagrange equation technique, the complete equations of motion for the inverted pendulum on a cart can be reached as

$$\begin{aligned}
(M + m)\ddot{x} + b\dot{x} + mL\ddot{\theta}\cos(\theta) - mL\dot{\theta}\sin(\theta) &= F, \\
(I + mL^2)\ddot{\theta} + mgL\sin(\theta) + mL\ddot{x}\cos(\theta) &= 0.
\end{aligned} \quad (34)$$

The following system parameters are assumed:

$$\begin{aligned}
&M = 0.5\,\mathrm{kg},\ m = 0.5\,\mathrm{kg},\ b = 0.1\mathrm{N}\frac{s}{m}, \\
&L = 0.3\,\mathrm{m},\ I = 0.06\,\mathrm{kg\,m^2}, \\
&\text{Sampling time: } T = 0.01\,\mathrm{s}.
\end{aligned}$$

Denote the following state variables:

$$x_1 = x(t), x_2 = \dot{x}(t), x_3 = \theta(t), x_4 = \dot{\theta}(t).$$

The following initial conditions are assumed:

$$x_1 = 1, x_2 = 0, x_3 = \pi/4, x_4 = 0.$$

The following design parameters are chosen to satisfy different mixed performance criteria:

Mixed NLQR–H_∞ design (predominant NLQR)

$$C = \begin{bmatrix} 0.01 & 0.01 & 0.01 & 0.01 \end{bmatrix},\ D = [0.01],\ M = I_4,\ N = 1,\ Q = 1,\ S = 0,\ R = -1.$$

Mixed NLQR–H_∞ design (predominant H_∞)

$$C = \begin{bmatrix} 1 & 1 & 1 & 1 \end{bmatrix},\ D = [1],\ M = 0.01 \times I_4,\ N = 0.01,\ Q = 1,\ S = 0,\ R = -10.$$

Mixed NLQR–H_∞–positive real design (NLQR passivity)

$$C = \begin{bmatrix} 1 & 1 & 1 & 1 \end{bmatrix},\ D = [1],\ M = I_4,\ N = 1,\ Q = 0.01,\ S = 1,\ R = 0.01.$$

All of the above mixed criteria control performance results are shown in Figures 1–5, in comparison with the traditional LQR technique based on linearization. From these figures, we find that the novel state-dependent LMI control has better performance compared with the traditional LQR technique based on linearization. Especially, Figures 3 and 4 show that the traditional LQR technique loses control of the angle and angular velocities of the pendulum, respectively. Figure 5 shows that the highest magnitude of control is needed by the predominant H_∞ control and the lowest control magnitude is needed by the linearization-based LQR technique.

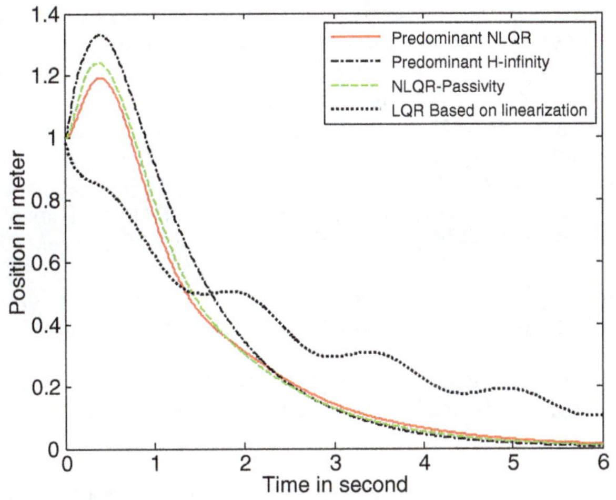

Figure 1. Position trajectory of the inverted pendulum.

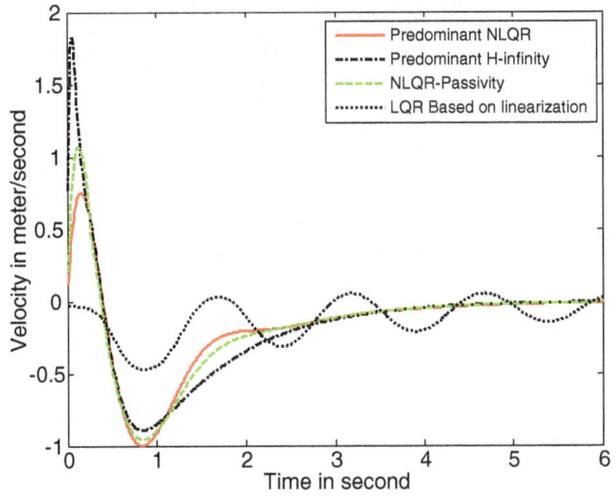

Figure 2. Velocity trajectory of the inverted pendulum.

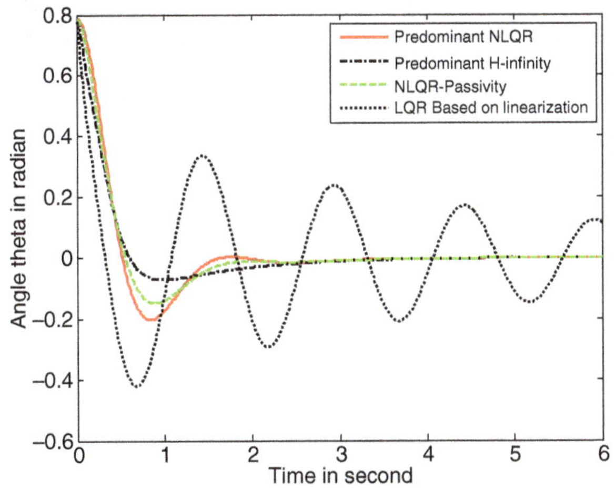

Figure 3. Angle "theta" trajectory of the inverted pendulum.

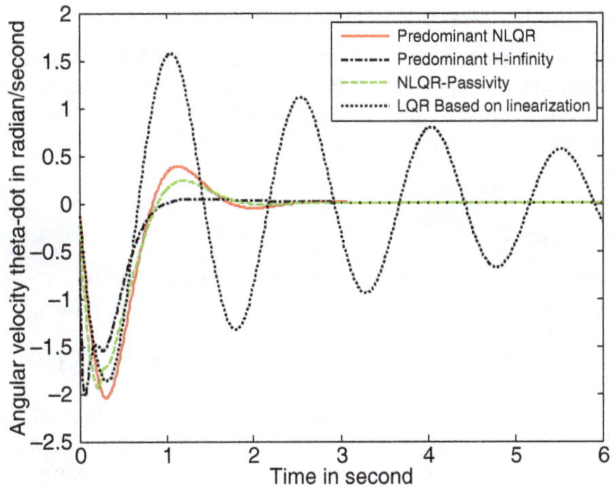

Figure 4. Angular velocity trajectory of the inverted pendulum.

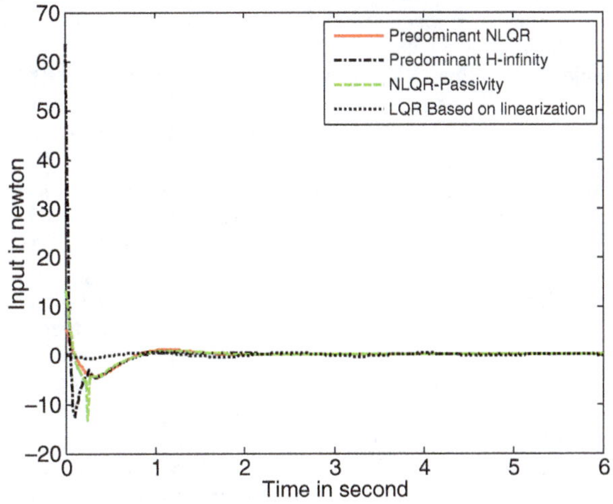

Figure 5. Control input.

5. Conclusions

This paper addresses nonlinear system control design with general NLQR and quadratic dissipative criteria to achieve asymptotic stability, quadratic optimality and strict quadratic dissipativeness. For systems with unstructured but bounded uncertainty, the LMI-based sufficient conditions are derived for the solution of general performance criteria control. Our results unify the existing results on SDRE control, robust H_∞, positive real control and sector-bounded control. The relative weighting matrices of these criteria can be achieved by choosing different weighing coefficient matrices. The optimal control can be obtained by solving LMI at each time step. The inverted pendulum on a cart control, which is a benchmark under-actuated nonlinear control system, is used as an example to demonstrate its effectiveness and robustness of the proposed method. The simulation studies show that the proposed method provides a satisfactory alternative to the existing nonlinear control approaches.

Acknowledgements

We thank the reviewers for their remarks, which have improved the quality of the paper. This research has been financially supported by a grant from NITC (National Institute for Transportation and Communities), which is gratefully acknowledged.

References

Basar, T., & Bernhard, P. (1995). *H-infinity optimal control and related minimax design problems – A dynamic game approach* (2nd ed.). Boston: Birkhauser.

Boyd, S., Ghaoui, L. E., Feron, E., & Balakrishnan, V. (1994). *Linear matrix inequalities in system and control theory, SIAM Studies in Applied Mathematics*. Philadelphia, PA: SIAM.

Cloutier, J. R. (1997). *State-dependent Riccati equation techniques: An overview*. Proceedings of the 1997 American Control Conference, pp. 932–936, Albuquerque, MN.

Cloutier, J. R., D'Souza, C. N., & Mracek, C. P. (1996). *Nonlinear regulation and nonlinear H-infinity control via the state-dependent Riccati equation technique: Part 1 theory, part 2 examples*. Proceedings of 1st international conference on nonlinear problems in aviation and aerospace, pp. 117–141, Daytona Beach, FL.

Doyle, J. C., Glover, K., Khargonekar, P. P., & Francis, B. A. (1989). State space solution to the standard H2 and H-infinity control problems. *IEEE Transactions on Automatic Control*, *34*, 831–847.

Gupta, S., & Joshi, S. M. (1994). *Some properties and stability results for sector bounded LTI systems*. Proceedings of 33rd IEEE conference on decision and control, pp. 2973–2978, Orlando, USA.

Haddad, W. M., & Bernstein, D. S. (1991). Robust stabilization with positive real uncertainty, beyond the small gain theorem. *Systems and Control Letters*, *17*, 191–208.

Hill, D. J., & Moylan, P. J. (1975). *Cyclo-dissipativeness, dissipativeness and losslessness for nonlinear dynamical systems* (Technical Report EE7526). Department of Electrical and Computer Engineering, University of Newcastle, Australia.

Hill, D. J., & Moylan, P. J. (1976). Stability of nonlinear dissipative systems. *IEEE Transactions on Automatic Control*, *21*, 708–711.

Hill, D. J., & Moylan, P. J. (1980). Dissipative dynamical systems: Basic input-output and state properties. *Journal of the Franklin Institute*, *309*, 327–357.

Huang, Y., & Lu, W-M. (1996). *Nonlinear optimal control: alternatives to Hamilton-Jacobi equation*. Proceedings of 35th conference on decision and control, pp. 3942–3947, Kobe, Japan.

Khargonekar, P. P., Petersen, I. R., & Zhou, K. (1990). Robust stabilization of uncertain linear systems: quadratic stabilizability and H-infinity control theory. *IEEE Transactions on Automatic Control*, 35, 356–361.

Mohseni, J., Yaz, E., & Olejniczak, K. (1998). *State dependent LMI control of discrete-time nonlinear systems*. Proceedings of the 37th IEEE conference on decision and control, pp. 4626–4627, Tampa, FL.

Petersen, I. R. (1987). A stabilization algorithm for a class of uncertain systems. *Systems and Control Letters*, *8*, 181–188.

Safonov, M. G., Jonckheere, E. A., Verma, M., & Limebeer, D. J. N. (1987). Synthesis of positive real multivariable feedback systems. *International Journal of Control*, *45*, 817–842.

Shim, D. (1996). Equivalence between positive real and norm-bounded uncertainty. *IEEE Transactions on Automatic Control*, *41*, 1190–1193.

Sun, W., Khargonekar, P. P., & Shim, D. (1994). Solution to the positive real control problem for linear time-invariant systems. *IEEE Transactions on Automatic Control*, *39*, 2034–2046.

Tan, Z., Soh, Y. C., & Xie, L. (2000). Dissipative control of linear discrete-time systems with dissipative uncertainty. *International Journal of Control*, *73*(4), 317–328.

Van der Shaft, A. J. (1993). Nonlinear state space H-infinity control theory. In H. J. Trentelman & J. C. Willems (Eds.), *Perspectives in control*, pp. 153–190. Groningen: Birkhauser.

Wang, X., & Yaz, E. E. (2009). *The state dependent control of continuous time nonlinear systems with mixed performance criteria*. Proceedings of IASTED international conference on identification and control applications, pp. 98–102, Honolulu, HI.

Wang, X., Yaz, E. E., & Jeong, C. S. (2010). *Robust nonlinear feedback control of discrete-time systems with mixed performance criteria*. Proceedings of American control conference, pp. 6357–6362, Baltimore, MD.

Wang, X., Yaz, E. E., & Yaz, Y. I. (2010). *Robust and resilient state dependent control of continuous time nonlinear systems with general performance criteria*. Proceedings of the 49th IEEE conference on decision and control, pp. 603–608, Atlanta, GA.

Wang, X., Yaz, E. E., & Yaz, Y. I. (2011). *Robust and resilient state dependent control of discrete time nonlinear systems with general performance criteria*. Proceedings of the 18th IFAC congress, pp. 10904–10909, Milano, Italy.

Willems, J. C. (1972a). Dissipative dynamical systems – Part 1: General theory. *Archives for Rational Mechanics and Analysis*, *45*, 321–351.

Willems, J. C. (1972b). Dissipative dynamical systems – Part 2: Linear systems with quadratic supply rates'. *Archives for Rational Mechanics and Analysis*, *45*, 352–393.

Xie, S., Xie, L., & De Souza, C. E. (1998). Robust dissipative control for linear systems with dissipative uncertainty. *International Journal of Control*, *70*(2), 169–191.

Zhou, K., & Khargonekar, P. P. (1988). An algebraic Riccati equation approach to H-infinity optimization. *Systems and Control Letters*, *11*, 85–91.

16

Impulsive synchronization of a class of chaotic systems

Yang Fang[a], Kang Yan[a] and Kelin Li[b*]

[a]*School of Automation and Electronic Information, Sichuan University of Science & Engineering, Sichuan 643000, People's Republic of China;* [b]*Institute of Nonlinear Science and Engineering Computing, Sichuan University of Science & Engineering, Sichuan 643000, People's Republic of China*

This paper deals with the impulsive synchronization problem of a class of chaotic systems. By employing the comparison principle and the linear matrix inequalities approach, some less conservative and easily verified sufficient conditions for impulsive synchronization of this class of chaotic systems are derived, these new sufficient conditions can be applied to analyze the impulsive synchronization of the Chua's oscillators. Moreover, the numerical simulation of Chua's oscillators under impulsive control shows the effectiveness of the proposed theory, and obtains better estimation of the boundary of the stable region than the existing approaches.

Keywords: impulsive synchronization; chaotic system; linear matrix inequality; Chua's oscillators

1. Introduction

Chaotic systems have complex dynamical behaviors which own some special performances, such as being extremely sensitive to tiny variations of initial conditions, having bounded trajectories in the phase space with a positive leading Lyapunov exponent, and so on. In recent decades, synchronization of coupled chaotic systems has been of ongoing interest due to its potential applications for secure communication. The phenomenon of chaos synchronization was first revealed by Pecora and Carroll (1990). Since then, numerous methods have been proposed for synchronization of chaotic systems, which include complete synchronization (Mahmoud & Mahmoud, 2010), phase synchronization (Ge & Chen, 2004), generalized synchronization (He & Cao, 2009), adaptive synchronization (Chen & Chang, 2006), lag synchronization (Bhowmick, Pal, Roy & Dana, 2012), impulsive control (He, Qian, Cao & Han, 2011; Li, Liao & Zhang, 2005; Sun, Zhang & Wu, 2002; Yang & Chua, 1997; Yang, Yang & Yang, 1997), etc.

It is common knowledge that impulsive control (Sun, Zhang & Wu, 2003; Yang, 1999, 2001) is characterized by the abrupt changes in the system dynamics at certain instants, which is an advantage in reducing the amount of information transmission and improving the security and robustness against disturbances especially in telecommunication network and power grid, orbital transfer of satellite. In addition, impulsive control allows the stabilization and synchronization of chaotic systems using only small control impulses. Thus, it has been widely used to stabilize and synchronize chaotic systems. Recently, several impulsive synchronization schemes have been reported in the literature (He et al., 2011; Li et al., 2005; Sun et al., 2002; Yang & Chua, 1997; Yang et al., 1997). Wang-Li He, etc in (He et al., 2011) have obtained some sufficient conditions for the synchronization of two nonidentical chaotic systems with time-varying delay via impulsive control. In Li et al. (2005), the authors used impulsive theory and the linear matrix inequality (LMI) technique derived from some less conservative and easily verified criteria for impulsive synchronization of chaotic systems. The stabilization and synchronization of Lorenz systems (Lorenz, 1995) via impulsive control are studied in Sun et al. (2002).

Inspired by the former, we further study the impulsive synchronization of a class of chaotic systems in this paper. Some less conservative and easily verified criteria for impulsive synchronization of chaotic systems are derived via impulsive control theory and LMI. We then use the LMI toolbox in MATLAB to obtain the synchronization conditions and estimate the stable region of synchronized systems. The proposed method is also applied to the original Chua's oscillators to demonstrate the effectiveness.

The organization of this paper is as follows. In Section 2, the theory of impulsive differential equations is given. Some new synchronization criteria are obtained in Section 3. In Section 4, the simulation results of the impulsive synchronization of the original Chua's oscillators are presented. Finally, conclusions are given in Section 5.

---

*Corresponding author. Email: lkl@suse.edu.cn

2. Model description and preliminaries

Consider a class of chaotic systems which can be described by

$$\dot{x}(t) = Ax(t) + f(x(t)) + J, \tag{1}$$

where $x(t) = (x_1(t), \ldots, x_n(t))^{\mathrm{T}}$ is the state variable, $A \in R^{n\times n}$, $J \in R^n$ is the constant input vector, and $f(x) = (f_1(x_1), \ldots, f_n(x_n))^{\mathrm{T}}$ is the nonlinear vector-value function.

Throughout the paper, we assume that:

(H1) Each function f_i is continuous, and there exist scalars l_i^- and l_i^+ such that

$$l_i^- \le \frac{f_i(a) - f_i(b)}{a - b} \le l_i^+$$

for any $a, b \in R, a \neq b$, where l_i^+ and l_i^- can be positive, negative or zero.

We set

$$L_1 = \mathrm{diag}(l_1^+ + l_1^-, l_2^+ + l_2^-, \ldots, l_n^+ + l_n^-),$$
$$L_2 = \mathrm{diag}(l_1^+ l_1^-, l_2^+ l_2^-, \ldots, l_n^+ l_n^-).$$

Remark 1 In the usual Lipschitz condition, it is assumed $l_i^- = -l_i^+$. Clearly, the condition (H1) is more general than the usual Lipschitz condition and it has been adopted in Khalil (1996).

We take system (1) as the drive system, and the corresponding response system is given by

$$\dot{y}(t) = Ay(t) + f(y(t)) + J, \tag{2}$$

where $y(t)$ is the state variable of the response system.

At discrete time t_k, the state variables of the drive system are transmitted to the response system as the control input such that the state variables of the response system are suddenly changed at these instants. Therefore, the impulsive controlled response system can be written as (3),

$$\begin{aligned} &\dot{y}(t) = Ay(t) + f(y(t)) + J, \quad t \neq t_k,\\ &\Delta y(t_k) = y(t_k^+) - y(t_k^-) = H(y(t_k^-) - x(t_k^-)),\\ &\qquad t = t_k, k = 1, 2, \ldots, \end{aligned} \tag{3}$$

where $\Delta y(t_k)$ denotes the state jumping at impulsive time instant $t = t_k, y(t_k^+), y(t_k^-)$ and $x(t_k^+), x(t_k^-)$ are the right-hand and left-hand limits of the functions $y(t)$ and $x(t)$ at t_k, respectively. Moreover, $y(t)$ and $x(t)$ are both left-hand continuous at $t = t_k$, i.e. $y(t_k) = y(t_k^-), x(t_k) = x(t_k^-)$. H is the impulsive matrix. Suppose that the discrete instant set $\{t_k\}$ satisfies $0 < t_1 < t_2 < \cdots$ and $\lim_{k\to\infty} t_k = \infty$. For simplicity, it is assumed that $y(t)$ is continuous at t_0, and $y(t_0^+) = y_0$.

Defining the synchronization error as $e(t) = y(t) - x(t)$, then we have the following error system:

$$\begin{aligned} &\dot{e}(t) = Ae(t) + g(e(t)), \quad t \neq t_k,\\ &\Delta e(t_k) = e(t_k^+) - e(t_k^-) = He(t_k^-), \quad t = t_k, k = 1, 2, \ldots,\\ &e(t_0^+) = e_0, \quad t_0 = 0, \end{aligned} \tag{4}$$

where $g(e(t)) = f(y(t)) - f(x(t))$.

To begin with, we introduce some notation and recall some basic definitions.

In general, the impulsive functional differential equation can be described by

$$\begin{aligned} &\dot{x}(t) = f(t, x(t)), \quad t \neq t_k, t > 0,\\ &\Delta x(t_k) = x(t_k^+) - x(t_k^-) = U(k, x(t_k^-)),\\ &\qquad t = t_k, k = 1, 2, \ldots,\\ &x(t_0) = x_0, \quad t_0 = 0, \end{aligned} \tag{5}$$

where $f : R_+ \times R^n \to R^n$ is continuous with $f(t, 0) \equiv 0$, $x \in R^n$ is the state variable, $U(k, x(t_k^-)) : S_\rho \to R^n$ with $U(k, 0) \equiv 0$.

Definition 1 (Yang, 2001) *Let $V : R_+ \times R^n \to R_+$, then V is said to belong to class $\mathcal{V}_0$, if*

(i) *V is continuous in $(t_{k-1}, t_k] \times R^n$ and for each $x \in R^n, k = 1, 2, \ldots, \lim_{(t,y)\to(t_k^+, x)} V(t, y) = V(t_k^+, x)$ exists;*
(ii) *V is locally Lipschitzian in x.*

Definition 2 *For $(t, x) \in (t_{k-1}, t_k] \times R^n$, we define the right and upper Dini's derivative of $V \in \mathcal{V}_0$ with respect to the time variable*

$$D^+V(t, x) \equiv \lim_{h\to 0^+} \sup \frac{1}{h}\{V[t + h, x + hf(t, x)] - V(t, x)\}. \tag{6}$$

Definition 3 (Yang, 2001) *Let $V \in \mathcal{V}_0$ and assume that*

$$\begin{aligned} &D^+V(t, x) \le g[t, V(t, x)], \quad t \neq t_k,\\ &V[t, x + U(k, x)] \le \psi_k[V(t, x)], \quad t = t_k, k = 1, 2, \ldots, \end{aligned} \tag{7}$$

where $g : R_+ \times R_+ \to R$ is continuous and $g(t, 0) = 0$, $\psi_k : R_+ \to R_+$ is nondecreasing. Then, the system

$$\begin{aligned} &\dot{\omega} = g(t, \omega), \quad t \neq t_k,\\ &\omega(t_k^+) = \psi_k(\omega(t_k)), \quad k = 1, 2, \ldots,\\ &\omega(t_0^+) = \omega_0 \ge 0. \end{aligned} \tag{8}$$

is called the comparison system of system (5).

DEFINITION 4 $S_\rho = \{x \in R^n | \|x\| < \rho\}$, *where* $\|\cdot\|$ *denotes the Euclidean norm on* R^n.

DEFINITION 5 *A function* α *is said to belong to* $\mathcal{K}$, *if* $\alpha \in [R_+, R_+], \alpha(0) = 0$, *and* $\alpha(x)$ *is strictly increasing in* x.

LEMMA 1 (Li, Wen & Soh, 2001) *Let* $g(t,\omega) = \dot{\lambda}(t)\omega, \lambda \in C^1[R_+, R_+], \psi_k(\omega) = d_k\omega, d_k \geq 0, k = 1, 2, \ldots$. *Then, the origin of system* (5) *is asymptotically stable if the following conditions hold*:

(i)

$$\sup_k \{d_k \exp[\lambda(t_{k+1}) - \lambda(t_k)]\} = \epsilon_0 < \infty; \quad (9)$$

(ii) *there exists a* $r > 1$ *such that*

$$\lambda(t_{2k+3}) + \ln(r^2 d_{2k+2} d_{2k+1}) \leq \lambda(t_{2k+1}) \quad (10)$$

holds for all $d_{2k+2}d_{2k+1} \neq 0, k = 0, 1, \ldots$;

(iii) $\dot{\lambda}(t) \geq 0$;

(iv) *there exists* $\alpha(\cdot)$ *and* $\beta(\cdot)$ *in class* $\mathcal{K}$ *such that* $\beta(\|x\|) \leq V(t,x) \leq \alpha(\|x\|)$.

3. Main results

In the section, we investigate the impulsive synchronization of the chaotic system by the stability analysis of impulsive functional differential equation.

THEOREM 1 *Assume that hypothesis* (H1) *holds. For any constant scalars* $\theta > 0, 0 < \kappa < 2$, *the origin of the error system* (4) *is asymptotically stable if there exist a symmetric and positive definite matrix* $P \in R^{n\times n}$, *and constant scalars* $0 < \mu < 1, \xi > 0, \gamma > 1$ *such that*

(C1)

$$\begin{bmatrix} A^{\mathrm{T}}P + PA - \dfrac{2\theta}{2-\kappa}L_2 + \dfrac{\theta}{2\kappa - \kappa^2}L_1^2 - \xi P & \sqrt{\dfrac{1}{\theta}}P \\ * & -I \end{bmatrix} \leq 0,$$

(C2)

$$\begin{bmatrix} \mu P & (I+H)^{\mathrm{T}}P \\ * & P \end{bmatrix} \geq 0,$$

(C3) $\max_k\{t_k - t_{k-1}\} \leq -\ln \gamma\mu/\xi, k = 1, 2, \ldots,$

where the notation $*$ *always denotes the symmetric block in one symmetric matrix.*

Proof Construct a Lyapunov functional as follows:

$$V(t, e(t)) = e^{\mathrm{T}}(t)Pe(t) \quad (11)$$

it follows (11) that $\lambda_m(P)e^{\mathrm{T}}(t)e(t), \lambda_M(P)e^{\mathrm{T}}(t)e(t) \in \mathcal{K}, \lambda_m(P)e^{\mathrm{T}}(t)e(t) \leq V(t,e(t)) \leq \lambda_M(P)e^{\mathrm{T}}(t)e(t)$, $\lambda_m(P)$ and $\lambda_M(P)$ denote the smallest and largest eigenvalues of P, respectively.

According to (H1), it is easy to see that

$$\begin{aligned} 0 &\leq \sum_{i=1}^{n}(l_i^+ e_i(t) - g_i(e_i(t)))(g_i(e_i(t)) - l_i^- e_i(t)) \\ &= e^{\mathrm{T}}(t)L_1g(e(t)) - g^{\mathrm{T}}(e(t))g(e(t)) - e^{\mathrm{T}}(t)L_2e(t), \end{aligned}$$

For $t \neq t_k, k = 1, 2, \ldots$, the derivative of $V(t, e(t))$ along the solution of (4) is

$$\begin{aligned} D^+V(t,e(t)) &= e^{\mathrm{T}}(t)(A^{\mathrm{T}}P + PA)e(t) + g^{\mathrm{T}}(e(t))Pe(t) \\ &\quad + e^{\mathrm{T}}(t)Pg(e(t)) \\ &\leq e^{\mathrm{T}}(t)(A^{\mathrm{T}}P + PA)e(t) + \theta g^{\mathrm{T}}(e(t))g(e(t)) \\ &\quad + \frac{1}{\theta}e^{\mathrm{T}}(t)P^{\mathrm{T}}Pe(t) \\ &\leq e^{\mathrm{T}}(t)\left(A^{\mathrm{T}}P + PA + \frac{1}{\theta}P^2\right)e(t) \\ &\quad + \theta(e^{\mathrm{T}}(t)L_1g(e(t)) - e^{\mathrm{T}}(t)L_2e(t)) \\ &= e^{\mathrm{T}}(t)\left[A^{\mathrm{T}}P + PA + \frac{1}{\theta}P^2 - \theta L_2\right]e(t) \\ &\quad + \theta e^{\mathrm{T}}(t)L_1g(e(t)), \end{aligned}$$

notice that

$$\begin{aligned} e^{\mathrm{T}}(t)L_1g(e(t)) &\leq \frac{1}{2\kappa}e^{\mathrm{T}}(t)L_1L_1^{\mathrm{T}}e(t) + \frac{\kappa}{2}g^{\mathrm{T}}(e(t))g(e(t)) \\ &\leq \frac{1}{2\kappa}e^{\mathrm{T}}(t)L_1^2e(t) + \frac{\kappa}{2}(e^{\mathrm{T}}(t)L_1g(e(t)) \\ &\quad - e^{\mathrm{T}}(t)L_2e(t)), \end{aligned}$$

which implies

$$e^{\mathrm{T}}(t)L_1g(e(t)) \leq e^{\mathrm{T}}(t)\left(\frac{1}{2\kappa - \kappa^2}L_1^2 - \frac{\kappa}{2-\kappa}L_2\right)e(t),$$

then

$$\begin{aligned} D^+V(t,e(t)) &\leq e^{\mathrm{T}}(t)\left[A^{\mathrm{T}}P + PA + \frac{1}{\theta}P^2 - \theta L_2\right]e(t) \\ &\quad + \theta e^{\mathrm{T}}(t)\left(\frac{1}{2\kappa-\kappa^2}L_1^2 - \frac{\kappa}{2-\kappa}L_2\right)e(t) \\ &= e^{\mathrm{T}}(t)\left[A^{\mathrm{T}}P + PA + \frac{1}{\theta}P^2 - \frac{2\theta}{2-\kappa}L_2 \right. \\ &\quad \left. + \frac{\theta}{2\kappa-\kappa^2}L_1^2\right]e(t) \\ &= e^{\mathrm{T}}(t)\left[A^{\mathrm{T}}P + PA + \frac{1}{\theta}P^2 - \frac{2\theta}{2-\kappa}L_2 \right. \\ &\quad \left. + \frac{\theta}{2\kappa-\kappa^2}L_1^2 - \xi P\right]e(t) + \xi V(t,e(t)), \end{aligned}$$

From condition (C1) and Schur complement (Boyd, Ghaoui, Feron & Balakrishnan, 1994), we obtain $D^+ V(t,e(t)) \leq \xi V(t,e(t))$.

For $t = t_k, k = 1,2,\ldots$,

$$\begin{aligned} V(t_k, e(t_k) + \Delta e(t_k)) &= e^{\mathrm{T}}(t_k^-)(I+H)^{\mathrm{T}}P(I+H)e(t_k^-) \\ &= e^{\mathrm{T}}(t_k)(I+H)^{\mathrm{T}}P(I+H)e(t_k). \end{aligned}$$

From condition (C2), i.e. $\begin{bmatrix} \mu P & (I+H)^{\mathrm{T}}P \\ * & P \end{bmatrix} \geq 0$, we have

$$\begin{bmatrix} I & -(I+H)^{\mathrm{T}} \\ 0 & I \end{bmatrix} \begin{bmatrix} \mu P & (I+H)^{\mathrm{T}}P \\ * & P \end{bmatrix} \begin{bmatrix} I & 0 \\ -(I+H) & I \end{bmatrix} = \begin{bmatrix} \mu P - (I+H)^{\mathrm{T}}P(I+H) & 0 \\ * & P \end{bmatrix} \geq 0$$

it follows that

$$\mu P - (I+H)^{\mathrm{T}}P(I+H) \geq 0,$$

which yields $V(t_k, e(t_k) + \Delta e(t_k)) \leq \mu V(t_k, e(t_k))$.

Let $\dot{\lambda}(t) = \xi, d_k = \mu, k = 1,2,\ldots$. From Lemma 1, the origin of the error system (4) is asymptotically stable. The proof of Theorem 1 is completed. ■

Remark 2 When θ, κ, ξ, μ are chosen, conditions (C1) and (C2) in Theorem 1 are LMIs, which can be solved numerically and very efficiently using the interior point algorithms (Boyd et al., 1994).

When $l_i^- = -l_i^+ = -L < 0$ in (H1), we have the following corollary holds.

COROLLARY 1 *Assume that hypothesis* (H1) *holds. For any constant scalars* $\theta > 0, 0 < \kappa < 2$, *if there exist a symmetric and positive definite matrix* $P \in R^{n\times n}$ *and constant scalars* $0 < \mu < 1, \xi > 0, \delta > 0, \gamma > 1$ *such that*

(C1′)

$$\begin{bmatrix} A^{\mathrm{T}}P + PA + \dfrac{2\theta}{2-\kappa}L^2 I - \xi P & \sqrt{\dfrac{1}{\theta}}P \\ * & -I \end{bmatrix} \leq 0;$$

(C2)

$$\begin{bmatrix} \mu P & (I+H)^{\mathrm{T}}P \\ * & P \end{bmatrix} \geq 0;$$

(C3) $\delta = \max_k\{t_k - t_{k-1}\} \leq -\ln \gamma\mu/\xi,\ k = 1,2,\ldots;$

then, the origin of the error system (4) is asymptotically stable.

Remark 3 It is worth noting that the Theorem 3 in Li et al. (2005) is a special form of Corollary 1 with appropriate θ, κ, thus our method is more general than those given out in Li et al. (2005).

In particular, when $l_i^- = 0, l_i^+ > 0$ in (H1), the following corollary holds

COROLLARY 2 *Assume that hypothesis* (H1) *holds. For any constant scalars* $\theta > 0, 0 < \kappa < 2$, *if there exist a symmetric and positive definite matrix* $P \in R^{n\times n}$ *and constant scalars* $0 < \mu < 1, \xi > 0, \delta > 0, \gamma > 1$ *such that*

(C1″)

$$\begin{bmatrix} A^{\mathrm{T}}P + AP + \dfrac{\theta}{2\kappa - \kappa^2}L_1^2 - \xi P & \sqrt{\dfrac{1}{\theta}}P \\ \sqrt{\dfrac{1}{\theta}}P & -I \end{bmatrix} \leq 0;$$

(C2) $(I+H)^{\mathrm{T}}P(I+H) - \mu P \leq 0;$

(C3) $\delta = \max_k\{t_k - t_{k-1}\} \leq -\ln \gamma\mu/\xi,\ k = 1,2,\ldots;$

then, the origin of the error system (4) is asymptotically stable.

The proof of Corollaries 1 and 2 is direct, so it is omitted.

4. Numerical experimental results

In this section, we use the proposed method to study the impulsive synchronization of Chua's oscillators. The dimensionless form of a Chua's oscillator is given by

$$\begin{aligned} \dot{x_1} &= \alpha[x_2 - x_1 - g(x_1)] \\ \dot{x_2} &= x_1 - x_2 + x_3 \qquad (12) \\ \dot{x_3} &= -\beta x_2, \end{aligned}$$

where $g(x_1)$ is the piecewise-linear characteristics of the Chua's diode, which is given by

$$g(x_1) = b_1 x_1 + 0.5(a_1 - b_1)(|x_1 + 1| - |x_1 - 1|), \quad (13)$$

where $a_1 < b_1 < 0$ are two constants.

Let $x^{\mathrm{T}} = (x_1, x_2, x_3)$, then we can rewrite the Chua's oscillator equation in the form as

$$\dot{x} = Ax + f(x) + J, \quad (14)$$

where $J = (0,0,0)^{\mathrm{T}}$, $A = \begin{bmatrix} -\alpha - \alpha b_1 & \alpha & 0 \\ 1 & -1 & 1 \\ 0 & -\beta & 0 \end{bmatrix}$,

$$f(x) = \begin{bmatrix} -0.5\alpha(a_1 - b_1)(|x_1+1| - |x_1-1|) \\ 0 \\ 0 \end{bmatrix}.$$

In the following numerical simulation, we choose the parameters of system (12) as $\alpha = 9.2156$, $\beta = 15.9946$, $a_1 = -1.24905$, $b_1 = -0.75735$. The initial conditions for drive and response systems are given by $x(0) = (0.15, 0.1, 0.2)^{\mathrm{T}}$ and $y(0) = (0.5, 0.3, -0.5)^{\mathrm{T}}$, respectively. The phase diagram of the drive system is shown in Figure 1, which is the Chua's double-scroll attractor.

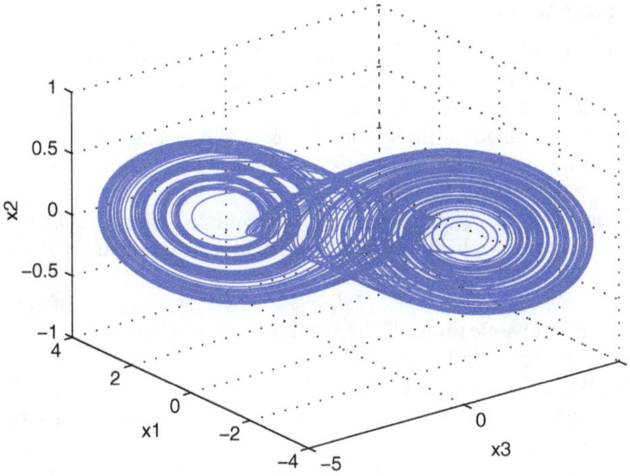

Figure 1. The double-scroll attractor of the origin Chua's oscillator.

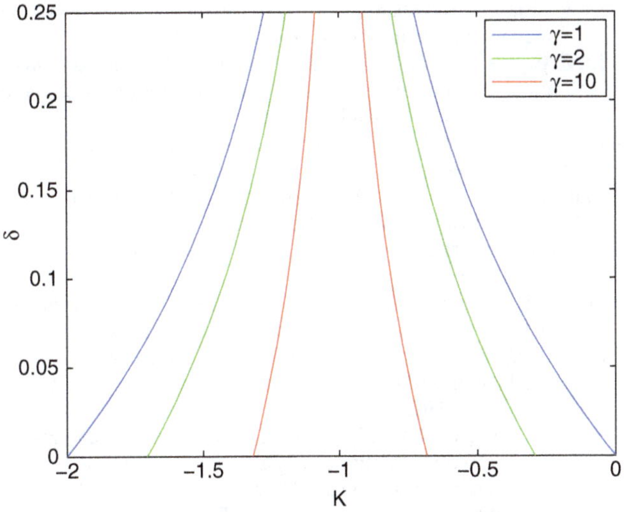

Figure 2. Estimate of the boundaries of stable regions with different $\gamma's$.

Choose $\theta = 1, \kappa = 1, H = \mathrm{diag}(K, K, K)$, then $-2 < K < 0$ and note that $L_1 = \mathrm{diag}(4.5313, 0, 0)$, $L_2 = \mathrm{diag}(0, 0, 0)$. From condition 1 in Theorem 1, we find the approximate smallest value of control gain, $\xi = 10.4053$. Then using Matlab LMI toolbox, we can obtain the following feasible solution to LMIs in Theorem 1:

$$P = \begin{bmatrix} 7.0484 & -0.5010 & 2.7907 \\ -0.5010 & 9.9859 & 0.7248 \\ 2.7907 & 0.7248 & 1.9347 \end{bmatrix}$$

with eigenvalues $\mathrm{eigen}(P) = [0.6266, 8.2613, 10.0811]^{\mathrm{T}}$. Let δ be the impulsive interval, then estimates of bounds of stable regions are given by

$$0 \leq \delta \leq -\frac{\ln \gamma + \ln(K+1)^2}{10.4053}, -2 < K < 0.$$

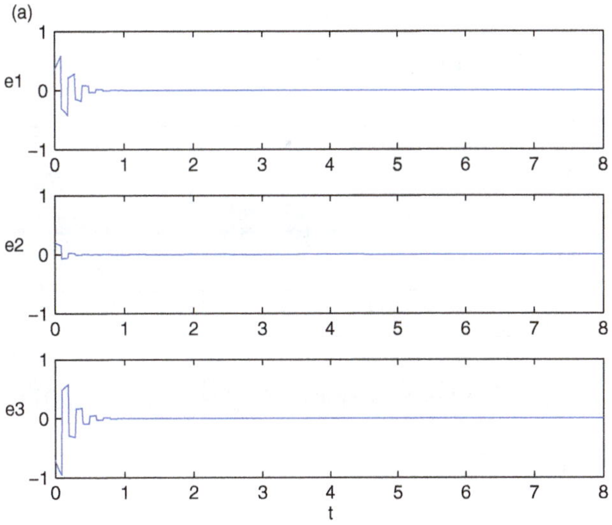

Figure 3. Time response of the synchronization error system with $K = -1.5, \delta = 0.1$.

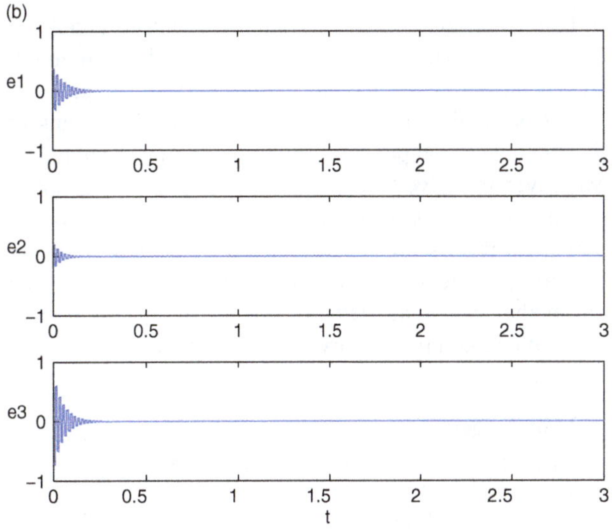

Figure 4. Time response of the synchronization error system with $K = -1.8, \delta = 0.01$.

Figure 2 shows the stable region for different $\gamma's$. The whole region under the curve of $\gamma = 1$ is the predicted stable region. When $\gamma \to \infty$, the stable region approaches a vertical line $K = -1$. Figures 3 and 4 show the stable results within the stable region for $K = -1.5$, $\delta = 0.1$ and for $K = -1.8$, $\delta = 0.01$, respectively. Figure 5 shows the unstable results outside the stable region for $K = -1.99$, $\delta = 0.1$.

Remark 4 When $(K, \delta) = (-1.5, 0.1)$, we note that the point $(-1.5, 0.1)$ falls outside of the stable region predicted in Yang and Chua (1997) and Li et al. (2005), respectively. Hence, our method is less conservative than those in Yang and Chua (1997) and Li et al. (2005).

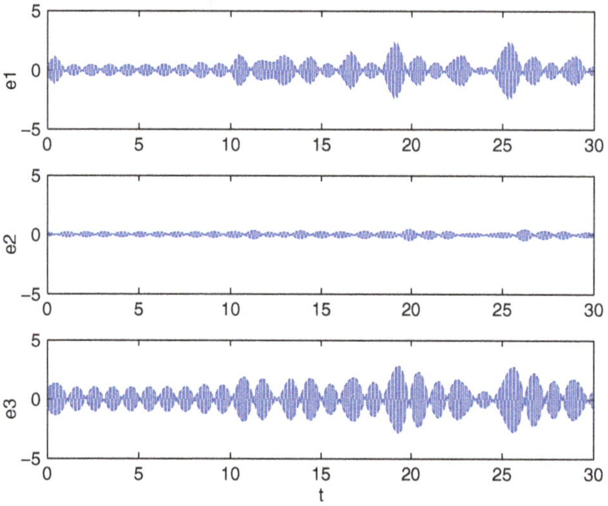

Figure 5. Time response of the synchronization error system with $K = -1.99, \delta = 0.1$.

5. Conclusion

In this paper, the impulsive synchronization of a class of chaotic systems have been investigated, in which an impulsive control scheme has been proposed and some new sufficient conditions have been established by means of the impulsive comparison principle and LMIs approach. Moreover, the drive system and the response system can be synchronized to within a desired control matrix. Finally, the numerical results have confirmed that these new sufficient conditions are less conservative than the existing ones, and our method can obtain better estimation of the boundary of the stable region than the existing approaches.

Acknowledgements

The authors thank the executive editor and the anonymous reviewers for their carefully comments and suggestions to improve the quality of the paper. This work is jointly supported by the Opening Fund of Artificial Intelligence Key Laboratory of the Sichuan Province under Grant 2011RK01, Opening Fund of Geomathematics Key Laboratory of the Sichuan Province under Grant scsxdz2011010, the scientific research fund of Sichuan University of Science and Engineering under Grant 2011PY08 and the Graduate innovation fundation of Sichuan University of Science and Engineering under Grant y2013021.

References

Bhowmick, S. K., Pal, P., Roy, P. K., & Dana, S. K. (2012). Lag synchronization and scaling of chaotic attractor in coupled system. *Chaos, 22*(2), 023151.

Boyd, S., Ghaoui, L., Feron, E., & Balakrishnan, V. (1994). *Linear matrix inequalities in system and control theory*. Philadelphia, PA: SIAM.

Chen, Y.-S., & Chang, C.-C. (2006). Simple adaptive synchronization of chaotic systems with random components. *Chaos, 16*(2), 023128.

Ge, Z.-M., & Chen, C.-C. (2004). Phase synchronization of coupled chaotic multiple time scales systems. *Chaos, Solitons & Fractals, 20*(3), 639–647.

He, W., & Cao, J. (2009). Generalized synchronization of chaotic systems: An auxiliary system approach via matrix measure. *Chaos, 19*(1), 013118.

He, W., Qian, F., Cao, J., & Han, H. (2011). Impulsive synchronization of two nonidentical chaotic systems with time-varying delay. *Physics Letters A, 375*(3), 498–504.

Khalil, H. K. (1996). *Nonlinear systems*. Upper Saddle River, NJ: Prentice-Hall.

Li, C., Liao, X., & Zhang, X. (2005). Impulsive synchronization of chaotic systems. *Chaos, 15*(2), 023104.

Li, Z. G., Wen, C. Y., & Soh, Y. C. (2001). Analysis and design of impulsive control systems. *IEEE Transactions on Automatic Control, 46*(6), 894–897.

Lorenz, E. N. (1995). *The essence of chaos*. Seattle: University of Washington Press.

Mahmoud, G. M., & Mahmoud, E. E. (2010). Complete synchronization of chaotic complex nonlinear systems with uncertain parameters. *Nonlinear Dynamics, 62*(4), 875–882.

Pecora, L. M., & Carroll, T. L. (1990). Synchronization in chaotic systems. *Physical Review Letters, 64*(8), 821–824.

Sun, J., Zhang, Y., & Wu, Q. (2002). Impulsive control for the stabilization and synchronization of Lorenz systems. *Physics Letters A, 298*(2), 153–160.

Sun, J., Zhang, Y., & Wu, Q. (2003). Less conservative conditions for asymptotic stability of impulsive control systems. *IEEE Transactions on Automatic Control, 48*(5), 829–831.

Yang, T. (1999). Impulsive control. *IEEE Transactions on Automatic Control, 44*(5), 1081–1083.

Yang, T. (2001). *Impulsive control theory*. Berlin: Springer.

Yang, T., & Chua, L. O. (1997). Impulsive stabilization for control and synchronization of chaotic systems: Theory and application to secure communication. *IEEE Transactions on Circuits and Systems I: Fundamental Theory and Applications, 44*(10), 976–988.

Yang, T., Yang, L.-B., & Yang, C.-M. (1997). Impulsive synchronization of Lorenz systems. *Physics Letters A, 226*(6), 349–354.

17

Hopf bifurcation in a partial dependent predator–prey system with multiple delays

Qingsong Liu and Yiping Lin*

Department of Applied Mathematics, Kunming University of Science and Technology, Kunming, Yunnan 650500, People's Republic of China

In this paper, a partial dependent predator–prey system with multiple delays is investigated. By choosing τ_1, τ_2 and τ_3 as bifurcating parameters, we show that Hopf bifurcations occur. In addition, by using theory of functional differential equation and Hassard's method, explicit algorithms for determining the direction of the Hopf bifurcation and the stability of bifurcating periodic solutions are derived. Finally, numerical simulations are performed to support the analytical results, and the chaotic behaviors are observed.

Keywords: multiple delays; Hopf bifurcation; chaos; predator–prey system

1. Introduction

Since the work of Volterra, the Lotka–Volterra system has been extensively investigated. A classical Lotka–Volterra system can be modeled by the following system:

$$\begin{aligned} \dot{x} &= x(r_1 + a_{11}x + a_{12}y), \\ \dot{y} &= y(r_2 + a_{21}x + a_{22}y), \end{aligned} \tag{1}$$

where x and y can be interpreted as the population densities of prey and predator at time t, respectively (for example, Jin & Ma, 2002; Saito, 2002; Tang & Zhou, 2003). For a long time, it has been recognized that delays can have a very complicated impact on the dynamics of a system, which cannot only cause the loss of stability, but also induce various oscillations and periodic solutions (for example, Faria & Magalhaes, 1995; Song, Han, & Peng, 2004; Song, Peng, & Wei, 2008; Sun, Lin, & Han, 2006; Yan & Chu, 2006). Recently, Zhang, Jin, Yuan, & Sun (2009) investigated the competition system with a single delay:

$$\begin{aligned} \dot{x}(t) &= x(t)(r_1 - a_{11}x(t-\tau) - a_{12}y(t-\tau)), \\ \dot{y}(t) &= y(t)(r_2 - a_{21}x(t-\tau) - a_{22}y(t-\tau)). \end{aligned} \tag{2}$$

Choosing the delay τ as the bifurcation parameter, they analyzed the stability of the interior positive equilibrium and the existence of the local Hopf bifurcation for system (2). Their results show that there exist critical values of the delay, as the delay passes through the first critical value, the positive equilibrium loses its stability and the Hopf bifurcation occurs. Further increasing the delay beyond the first critical value, the system goes into oscillations.

In the present paper, we devote our attention to the bifurcating phenomenons of the predator–prey system with three delays. So the system is not only more complicated, but also more close to the actuality. We described the system by

$$\begin{aligned} \dot{x}(t) &= x(t)(r_1 - a_{11}x(t)) - a_{12}x(t-\tau_1)y(t-\tau_2), \\ \dot{y}(t) &= y(t)(r_2 - a_{22}y(t)) + a_{21}x(t-\tau_1)y(t-\tau_3), \end{aligned} \tag{3}$$

where $r_1, r_2, a_{11}, a_{12}, a_{21}$ and a_{22} are all positive constants. r_2 is positive, which denotes that the food of the predator is partially dependent on the prey of the system. In this paper, assuming that the predator not only takes time to hunt prey, but also takes time to digest and the predator only feeds on the mature prey, then τ_1 denotes the time of the prey maturation, while τ_2 denotes the time taken for hunting of the maturation prey, called hunting delay, finally, τ_3 denotes the time predator use for digestion.

This paper is organized as follows: in Section 2, we investigate the effect of the time delays τ_1, τ_2 and τ_3 on the stability of the positive equilibrium of system (3). In Section 3, we derive the direction and stability of the Hopf bifurcation by using normal form and central manifold theory. Finally in Section 4, numerical simulations are carried out to illustrate the theoretical prediction and to explore the complex dynamics including chaos.

2. Stability analysis and Hopf bifurcation

It is easy to see that system (3) has a unique positive equilibrium $E_*(x_*, y_*)$ provided that the following condition

*Corresponding author. Email: linyiping689@163.com

is satisfied:

$$(H_*)\quad a_{22}r_1 - a_{12}r_2 > 0,$$

where

$$x_* = \frac{a_{22}r_1 - a_{12}r_2}{a_{11}a_{22} + a_{12}a_{21}}, \quad y_* = \frac{a_{11}r_2 + a_{21}r_1}{a_{11}a_{22} + a_{12}a_{21}}.$$

Let $\bar{x} = x - x_*, \bar{y} = y - y_*$, and still denote by $\bar{x} = x, \bar{y} = y$, system (3) can be written as

$$\dot{x} = \alpha_1 x(t) + \alpha_2 x(t-\tau_1) + \alpha_3 y(t-\tau_2) + f_1',$$
$$\dot{y} = \alpha_4 y(t) + \alpha_5 x(t-\tau_1) + \alpha_6 y(t-\tau_3) + f_2',$$

where

$$\alpha_1 = r_1 + 2a_{11}x_*, \quad \alpha_2 = a_{12}y_*, \quad \alpha_3 = a_{12}x_*,$$
$$\alpha_4 = r_2 + 2a_{22}y_*, \quad \alpha_5 = -a_{21}y_*, \quad \alpha_6 = -a_{21}x_*,$$
$$f_1' = -a_{11}x^2 - a_{12}x(t-\tau_1)y(t-\tau_2),$$
$$f_2' = -a_{22}y^2 + a_{21}x(t-\tau_1)y(t-\tau_3).$$

We then obtain the linearized system

$$\dot{x} = \alpha_1 x(t) + \alpha_2 x(t-\tau_1) + \alpha_3 y(t-\tau_2),$$
$$\dot{y} = \alpha_4 y(t) + \alpha_5 x(t-\tau_1) + \alpha_6 y(t-\tau_3).$$

The corresponding characteristic equation is

$$\lambda^2 - A\lambda + B + (C\lambda + D)\,e^{-\lambda\tau_1} + (E\lambda + F)\,e^{-\lambda\tau_3} + G\,e^{-\lambda(\tau_1+\tau_3)} - H\,e^{-\lambda(\tau_1+\tau_2)} = 0, \qquad (4)$$

where

$$A = \alpha_1 + \alpha_4, \quad B = \alpha_1\alpha_4, \quad C = -\alpha_2,$$
$$D = \alpha_2\alpha_4, \quad E = -\alpha_6, \quad F = \alpha_1\alpha_6,$$
$$G = \alpha_2\alpha_6, \quad H = \alpha_3\alpha_5.$$

Case (a1) $\tau_1 = \tau_2 = \tau_3 = 0$, Equation (4) becomes

$$\lambda^2 - (A - C - E)\lambda + B + D + F + G - H = 0.$$

All roots have negative real parts if and only if

$$A - C - E < 0, \quad B + D + F + G - H > 0. \qquad (5)$$

THEOREM 2.1 *For $\tau_1 = \tau_2 = \tau_3 = 0$, the interior equilibrium point E_* is locally asymptotically stable if conditions (H_*) and (5) hold.*

Case (b1) $\tau_2 = \tau_3 = 0, \tau_1 > 0$.

THEOREM 2.2 *For $\tau_2 = \tau_3 = 0$, assume that (H_*) and $(B+F)^2 < (D+G-H)^2$ hold, the interior equilibrium point E_* is locally asymptotically stable for $0 < \tau_1 < \tau_{11_0}$ and it undergoes the Hopf bifurcation at $\tau_1 = \tau_{11_0}$ given by*

$$\tau_{11_0} = \frac{1}{\omega_{11}}\arccos \times \left\{\frac{[\omega_{11}^2 - (B+F)](D+G-H) + C(A-E)\omega_{11}^2}{(D+G-H)^2 + C^2\omega_{11}^2}\right\}. \qquad (6)$$

Proof On substituting $\tau_2 = \tau_3 = 0$, the characteristic equation (4) becomes

$$\lambda^2 - (A-E)\lambda + B + F + (C\lambda + D + G - H)\,e^{-\lambda\tau_1} = 0. \qquad (7)$$

Let $i\omega\ (\omega > 0)$ be a purely imaginary root of Equation (7), then it follows that

$$C\omega \sin\omega\tau_1 + (D+G-H)\cos\omega\tau_1 = -\omega^2 - (B+F),$$
$$C\omega\cos\omega\tau_1 - (D+G-H)\sin\omega\tau_2 = (A-E)\omega.$$

Squaring both sides and adding them up, we get the following polynomial equation:

$$\omega^4 + [(A-E)^2 - C^2 - 2(B+F)]\omega^2 + [(B+F)^2 - (D+G-H)^2] = 0. \qquad (8)$$

Equation (8) has unique positive root ω_{11}^2 if

$$(B+F)^2 < (D+G-H)^2.$$

The corresponding critical value of time delay τ_{11_n} is

$$\tau_{11_n} = \frac{1}{\omega_{11}}\arccos \times \left\{\frac{[\omega_{11}^2 - (B+F)](D+G-H) + C(A-E)\omega_{11}^2}{(D+G-H)^2 + C^2\omega_{11}^2}\right\} + \frac{2n\pi}{\omega_{11}}, \quad n = 0, 1, 2, \ldots$$

Let $\lambda(\tau_{11_n}) = \pm i\omega_{11}$ be the root of Equation (7), then the transversal condition can be obtained as

$$\left(\frac{d\lambda}{d\tau_1}\right)^{-1}_{\tau_1 = \tau_{11_n}} = \frac{(2\lambda - (A-E))\,e^{\lambda\tau_1}}{\lambda(C\lambda + D + G - H)} + \frac{C}{\lambda(C\lambda + D + G - H)} - \frac{\tau_1}{\lambda}.$$

Since

$$\operatorname{Sign}\left\{\frac{d(\operatorname{Re}\lambda(\tau_1))}{d\tau_1}\right\}^{-1} = \operatorname{Sign}\left\{\left[\operatorname{Re}\frac{(2\lambda - (A-E))\,e^{\lambda\tau_1}}{\lambda(C\lambda + D + G - H)}\right] + \left[\operatorname{Re}\frac{C}{\lambda(C\lambda + D + G - H)}\right]\right\},$$

we can obtain

$$\begin{aligned}
&\mathrm{Sign}\left\{\frac{\mathrm{d}(\mathrm{Re}\lambda(\tau_1))}{\mathrm{d}\tau_1}\right\}^{-1}\\
&= \mathrm{Sign}\left\{\mathrm{Re}\left[\frac{(E_A)\cos\omega_{11}\tau_1 - 2\omega_{11}\sin\omega_{11}\tau_1}{-C\omega_{11}^2 + \mathrm{i}(D+G-H)\omega_{11}}\right]\right.\\
&\quad +\mathrm{Re}\left[\mathrm{i}\frac{(E_A)\sin\omega_{11}\tau_1 + 2\omega_{11}\cos\omega_{11}\tau_1}{-C\omega_{11}^2 + \mathrm{i}(D+G-H)\omega_{11}}\right]\\
&\quad \left.+\mathrm{Re}\left[\frac{C}{-C\omega_{11}^2 + \mathrm{i}(D+G-H)\omega_{11}}\right]\right\}\\
&= \mathrm{Sign}\left\{\frac{(A-E)^2 - C^2 - 2(B+F) + 2\omega_{11}^2}{C^2\omega_{11}^2 + (D+G-H)^2}\right\}\\
&= \mathrm{Sign}\left\{\frac{\sqrt{\begin{array}{c}[(A-E)^2 - C^2 - 2(B+F)]^2\\ -4[(B+F)^2 - (D+G-H)^2]\end{array}}}{C^2\omega_{11}^2 + (D+G-H)^2}\right\} > 0,
\end{aligned}$$

then we have

$$\left.\frac{\mathrm{d}(\mathrm{Re}\lambda)}{\mathrm{d}\tau_1}\right|_{\tau_1=\tau_{11_0}} > 0.$$

■

Case (b2) $\tau_1 = \tau_3 = 0,\ \tau_2 > 0.$

THEOREM 2.3 *For $\tau_1 = \tau_3 = 0$, assume that (H_*) and $(B+D+G+F)^2 < H^2$ hold, the interior equilibrium point E_* is locally asymptotically stable for $0 < \tau_2 < \tau_{21_0}$ and it undergoes the Hopf bifurcation at $\tau_2 = \tau_{21_0}$ given by*

$$\tau_{21_0} = \frac{1}{\omega_{21}}\arccos\left(\frac{-\omega_{21}^2 + B + D + F + G}{H}\right), \tag{9}$$

where $\mathrm{i}\omega_{21}$ $(\omega_{21} > 0)$ is a root of corresponding characteristic equation.

The proof is similar as in case (b1).

Case (b3) $\tau_1 = \tau_2 = 0,\ \tau_3 > 0.$

THEOREM 2.4 *For $\tau_1 = \tau_2 = 0$, assume that (H_*) and $(B+D-H)^2 < (F+G)^2$ hold, the interior equilibrium point E_* is locally asymptotically stable for $0 < \tau_3 < \tau_{31_0}$ and it undergoes the Hopf bifurcation at $\tau_3 = \tau_{31_0}$ given by*

$$\tau_{31_0} = \frac{1}{\omega_{31}}\arccos \times\left[\frac{(\omega_{31}^2 - B - D + H)(F+G) + E(A-C)\omega_{31}^2}{(F+G)^2 + E^2\omega_{31}^2}\right], \tag{10}$$

where $\mathrm{i}\omega_{31}$ $(\omega_{31} > 0)$ is a root of corresponding characteristic equation.

The proof is similar as in case (b1).

Case (c1) $\tau_3 = 0$, τ_2 is fixed in the interval $(0, \tau_{21_0})$ and $\tau_1 > 0$.

THEOREM 2.5 *Let $(B+F)^2 < (D-H)^2 + G^2 - 2HG$, $\tau_3 = 0$ and $\tau_2 \in (0, \tau_{21_0})$, if (H_*) holds, then the equilibrium E_* is asymptotically stable for $\tau_1 \in (0, \tau_{12_0})$, and system (3) undergoes Hopf bifurcation at E_* when $\tau_1 = \tau_{12_0}$, where*

$$\tau_{12_0} = \frac{1}{\omega_{12}}\arccos\left(\frac{B_1D_1 - A_1C_1}{B_1^2 + C_1^2}\right). \tag{11}$$

Proof We know $\tau_3 = 0$, τ_2 is in its stable interval and τ_1 is considered as a parameter. Let $\mathrm{i}\omega$ $(\omega > 0)$ be a root of Equation (4). Separating real and imaginary parts leads to

$$\begin{aligned}
\omega^2 - (B+F) &= (C\omega + H\sin\omega\tau_2)\sin\omega\tau_1\\
&\quad - (H\cos\omega\tau_2 - D - G)\cos\omega\tau_1,\\
(A-E)\omega &= (C\omega + H\sin\omega\tau_2)\cos\omega\tau_1\\
&\quad + (H\cos\omega\tau_2 - D - G)\sin\omega\tau_1.
\end{aligned} \tag{12}$$

It can give

$$H(\omega) = \omega^4 + l_1\omega^2 + l_2\omega + l_3 = 0, \tag{13}$$

where

$$\begin{aligned}
l_1 &= (A-E)^2 - 2(B+F) - C^2,\\
l_2 &= -2HC\sin\omega\tau_2,\\
l_3 &= (B+F)^2 - H^2 - D^2 - G^2 + 2HD\cos\omega\tau_2\\
&\quad + 2HG\cos\omega\tau_2.
\end{aligned}$$

We assumed that

$$(B+F)^2 < (D-H)^2 + G^2 - 2HG,$$

then $H(0) < 0$ and $H(\infty) = \infty$.

Without going into detailed analysis with Equation (13), it is assumed there exists at least one real positive root ω_{12}. Now Equation (12) can be written as

$$\begin{aligned}
A_1 &= B_1\sin\omega_{12}\tau_1 - C_1\cos\omega_{12}\tau_1,\\
D_1 &= B_1\cos\omega_{12}\tau_1 + C_1\sin\omega_{12}\tau_1,
\end{aligned} \tag{14}$$

where

$$\begin{aligned}
A_1 &= \omega_{12}^2 - (B+F), \quad B_1 = C\omega_{12} + H\sin\omega_{12}\tau_2,\\
C_1 &= H\cos\omega_{12}\tau_2 - D - G, \quad D_1 = (A-E)\omega_{12}.
\end{aligned}$$

Equations (14) are simplified to give

$$\tau_{12_j} = \frac{1}{\omega_{12}}\arccos\frac{B_1D_1 - A_1C_1}{B_1^2 + C_1^2} + \frac{2j\pi}{\omega_{12}}, \quad j = 0, 1, 2, \ldots$$

and $\pm\mathrm{i}\omega_{12}$ is purely imaginary root of Equation (4) for $\tau_2 \in [0, \tau_{21_0}]$. Now verify the transversal condition of the Hopf bifurcation. Differentiating Equation (4) with respect to τ_1, we obtain the following:

$$\left(\frac{d\lambda}{d\tau_1}\right)_{\tau_1=\tau_{12_j}} = \frac{P_1 + iQ_1}{R_1 + iS_1},$$

where

$$\begin{aligned}
P_1 &= -A + E - (D+G)\tau_{12_j}\cos\omega_{12}\tau_{12_j} \\
&\quad + C\omega_{12}\sin\omega_{12}\tau_{12_j}(1-\tau_{12_j}) \\
&\quad + H(\tau_{12_j}+\tau_2)\cos\omega_{12}(\tau_{12_j}+\tau_2), \\
Q_1 &= \omega_{12}(C\tau_{12_j})\cos\omega_{12}\tau_{12_j}) \\
&\quad + H(\tau_{12_j}+\tau_2)\sin\omega_{12}(\tau_{12}+\tau_2) \\
&\quad - (C - (D+G)\tau_{12_j}\sin\omega_{12}\tau_{12_j}), \\
R_1 &= -C\omega_{12}^2\cos\omega_{12}\tau_{12_j} + (D+G)\omega_{12}\sin\omega_{12}\tau_{12_j} \\
&\quad - H\omega_{12}\sin\omega_{12}(\tau_{12_j+\tau_2}), \\
S_1 &= \omega_{12}(C\omega_{12}\sin\omega_{12}\tau_{12_j} - (D+G)\cos\omega_{12}\tau_{12_j}) \\
&\quad - H\omega_{12}\cos\omega_{12}(\tau_{12_j+\tau_2}),
\end{aligned}$$

then

$$(R_1^2 + S_1^2)\left[\mathrm{Re}\frac{d(\lambda)}{d\tau_1}\right]_{\tau_1=\tau_{11_j}} = P_1R_1 + Q_1S_1 \neq 0.$$

Noting that

$$\mathrm{Sign}\left[\mathrm{Re}\left(\frac{d\lambda}{d\tau_1}\right)\right]_{\tau_1=\tau_{11_0}} = \mathrm{Sign}\left[\mathrm{Re}\left(\frac{d\lambda}{d\tau_1}\right)^{-1}\right]_{\tau_1=\tau_{11_0}}.$$

■

Case (c2) $\tau_3 = 0$, τ_1 is fixed in the interval $(0, \tau_{11_0})$ and $\tau_2 > 0$.

THEOREM 2.6 *Let* $(B+F+D+G)^2 < H^2$, $\tau_3 = 0$ *and* $\tau_1 \in (0, \tau_{11_0})$, *if* (H_*) *holds, then the equilibrium* E_* *is asymptotically stable for* $\tau_2 \in (0, \tau_{22_0})$, *and system* (3) *undergoes Hopf bifurcation at* E_* *when* $\tau_2 = \tau_{22_0}$, *where*

$$\tau_{22_0} = \frac{1}{\omega_{22}}\arccos\left[\frac{(A_2 - B_2 - C_2)H_2 + (D_2 - E_2 + F_2)G_2}{G_2^2 + H_2^2}\right], \tag{15}$$

$i\omega_{22}$ $(\omega_{22} > 0)$ *is a root of the corresponding characteristic equation and*

$$\begin{aligned}
A_2 &= \omega_{22}^2 - (B+F), \quad B_2 = C\omega_{22}\sin\omega_{22}\tau_1, \\
C_2 &= (D+G)\cos\omega_{22}\tau_1, \quad D_2 = (A-E)\omega_{22}, \\
E_2 &= C\omega_{22}\cos\omega_{22}\tau_1, \quad F_2 = (D+G)\sin\omega_{22}\tau_1, \\
G_2 &= H\sin\omega_{22}\tau_1, \quad H_2 = H\cos\omega_{22}\tau_1.
\end{aligned}$$

The proof is similar as in case (c1).

Case (c3) $\tau_2 = 0$, τ_3 is fixed in the interval $(0, \tau_{31_0})$ and $\tau_1 > 0$.

THEOREM 2.7 *Let* $(B+F)^2 < D^2 + G^2$, $\tau_2 = 0$ *and* $\tau_3 \in (0, \tau_{31_0})$, *if* (H_*) *holds, then the equilibrium* E_* *is asymptotically stable for* $\tau_1 \in (0, \tau_{13_0})$, *and system* (3) *undergoes Hopf bifurcation at* E_* *when* $\tau_1 = \tau_{13_0}$, *where*

$$\tau_{13_0} = \frac{1}{\omega_{13}}\arccos\left[\frac{(A_3 - B_3 - C_3)H_3 + (D_3 - E_3 + F_3)G_3}{G_3^2 + H_3^2}\right], \tag{16}$$

$i\omega_{13}$ $(\omega_{13} > 0)$ *is a root of the corresponding characteristic equation and*

$$\begin{aligned}
A_3 &= \omega_{13}^2 - B, \quad B_3 = E\omega_{13}\sin\omega_{13}\tau_3, \\
C_3 &= F\cos\omega_{13}\tau_3, \quad D_3 = A\omega_{13}, \\
E_3 &= E\omega_{13}\cos\omega_{13}\tau_3, \quad F_3 = F\sin\omega_{13}\tau_3, \\
G_3 &= C\omega_{13} - G\sin\omega_{13}\tau_3, \quad H_3 = D - H + G\cos\omega_{13}\tau_3.
\end{aligned}$$

The proof is similar as in case (c1).

Case (c4) $\tau_2 = 0$, τ_1 is fixed in the interval $(0, \tau_{11_0})$ and $\tau_3 > 0$.

THEOREM 2.8 *Let* $(B+D-H)^2 < F^2 + G^2, \tau_2 = 0$ *and* $\tau_1 \in (0, \tau_{11_0})$, *if* (H_*) *holds, then the equilibrium* E_* *is asymptotically stable for* $\tau_3 \in (0, \tau_{32_0})$, *and system* (3) *undergoes Hopf bifurcation at* E_* *when* $\tau_3 = \tau_{32_0}$, *where*

$$\tau_{32_0} = \frac{1}{\omega_{32}}\arccos\left[\frac{(A_4 - B_4 - C_4)H_4 + (D_4 - E_4 + F_4)G_4}{G_4^2 + H_4^2}\right], \tag{17}$$

$i\omega_{32}$ $(\omega_{32} > 0)$ *is a root of corresponding the characteristic equation and*

$$\begin{aligned}
A_4 &= \omega_{32}^2 - B, \quad B_4 = C\omega_{32}\sin\omega_{32}\tau_1, \\
C_4 &= (D-H)\cos\omega_{32}, \quad E_4 = C\omega_{32}\cos\omega_{32}\tau_1, \\
F_4 &= (D-H)\sin\omega_{32}\tau_1, \\
G_4 &= E\omega_{32} - G\sin\omega_{32}\tau_1, \\
H_4 &= F + G\cos\omega_{32}\tau_1.
\end{aligned}$$

The proof is similar as in case (c1).

Case (c5) $\tau_1 = 0$, τ_3 is fixed in the interval $(0, \tau_{31_0})$ and $\tau_2 > 0$.

THEOREM 2.9 *Let* $(B+D+F+G)^2 < H^2, \tau_1 = 0$ *and* $\tau_3 \in (0, \tau_{31_0})$, *if* (H_*) *holds, then the equilibrium* E_* *is asymptotically stable for* $\tau_2 \in (0, \tau_{23_0})$, *and system* (3) *undergoes Hopf bifurcation at* E_* *when* $\tau_2 = \tau_{23_0}$, *where*

$$\tau_{23_0} = \frac{1}{\omega_{23}}\arccos\left(\frac{A_5 + B_5 + C_5}{H}\right), \tag{18}$$

$i\omega_{23}$ $(\omega_{23} > 0)$ *is a root of the corresponding characteristic equation and*

$$A_5 = -\omega_{23}^2 + B + D, \quad B_5 = E\omega_{23}\sin\omega_{23}\tau_3,$$
$$C_5 = (F+G)\cos\omega_{23}\tau_3.$$

The proof is similar as in case (c1).

Case (c6) $\tau_1 = 0$, τ_2 is fixed in the interval $(0, \tau_{21_0})$ and $\tau_3 > 0$.

THEOREM 2.10 *Let* $(B+D-H)^2 < (F+G)^2$, $\tau_1 = 0$ *and* $\tau_2 \in (0, \tau_{21_0})$, *if* (H_*) *holds, then the equilibrium* E_* *is asymptotically stable for* $\tau_3 \in (0, \tau_{33_0})$, *and system* (3) *undergoes Hopf bifurcation at* E_* *when* $\tau_3 = \tau_{33_0}$, *where*

$$\tau_{33_0} = \frac{1}{\omega_{33}} \arccos\left[\frac{(A_6+B_6)H_6 + (D_6-E_6)G_6}{G_6^2 + H_6^2}\right], \quad (19)$$

$i\omega_{33}$ $(\omega_{33} > 0)$ *is a root of the corresponding characteristic equation and*

$$A_6 = \omega_{33}^2 - (B+D), \quad B_6 = H\cos\omega_{33}\tau_2,$$
$$D_6 = (A-C)\omega_{33}, \quad E_6 = H\sin\omega_{33}\tau_2,$$
$$G_6 = E\omega_{33}, \quad H_6 = (F+G).$$

The proof is similar as in case (c1).

Case (d1) τ_2 and τ_3 are fixed in the interval $(0, \tau_{21_0})$ and $(0, \tau_{31_0})$, $\tau_1 > 0$.

THEOREM 2.11 *Let* $(B+F)^2 < (D+G)^2 + H^2 - 2HG - 2HD$, $\tau_2 \in (0, \tau_{21_0})$ and $\tau_3 \in (0, \tau_{31_0})$, *if* (H_*) *holds, then the equilibrium* E_* *is asymptotically stable for* $\tau_1 \in (0, \tau_{14_0})$, *and system* (3) *undergoes Hopf bifurcation at* E_* *when* $\tau_1 = \tau_{14_0}$, *where*

$$\tau_{14_0} = \frac{1}{\omega_{14}} \arccos\frac{(A_7-B_7-C_7)H_7 + (D_7-E_7+F_7)G_7}{G_7^2 + H_7^2}. \quad (20)$$

Proof We know τ_2, τ_3 are in its stable interval and τ_1 is considered as a parameter. Let $i\omega$ $(\omega > 0)$ be a root of Equation (4). Separating real and imaginary parts leads to

$$\begin{aligned} &\omega^2 - B - E\omega\sin\omega\tau_3 - F\cos\omega\tau_3 \\ &\quad = (C\omega - G\sin\omega\tau_3 + H\sin\omega\tau_2)\sin\omega\tau_1 \\ &\qquad + (D - H\cos\omega\tau_2 + G\cos\omega\tau_3)\cos\omega\tau_1, \\ &A\omega - E\omega\cos\omega\tau_3 + F\sin\omega\tau_3 \\ &\quad = (C\omega - G\sin\omega\tau_3 + H\sin\omega\tau_2)\cos\omega\tau_1 \\ &\qquad - (D - H\cos\omega\tau_2 + G\cos\omega\tau_3)\sin\omega\tau_1. \end{aligned} \quad (21)$$

It gives

$$H(\omega) = \omega^4 + k_1\omega^3 + k_2\omega^2 + k_3\omega + k_4 = 0, \quad (22)$$

where

$$\begin{aligned} k_1 &= -2E\sin\omega\tau_3, \\ k_2 &= -2B - C + A^2 - 2(F+AE)\cos\omega\tau_3, \\ k_3 &= 2(BE + CG + AF)\sin\omega\tau_3 - 2CH\sin\omega\tau_2, \\ k_4 &= F^2 + B^2 - D^2 - G^2 - H^2 + 2(BF - DG)\cos\omega\tau_3 \\ &\quad + 2DH\cos\omega\tau_2 + 2GH\cos\omega(\tau_2 - \tau_3). \end{aligned}$$

We have assumed that

$$(B+F)^2 < (D+G)^2 + H^2 - 2HG - 2HD,$$

then $H(0) < 0$ and $H(\infty) = \infty$.

Without going into detailed analysis with Equation (22), it is assumed there exists at least one real positive root ω_{14}. Now Equation (21) can be written as

$$\begin{aligned} A_7 - B_7 - C_7 &= G_7\sin\omega_{14}\tau_1 + H_7\cos\omega_{14}\tau_1, \\ D_7 - E_7 + F_7 &= G_7\cos\omega_{14}\tau_1 - H_7\sin\omega_{14}\tau_1, \end{aligned} \quad (23)$$

where

$$\begin{aligned} &A_7 = \omega_{14}^2 - B, \quad B_7 = E\omega_{14}\sin\omega_{14}\tau_3, \\ &C_7 = F\cos\omega_{14}\tau_3, \quad D_7 = (A-E)\omega_{12}, \\ &E_7 = E\omega_{14}\cos\omega_{14}\tau_3, \quad F_7 = F\sin\omega_{14}\tau_3, \\ &G_7 = C\omega_{14} - G\sin\omega_{14}\tau_3 + H\sin\omega_{12}\tau_2, \\ &H_7 = D - H\cos\omega_{14}\tau_2 + G\cos\omega_{14}\tau_3. \end{aligned}$$

Equations (23) are simplified to give

$$\tau_{14_j} = \frac{1}{\omega_{14}}\arccos\frac{(A_7-B_7-C_7)H_7 + (D_7-E_7+F_7)G_7}{G_7^2+H_7^2} + \frac{2j\pi}{\omega_*}, \quad j = 0, 1, 2, \ldots$$

and $\pm i\omega_{14}$ are purely imaginary root of Equation (4) for $\tau_2 \in [0, \tau_{21_0}]$, $\tau_3 \in [0, \tau_{31_0}]$. Now verify the transversal condition of the Hopf bifurcation, differentiating Equation (4) with respect to τ_1, it is obtained that

$$\left(\frac{d\lambda}{d\tau_1}\right)_{\tau_1 = \tau_{14_j}} = \frac{P_2 + iQ_2}{R_2 + iS_2},$$

where

$$\begin{aligned} P_2 &= (C - G\tau_3\cos\omega_{14}\tau_3 + H\tau_2\cos\omega_{14}\tau_2)\cos\omega_{14}\tau_{14_j} \\ &\quad - (C\omega_{14} - G\tau_3\sin\omega_{14}\tau_3 + H\tau_2\sin\omega_{14}\tau_2) \\ &\quad \times \tau_{14_j}\sin\omega_{14}\tau_{14_j} - (H\tau_2\sin\omega_{14}\tau_2 - G\tau_3\sin\omega_{14}\tau_3) \\ &\quad \times \sin\omega_{14}\tau_{14_j} - (D + G\cos\omega_{14}\tau_3 - H\cos\omega_{14}\tau_2) \\ &\quad \times \tau_{14_j}\cos\omega_{14}\tau_{14_j} - (A - E\cos\omega_{14}\tau_3 \\ &\quad + E\tau_3\omega_{14}\sin\omega_{14}\tau_3 + F\tau_3\cos\omega_{14}\tau_3), \end{aligned}$$

$$\begin{aligned} Q_2 = &(C - G\tau_3 \cos\omega_{14}\tau_3 + H\tau_2 \cos\omega_{14}\tau_2) \sin\omega_{14}\tau_{14_j} \\ &+ (C\omega_{14} - G\tau_3 \sin\omega_{14}\tau_3 + H\tau_2 \sin\omega_{14}\tau_2)\tau_{14_j} \\ &\times \cos\omega_{14}\tau_{14_j} + (H\tau_2 \sin\omega_{14}\tau_2 - G\tau_3 \sin\omega_{14}\tau_3) \\ &\times \cos\omega_{14}\tau_{14_j} - (D + G\cos\omega_{14}\tau_3 - H\cos\omega_{14}\tau_2)\tau_{14_j} \\ &\times \sin\omega_{14}\tau_{14_j} - (2\omega_{14} - E\sin\omega_{14}\tau_3 \\ &- E\tau_3\omega_{14}\sin\omega_{14}\tau_3 + F\tau_3 \sin\omega_{14}\tau_3), \\ R_2 = &-(C\omega_{14} - G\sin\omega_{14}\tau_3 + H\sin\omega_{14}\tau_2)\omega_{14}\cos\omega_{14}\tau_{14_j} \\ &+ (D + G\cos\omega_{14}\tau_3 - H\cos\omega_{14}\tau_2)\omega_{14}\sin\omega_{14}\tau_{14_j}, \\ S_2 = &(C\omega_{14} - G\sin\omega_{14}\tau_3 + H\sin\omega_{14}\tau_2)\omega_{14}\sin\omega_{14}\tau_1 \\ &+ (D + G\cos\omega_{14}\tau_3 - H\cos\omega_{14}\tau_2)\omega_{14}\cos\omega_{14}\tau_{14_j}, \end{aligned}$$

then

$$(R_2^2 + S_2^2)\left[\mathrm{Re}\frac{d(\lambda)}{d\tau_1}\right]_{\tau_1=\tau_{14_j}} = P_2R_2 + Q_2S_2 \neq 0.$$

Noting that

$$\mathrm{Sign}\left[\mathrm{Re}\left(\frac{d\lambda}{d\tau_1}\right)\right]_{\tau_1=\tau_{14_0}} = \mathrm{Sign}\left[\mathrm{Re}\left(\frac{d\lambda}{d\tau_1}\right)^{-1}\right]_{\tau_1=\tau_{14_0}}.$$

■

Case (d2) τ_1 and τ_3 are fixed in the interval $(0, \tau_{11_0})$, $(0, \tau_{31_0})$ and $\tau_2 > 0$.

THEOREM 2.12 *Let* $(B+F)^2 + (D+F)^2 + G^2 + 2FG < H^2$, $\tau_1 \in (0, \tau_{11_0})$ *and* $\tau_3 \in (0, \tau_{31_0})$, *if* (H_*) *holds, then the equilibrium* E_* *is asymptotically stable for* $\tau_2 \in (0, \tau_{24_0})$, *and system* (3) *undergoes Hopf bifurcation at* E_* *when* $\tau_2 = \tau_{24_0}$, *where*

$$\tau_{24_0} = \frac{1}{\omega_{24}} \arccos\left[\frac{(D_8 - E_8 - F_8)G_8 - (A_8 - B_8 - C_8)H_8}{G_8^2 + H_8^2}\right], \tag{24}$$

$i\omega_{24}$ $(\omega_{24} > 0)$ *is a root of the corresponding characteristic equation and*

$$\begin{aligned} A_8 &= \omega_{24}^2 - B, \\ B_8 &= (C\omega_{24} - G\sin\omega_{24}\tau_3)\sin\omega_{24}\tau_1 \\ &\quad + (D + G\cos\omega_{24}\tau_3)\cos\omega_{24}\tau_1, \\ C_8 &= E\omega_{24}\sin\omega_{24}\tau_3 + F\cos\omega_{24}\tau_3, \quad D_8 = A\omega_{24}, \\ E_8 &= (C\omega_{24} - G\sin\omega_{24}\tau_3)\cos\omega_{24}\tau_1 \\ &\quad + (D + G\cos\omega_{24}\tau_3)\sin\omega_{24}\tau_1, \\ F_8 &= E\omega_{24}\cos\omega_{24}\tau_3 - F\sin\omega_{24}\tau_3, \\ G_8 &= H\sin\omega_{24}\tau_1, \quad H_8 = H\cos\omega_{24}\tau_1. \end{aligned}$$

The proof is similar as in case (d1).

Case (d3) τ_1 and τ_2 are fixed in the interval $(0, \tau_{11_0})$, $(0, \tau_{21_0})$ and $\tau_3 > 0$.

THEOREM 2.13 *Let* $(B+D)^2 + H^2 < 2BH + 2DH - G^2 - F^2$, $\tau_1 \in (0, \tau_{11_0})$ *and* $\tau_2 \in (0, \tau_{21_0})$, *if* (H_*) *holds, then the equilibrium* E_* *is asymptotically stable for* $\tau_3 \in (0, \tau_{34_0})$, *and system* (3) *undergoes the Hopf bifurcation at* E_* *when* $\tau_3 = \tau_{34_0}$, *where*

$$\tau_{34_0} = \frac{1}{\omega_{34}} \arccos\left[\frac{(A_9 - B_9 + C_9)H_9 + (D_9 - E_9 - F_9)G_9}{G_9^2 + H_9^2}\right], \tag{25}$$

$i\omega_{34}$ $(\omega_{34} > 0)$ *is a root of the corresponding characteristic equation and*

$$\begin{aligned} A_9 &= \omega_{34}^2 - B, \\ B_9 &= C\omega_{34}\sin\omega_{34}\tau_1 + D\cos\omega_{34}\tau_1, \\ C_9 &= H\cos\omega_{34}(\tau_1 + \tau_2), \quad D_9 = A\omega_{34}, \\ E_9 &= C\omega_{34}\cos\omega_{34}\tau_1 - D\sin\omega_{34}\tau_1, \\ F_9 &= H\sin\omega_{34}(\tau_1 + \tau_2), \\ G_9 &= E\omega_{34} - G\sin\omega_{34}\tau_1, \\ H_9 &= F + G\cos\omega_{34}\tau_1. \end{aligned}$$

The proof is similar as in case (d1).

3. Direction and stability of the Hopf bifurcation

In this section, we show that system (3) undergoes a Hopf bifurcation for different combinations of τ_1, τ_2 and τ_3 satisfying sufficient conditions as described. By using the method based on the normal form theory and the center manifold theory introduced by Hassard, Kazarinoff, and Wan (1981), we study the direction of bifurcations and the stability of bifurcating periodic solutions. Without loss of generality, these properties are studied for variable τ_2 as parameter and $\tau_1 \in (0, \tau_{11_0})$, $\tau_3 \in (0, \tau_{31_0})$ are fixed. Let $\tau_2 = \tau_{24_0} + \mu$, $\mu \in R$, τ_{24_0} is described in Equation (18), then the Hopf bifurcation occurs at $\mu = 0$. Assume that $\tau_3^* < \tau_1^* < \tau_{24_0}$ where $\tau_3^* \in (0, \tau_{31_0})$, $\tau_1^* \in (0, \tau_{11_0})$. Now we rescale the time by $t = t/\tau_2$, $\bar{x}(t) = x - x_*$, $\bar{y}(t) = y - y_*$, for convenience, $\bar{x}(t), \bar{y}(t)$ are still written as $x(t), y(t)$, then system (3) can be written as

$$\begin{aligned} \dot{U}(t) = (\tau_{24_0} + \mu)\bigg(&B'U(t) + C'U\left(t - \frac{\tau_3^*}{\tau_2}\right) \\ &+ D'U\left(t - \frac{\tau_1^*}{\tau_2}\right) + E'U(t-1) + f(x,y)\bigg), \end{aligned}$$

where

$$U(t) = (x(t), y(t))^{\mathrm{T}},$$

$$B' = \begin{pmatrix} \alpha_1 & 0 \\ 0 & \alpha_4 \end{pmatrix}, \quad C' = \begin{pmatrix} 0 & 0 \\ 0 & \alpha_6 \end{pmatrix},$$

$$D' = \begin{pmatrix} \alpha_2 & 0 \\ \alpha_5 & 0 \end{pmatrix}, \quad E' = \begin{pmatrix} 0 & \alpha_3 \\ 0 & 0 \end{pmatrix}, \quad f = (f_1, f_2)^{\mathrm{T}},$$

respectively, the nonlinear terms f_1 and f_2 are

$$f_1 = -a_{11}x^2(t) - a_{12}x(t)y\left(t - \frac{\tau_3^*}{\tau_2}\right),$$

$$f_2 = -a_{22}y^2(t) + a_{21}x(t-1)y\left(t - \frac{\tau_1^*}{\tau_2}\right).$$

The delayed system can be written in the functional form as

$$L_\mu\varphi = (\tau_{24_0} + \mu) \times \left[B'\varphi(0) + C'\varphi\left(-\frac{\tau_3^*}{\tau_2}\right) + D'\varphi(-\frac{\tau_1^*}{\tau_2}) + E'\varphi(-1)\right],$$

$$\varphi = (\varphi_1, \varphi_2)^{\mathrm{T}} \in C([-1,0], R^2).$$

By the Riesz representation theorem, there exists a matrix whose components are bounded variation functions $\eta(\theta,\mu) : [-1,0] \to R^2$ such that

$$L_\mu\varphi = \int_{-1}^{0} \mathrm{d}\eta(\theta,\mu)\varphi(\theta),$$

where choosing

$$\eta(\theta,\mu) = \begin{cases} (\tau_{24_0} + \mu)(B' + C' + D' + E'), & \theta = 0, \\ (\tau_{24_0} + \mu)(B' + C' + D'), & \theta \in \left[-\frac{\tau_3^*}{\tau_2}, 0\right), \\ (\tau_{24_0} + \mu)(B' + C'), & \theta \in \left[-\frac{\tau_1^*}{\tau_2}, -\frac{\tau_3^*}{\tau_2}\right), \\ (\tau_{24_0} + \mu)B', & \theta \in \left[-1, -\frac{\tau_1^*}{\tau_2}\right), \\ 0, & \theta = -1. \end{cases}$$

For $\varphi = (\varphi_1, \varphi_2)^{\mathrm{T}} \in C([-1,0], R^2)$, define

$$A(\mu)\varphi = \begin{cases} \dfrac{\mathrm{d}\varphi(\theta)}{\mathrm{d}\theta}, & \theta \in [-1, 0), \\ \displaystyle\int_{-1}^{0} d\eta(s,\mu)\varphi(s), & \theta = 0, \end{cases}$$

and

$$R(\mu)\varphi = \begin{cases} 0, & \theta \in [-1,0), \\ h(\mu,\varphi), & \theta = 0, \end{cases}$$

where

$$h(\mu,\varphi) = (\tau_{24_0} + \mu)\begin{pmatrix} h_1 \\ h_2 \end{pmatrix},$$

$$\varphi = (\varphi_1, \varphi_2)^{\mathrm{T}} \in C([-1,0], R^2),$$

$$h_1 = -a_{11}\varphi_1^2(0) - a_{12}\varphi_1\left(-\frac{\tau_1^*}{\tau_2}\right)\varphi_2(-1),$$

$$h_2 = -a_{22}\varphi_2^2(0) + a_{21}\varphi_1\left(-\frac{\tau_1^*}{\tau_2}\right)\varphi_2\left(-\frac{\tau_3^*}{\tau_2}\right).$$

Hence, Equation (3) can be rewritten as

$$\dot{U}_t = A(\mu)U_t + R(\mu)U_t,$$

where $U = (x(t), y(t))^{\mathrm{T}}$ and $U_t(\theta) = U(t+\theta), \theta \in [-1,0]$. For $\psi \in C([-1,0], (R^2)^*)$, define $A(0) = A$ and the adjoint operator A^* of A as

$$A^*\psi(s) = \begin{cases} -\dfrac{\mathrm{d}\psi(s)}{\mathrm{d}s}, & s \in (0,1], \\ \displaystyle\int_{-1}^{0} \mathrm{d}\eta^{\mathrm{T}}(t,0)\psi(-t), & s = 0, \end{cases}$$

where η^{T} is the transpose of the matrix η.

For $\varphi \in C([-1,0], R^2)$ and $\psi \in C([-1,0], (R^2)^*)$, in order to normalize the eigenvectors of operator A and adjoint operator A^*, we define a bilinear inner product

$$\langle\psi(s), \varphi(\theta)\rangle = \bar{\psi}(0)\varphi(0) - \int_{-1}^{0}\int_{\xi=0}^{\theta} \bar{\psi}(\xi - \theta)\,\mathrm{d}\eta(\theta)\varphi(\xi)\,\mathrm{d}\xi,$$

where $\eta(\theta) = \eta(\theta, 0)$.

Since $\pm \mathrm{i}\omega_{24}\tau_{24_0}$ are eigenvalues of A, they will also be the eigenvalues of A^*. The eigenvectors of A and A^* are calculated corresponding to the eigenvalues $+\mathrm{i}\omega_{24}\tau_{24_0}$ and $-\mathrm{i}\omega_{24}\tau_{24_0}$.

LEMMA 3.1 *$q(\theta) = (1,\rho)^{\mathrm{T}}\mathrm{e}^{\mathrm{i}\omega_{24}\tau_{24_0}\theta}$ is the eigenvector of A corresponding to $+\mathrm{i}\omega_{24}\tau_{24_0}$; $q^*(s) = (1/K)(1,\sigma)^{\mathrm{T}}\mathrm{e}^{-\mathrm{i}\omega_{24}\tau_{24_0}s}$ is the eigenvector of A^* corresponding to $-\mathrm{i}\omega_{24}\tau_{24_0}$ and*

$$\langle q^*(s), q(\theta)\rangle = 1, \quad \langle q^*(s), \bar{q}(\theta)\rangle = 0,$$

where

$$\rho = (\mathrm{i}\omega_{24} - \alpha_1 - \alpha_2\,\mathrm{e}^{-\mathrm{i}\omega_{24}\tau_{24_0}})(\alpha_3\,\mathrm{e}^{-\mathrm{i}\omega_{24}(\tau_3^*/\tau_{24_0})})^{-1},$$

$$\sigma = -(\mathrm{i}\omega_{24} + \alpha_1 + \alpha_2\,\mathrm{e}^{\mathrm{i}\omega_{24}\tau_{24_0}})(\alpha_5\,\mathrm{e}^{\mathrm{i}\omega_{24}\tau_{24_0}})^{-1},$$

$$\bar{K} = 1 + \rho\bar{\sigma} + \tau_{24_0}\left[(\alpha_2 + \alpha_5\rho\bar{\sigma})\,\mathrm{e}^{-\mathrm{i}\omega_{24}\tau_{24_0}} + \frac{\tau_3^*}{\tau_{24_0}}\rho\alpha_6\bar{\sigma}\,\mathrm{e}^{-\mathrm{i}\omega_{24}(\tau_1^*/\tau_{24_0})} + \frac{\tau_3^*}{\tau_{24_0}}\rho\alpha_3\,\mathrm{e}^{-\mathrm{i}\omega_{14}(\tau_3^*/\tau_{24_0})}\right].$$

Following the algorithms explained in Hassard, Kazarinoff, and Wan (1981), which is used to obtain the properties of Hopf bifurcation:

$$g_{20} = \frac{2\tau_{24_0}}{\bar{K}}[-a_{11} - a_{12}\rho\,\mathrm{e}^{-\mathrm{i}\omega_{24}\tau_3^*} + \bar{\sigma}(-a_{22}\rho^2 + a_{21}\rho\,\mathrm{e}^{-\mathrm{i}\omega_{24}(\tau_{24_0}+\tau_1^*)})],$$

$$g_{11} = \frac{\tau_{24_0}}{\bar{K}}[-2a_{11} - a_{12}\bar{\rho}\,\mathrm{e}^{\mathrm{i}\omega_{24}\tau_3^*} - a_{12}\rho\,\mathrm{e}^{-\mathrm{i}\omega_{24}\tau_3^*} + \bar{\sigma}(-2a_{22}\rho\bar{\rho} + a_{21}\bar{\rho}\,\mathrm{e}^{-\mathrm{i}\omega_{24}(\tau_{24_0}-\tau_1^*)} + a_{21}\rho\,\mathrm{e}^{\mathrm{i}\omega_{24}(\tau_{24_0}-\tau_1^*)})],$$

$$g_{02} = \frac{2\tau_{24_0}}{\bar{K}}[-a_{11} - a_{12}\bar{\rho}\, e^{i\omega_{24}\tau_3^*} + \bar{\sigma}(-a_{22}\bar{\rho}^2 + a_{21}\bar{\rho}\, e^{i\omega_{24}(\tau_{24_0}+\tau_1^*)})],$$

$$g_{21} = \frac{2\tau_{24_0}}{\bar{K}}\Bigg[-2a_{11}W_{11}^{(1)}(0) - a_{11}W_{20}^{(1)}(0) - a_{12}W_{11}^{(2)}\left(-\frac{\tau_3^*}{\tau_{24_0}}\right) - \frac{1}{2}a_{12}W_{20}^{(2)}\left(-\frac{\tau_3^*}{\tau_{24_0}}\right) - \frac{1}{2}a_{12}W_{20}^{(1)}(0)\, e^{i\omega_{24}\tau_3^*} - a_{12}\rho W_{11}^{(1)}(0)\, e^{i\omega_{24}\tau_3^*} + \bar{\sigma}\Bigg(-a_{22}\rho W_{11}^{(2)}(0) - a_{22}\bar{\rho}W_{20}^{(2)}(0) - a_{22}\rho W_{11}^{(2)}(0) + a_{21}W_{11}^{(2)}\left(-\frac{\tau_1^*}{\tau_{24_0}}\right)e^{-i\omega_{24}\tau_{24_0}} + \frac{1}{2}a_{21}W_{20}^{(2)}\left(-\frac{\tau_1^*}{\tau_{24_0}}\right)e^{i\omega_{24}\tau_{24_0}} + \frac{1}{2}a_{21}\bar{\rho}W_{20}^{(1)}(-1) \times e^{i\omega_{24}\tau_1^*} + a_{21}\rho W_{11}^{(1)}(-1)\, e^{-i\omega_{24}\tau_1^*}\Bigg)\Bigg],$$

where

$$W_{20}(\theta) = \frac{ig_{20}}{\omega_{24}\tau_{24_0}}q(0)\, e^{i\theta\omega_{24}\tau_{24_0}} + \frac{i\bar{g}_{02}}{3\tau_{24_0}\omega_{24}}\bar{q}(0)\, e^{-i\theta\omega_{24}\tau_{24_0}} + R\, e^{2i\theta\omega_{24}\tau_{24_0}},$$

$$W_{11}(\theta) = -\frac{ig_{11}}{\tau_{24_0}\omega_{24}}q(0)\, e^{i\theta\omega_{24}\tau_{24_0}} + \frac{i\bar{g}_{11}}{\tau_{24_0}\omega_{24}}\bar{q}(0)\, e^{-i\theta\omega_{24}\tau_{24_0}} + S.$$

We know $R = (R^{(1)}, R^{(2)}) \in R^2$ and $S = (S^{(1)}, S^{(2)}) \in R^2$ are constant vectors.

Thus, we can compute the following quantities:

$$c_1(0) = \frac{i}{2\omega_{24}\tau_{24_0}}\left(g_{20}g_{11} - 2|g_{11}|^2 - \frac{1}{3}|g_{02}|^2\right) + \frac{g_{21}}{2},$$

$$\mu_2 = -\frac{\mathrm{Re}\{c_1(0)\}}{\mathrm{Re}\{\lambda'(\tau_{24_0})\}},$$

$$\beta_2 = 2\,\mathrm{Re}\{c_1(0)\},$$

$$T_2 = -\frac{\mathrm{Im}\{c_1(0)\} + \mu_2\,\mathrm{Im}\{\lambda'(\tau_{24_0})\}}{\omega_{24}\tau_{24_0}}.$$

These expressions give a description of the bifurcating periodic solutions in the center manifold of system (3) at critical values $\tau_2 = \tau_{24_0}$, whereas, $\mathrm{Re}\{\lambda'(\tau_{24_0})\} > 0$ which can be stated as follows:

(i) μ_2 gives the direction of the Hopf bifurcation: if $\mu_2 > 0\,(\mu_2 < 0)$, the Hopf bifurcation is supercritical (subcritical).
(ii) β_2 determines the stability of the bifurcating periodic solution, the periodic solution is stable (unstable) if $\beta_2 < 0\,(\beta_2 > 0)$.
(iii) T_2 denotes the period of bifurcating period solutions, if $T_2 > 0\,(T_2 < 0)$, the period increases (decrease).

4. Numerical simulations

To demonstrate the algorithm for determining the existence of the Hopf bifurcation in Section 2 and the direction and stability of the Hopf bifurcation in Section 3, we carry out numerical simulations on a particular case of Equation (3) in the following form.

$$\begin{aligned}\dot{x}(t) &= x(t)(1.9 - 0.2x(t)) - 1.1x(t-\tau_1)y(t-\tau_2),\\ \dot{y}(t) &= y(t)(2.8 - 17.5y(t)) + 17x(t-\tau_1)y(t-\tau_3),\end{aligned} \tag{26}$$

where $r_1 = 1.9$, $r_2 = 2.8$, $a_{11} = 0.2$, $a_{12} = 1.1$, $a_{21} = 17$ and $a_{22} = 17.5$. It is easy to show that system (26) has a unique coexistence equilibrium $E_*(1.359, 1.4802)$. By calculation, when $\tau_2 = 0$ and $\tau_3 = 0$, the critical delay for τ_1 is obtained as $\tau_{11_0} = 1.7601$, while $\tau_{21_0} = 3.672$ when $\tau_1 = 0$ and $\tau_3 = 0$, whereas, $\tau_{31_0} = 0.1065$ if $\tau_1 = 0$ and $\tau_2 = 0$.

We can see from Figure 1(a) that E_* is asymptotically stable at $\tau_1 = 1.3 < \tau_{11_0} = 1.7601$, $\tau_2 = 2.6 < \tau_{21_0} = 3.672$ and $\tau_3 = 0.085 < \tau_{31_0} = 0.1065$, while E_* loses its stability and the Hopf bifurcation occurs at $\tau_1 = 2.1 > \tau_{11_0}$, $\tau_2 = 4.1 > \tau_{21_0}$ and $\tau_3 = 0.3 > \tau_{31_0}$, see

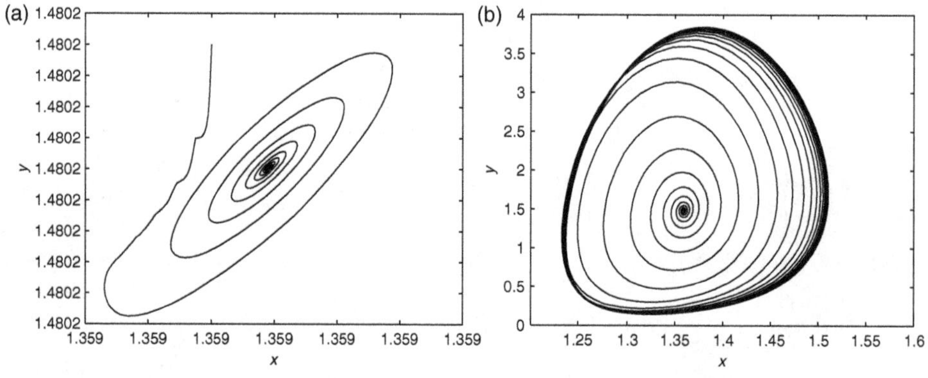

Figure 1. (a) E_* is asymptotically stable equilibrium at $\tau_1 = 1.3$, $\tau_2 = 2.6$ and $\tau_3 = 0.085$; (b) E_* loses stability and Hopf bifurcation occurs at $\tau_1 = 2.1$, $\tau_2 = 4.1$ and $\tau_3 = 0.3$.

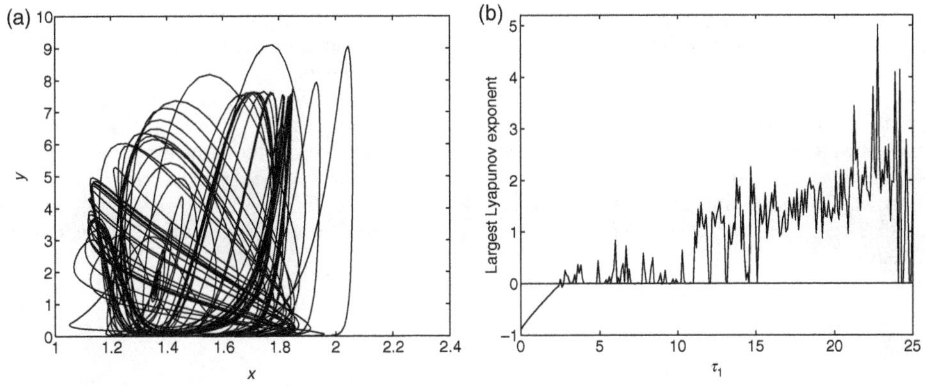

Figure 2. (a) E_* loses stability and a chaotic solution occurs at $\tau_1 = 2.7, \tau_2 = 5.0$ and $\tau_3 = 2.1$; (b) the largest Lyapunov exponent diagram of system (4.1) for variable τ_1 at $\tau_2 = 5.0$ and $\tau_3 = 2.1$.

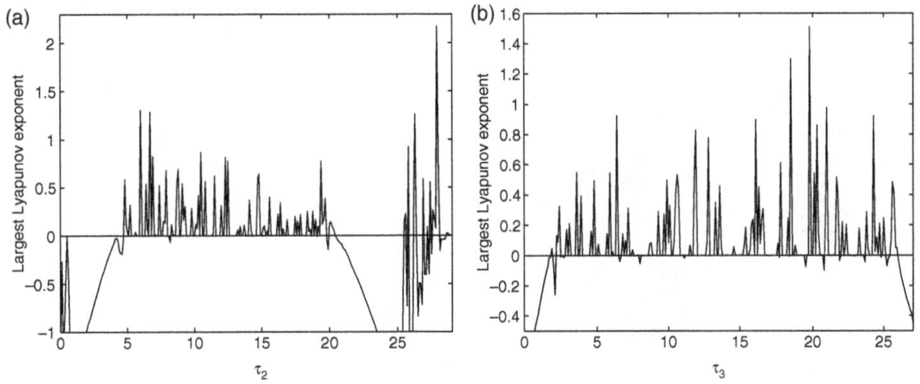

Figure 3. (a) The largest Lyapunov exponent diagram of system (4.1) for variable τ_2 at $\tau_1 = 2.7$ and $\tau_3 = 2.1$; (b) for a variable τ_3 at $\tau_1 = 2.7$ and $\tau_2 = 5.0$.

Figure 1(b). Then using the algorithm derived in Section 3, we obtain that $c_1(0) = -283.67 - 392.23i$, $\mu_2 = 1056.7$, $\beta_2 = -567.34$, $T_2 = 173.54$, we know that the Hopf bifurcation is supercritical, the bifurcating periodic solution is stable and the period increases. Whereas, at $\tau_1 = 2.7 > \tau_{11_0}$, $\tau_2 = 5.0 > \tau_{21_0}$ and $\tau_3 = 2.1 > \tau_{31_0}$ it loses stability and a chaotic solution occurs (see Figure 2(a)). In Figure 2(b), the largest Lyapunov exponent diagram is plotted for variable τ_1 when $\tau_2 = 5.0$ and $\tau_3 = 2.1$, it is easy to know when $\tau_1 > 2.6$, the Lyapunov exponent is almost positive, then the chaotic solutions occur. From Figure 3(a), the largest Lyapunov exponent diagram is plotted for variable τ_2 when $\tau_1 = 2.7$ and $\tau_3 = 2.1$, it is easy to know when $4.85 < \tau_2 < 20.0$, the Lyapunov exponent is almost positive, then the chaotic solutions occur. Similarity, from Figure 3(b), the largest Lyapunov exponent diagram is plotted for variable τ_3 when $\tau_1 = 2.7$ and $\tau_2 = 5.0$, it is easy to know when $2.0 < \tau_3 < 25.8$ the Lyapunov exponent is almost positive, then the chaotic solutions occur.

5. Conclusions

In this paper, we investigate the effect of the time delays τ_1, τ_2 and τ_3 on the stability of the positive equilibrium of the system (3), and derive the direction and stability of the Hopf bifurcation. Numerical simulations are carried out to illustrate the theoretical prediction and to explore the complex dynamics including chaos.

Acknowledgements

We are grateful to the reviewers for their valuable comments and suggestions which have led to an improvement of this paper. This research is supported by the National Natural Science Foundation of China (No. 11061016).

References

Faria, T., & Magalhaes, L. (1995). Normal form for retarded functional differential equations and applications to Bogdanov–Takens singularity. *Journal of Differential Equations*, *122*, 201–224.

Hassard, B., Kazarinoff, N., & Wan, Y. H. (1981). *Theory and applications of Hopf bifurcation*. London mathematical society lecture note series (Vol. 41). Cambridge: Cambridge University Press.

Jin, Z., & Ma, Z. (2002). Stability for a competitive Lotka–Volterra system with delays. *Nonlinear Analysis*, *52*, 1131–1142.

Saito, Y. (2002). The necessary and sufficient condition for global stability of a Lotka–Volterra cooperative or competition system with delays. *Journal of Mathematical Analysis and Applications*, *268*, 109–124.

Song, Y., Han, M., & Peng, Y. (2004). Stability and Hopf bifurcation in a competitive Lotka–Volterra system with two delays. *Chaos, Solitons and Fractals*, *22*, 1139–1148.

Song, Y., Peng, Y., & Wei, J. (2008). Bifurcation for a predator–prey system with two delays. *Journal of Mathematical Analysis and Applications*, *337*, 446–479.

Sun, C., Lin, Y., & Han, M. (2006). Stability and Hopf bifurcation for an epidemic disease model with delay. *Chaos, Solitons and Fractals*, *30*, 204–216.

Tang, X., & Zhou, X. (2003). Global attractivity of non-autonomous Lotka–Volterra competition system without instantaneous negative feedback. *Journal of Differential Equations*, *192*, 502–535.

Yan, X., & Chu, Y. (2006). Stability and bifurcation analysis for a delayed Lotka–Volterra predator–prey system. *Journal of Computational and Applied Mathematics*, *196*, 198–210.

Zhang, J., Jin, Z., Yan, J., & Sun, G. (2009). Stability and Hopf bifurcation in a delayed competition system. *Nonlinear Analysis*, *70*, 658–670.

18

Neural network and support vector machine predictive control of *tert*-amyl methyl ether reactive distillation column

Neha Sharma* and Kailash Singh

Department of Chemical Engineering, Malaviya National Institute of Technology, JLN Marg, Malaviya Nagar, Jaipur 302017, India

An algorithm of model predictive control based on artificial neural network and least-square support vector machine method is presented for a class of industrial process with strong nonlinearity such as *tert*-amyl methyl ether (TAME). Integral constant is added to improve the performance of the controller. In the present work, two different control methodologies neural network predictive control (NNPC) and support vector machine-based predictive control (SVMPC) are implemented and compared with a conventional proportional-integral-derivative (PID) control methodology to a TAME reactive distillation column. The simulation result shows that both NNPC and SVMPC gives better control performance than PID for set-point change as well as for load change of ±10% in methanol feed flow rate and molar ratio of methanol to isoamylene in reactor effluent feed.

Keywords: reactive distillation column; neural network predictive control; SVMPC; TAME

1. Introduction

Model predictive control (MPC) has become one of the most successful control algorithms in process industries such as chemical plants and oil refineries (Garcia & Morshedi, 1986; Qin & Badgwell, 1997, 2003; Rewagad & Kiss, 2011; Tatjewski & Lawrynczuk, 2006). Predictive model is a basic element of MPC. Most of the MPC algorithms are based on a linear model of the processes. The major drawback associated with these linear controllers is that they do not perform well over the wide range of operating conditions and with large disturbances. Since most industrial processes exhibit severe nonlinearity, the studies on the use of artificial intelligence techniques such as artificial neural network (ANN), support vector machine (SVM) and fuzzy logic have drawn increasing attention in recent years because of their ability to represent nonlinear systems and their self-learning capabilities.

tert-amyl methyl ether (TAME) is a widely studied model system to understand the complex behavior of reactive distillation column (Katariya, Kamath, Moudgalya, & Mahajani, 2008). But very limited research has been found in the area of the control, due to the complex dynamics of the system. Al-Arfaj and Luyben (2004) studied the plant-wide control of the TAME process using the proportional-integral (PI) control methodology. In the present work, we studied control of reactive distillation column using artificial intelligence-based control methodology and compare these techniques with the conventional control methodology (PI).

Synthesis of multiple reaction system TAME exhibits highly nonlinear behavior, i.e. steady-state multiplicity, strong interactions between process variables, etc. (Katariya, Moudgalya, & Mahajani, 2006; Mohl et al., 1997, 1999; Sharma & Singh, 2010). This complex dynamics makes process control of the reactive distillation column a challenging task. The need to handle such difficult control problem has led to use ANN and SVM in MPC and has recently attracted a great deal of attention. The attractive advantage of the neural network approach is that an accurate representation of the process can be obtained by training the network. Neural networks are capable of handling complex and nonlinear problems and can reduce the engineering efforts required in controller model development. In the field of chemical engineering, these have been successfully implemented in distillation control, for example. The papers by ZareNezhad and Aminian (2011), Hui, Hui, Aziz, & Ahmad (2011), Lawrynczuk (2010), Konakom, Kittisupakorn, Saengchan, and Mujtaba (2010), Arumugasamy and Ahmad (2009), Ahmad and Mat Noor (2009), Hussain (1999), Ramachandran and Russell Rhinehart (1995), Thibault and Grandjean (1991) and Åström and McAvoy (1992), Bhat and McAvoy (1990) provide in-depth reviews on neural network application in chemical process control.

Least-square support vector machine (LS-SVM) has been proposed by Syukens and Vandewalle (1999) and has been successfully applied to many applications

*Corresponding author. Email: nsharma.mnit@gmail.com

(Suykens, Van Gestel, De Brabanter, De Moor, & Vandewalle, 2002). SVM has excellent performance in function regression so it can be used for nonlinear system identification and system control. Several papers have appeared on SVM-based MPC (Bao, Pi, & Sun, 2007; Basak, Pal, & Patranabis, 2007; Deng et al., 2010; Iplikci, 2010; Li, Su, & Chu, 2007; Liu, Wang, & Li, 2008; Xi, Poo, & Chou, 2007; Zhang & Wang, 2008; Zhang, Xue, & Wang, 2011). The degree of nonlinearity is controlled via hidden nodes in neural network and support vectors in SVM.

Artificial Intelligence (AI) techniques are able to deal with nonlinear problems, and once trained can perform prediction and generalization at high speed (Kalogirou, 2003). They are widely used in system modeling. In this paper, we have used ANN and LS-SVM as a predictive model in MPC, namely neural network predictive control (NNPC) and support vector machine predictive control (SVMPC). MATLAB® was used to implement NNPC, SVMPC and proportional-integral-derivative (PID) in the TAME reactive distillation column. The paper is organized as follows: in Section 2, we briefly describe the synthesis of the process TAME, Section 3 introduces the ANN and LS-SVM as a model and describes the steps for model development and its statistical analysis. Section 4 describes the design methodology for artificial intelligence-based model predictive controller, namely NNPC and SVMPC. Section 5 gives optimum values of parameters of NNPC and SVMPC. Section 6 compares both control methodologies for disturbance rejection and set-point tracking which is justified by several performance criteria such as integral of time absolute error (ITAE), integral of squared error (ISE), integrated absolute error (IAE) and integral of time squared error (ITSE).

2. Process description

TAME is one of most possible antiknock additives to gasoline. It is added both to enhance octane number to replace banned tetraethyl lead and to raise the oxygen content in gasoline. The largest volume component in the past was MTBE but it is being phased out because of groundwater contamination problems. The TAME reactive distillation is an etherification process that is similar to ETBE. Therefore, TAME is becoming more important. TAME is formed by reaction of isoamylenes (IAs) (2M1B and 2M2B), which is coming from C5-stream of the refinery, with methanol (MeOH) in the presence of inert components (i.e. isopentane). Three reactions take place simultaneously in TAME synthesis: etherification of the two methylbutenes and their isomerization. The TAME reactions have been shown to be reversible and fairly exothermic. Besides TAME formation, several side reactions such as isomerization of reactive amylenes and hydration of IAs to *tert*-amyl alcohol. also take place, among which the isomerization reaction between the two IAs is the most important (Sharma & Singh, 2012):

$$\text{MeOH} + 2M1B \leftrightarrow \text{TAME}\ R_1 = M_{\text{cat}}(k_{F1}x_{2M1B}x_{\text{MeOH}} - k_{B1}x_{\text{TAME}}), \quad (1)$$

$$\text{MeOH} + 2M2B \leftrightarrow \text{TAME}\ R_2 = M_{\text{cat}}(k_{F2}x_{2M2B}x_{\text{MeOH}} - k_{B2}x_{\text{TAME}}), \quad (2)$$

$$2M1B \leftrightarrow 2M2B\ R_3 = M_{\text{cat}}(k_{F3}x_{2M1B} - k_{B3}x_{2M2B}). \quad (3)$$

The chemical reaction kinetic model is adopted from Luyben and Yu (2008):

$$k_{F1} = 1.19367 \times 10^{14} \exp\left(\frac{-76103.737}{RT}\right), \quad (4)$$

$$k_{B1} = 2.118 \times 10^{17} \exp\left(\frac{-110540.899}{RT}\right), \quad (5)$$

$$k_{F2} = 1.23462 \times 10^{17} \exp\left(\frac{-98230.2176}{RT}\right), \quad (6)$$

$$k_{B2} = 1.38726 \times 10^{20} \exp\left(\frac{-124993.965}{RT}\right), \quad (7)$$

$$k_{F3} = 2044683 \times 10^{16} \exp\left(\frac{-965122.6384}{RT}\right), \quad (8)$$

$$k_{B3} = 3.86397 \times 10^{16} \exp\left(\frac{-104196.053}{RT}\right). \quad (9)$$

where the kinetic constants k's are in mol/L-s, R in J/mol-K and T in K. The TAME synthesis process flow sheet using reactive distillation in MATLAB® is presented in Figure 1. An equilibrium stage dynamic model for the synthesis of TAME of the RDC has been developed and described in our previous paper (Sharma & Singh, 2012). The column has a total condenser and reboiler. The theoretical stages are numbered from top to bottom. For this purpose, continuous mode of operation has been assumed. The equations for 15 stages for the system and five-component system were solved in MATLAB® by *ODE15S* solver. TAME is the heaviest component, which leaves the column in the

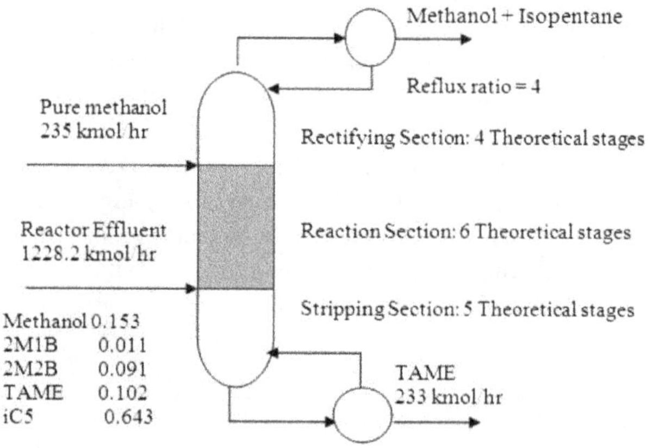

Figure 1. Synthesis of TAME in reactive distillation column.

Table 1. Column specification.

Parameter	Value
Feed flow rates:	
Fresh methanol	65.3 mol/s
Reactor effluent	341.2 mol/s
Pressure	4 bar
Reflux ratio	4
Number of stages (N)	15
Reactive zone	Tray number 5–10
Feed stage location:	
Fresh methanol	Tray number 5
Reactor effluent	Tray number 10
Volume of each tray	1220 l
Initial volume of reboiler	1220 l
Reboiler heat duty	15.727 MW

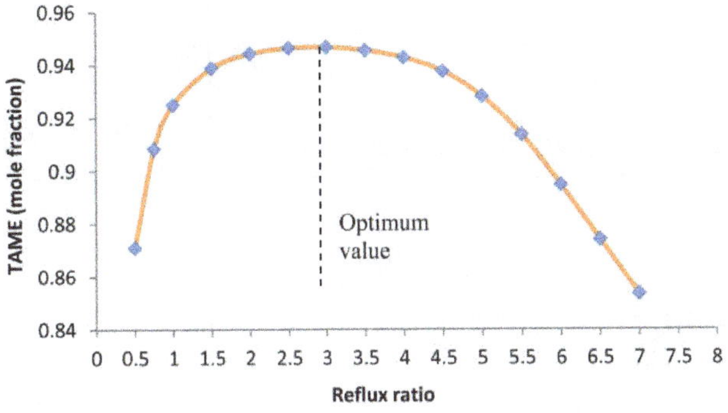

Figure 2. Effect of reflux ratio on TAME purity in bottoms.

bottoms. Product composition was chosen as a controlled variable and reboiler heat duty was chosen as a manipulating variable. Input multiplicities are also observed in the TAME system. The column specification is given in Table 1 (Sharma & Singh, 2012).

Input multiplicity is common in reactive distillation due to its nonlinear dynamics, which places substantial restrictions on the selection of controlled variables. It occurs when two or more unique sets of input variables produce the same output predictions. As shown in Figure 2, input multiplicity was found with different reflux ratios. Reflux ratios of 0.5 and 6.57 both produce a TAME purity of 0.8710 and reflux ratios of 0.75 and 5.648 both produce a TAME purity of 0.9083. Similarly, the same value of TAME product purity was obtained at different reflux ratios. Below the optimum reflux ratio (~3), the operating conditions are favorable for the TAME reaction (as indicated by the relatively high conversions) and above the optimum reflux ratio, TAME purity decreases. Input multiplicity is caused by two (or more) conflicting effects operating via the same manipulated variable. As reactive distillation usually always involves a compromise between reaction and separation effects, there are many variables and operating conditions that have the potential to result in input multiplicities.

3. Methodologies used for model development

3.1. *Artificial neural network*

Steps for the neural network model are as follows:

(i) *Data generation*: At first, the fundamental model was used as the real process; it was simulated open-loop in order to obtain two sets of random number data, namely training and test sets. Training data sets contained 5000 samples and test data set contained 2001 samples with 100 s sampling interval. These data sets were used to generate an input matrix consisting of an output variable and manipulated variable.

(ii) *Configuration of neural network*: Feed-forward back propagation network was created, which consists of number of layers using the weight function, net input function and the specified transfer functions. The first layer has weights coming from the input. Each subsequent layer has a weight coming from the previous layer. The last layer is the network output. Network contains two bias vectors (values given in Table 2), one input weight matrix and one layer weight matrix (values of weight matrices are

Table 2. Values of bias vectors.

Bias vectors	
Hidden layer	Output layer
−0.1781	−0.3801
0.2255	
−0.2014	
0.2655	
0.4394	
−0.4617	
−0.4475	
0.1295	
0.0435	
0.4256	
−0.6434	
1.5630	
−0.9224	
0.0108	
−0.4806	
0.0519	
0.1085	
0.6328	
0.0869	
−0.2992	
−0.2281	
1.5911	
0.0070	
−0.8065	
−0.0724	
−0.3683	
−0.3659	
0.0021	
−0.2727	
0.2461	

Table 3. Weight matrices of neural network.

Neuron in hidden layer	u_{k-1}	u_{k-2}	u_{k-3}	u_{k-4}	u_{k-5}	u_{k-6}	u_{k-7}	u_{k-8}	u_{k-9}	u_{k-10}
Input-hidden layer weight matrix										
1 Neuron	−1.456	1.032	0.407	−0.045	0.033	0.438	−0.429	−0.435	−0.22	−0.023
2 Neuron	0.626	−0.477	0.022	−0.427	−0.429	−0.445	0.125	0.104	0.465	0.459
3 Neuron	0.066	0.253	0.002	0.164	0.479	0.492	0.222	0.148	−0.02	−0.199
4 Neuron	−0.698	−0.177	0.127	0.217	0.007	0.038	−0.277	−0.768	−0.334	−0.577
5 Neuron	−0.466	0.05	−0.006	0.083	−0.092	−0.381	0.456	0.176	0.278	−0.221
6 Neuron	−0.688	0.544	−0.234	−0.231	−0.213	−0.032	−0.621	0.025	0.016	−0.226
7 Neuron	0.517	−0.269	−0.086	−0.311	0.79	0.493	−0.16	0.201	−0.098	0.044
8 Neuron	0.241	−0.065	0.185	−0.492	−0.327	−0.28	−0.37	−0.272	−0.559	−0.357
9 Neuron	0.689	−0.338	0.307	−0.352	0.282	0.407	0.15	−0.183	0.429	−0.059
10 Neuron	0.155	−0.141	0.076	0.693	−0.195	0.479	0.291	0.105	0.076	−0.179
11 Neuron	−1.163	0.401	0.452	0.28	0.361	0.11	−0.317	−0.014	−0.517	−0.562
12 Neuron	1.247	−0.492	−0.184	−0.245	0.164	−0.977	0.345	0.01	0.372	0.231
13 Neuron	−0.77	0.605	−0.117	−0.598	−0.407	−0.456	0.048	−0.157	−0.143	−0.347
14 Neuron	−0.306	−0.767	0.259	0.023	−0.011	0.244	−0.02	−0.17	−0.184	−0.141
15 Neuron	−0.154	0.219	−0.528	−0.493	−0.182	0.164	−0.271	0.038	−0.331	−0.201
16 Neuron	−0.233	0.485	0.179	0.779	−0.896	0.498	0.352	−0.418	0.412	0.02
17 Neuron	0.455	0.097	0.25	−0.068	0.366	−0.176	0.297	0.218	0.174	−0.081
18 Neuron	−0.691	−0.2	0.114	0.351	−0.236	−0.044	0.288	0.136	0.015	−0.53
19 Neuron	−0.501	0.032	0.066	0.35	0.062	0.258	0.47	0.107	−0.378	0.366
20 Neuron	−1.347	0.988	−0.177	0.019	0.263	0.711	0.206	0.053	−0.42	0.112
21 Neuron	−0.07	−0.262	0.136	0.34	0.128	0.283	−0.181	−0.284	−0.443	−0.292
22 Neuron	1.621	0.029	−0.327	−0.384	−0.033	−0.434	0.228	0.238	0.357	0.254
23 Neuron	0.325	−0.013	−0.234	−0.151	0.261	0.011	−0.548	−0.2	−0.108	0.208
24 Neuron	−0.836	0.491	−0.3	0.04	0.319	0.577	−0.137	0.174	0.005	0.099
25 Neuron	−0.155	0.35	−0.161	0.441	0.644	0.864	−0.591	−0.287	−0.348	−0.459
26 Neuron	−0.628	−0.031	0.358	0.263	0.195	0.177	−0.244	−0.364	−0.407	−0.279
27 Neuron	−1.376	0.795	0.255	−0.026	−0.175	0.044	−0.013	−0.528	−0.514	0.288
28 Neuron	0.213	0.394	−0.264	0.057	0.141	0.145	0.279	0.708	0.562	0.434
29 Neuron	−1.159	0.839	−0.391	−0.327	−0.12	−0.085	0.024	0.052	−0.339	−0.088
30 Neuron	0.457	−0.152	0.276	0.268	0.13	0.833	−0.627	−0.126	−0.064	−0.387
Hidden layer–output weight matrix										
−1.27	1.45	0.73	0.76	1.18	1.17	−0.92	0.85	1.1	1.14	0.67

(Continued).

Table 3. Continued.

Neuron in hidden layer	y_{k-1}	y_{k-2}	y_{k-3}	y_{k-4}	y_{k-5}	y_{k-6}	y_{k-7}	y_{k-8}	y_{k-9}	y_{k-10}
Input-hidden layer weight matrix										
1 Neuron	0.331	−0.587	−0.137	−0.244	−0.885	0.944	0.571	−0.087	−0.29	−0.275
2 Neuron	−0.015	0.32	0.05	0.084	0.011	0.368	0.14	−0.458	−0.55	0.809
3 Neuron	0.329	0.649	−0.329	0.063	−0.384	−0.142	0.07	0.013	−0.33	0.599
4 Neuron	0.665	−0.039	−0.289	0.103	−0.695	0.292	0.301	0.384	0.299	−0.41
5 Neuron	−0.339	−0.345	0.139	−0.088	−0.215	−0.029	0.114	−0.08	0.062	0.012
6 Neuron	0.02	0.58	0.294	−0.418	0.59	−0.254	0.068	−0.765	−0.092	−0.162
7 Neuron	0.678	0.457	0.173	−0.11	−0.231	−0.334	0.207	0.087	−0.336	0.387
8 Neuron	−0.362	−0.298	0.446	0.2	0.221	−0.023	−0.083	0.359	0.068	−0.213
9 Neuron	0.438	−0.586	0.335	−0.12	0.18	−0.257	0.315	0.022	0.313	0.121
10 Neuron	−0.35	−0.039	0.492	0.34	−0.29	−0.091	0.248	0.375	−0.066	0.616
11 Neuron	0.483	−0.424	0.14	−0.202	0.489	−0.148	−0.236	0.081	0.5	−0.366
12 Neuron	−0.217	0.205	−0.25	−0.073	0.274	−0.129	0.137	0.028	−0.269	−0.688
13 Neuron	0.365	0.463	0.044	0.102	0.104	−0.069	−0.147	−0.295	0.317	−0.336
14 Neuron	0.014	−0.549	−0.353	0.042	−0.362	−0.146	0.243	0.205	0.302	0.028
15 Neuron	0.235	0.773	0.217	−0.337	0.416	−0.505	−0.22	−0.414	0.139	−0.313
16 Neuron	−0.546	0.474	0.095	0.163	0.216	−0.157	0.32	0.062	−0.073	0.016
17 Neuron	0.18	−0.075	−0.196	0.153	0.102	0.491	−0.102	−0.318	−0.326	0.439
18 Neuron	0.021	−0.221	−0.202	−0.121	−0.452	0.02	−0.103	0.011	0.601	0.058
19 Neuron	0.326	0.414	−0.636	−0.004	−0.348	0.338	−0.046	0.657	−0.128	0.048
20 Neuron	−0.218	0.056	−0.216	0.116	0.05	−0.21	−0.056	0.069	0.038	0.15
21 Neuron	−0.734	0.194	−0.201	−0.69	0.798	−0.167	0.022	−0.385	−0.611	0.738
22 Neuron	−0.267	0.257	0.534	−0.019	0.202	−0.003	−0.195	−0.227	−0.401	−0.175
23 Neuron	0.201	−0.186	−1.002	−0.412	0.267	−0.03	−0.377	−0.283	0.14	−0.66
24 Neuron	0.054	0.129	0.062	−0.015	0.211	0.16	−0.144	−0.171	−0.125	0.319
25 Neuron	−0.028	−0.543	−0.15	−0.342	−0.467	−0.014	0.201	0.51	0.5	−0.936
26 Neuron	−0.364	0.01	−0.127	−0.769	0.865	−0.07	0.259	−0.408	−0.573	0.633
27 Neuron	0.052	−0.154	−0.099	0.058	−0.711	0.669	0.652	0.165	−0.599	−0.384
28 Neuron	−0.444	0.206	0.029	−0.171	0.291	−0.102	−0.254	−0.704	−0.257	0.331
29 Neuron	0.203	0.009	−0.07	0.153	−0.318	−0.396	−0.285	0.4	0.476	−0.293
30 Neuron	0.256	−0.368	−0.299	−0.087	0.002	−0.213	−0.186	0.066	0.532	0.226

Hidden layer–output weight matrix

−0.96	1.34	−0.78	−1.38	−1.03	−1.07	−0.83	0.96	0.78	1.07	1.11	0.79	−1.21	0.94	−1.32	1.26	1.32	−0.88	0.76	

given in Table 3). For creating a feed-forward back propagation network, first we define the number of hidden layers and number of neurons in the hidden layer. In this study, Levenberg–Marquardt training algorithm with Bayesian regularization in single hidden layer with 30 neurons is used. *Tansigmoid* transfer function is used for the input layer and pure linear transfer function is used for the output layer. Figure 3 shows the resulting network.

(iii) *Training*: The neural network was trained using the Levenberg–Marquardt backpropagation (trainlm) method. The regression shows that the training is perfect as shown in Figure 4 having R^2 value as 99.94%.

(iv) *Validation*: Test data set were validated with the trained neural network. Figure 5 shows the validation with $R^2 = 0.9979$ and mean square error is 8.2745×10^{-4}. Minimum mean squared error (MSE) is used as the criterion for the network selection and also for the stopping of weights and bias adjustment. Table 4 shows the performance of ANN model on both training and testing data sets. These results indicate that the neural network was trained perfectly.

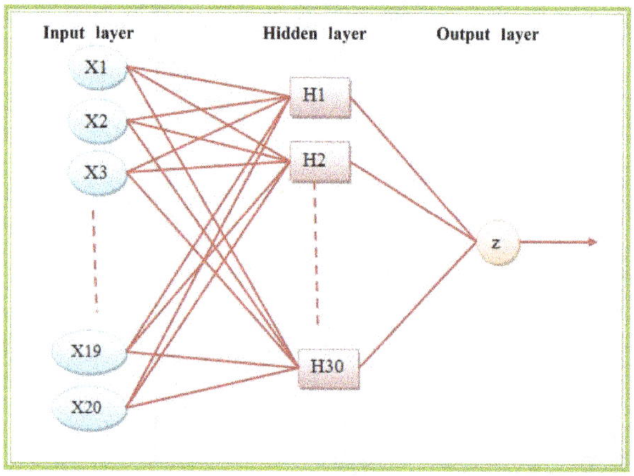

Figure 3. Structure of neural network.

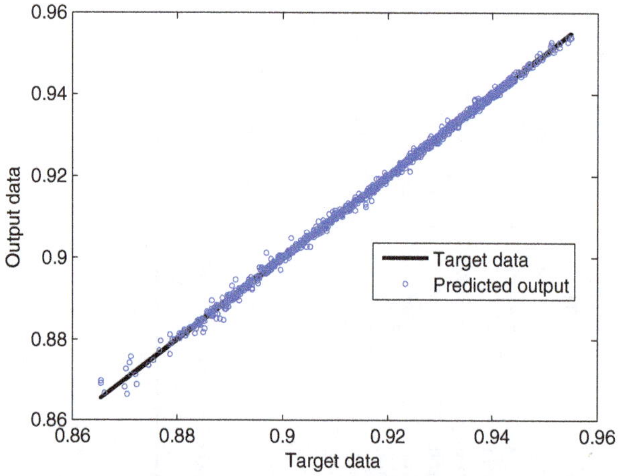

Figure 4. Process output prediction by trained neural network for training data.

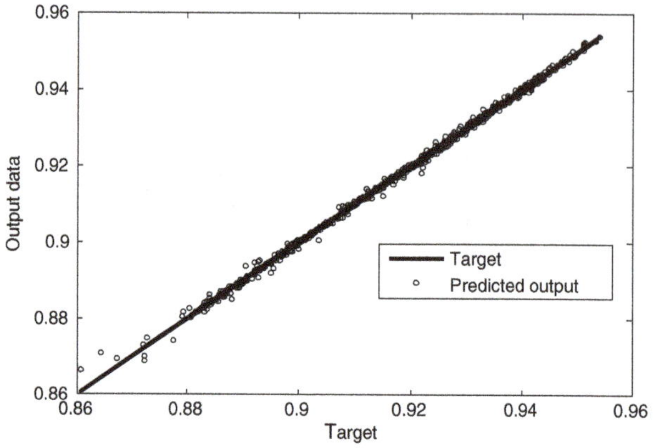

Figure 5. Process output prediction by trained neural network for testing data.

Table 4. Prediction error by ANN-based model.

	Prediction performance by ANN model	
	Training data sets	Testing data sets
RMSE	7.8258×10^{-4}	$802,745 \times 10^{-4}$
R^2	0.9981	0.9979
Optimum number of nodes	7	
Input activation function	Tansigmoid	
Output activation function	Purelin	
Best training algorithm	Levenberg–Marquardt	

3.2. *Support vector regression*

Steps for SVM model

(i) *Data generation*: Data generation in SVM is similar to the neural network model as described earlier by using fundamental model as the real process. Training data sets were chosen as 5001 samples and test data set contained 2001 samples with 100 s sampling interval. These data sets were used to generate an input matrix consisting of output variable and manipulated variable.

(ii) *Selection of kernel*: SVM handles nonlinear systems by using "kernel function." Different kernel types such as linear, polynomial and radial basis function (RBF) have been trained, validated and tested for best prediction using these data sets. There are two parameters to be evaluated to design a successful regression model. These parameters are kernel type and kernel parameter

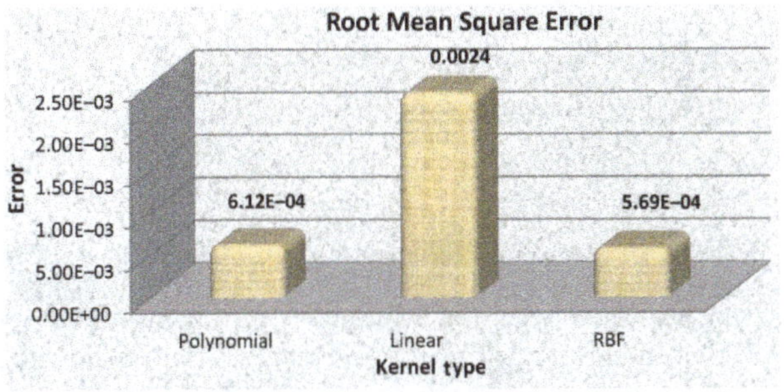

Figure 6. Performance of LS-SVM for different kernel type.

(hyper-parameters). All the kernels are tested since the prior knowledge is not there regarding the suitability of particular value of these parameters. Each run was exposed with same training and testing data and root mean square error (RMSE) was calculated for each run. It is evident from Figure 6 that RBF kernel produces the best results among all other algorithms.

(iii) *Tuning*: In the formulation of LS-SVM, the model includes two hyper-parameters, i.e. regularization parameter (γ) and kernel parameter (σ). The objective of tuning is to find the optimum values of the tuning parameters (γ and σ). Finding the hyper-parameters with a good generalization performance is crucial for the successful application of LS-SVM. A popular way to estimate the generalization performance of a model is cross-validation (CV). In this study, we are using 10-fold (k-fold) CV, which means the following:
   (a) It breaks data into 10 sets of size $n/10$,
   (b) Train on 9 data sets and test on 1 data set and
   (c) Repeat 10 times and take a mean accuracy.

SVM algorithm with CV cost function demands a high computational load. It finds the value of tuning parameters in approximately 40 min on a computer with Intel core i-5 processor and 4 GB RAM; the execution time was found to be much higher than that in the neural network (1 minute only) to train network. Low value of regularizing parameter increases the complexity of the model and high value of γ fitted good on training data points, whereas large value of kernel parameter (σ^2) indicates a stronger smoothing. Optimal value of regularization parameter (γ) is 1.3092×10^5 and kernel parameter (σ^2) is 90.5660 with RBF kernel (as given in Table 5).

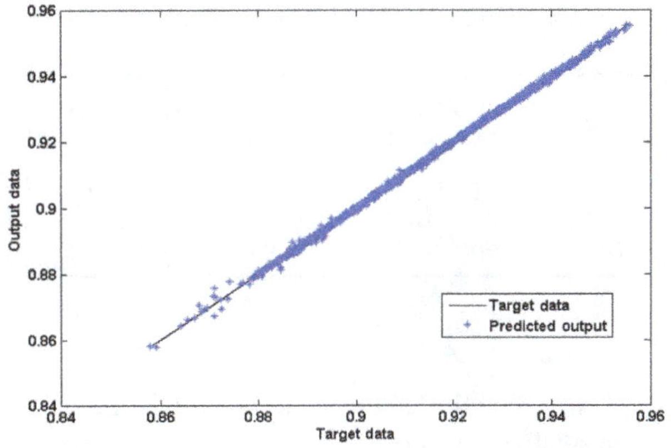

Figure 7. Process output prediction by trained SVM for testing data.

(iv) *Training*: Train the support values and the bias term of an LS-SVM for function approximation. The regression shows that the training is perfect as R^2 value is 99.94% and RMSE is 5.6875×10^{-4} on training data set.

(v) *Validation*: Test data set were validated with the trained SVM model. Figure 7 shows the validation with R^2 as 99.82% and mean square error as 7.5817×10^{-4}. Minimum MSE is used as the criterion for the model selection. These results indicate that SVM model was trained perfectly.

3.3. Statistical analysis of the model

The statistical analysis of the NN and SVR prediction is based on the following performance criteria:

(1) RMSE on test data set should be the minimum

$$\mathrm{RMSE} = \sqrt{\mathrm{MSE}(Y_{\mathrm{experimental}} - Y_{\mathrm{predicted}})}.$$

(2) The cross-correlation co-efficient (R) between the input and the output should be around unity

$$R^2 = \frac{\sum((Y_{\mathrm{experimental}} - \mathrm{mean}(Y_{\mathrm{experimental}}))(Y_{\mathrm{predicted}} - \mathrm{mean}\ (Y_{\mathrm{predicted}})))^2}{\sum((Y_{\mathrm{experimental}} - \mathrm{mean}(Y_{\mathrm{experimental}}))^2)\sum((Y_{\mathrm{predicted}} - \mathrm{mean}(Y_{\mathrm{predicted}}))^2)}.$$

Prediction performances of both the controllers are almost similar as given in Table 6 in terms of mean absolute percent error (MAPE), RMSE and R^2.

Table 5. Performance of SVM for different kernel types.

Type of Kernel	RMSE		R^2		Cost CV	Regularization parameter (σ^2)	Kernel parameter (σ^2)
	Training	Testing	Training	Testing			
Linear	0.0024	0.0024	0.9811	0.9881	6.0731×10^{-6}	6.2948	–
Polynomial	6.1157×10^{-4}	7.8614×10^{-4}	0.9988	0.9981	5.1405×10^{-7}	0.0340	18.6222
RBF	5.6875×10^{-4}	7.5817×10^{-4}	0.999	0.9982	4.8789×10^{-7}	1.3092×10^5	90.5660

Table 6. Comparison of performance of SVM and ANN models.

	Prediction performance by ANN	Prediction performance by SVM
RMSE	8.2745×10^{-4}	7.5019×10^{-4}
MAPE	0.0528	0.0445
R^2	0.9978	0.9982

4. Design of artificial intelligence-based model predictive controller

NNPC and SVMPC are basically a model-based predictive control where the model for predictions is the neural network and SVM model. It offers an alternative approach to modeling process behavior as they do not necessarily require a previous knowledge of the process phenomena. They are taught to emulate a process by "training" them, in which they are exposed to sets of input–output data and a least-squares optimization is performed. During this optimization, the neural network and SVM forms its own model of the process which can be used to predict output(s) for a given set of inputs. This model is then used for prediction in MPC algorithm. Figure 8 represents the block schemes of the control configuration. NNPC and SVMPC uses a neural network and SVM model of the process, a history of past control moves and an optimization cost function over the

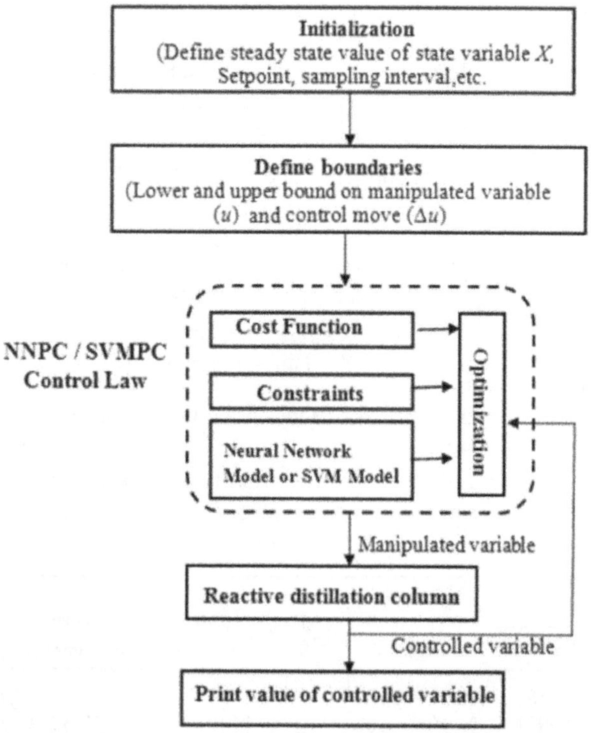

Figure 8. Algorithm of the neural network and SVMPC.

receding prediction horizon to calculate the optimal control moves (Sharma & Singh, 2012). The control evaluation consists of minimization of cost function given by

$$f = Q\sum_{i=1}^{N_p}(y_{c,i} - y_{sp})^2 + \sum_{i=1}^{N_c} S(\Delta u_i)^2,$$

where y_{sp} is the set-point value of controlled variable, Q is the weighting co-efficient for the relative importance of error between set-point and actual values and S is the weighting co-efficient penalizing relative big changes in the manipulated variable.

The objective function is subject to the following constraints:

$$u_{min} \le u_i \le u_{max} (i = 1, \ldots, N_c - 1)$$

(Constraint on manipulated variable),

$$\Delta u_{min} \le \Delta u_i \le \Delta u_{max} (i = 1, \ldots, N_c - 1)$$

(Constraint on control move),

$$y_{min} \le y_i \le y_{max} (i = 1, \ldots, N_p - 1)$$

(Constraint on process variable),

where y_{max} and y_{min}, u_{max}, andu_{min}, Δu_{max} and Δu_{min} are upper and lower bounds for the vectors y, u and Δu, respectively. N_c and N_p are the control horizon and prediction horizon. The optimization algorithm calculates the values of Δu to minimize the cost objective function. The predictions of the neural networks and support vectors are corrected by a process/model error:

$$y_d = y_k - y''$$

$$y_c = y_m + y_d,$$

where y_k is the vector of measured outputs at the present sampling interval, y'' is the vector of the networks predictions (calculated at the previous sampling interval), y_c is the vector of the corrected model predictions and y_m is the value predicted by the neural network and SVM.

As the neural networks and SVM are trained to predict the controlled variables only one step ahead, in the control algorithm the networks are iterated to obtain a total of N_p future predictions by using the outputs of the neural networks and SVM as their own inputs in the next iteration.

Although the optimization is based on a control horizon, only the first control action (for each manipulated variable) is implemented in the process and the optimization problem is solved again at the next sampling interval. Matlab® function *fmincon* was used to solve the optimization problem. The constraints on variables were taken as follows:

$$14.727 \le u \le 16.72,$$

$$-0.1 \le \Delta\Delta u \le +0.1,$$

$$0.88 \le y \le 0.95.$$

5. Parameter optimization of NNPC and SVMPC

5.1. *Weighting co-efficient*

Figure 9 shows the effect of weighting co-efficients Q and S on the NNPC methodology. Usually, the value of Q is selected to be much higher than the S factor. If we increase the value of S, then system becomes sluggish and takes more time to obtain steady-state condition (Figure 9(a)), whereas on increasing Q, system responds faster to reach to set point (Figure 9(b)). When we further increase the value of Q (beyond 1000), there is not much change in the response of both controllers. Table 7 shows the effect of Q and S in terms of performance indexes for set-point change from 0.92506 to 0.93506. For all these performance criteria, the error is minimum with Q value as 1000 and S as 0.03. Therefore, for further study of the NNPC for different set-point and load changes, the values of Q and S were specified as 1000 and 0.03, respectively.

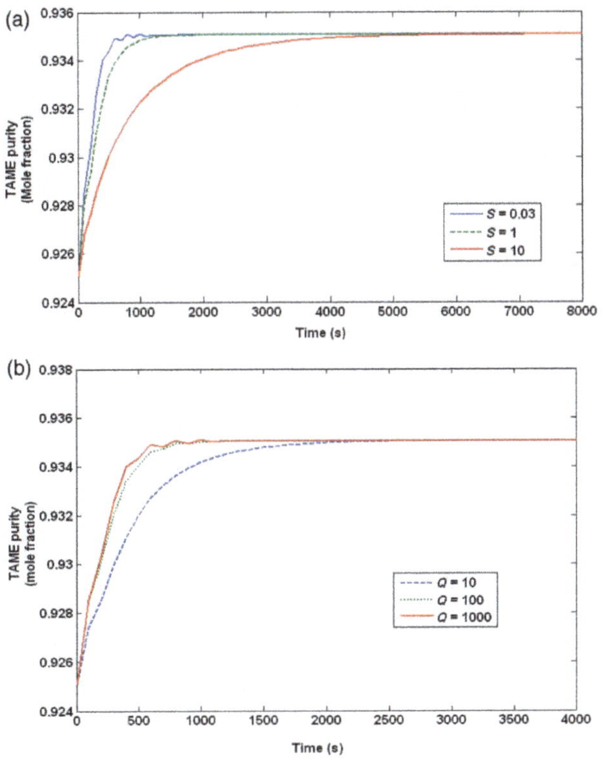

Figure 9. Effect of weighing co-efficients (a) S and (b) Q on performance of the NNPC methodology.

Table 7. Effect of weighing co-efficient Q and S in terms of performance indexes.

Parameter	ITAE	IAE	ISE	ITSE
$Q = 10$	1.6913×10^3	3.3351	0.01365	4.08469
$Q = 100$	4.1913×10^2	1.4981	0.00588	1.15512
$Q = 1000$	3.3472×10^2	1.2868	0.005113	0.91443
$S = 0.03$	3.3471×10^2	1.2868	0.005113	4.08469
$S = 1$	7.3009×10^2	2.1011	0.00858	1.15509
$S = 10$	7.5642×10^3	7.24745	0.028628	1.15509

5.2. *Integral action*

To improve the performance of the controller, we add integral action in the algorithm of the NNPC methodology. We observed that for large load changes in methanol feed flow rate such as from 65.3 to 50 mol/s at 1000 s, controller shows offset. As shown in Figure 10, dotted line shows the response of controller without integral action, whereas dashed and solid lines show the response of controller with integral action with different values of integral constant (0.001 and 0.1). When methanol feed flow rate is given a step decrease, then initially TAME composition increases and then after some time comes back to the set point. The result shows that controller gives better performance with integral constant of 0.001.

5.3. *Effect of prediction horizon*

Prediction horizon (N_P) represents the number of samples into the future over which NNPC computes the predicted process variable profile and minimizes the predicted error. Control horizon (N_c) is the number of manipulated variable moves that controller computes at a given sampling interval to eliminate the current predicted error. Figure 11 shows a comparison between the responses for different values of the prediction horizon. In both the cases (Figure 11), with higher values of the prediction horizon, controller shows a sluggish behavior. Lower value of N_p is suggested. The error is found to be minimum for $N_p = 3$ in the NNPC.

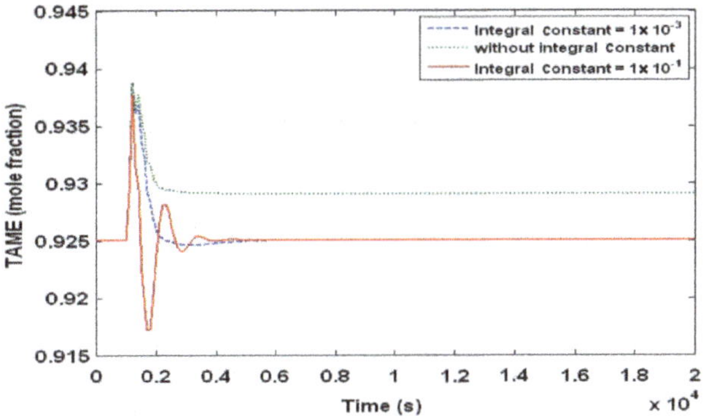

Figure 10. Effect of integral action on performance of the NNPC methodology for load change in flow rate of methanol from 65.3 to 50 mol/s.

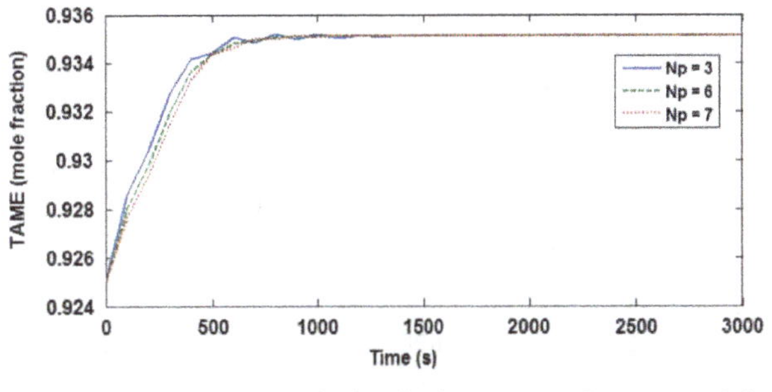

Figure 11. Effect of prediction horizon on performance of the NNPC methodology.

Due to similar prediction performance (as shown in Table 6), same weighting co-efficients (Q and S) and integral constant were taken with the SVMPC methodology. Response of prediction horizon implies a larger impact on the performance of SVMPC. As shown in Figure 12 as we increase the value of prediction horizon, by taking other parameters same as the NNPC, controller gives an offset. Value of prediction horizon has been taken as 1 with SVMPC.

6. Results and discussion

Responses of both control methodologies were compared with the PID control methodology. The values of K_c and τ_I were specified as 1 and 0.1, respectively, as discussed in our previous paper (Sharma & Singh, 2012).

6.1. Set-point change

As a base case, reboiler heat duty was fixed at 15.727 MW with a TAME composition of 0.92506 and reflux ratio fixed at 1. To observe the effect of NNPC and SVMPC, different simulation runs were performed at different step changes in set point as shown in Figure 13. Upper part of this figure shows NNPC, SVMPC and PID responses of reactive distillation column for step changes of +0.01 and −0.01 in the set point of the TAME mole fraction in the bottom product. Lower part of the figure shows the

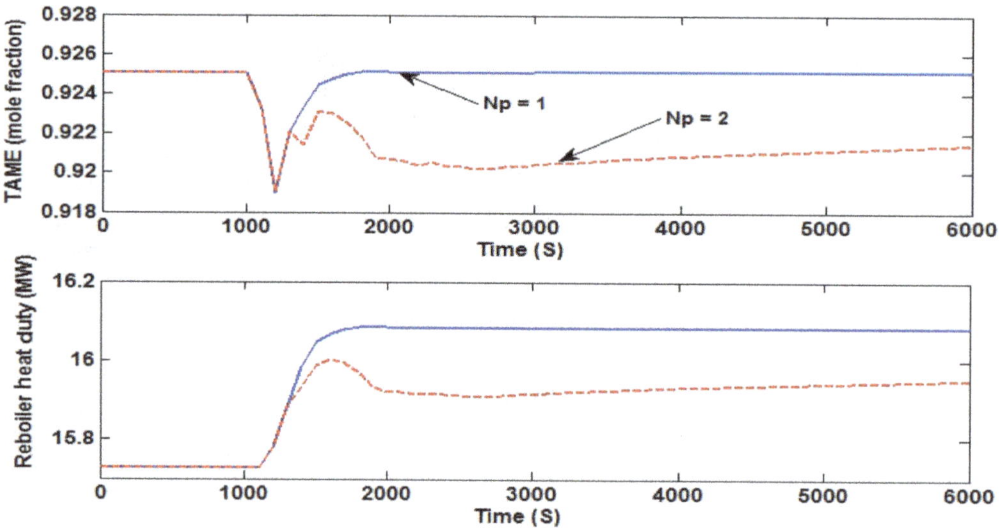

Figure 12. Effect of prediction horizon on performance of the SVMPC.

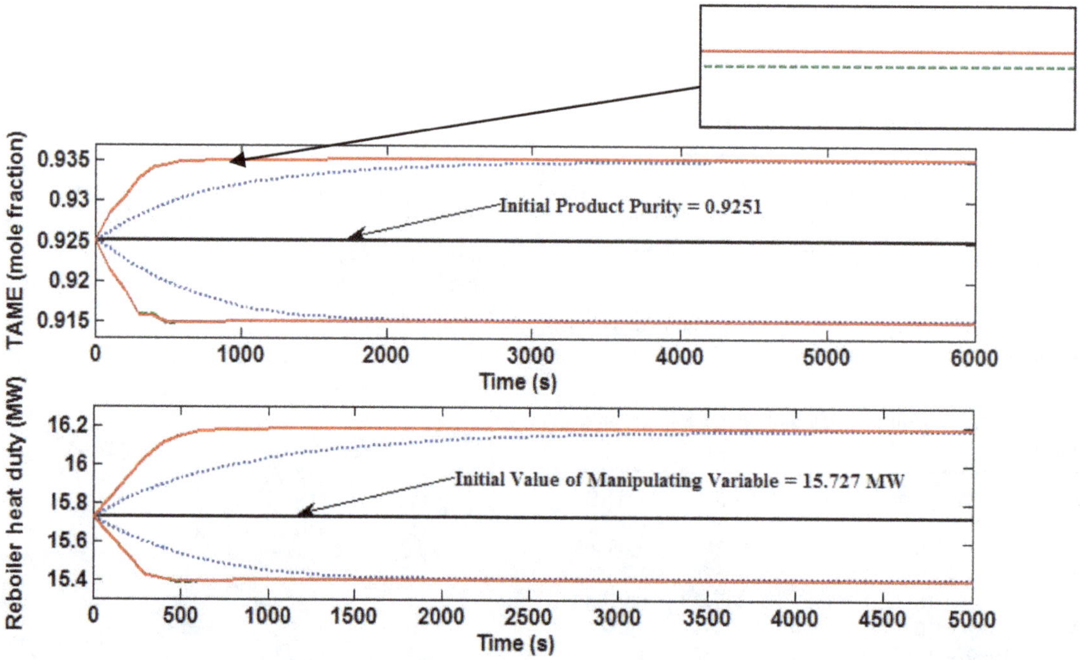

Figure 13. Responses of the SVMPC (green dashed lines), NNPC (red solid) and PID controller (dotted lines) for set-point change.

corresponding reboiler heat duty. It is depicted from the figure that both NNPC and SVMPC give exactly similar response, due to similar prediction performance, which can be seen in the expanded part of top-right corner of the figure. The result shows that both NNPC and SVMPC obtained steady state at around 1000 s, whereas the PID controller shows a sluggish response (obtained steady state at around 4000 s).

6.2. *Load change in methanol feed flow rate*

A load change was given in methanol feed flow rate from 65.3 to 71.83 (+10%) mol/s at 1000 s and another step change was given in the feed flow rate from 65.3 to 58.77 (−10%) mol/s at 1000 s. The corresponding responses of both controllers were compared in Figures 14 and 15. On giving a step increase in methanol feed flow rate, TAME composition decreases initially and after some time is brought back to the set point (Figure 14). Similarly, when methanol feed flow rate is given a step decrease, then initially TAME composition increases and then after some time is brought back to the set point (Figure 15). It is clear from these figures that for load change in methanol feed flow rate, NNPC and SVMPC work well in comparison to the PID controller.

6.3. *Load change in molar ratio*

Molar ratio of methanol to IA in reactor effluent feed implies a larger impact on TAME purity and IA conversion. The methanol feed to the reactive column is very important. Less methanol in feed will result in low conversion and purity. This occurs because methanol forms an azeotrope with isopentane, which remains in significant amount in distillate. When the methanol is reduced, the amount of methanol needed in the azeotrope does not change and less methanol is available for reaction. Further increase in methanol will result in higher conversion and purity (molar ratio 0.5–1.5). Further increasing methanol will result in (i) reducing TAME purity because most of the excess methanol will leave the column in the bottoms and (ii) reducing the IA conversion because some amount of product leave in the distillate. Therefore, both reactants must be used in correct proportions to maintain the high conversion and purity of the TAME. Initially, molar ratio of 1.5 was considered for the study.

A load change was given in molar ratio of methanol to IA in reactor effluent feed from 1.5 to 1.65 (+10%) at 1000 s and another load change was given in the molar ratio from 1.5 to 1.35 (−10%) at 1000 s. The corresponding responses of both controllers were compared in Figures 16 and17. On giving a step increase in molar ratio, TAME composition decreases initially and after some time is brought back to the set point (Figure 16). Similarly, when molar ratio is given a step decrease, then initially TAME composition increases and then after some time is brought back to the set point (Figure 17). It is depicted from the figure that both NNPC and SVMPC give exactly similar response, which can be seen in the expanded part of top-right corner of the figure. It is clear from these figures that for load change in molar ratio of methanol to IA in reactor, effluent feed machine learning-based control methodology works well in comparison to the conventional PID control methodology.

6.4. *Performance of the control system*

Model is said to be optimal with respect to cost criteria if the cost is the lowest when using that model. In general, there

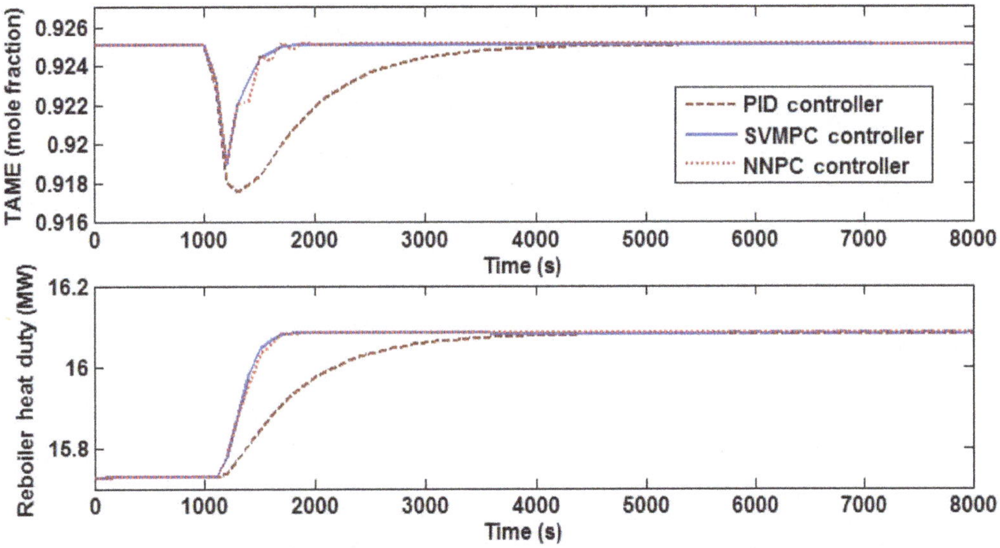

Figure 14. Responses of the SVMPC, NNPC and PID controllers for load change in flow rate of methanol from 65.3 to 71.83 (+10%) mol/s.

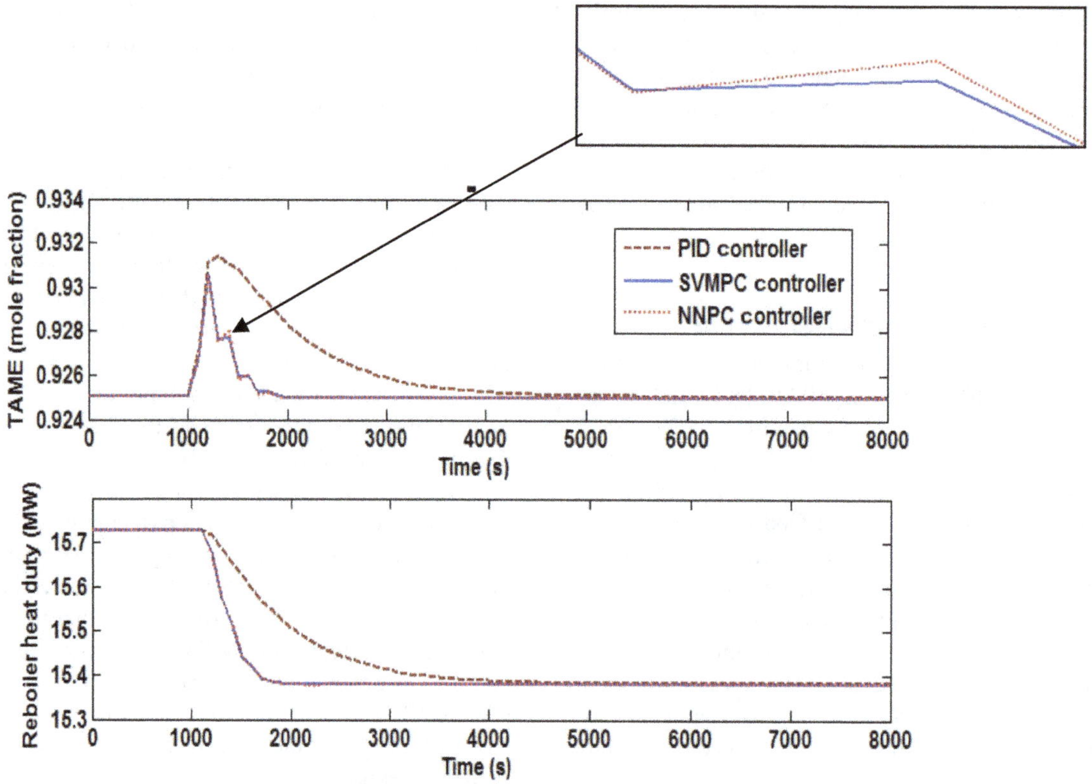

Figure 15. Responses of the SVMPC, NNPC and PID controllers for load change in flow rate of methanol from 65.3 58.77 (−10%) mol/s.

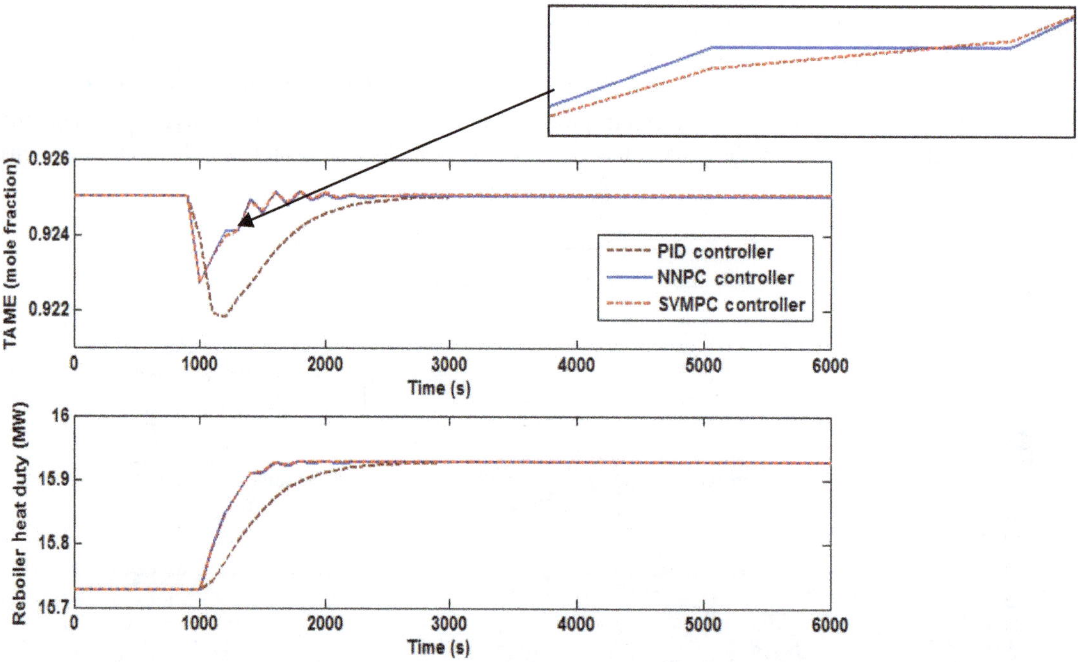

Figure 16. Responses of the SVMPC, NNPC and PID controllers for load change in molar ratio of methanol to IA in reactor effluent feed from 1.5 to 1.65 (+10%).

are different types of performance criteria of the controllers such as the IAE, ISE, ITAE and ITSE. The IAE and ISE cost criteria weight the initial values of the error more than the later value, while the time weighted criteria such as the ITAE and ITSE criteria weight the later values of the error more. Minimizing the integral constraint tries to keep the

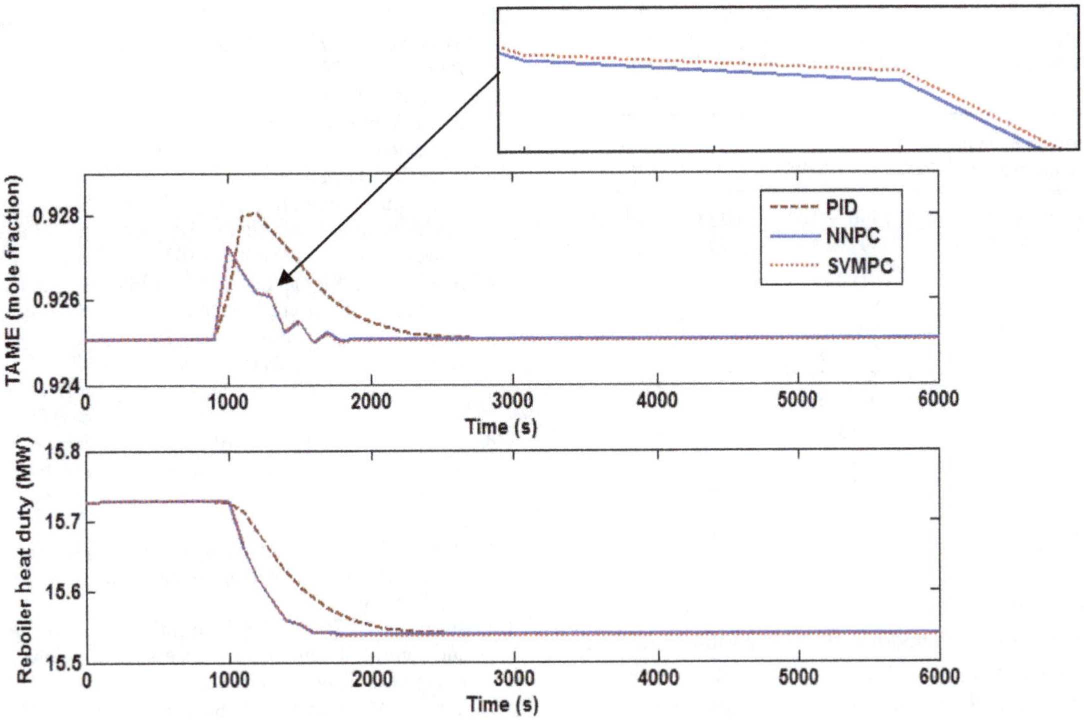

Figure 17. Responses of the SVMPC, NNPC and PID controllers for load change in molar ratio of methanol to IA in reactor effluent feed from 1.5 to 1.35 (−10%).

error small in general sense.

$$\mathrm{ITAE} = \int_0^t t|y - y_{\mathrm{sp}}|\,\mathrm{d}t,$$

$$\mathrm{IAE} = \int_0^t |y - y_{\mathrm{sp}}|\,\mathrm{d}t,$$

$$\mathrm{ISE} = \int_0^t (y - y_{\mathrm{sp}})^2\,\mathrm{d}t,$$

$$\mathrm{ITSE} = \int_0^t (y - y_{\mathrm{sp}})^2\,\mathrm{d}t.$$

In this paper, all these indexes were used to estimate the performance of the NNPC, SVMPC and PID. Tables 8–10 show the comparison between the responses of all the controllers for set-point change, load change in methanol feed flow rate and molar ratio of methanol to IA in reactor effluent feed. For all these performance criteria, the error is minimum with NNPC and SVMPC. Therefore, the implementation of machine learning-based controller to the present system has proved to be effective in maintaining the desired purity in the presence of disturbances and set-point changes.

Table 8. Performance criteria for positive (+0.01) and negative (−0.01) set-point change using SVMPC, NNPC and PID controllers.

Performance criteria and controller		Positive set-point change (+0.01)	Negative set-point change (−0.01)
ITAE	SVMPC	0.4178×10^3	0.3418×10^3
	NNPC	0.4142×10^3	0.3362×10^3
	PID	7.8178×10^3	3.0737×10^3
IAE	SVMPC	2.0959	1.8118
	NNPC	2.0918	1.8076
	PID	8.5252	6.0049
ISE	SVMPC	0.0120	0.0104
	NNPC	0.0120	0.0104
	PID	0.0409	0.0333
ITSE	SVMPC	1.0656	0.7395
	NNPC	1.0639	0.7371
	PID	17.2911	9.5635

Table 9. Performance criteria for positive (+10%) and negative (−10%) load change in methanol feed flow rate using SVMPC, NNPC and PID controllers.

Performance criteria and controller		Positive set-point change (+10%)	Negative set-point change (−10%)
ITAE	SVMPC	1.8679×10^3	1.6495×10^3
	NNPC	1.9873×10^3	1.9472×10^3
	PID	12.762×10^3	12.997×10^3
IAE	SVMPC	1.2893	1.4083
	NNPC	1.4909	1.4415
	PID	6.9745	6.7881
ISE	SVMPC	0.0051	0.0048
	NNPC	0.0055	0.0048
	PID	0.0331	0.0272
ITSE	SVMPC	6.2961	6.0036
	NNPC	6.8519	6.0818
	PID	51.1814	43.8539

Table 10. Performance criteria for positive (+10%) and negative (−10%) load change in feed ratio (methanol/IA) using SVMPC, NNPC and PID controllers.

Performance criteria and controller		Positive load change (+10%)	Negative load change (−10%)
ITAE	SVMPC	0.7590×10^3	0.7248×10^3
	NNPC	0.7597×10^3	0.7025×10^3
	PID	2.8246×10^3	2.6048×10^3
IAE	SVMPC	0.6023	0.5928
	NNPC	0.6018	0.5784
	PID	1.9548	1.8256
ISE	SVMPC	7.4753×10^{-4}	7.7341×10^{-4}
	NNPC	7.4742×10^{-4}	7.3745×10^{-4}
	PID	0.0042	0.0037
ITSE	SVMPC	0.8356	0.8725
	NNPC	0.8276	0.8422
	PID	5.4707	4.9085

7. Conclusions

NNPC and SVMPC methodologies were applied and compared to control the product purity in bottoms of reactive distillation column. Both controllers were compared for set-point changes as well as load changes in feed flow rate of methanol and molar ratio of methanol to IA in reactor effluent feed. Responses of both the controllers were then compared with the PID control methodology. SVM algorithm demands a high computational load due to the form of its optimization problem. Neural network provides good generalization results, if the structure is suitably chosen. In this study, it was found that for large disturbances, both NNPC and SVMPC shows offset. Addition of an integral action in these controllers improves the performance of the controllers, thus it is suggested that the machine learning-based control methodology with integral action could be advantageous for nonlinear processes. In comparison to PID, machine learning-based control methodologies, i.e. SVMPC and NNPC in the present study, show better response for both disturbance rejection and set-point tracking as justified by several performance criteria, namely ITAE, ISE, IAE and ITSE.

References

Ahmad, Z., & Mat Noor, R. (2009). Multiple neural networks modeling techniques in process control: A review. *Asia-Pacific Journal of Chemical Engineering, 4*, 403–419.

Al-Arfaj, M. A., & Luyben, W. L. (2004). Plantwide control for TAME production using reactive distillation. *AIChE Journal, 50*, 1462–1473.

Arumugasamy, S. K., & Ahmad, Z. (2009). Elevating model predictive control using feedforward artificial neural networks: A review. *Chemical Product and Process Modeling, 4*, 1–40.

Åström, K. J., & McAvoy, T. J. (1992). Intelligent control. *Journal of Process Control, 2*, 115–127.

Bao, Z., Pi, D., & Sun, Y. (2007). Nonlinear model predictive control based on support vector machine with multikernel. *Chinese Journal of Chemical Engineering, 15*, 691–697.

Basak, D., Pal, S., & Patranabis, D. C. (2007). Support vector regression. *Neural Information Processing – Letters and Reviews, 11*, 203–224.

Bhat, N., & McAvoy, T. J. (1990). Use of neural nets for dynamic modeling and control of chemical process systems. *Computers & Chemical Engineering, 14*, 573–582.

Deng, Z., Cao, H., Li, X., Jiang, J., Yang, J., & Qin, Y. (2010). Generalized predictive control for fractional order dynamic model of solid oxide fuel cell output power. *The Journal of Power Sources, 195*, 8097–8103.

Garcia, C. E., & Morshedi, A. M. (1986). Quadratic programming solution of dynamic matrix control (QDMC). *Chemical Engineering Communications, 46*, 73–87.

Hui, L. Q., Hui, L. W., Aziz, N., & Ahmad, Z. (2011). Nonlinear process modeling of "shell" heavy oil fractionator using neural network. *Journal of Applied Sciences, 11*, 2114–2124.

Hussain, A. (1999). Review of the applications of neural networks in chemical process control – simulation and online implementation. *Artificial Intelligence in Engineering, 13*, 55–68.

Iplikci, S. (2010). A support vector machine based control application to the experimental three-tank system. *ISA Transactions, 49*, 376–386.

Kalogirou, S. A. (2003). Artificial intelligence for the modeling and control of combustion processes: A review. *Progress in Energy and Combustion Science, 29*, 515–566.

Katariya, A. M., Kamath, R. S., Moudgalya, K. M., & Mahajani, S. M. (2008). Non-equilibrium stage modeling and non-linear dynamic effects in the synthesis of TAME by reactive distillation. *Computers & Chemical Engineering, 32*, 2243–2255.

Katariya, A. M., Moudgalya, K. M., & Mahajani, S. M. (2006). Nonlinear dynamic effects in reactive distillation for synthesis of TAME. *Journal of Industrial and Engineering Chemistry, 45*, 4233–4242.

Konakom, K., Kittisupakorn, P., Saengchan, A., & Mujtaba, I. M. (2010). *Optimal policy tracking of a batch reactive distillation by neural network-based model predictive control (NNMPC) strategy*. World congress on engineering and computer science (WCECS 2010), San Francisco.

Lawrynczuk, M. (2010). Training of neural models for predictive control. *Neurocomputing, 73*, 1332–1343.

Li, L. J., Su, H. Y., & Chu, J. (2007). Generalized predictive control with online least squares support vector machines. *Acta Automatica Sinica, 33*, 1182–1188.

Liu, Y., Wang, H., & Li, P. (2008). Kernel learning adaptive one-step-ahead predictive control for nonlinear processes. *Asia-Pacific Journal of Chemical Engineering, 3*, 673–679.

Luyben, W. L., & Yu, C.-C. (2008). *Reactive distillation design and control*. Hoboken, NJ: John Wiley & Sons.

Mohl, K. D., Kienle, A., Gilles, E. D., Rapmund, P., Sundmacher, K., & Hoffmann, U. (1997). Nonlinear dynamics of reactive distillation processes for the production of fuel ethers. *Computers & Chemical Engineering, 21*, S989–S994.

Mohl, K. D., Kienle, A., Gilles, E. D., Rapmund, P., Sundmacher, K., & Hoffmann, U. (1999). Steady-state multiplicities in reactive distillation columns for the production of fuel ethers MTBE and TAME: Theoretical analysis and experimental verification. *Chemical Engineering Science, 54*, 1029–1043.

Qin, S. J., & Badgwell, T. A. (1997). *An overview of industrial model predictive control technology* (pp. 232–256). New York, NY: American Institute of Chemical Engineers.

Qin, S. J., & Badgwell, T. A. (2003). A survey of industrial model predictive control technology. *Control Engineering Practice, 11*, 733–764.

Ramchandran, S., & Russell Rhinehart, R. (1995). A very simple structure for neural network control of distillation. *Journal of Process Control, 5*, 115–128.

Rewagad, R. R., & Kiss, A. A. (2011). Dynamic optimization of a dividing-wall column using model predictive control. *Chemical Engineering Science, 68*, 132–142.

Sharma, N., & Singh, K. (2010). Control of reactive distillation column: A review. *International Journal of Chemical Reactor Engineering, 8*, R5, 1–57.

Sharma, N., & Singh, K. (2012). Model predictive control and neural network predictive control of TAME reactive distillation column. *Chemical Engineering and Processing, 59*, 9–21.

Suykens, J. A. K., Van Gestel, T., De Brabanter, J., De Moor, B., & Vandewalle, J. (2002). *Least squares support vector machines*. Singapore: World Scientific.

Suykens, J. A. K., and Vandewalle, J. (1999). Least squares support vector machine classifiers. *Neural Processing Letters, 9*, 293–300.

Tatjewski, P., & Lawrynczuk, M. (2006). Soft computing in model-based predictive control. *International Journal of Applied Mathematics and Computer Science, 16*, 7–26.

Thibault, J., & Grandjean, B. P. (1991). *Neural networks in process control – a survey*. Advanced Control of Chemical Processes – ADCHEM'91, IFAC Symp., Toulouse, France, pp. 251–260.

Xi, X. C., Poo, A. N., & Chou, S. K. (2007). Support vector regression model predictive control on a HVAC plant. *Control Engineering Practice, 15*, 897–908.

ZareNezhad, B., & Aminian, A. (2011). Application of the neural network-based model predictive controllers in nonlinear industrial systems: Case study. *Journal of the University of Chemical Technology and Metallurgy, 46*, 67–74.

Zhang, R., & Wang, S. (2008). Support vector machine based predictive functional control design for output temperature of coking furnace. *Journal of Process Control, 18*, 439–448.

Zhang, R., Xue, A., & Wang, S. (2011). Modeling and nonlinear predictive functional control of liquid level in a coke fractionation tower. *Chemical Engineering Science, 66*, 6002–6013.

19

A multivariable adaptive controller for a quadrotor with guaranteed matching conditions

Justin M. Selfridge* and Gang Tao

Department of Electrical and Computer Engineering, University of Virginia, Charlottesville, VA 22904, USA

This paper develops an adaptive control system for a quadrotor unmanned aerial vehicle. It employs a state feedback output tracking design for multi-input multi-output systems, using a less restrictive matching condition than a state tracking design, and offers a simpler controller structure than an output feedback design. Some key characteristics of the quadrotor dynamics are derived for adaptive control design which deals with system uncertainties from changing operating points. The plant–model matching is ensured despite of system parameter uncertainties which cannot be handled by an existing state tracking design. The adaptive law is based on a parametrization using an LDS decomposition of the high-frequency gain matrix, which ensures closed-loop stability and asymptotic output tracking. A simulation study is carried out on the nonlinear quadrotor model, and results are presented to demonstrate the desired adaptive system performance.

Keywords: adaptive control; multivariable system; nonlinear model; state feedback; output tracking; quadrotor

1. Introduction

This research develops an adaptive control architecture for a quadrotor unmanned aerial vehicle (UAV). Traditional controllers require accurate models, but in practice the system is rarely completely known in advance or may change over time. Adaptive controllers are designed to accommodate these uncertainties by adapting to the changing system online. Model reference adaptive control (MRAC) is a fundamental adaptive control architecture, but established theory often requires a strict matching condition between the open and closed-loop systems, which may be difficult or impossible to obtain a priori.

Quadrotors are highly maneuverable and have a very simple structural design, so these vehicles have many military and civilian surveillance applications. As demand and expectations for these vehicles continue to increase, robust controllers designed around multiple operating points will be required. However, existing quadrotor controllers are often based on simplifying assumptions or single operating points, and cannot accommodate a wide range of conditions.

Background: This paper builds upon research on adaptive control theory and quadrotor controller architecture. Both areas of research are well documented in the literature.

MRAC is a fundamental adaptive control methodology with a rich literature including Elliott and Wolovich (1982), Goodwin and Sin (1984), Ioannou and Sun (1996), and Tao (2003a); which provide comprehensive material on parameter estimation and adaptive control theory. State feedback output tracking for nonlinear systems is documented in Isidori (1995), Krstic, Kanellakopoulos, and Kokotovic (1995), and Guo, Liu, & Tao (2009). High-frequency gain matrix decompositions, commonly used with output tracking designs, are presented in Guo, Tao, and Liu (2011), Tao (2003b), and Imai, Costa, Hsu, Tao, and Kokotovic (2001).

Quadrotor research has received considerable attention from numerous groups. A comprehensive quadrotor model is presented with proportional-derivative (PD) and linear quadratic regulator (LQR) controller designs in Bouabdallah, Murrieri, and Siegwart (2004). Stability and robustness under the presence of external disturbances is covered in Nicol, Macnab, and Ramirez-Serrano (2011). A large quadrotor with high fidelity model is presented in Dydek, Annaswamy, and Lavretsky (2013), and Pounds, Mahony, and Corke (2010) provide a comparison of many traditional controllers and then supplement the design with an adaptive control scheme. A nonlinear approach to quadrotor control is addressed in Diao, Xian, Yin, Zeng, Li, and Yang (2011), dynamic inversion is covered in Das, Subbarao, and Lewis (2009), and back-stepping control is presented in Madani and Benallegue (2006). Even the classic control problem of the "inverted pendulum" is applied to quadrotor vehicles in Hehn and D'Andrea (2011).

Motivation: Both state feedback and output feedback control designs can be applied to MRAC; however, state tracking requires a strict matching condition between the plant and model, and output feedback requires a more complicated controller structure. This work reviews state feedback with output tracking based on an LDS decomposition of the high-frequency gain matrix, which keeps the simple state feedback structure while relaxing the

*Corresponding author. Email: jms5gd@virginia.edu

required matching condition. Quadrotors have nonlinear coupling that is often simplified during the design process, time-varying parameters which move the model off the nominal design condition, and ancillary tasks which alter the vehicle parameters. This research offers a controller that adapts to changing system parameters and can accommodate different operating points. The contributions of this paper include:

(1) a characterization of the quadrotor system,
(2) a controller with ensured matching condition, and
(3) a system that adapts to many operating points.

Problem statement: Quadrotor dynamics are often simplified, unknown, or changing over time, and they may maintain several different operating points during a flight, so the controller must adapt to accommodate this uncertainty. State feedback offers a simple controller structure, but state tracking requires a strict matching condition between the plant and the reference model, so the controller should maintain the simple structure and guarantee a matching condition. The open-loop dynamics of the quadrotor are unstable, so the controller must ensure asymptotic stability, alter the system dynamics to some prescribed response characteristics, and prove that all output errors decay to zero exponentially.

Paper outline: This work investigates an adaptive controller applied to a quadrotor. The dynamics and equations of motion are discussed in Section 2, and the theory for nominal and adaptive control is reviewed in Section 3. The state feedback output tracking controller for the quadrotor is developed in Section 4, and the controller is evaluated through a simulation of the nonlinear system in Section 5. Finally, Section 6 closes the paper with the conclusions.

2. System model

This section presents the problem statement, describes the general quadrotor system, derives the dynamics of the vehicle, and outlines the linearization process.

2.1. System description

The quadrotor has four fixed-pitch props arranged in a symmetric "X" formation. A diagram of the vehicle geometry and coordinate system is provided in Figure 1. The vehicle is controlled entirely through the motor systems, where

- Roll is differential thrust between left/right motors.
- Pitch is differential thrust between front/rear motors.
- Yaw is differential torque between clockwise (CW) and counterclockwise (CCW) motors.
- Altitude ramps up/down the motors in unison.

The vehicle is an underactuated system, so independently manipulating all six degrees of freedom is not possible.

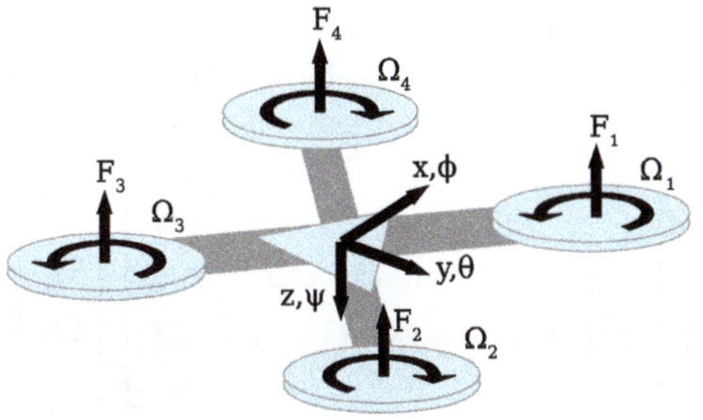

Figure 1. Quadrotor configuration.

Forward and lateral translations are coupled to pitch and roll of the vehicle, so it must pitch to move longitudinally, and roll to move laterally. Rotation matrix R, described by

$$R = \begin{bmatrix} c_\theta c_\psi & -c_\phi s_\psi + s_\phi s_\theta c_\psi & s_\phi s_\psi + c_\phi s_\theta c_\psi \\ c_\theta s_\psi & c_\phi c_\psi + s_\phi s_\theta s_\psi & -s_\phi c_\psi + c_\phi s_\theta s_\psi \\ -s_\theta & s_\phi c_\theta & c_\phi c_\theta \end{bmatrix}, \quad (1)$$

maps the body frame to the inertial frame, where sine and cosine (s_*, c_*) are applied to the vehicle attitude (ϕ, θ, ψ).

2.2. Quadrotor dynamics

The vehicle is modeled as a rigid body with external forces and moments acting on the body. Gravity, the gyroscopic effects from the propellers, and the control inputs from the motors are the primary external forces and moments on the vehicle (Bouabdallah et al. 2004). This section presents the rigid body dynamics and then addresses external forces and moments individually.

Rigid body dynamics: The dynamics of a rigid body under external forces and moments expressed in the inertial frame are governed by

$$m\dot{v} + \omega \times mv = \Sigma F, \quad J\dot{\omega} + \omega \times J\omega = \Sigma M, \quad (2)$$

where $\Sigma F \in \mathbb{R}^3$ and $\Sigma M \in \mathbb{R}^3$ represent the external forces and moments acting on the vehicle, v is the velocity in the body frame, ω the angular rate of the vehicle, m the quadrotor mass, and J the moment of inertia matrix.

Control inputs: The controller design is better suited for manipulating forces and moments on the vehicle, but the vehicle is physically controlled by adjusting motor speeds. Simplified propeller dynamics reduce to a linear relationship between the squared rotor angular rate and the force and moment of the propellers. The relationship is given by

$$F_{zi} = c_f \Omega_i^2, \quad M_{zi} = c_m \Omega_i^2, \quad (3)$$

where i is the motor index, c_f and c_m are thrust and drag coefficients, and Ω_i the rotor angular velocity. The motors are all aligned with the vertical axis of the body frame, so the force and moments only occur in the z-direction of the body frame; i.e. $F_{xi} = F_{yi} = 0$ and $M_{xi} = M_{yi} = 0$, where

F_z is negative to remain consistent with the downward z-axis convention. The relationship that maps motor angular velocities to forces and moments on the vehicle is

$$\begin{bmatrix} F_z \\ M_x \\ M_y \\ M_z \end{bmatrix} = \begin{bmatrix} -c_f & -c_f & -c_f & -c_f \\ -c_f d & -c_f d & c_f d & c_f d \\ c_f d & -c_f d & -c_f d & c_f d \\ c_m & -c_m & c_m & -c_m \end{bmatrix} \begin{bmatrix} \Omega_1^2 \\ \Omega_2^2 \\ \Omega_3^2 \\ \Omega_4^2 \end{bmatrix}, \quad (4)$$

where d is the distance between the motors and CG.

Gyroscopic moment: The spinning propellers create a gyroscopic torque when the quadrotor rotates in space. All propellers have the same moment of inertia, so the sum of the rotor angular velocities, $\Omega_r = -\Omega_1 + \Omega_2 - \Omega_3 + \Omega_4$, can be used to model gyroscopic effects. The rotational velocity of the vehicle ω and the net angular velocity of the rotors Ω_r determine the resultant gyroscopic moment: $M_g = \omega \times J_r \Omega_r$.

Complete equations of motion: The complete equations of motion for the quadrotor are obtained by expanding out the rigid body dynamics, and then adding the external forces and moments from the control inputs, the gyroscopic effects, and the gravitational force. After manipulating terms, the vehicle dynamics are determined to be

$$\begin{aligned}
\ddot{x} &= \left(s_\phi s_\psi + c_\phi s_\theta c_\psi\right) \frac{F_z}{m}, \\
\ddot{y} &= \left(-s_\phi c_\psi + c_\phi s_\theta s_\psi\right) \frac{F_z}{m}, \\
\ddot{z} &= g + \left(c_\phi c_\theta\right) \frac{F_z}{m}, \\
\ddot{\phi} &= \dot{\theta}\dot{\psi} \left(\frac{J_y - J_z}{J_x}\right) - \frac{J_r}{J_x}\dot{\theta}\Omega_r + \frac{M_x}{J_x}, \\
\ddot{\theta} &= \dot{\phi}\dot{\psi} \left(\frac{J_z - J_x}{J_y}\right) + \frac{J_r}{J_y}\dot{\phi}\Omega_r + \frac{M_y}{J_y}, \\
\ddot{\psi} &= \dot{\phi}\dot{\theta} \left(\frac{J_x - J_y}{J_z}\right) + \frac{M_z}{J_z},
\end{aligned} \quad (5)$$

which are expressed in the inertial frame. Sine and cosine terms (s_*, c_*) come from the rotation R between the body frame and the inertial frame, translation $(\ddot{x}, \ddot{y}, \ddot{z})$ is dictated by the vertical force from the props (F_z) and the attitude (ϕ, θ, ψ) of the quadrotor, and rotational acceleration $(\ddot{\phi}, \ddot{\theta}, \ddot{\psi})$ is dominated by the moments (M_x, M_y, M_z) from the props.

2.3. *Linearized system*

Linearization takes the nonlinear dynamic system, identifies meaningful operating points, and creates a linear model around those points. The nominal controller is developed around the linearized model, and the adaptive controller adapts to the linearized model during changing conditions.

The equations of motion of the plant are expressed in compact form as $\dot{x}_p(t) = f(x_p(t), u(t))$ where

$$\begin{aligned}
x_p(t) &= [x\; y\; z\; \phi\; \theta\; \psi\; \dot{x}\; \dot{y}\; \dot{z}\; \dot{\phi}\; \dot{\theta}\; \dot{\psi}]^{\mathrm{T}} \in \mathbb{R}^{12}, \\
u(t) &= [F_z\; M_x\; M_y\; M_z]^{\mathrm{T}} \in \mathbb{R}^4,
\end{aligned} \quad (6)$$

which has $n = 12$ states and $m = 4$ control inputs. Denote an operating point as $\hat{x}_p$ and $\hat{u}$, so that perturbations are described by $\Delta x_p = x_p - \hat{x}_p$ and $\Delta u = u - \hat{u}$. The Taylor series expansion of the nonlinear system yields

$$\begin{aligned}
\dot{x}_p = f(x_p, u) &\cong f(\hat{x}_p, \hat{u}) + \left.\frac{\partial f}{\partial x_p}\right|_{(\hat{x}_p,\hat{u})} (x_p - \hat{x}_p) \\
&+ \left.\frac{\partial f}{\partial u}\right|_{(\hat{x}_p,\hat{u})} (u - \hat{u}) + \mathrm{HOT},
\end{aligned} \quad (7)$$

and after disregarding higher order terms (HOT), becomes

$$\Delta\dot{x}_p = f(\hat{x}_p, \hat{u}) + A_p \Delta x_p + B_p \Delta u. \quad (8)$$

When an operating point is also an equilibrium point (derivatives are equal to zero), then $f(\hat{x}_p, \hat{u}) = 0$ and the system reduces to $\Delta\dot{x}_p = A_p \Delta x_p + B_p \Delta u$, where future equations omit the Δ symbol for readability.

Linearization is accurate within a small region, and most quadrotor controller research limits the flight envelop to stay within this trusted region. The proposed controller adapts to changes in the high-frequency gain matrix, K_p, so the vehicle can accommodate different and changing operating points during a flight.

General form: The general form of the linearized system uses arbitrary values for all the states. The state matrix A_p is filled with the partial derivatives with respect to each state

$$A_p = \begin{bmatrix} 0_{3\times3} & 0_{3\times3} & I_{3\times3} & 0_{3\times3} \\ 0_{3\times3} & 0_{3\times3} & 0_{3\times3} & I_{3\times3} \\ 0_{3\times3} & A_t & 0_{3\times3} & 0_{3\times3} \\ 0_{3\times3} & 0_{3\times3} & 0_{3\times3} & A_r \end{bmatrix} \in \mathbb{R}^{12\times12}, \quad (9)$$

and the quadrotor's B_p matrix is populated with the partial derivatives with respect to each control input

$$B_p = \begin{bmatrix} 0_{3\times1} & 0_{3\times3} \\ 0_{3\times1} & 0_{3\times3} \\ B_t & 0_{3\times3} \\ 0_{3\times1} & B_r \end{bmatrix} \in \mathbb{R}^{12\times4}, \quad (10)$$

where submatrices A_t, A_r, B_t, and B_r are given by

$$A_t = \begin{bmatrix} c_\phi s_\psi - s_\phi s_\theta c_\psi & c_\phi c_\theta c_\psi & s_\phi c_\psi - c_\phi s_\theta s_\psi \\ -c_\phi c_\psi - s_\phi s_\theta s_\psi & c_\phi c_\theta s_\psi & s_\phi s_\psi + c_\phi s_\theta c_\psi \\ -s_\phi c_\theta & -c_\phi s_\theta & 0 \end{bmatrix} \frac{F_z}{m}, \quad (11)$$

$$A_r = \begin{bmatrix} 0 & \dot{\psi}\left(\frac{J_y - J_z}{J_x}\right) - \frac{J_r}{J_x}\Omega_r & \dot{\theta}\left(\frac{J_y - J_z}{J_x}\right) \\ \dot{\psi}\left(\frac{J_z - J_x}{J_y}\right) + \frac{J_r}{J_y}\Omega_r & 0 & \dot{\phi}\left(\frac{J_z - J_x}{J_y}\right) \\ \dot{\theta}\left(\frac{J_x - J_y}{J_z}\right) & \dot{\phi}\left(\frac{J_x - J_y}{J_z}\right) & 0 \end{bmatrix}, \tag{12}$$

$$B_t = \begin{bmatrix} s_\phi s_\psi + c_\phi s_\theta c_\psi \\ -s_\phi c_\psi + c_\phi s_\theta s_\psi \\ c_\phi c_\theta \end{bmatrix} \frac{1}{m}, \tag{13}$$

$$B_r = \operatorname{diag}\left\{\frac{1}{J_x}, \frac{1}{J_y}, \frac{1}{J_z}\right\}. \tag{14}$$

The controller uses diagonal matrices to decouple the inputs and outputs, so the system outputs must closely match the control inputs. With $u(t) = [F_z\, M_x\, M_y\, M_z]^{\mathrm{T}}$, the system output is selected as $y = [z\, y\, x\, \psi]^{\mathrm{T}}$. Nominal control inputs are needed to complete the linearization process. Maintaining a zero rotational acceleration implies $M_x = M_y = M_z = 0$, whereas maintaining a zero vertical acceleration necessitates $\ddot{z} = g + (c_\phi c_\theta)(F_z/m) = 0 \Rightarrow F_z = -mg/(c_\phi c_\theta)$.

Hover condition: The logical starting point for linearization is around the hover condition. During hover the vehicle has zero tilt about the roll and pitch axes, the heading is arbitrary, and the angular rates must all be equal to zero. The position in space is arbitrary, but the velocities must all be zero. With this description, the nominal state is given by $\hat{x}_p = [x\ y\ z\ 0\ 0\ \psi\ 0\ 0\ 0\ 0\ 0\ 0]^{\mathrm{T}}$, and the nominal control input is given by $\hat{u} = [-mg\ 0\ 0\ 0]^{\mathrm{T}}$. This state represents an equilibrium point, so $f(\hat{x}_p, \hat{u}) = 0$. Evaluating A_p at the operating point yields

$$A_t = \begin{bmatrix} -gs_\psi & -gc_\psi & 0 \\ gc_\psi & -gs_\psi & 0 \\ 0 & 0 & 0 \end{bmatrix},$$

$$A_r = \begin{bmatrix} 0 & -\frac{J_r}{J_x}\Omega_r & 0 \\ \frac{J_r}{J_y}\Omega_r & 0 & 0 \\ 0 & 0 & 0 \end{bmatrix}, \tag{15}$$

and evaluating B_p at the operating point yields

$$B_t = \left[0\ 0\ \frac{1}{m}\right]^{\mathrm{T}}, \quad B_r = \operatorname{diag}\left\{\frac{1}{J_x}\frac{1}{J_y}\frac{1}{J_z}\right\}. \tag{16}$$

A similar procedure can be used to evaluate operating points other than the hover condition.

3. Control system design

This section addresses the conditions to implement the controller, describes the theory to develop a nominal controller for a multi-input multi-output (MIMO) system, and then builds upon that foundation to develop the adaptive controller for the quadrotor.

3.1. *Plant assumptions*

Consider a linearized MIMO system described by

$$\dot{x}_p(t) = A_p x_p(t) + B_p u(t), \quad y_p(t) = Cx_p(t) \tag{17}$$

with transfer matrix $G(s) = C(sI - A_p)^{-1}B_p$, where the parameters in A_p, B_p, and C are unknown. From Tao (2003a), we know that an $m \times m$ strictly proper and full rank rational transfer matrix $G(s)$ has an interactor matrix $\xi(s)$ such that the high-frequency gain matrix of $G(s)$, defined as $K_p = \lim_{s\to\infty} \xi(s)G(s)$, is finite and nonsingular. Also, the high-frequency gain matrix $K_p \in \mathbb{R}^{m\times m}$, with all its leading principle minors Δ_i being nonzero, has a non-unique decomposition $K_p = \mathrm{LDS}$, where $S = S^{\mathrm{T}} > 0$, $L \in \mathbb{R}^{m\times m}$ is a unity lower triangular matrix, and

$$D = \operatorname{diag}\left\{\operatorname{sign}[\Delta_1]\gamma_1, \ldots, \operatorname{sign}\left[\frac{\Delta_m}{\Delta_{m-1}}\right]\gamma_m\right\}, \tag{18}$$

such that $\gamma_i > 0$, $i = 1, \ldots, m$, is arbitrary.

For a MIMO state feedback controller, the following conditions must be satisfied (Tao 2003a):

(1) (A_p, B_p) controllable and (A_p, C) observable;
(2) all zeros of $G(s)$ have negative real parts;
(3) $G(s)$ has full rank with known $\xi(s)$; and
(4) all leading principle minors Δ_i, $i = 1, \ldots, m$, of K_p are nonzero and their signs are known.

Condition (1) is needed for stable plant–model matching. Zeros of a MIMO system cause $G(s)$ to lose rank, so that a control input $u(t) \neq 0$ exists where $y(t) = G(s)u(t) = 0$; meaning the control input has no influence on the system output. Condition (2) ensures that no transmission zeros exist which can cause the system to become uncontrollable. Condition (3) is needed to select a viable reference model, and Condition (4) ensures that the adaptation laws converge in the correct direction.

3.2. *Nominal controller design*

All states are available for measurement, so state feedback control is used. This section describes state tracking control, and illustrates the limitations with the matching condition when plant matrices are unknown. Then output tracking is presented because it offers a simple control structure and alleviates the matching condition requirement.

State tracking: The goal of state feedback state tracking control is to influence the system (17), by designing a control input $u(t)$ such that all the signals in the closed-loop system are bounded, and the state vector signal $x_p(t)$ asymptotically tracks a reference state vector signal $x_m(t)$.

A bounded reference signal $r(t) \in \mathbb{R}^m$ is applied to the reference system

$$\dot{x}_m(t) = A_m x_m(t) + B_m r(t), \tag{19}$$

where stable $A_m \in \mathbb{R}^{n\times n}$ and $B_m \in \mathbb{R}^{n\times m}$ describe the desired system characteristics. The control input

$$u(t) = K_x x_p(t) + K_r r(t), \tag{20}$$

with $K_x \in \mathbb{R}^{m\times n}$ and $K_r \in \mathbb{R}^{m\times m}$, achieves the desired control objective by producing a closed-loop system governed by $\dot{x}_p(t) = A_m x_p(t) + B_m r(t)$. The gains must be selected so $A_m = A_p + B_p K_x$ and $B_m = B_p K_r$, which describe the necessary matching condition (Franklin, Powell, & Emami-Naeini 2009).

Matching condition limitations: Consider the general structure of the quadrotor plant matrices A_p and B_p, given in Equations (9) and (10). Denote the submatrices as $A_t = \{a_t\}_{ij}$ and $A_r = \{a_r\}_{ij}$ for $i,j = 1,2,3$, the column vector as $B_t = [b_1\, b_2\, b_3]^{\mathrm{T}}$, and the diagonal matrix as $B_r = \mathrm{diag}\{b_4\, b_5\, b_6\}$. The matching condition becomes

$$A_m = \begin{bmatrix} 0_{3\times3} & 0_{3\times3} & I_{3\times3} & 0_{3\times3} \\ 0_{3\times3} & 0_{3\times3} & 0_{3\times3} & I_{3\times3} \\ & A_{mt} \in \mathbb{R}^{3\times12} & & \\ & A_{mr} \in \mathbb{R}^{3\times12} & & \end{bmatrix},$$

$$B_m = \begin{bmatrix} 0_{6\times4} \\ B_{mt} \in \mathbb{R}^{3\times4} \\ B_{mr} \in \mathbb{R}^{3\times4} \end{bmatrix}, \tag{21}$$

where

$$a_{mt_{ij}} = \begin{cases} b_i k_{x1j} + a_{t_{ij-3}} & \text{for } j = 4,5,6, \\ b_i k_{x1j} & \text{otherwise,} \end{cases}$$

$$a_{mr_{ij}} = \begin{cases} b_{i+3} k_{xi+1j} + a_{r_{ij-9}} & \text{for } j = 10,11,12, \\ b_{i+3} k_{xi+1j} & \text{otherwise,} \end{cases} \tag{22}$$

for $i = 1,2,3$ and $j = 1,2,\ldots,12$, and

$$b_{mt_{ij}} = b_i k_{r1j}, \quad b_{mr_{ij}} = b_{i+3} k_{ri+1j}, \tag{23}$$

for $i = 1,2,3$ and $j = 1,2,3,4$. The sparse nature of the required A_m and B_m matrices, and the limited options for elements within those matrices, make the matching condition difficult or impossible to satisfy.

Output tracking: To alleviate the state tracking matching condition, an output tracking controller is presented. The goal of state feedback output tracking control is to influence the system (17), by designing a control input $u(t)$ such that all the signals in the closed-loop system are bounded, and the output $y_p(t)$ asymptotically tracks a reference output $y_m(t)$; meaning $\lim_{t\to\infty}(y_p(t) - y_m(t)) = 0$. A bounded reference signal $r(t) \in \mathbb{R}^m$ is applied to the reference system

$$y_m(t) = G_c(s)[r](t), \quad G_c(s) = \xi_m^{-1}(s), \tag{24}$$

where the closed-loop transfer matrix G_c is described by

$$G_c(s) = C(sI - A_p - B_p K_x)^{-1} B_p K_r, \tag{25}$$

and $\xi_m = \mathrm{diag}\{d_1(s), \ldots, d_m(s)\}$ is the modified left interactor matrix, where $d_i(s)$ are the desired closed-loop characteristic polynomials of degree l_i. The high-frequency gain matrix K_p is related to the control gains K_x and K_r through the relationships

$$K_x = K_p^{-1} K_0, \quad K_r = K_p^{-1}, \tag{26}$$

where the ith row of K_0 is described by

$$K_{0i} = -c_i A_p^{l_i} - d_{i1} c_i A_p^{l_i - 1} - \cdots - d_{il_i-1} c_i A_p - d_{il_i} c_i, \tag{27}$$

where c_i is the ith row of the C matrix, and d_{ij} are coefficients of the desired characteristic polynomials (Goodwin, Graebe, & Salgado 2001).

Relaxed matching condition: Whereas, state tracking control requires a strict matching condition between $A_m = A_p + B_p K_x$ and $B_m = B_p K_r$, the structure of the output tracking controller is more flexible. The gain matrices K_x and K_r are defined in terms of K_0, which encompasses the terms for the desired closed-loop characteristic polynomials $d_i(s)$. These polynomials can be arbitrarily tailored to suit any desired system response. The only restriction on the selection of $d_i(s)$ is the degree of each polynomial must match the degree of the plant interactor matrix $\xi(s)$, which is common in any traditional pole placement controller design.

3.3. *Adaptive controller design*

When the state and input matrices A_p and B_p are known, the state feedback gains K_x and K_r can be uniquely determined (Chen 2013). However, when A_p and B_p are either unknown or changing, then static values for K_x and K_r are not appropriate. The uncertainty associated with the nominal controller motivates the development of the adaptive controller, where the parameters of A_p and B_p are estimated to determine the appropriate feedback gains.

The goal of the adaptive control algorithm is to have the system output $y_p(t)$ track a desired output $y_m(t)$ given by Equation (24), and to have the tracking error, $e(t) = y_p(t) - y_m(t)$, decay to zero exponentially. Let K_x^* and K_r^* denote the true (unknown) feedback gains, and $K_x(t)$ and $K_r(t)$ be their estimates. The feedback control law becomes

$$u(t) = K_x(t) x_p(t) + K_r(t) r(t), \tag{28}$$

where $K_x(t)$ and $K_r(t)$ are updated from adaptive laws. Uncertainties in A_p and B_p are accounted for in the high-frequency gain matrix K_p, and the LDS decomposition of K_p is included within the adaptation process.

Control structure: Substituting the estimates of the gain matrices $K_x(t)$ and $K_r(t)$ for the control law yields

$$\dot{x}_p(t) = (A_p + B_p K_x^*) x_p(t) + B_p K_r^* r(t) + B_p[(K_x(t) - K_x^*) x_p(t) + (K_r(t) - K_r^*) r(t)], \quad (29)$$

$$y_p(t) = C x_p(t),$$

so the output tracking error is expressed as

$$e(t) = G_c(s) K_p [\tilde{\Theta}^{\mathrm{T}} \omega](t) + C e^{(A_p + B_p K_x^* t)} x_p(0), \quad (30)$$

where

$$\begin{aligned} \Theta^* &= [K_x^{*\mathrm{T}}, K_r^*]^{\mathrm{T}}, \\ \Theta(t) &= [K_x^{\mathrm{T}}(t), K_r(t)]^{\mathrm{T}}, \\ \tilde{\Theta}(t) &= \Theta(t) - \Theta^*, \quad (31) \\ \omega &= \left[x_p^{\mathrm{T}}(t), r^{\mathrm{T}}(t)\right]^{\mathrm{T}}. \end{aligned}$$

When there is no estimation error, $K_x(t) = K_x^*$ and $K_r(t) = K_r^*$. This indicates $\lim_{t\to\infty} e(t) = 0$, because the estimation error $\tilde{\Theta}(t)$ is zero, and the stability of $A_p + B_p K_x^*$ ensures $C e^{(A_p + B_p K_x^* t)} x(0)$ converges to zero exponentially.

Error parametrization: Following Guo et al. (2011), we ignore the exponentially decaying term, and substitute the LDS decomposition for K_p, so the tracking error is expressed as

$$L^{-1} \xi_m(s)[e](t) = D S \tilde{\Theta}^{\mathrm{T}}(t) \omega(t). \quad (32)$$

Eliminate the unity on the diagonal in L^{-1} by introducing $\Lambda^* = L^{-1} - I = \{\lambda_{ij}^*\}$, so that $\lambda_{ij}^* = 0$ on its diagonal and upper triangle, and then substituting in Λ^* yields

$$\Lambda^* \xi_m(s)[e](t) + \xi_m(s)[e](t) = D S \tilde{\Theta}^{\mathrm{T}}(t) \omega(t). \quad (33)$$

Introduce a stable filter $F(s) = 1/f(s)$, where $f(s)$ is a stable monic polynomial, where the degree of $f(s)$ matches the degree of $\xi_m(s)$. Operate on both sides by $F(s)$, so

$$\Lambda^* \bar{e}(t) + \bar{e}(t) = D S F(s)[\tilde{\Theta}^{\mathrm{T}} \omega](t), \quad (34)$$

where the transformed error signal $\bar{e}(t)$ is given by

$$\bar{e}(t) = \xi_m(s) F(s)[e](t) = [\bar{e}_1(t), \bar{e}_2(t), \ldots, \bar{e}_m(t)]^{\mathrm{T}}. \quad (35)$$

The $\Lambda^* \bar{e}(t)$ term can be expressed as

$$\Lambda^* \bar{e}(t) = [0,\ \lambda_2^{*\mathrm{T}} \eta_2(t),\ \lambda_3^{*\mathrm{T}} \eta_3(t), \ldots, \lambda_m^{*\mathrm{T}} \eta_m(t)]^{\mathrm{T}}, \quad (36)$$

where $\lambda_i^* = [\lambda_{i1}^*, \ldots, \lambda_{ii-1}^*]^{\mathrm{T}}$ and $\eta_i = [\bar{e}_1(t), \ldots, \bar{e}_{i-1}]^{\mathrm{T}}$, for $i = 2, \ldots, m$, and isolate the transformed error $\bar{e}(t)$ so

$$\bar{e}(t) = -\left[0, \lambda_2^{*\mathrm{T}} \eta_2(t), \lambda_3^{*\mathrm{T}} \eta_3(t), \ldots, \lambda_m^{*\mathrm{T}} \eta_m(t)\right]^{\mathrm{T}} + D S F(s)[\tilde{\Theta}^{\mathrm{T}} \omega](t). \quad (37)$$

Let $\lambda_{ij}(t)$ be the estimate of λ_{ij}^*, and denoting $\Psi^* = DS$, let $\Psi(t)$ be the estimate of Ψ^*. Estimation errors are the difference between the estimates and their true values, so

$$\tilde{\Psi}(t) = \Psi(t) - \Psi^* \quad \text{and} \quad \tilde{\lambda}_i(t) = \lambda_i(t) - \lambda_i^*. \quad (38)$$

Define $\epsilon(t)$ as the summation of all the parameter estimates

$$\epsilon(t) = [0, \lambda_2^{\mathrm{T}}(t) \eta_2(t), \ldots, \lambda_m^{\mathrm{T}}(t) \eta_m(t)]^{\mathrm{T}} + \bar{e}(t) + \Psi(t) \rho(t), \quad (39)$$

where $\rho(t) = \Theta^{\mathrm{T}}(t) \zeta(t) - F(s)[\Theta^{\mathrm{T}} \omega](t)$ and $\zeta(t) = F(s)[\omega](t)$. Substituting in the transformed tracking error $\bar{e}(t)$ yields

$$\epsilon(t) = [0, \tilde{\lambda}_2^{\mathrm{T}}(t) \eta_2(t), \ldots, \tilde{\lambda}_m^{\mathrm{T}}(t) \eta_m(t)]^{\mathrm{T}} + \tilde{\Psi} \rho(t) + D S \tilde{\Theta}^{\mathrm{T}}(t) \zeta(t), \quad (40)$$

which puts $\epsilon(t)$ completely in terms of parameter errors.

Adaptive laws: The estimates do not always equal their true values, so $K_x(t)$ and $K_r(t)$ are updated from adaptive laws. Select a normalizing signal $m(t)$ described by

$$m^2(t) = 1 + \zeta^{\mathrm{T}}(t) \zeta(t) + \rho^{\mathrm{T}}(t) \rho(t) + \sum_{i=2}^{m} \eta_i^{\mathrm{T}}(t) \eta_i(t), \quad (41)$$

then the expression for the estimation error $\epsilon(t)$ suggests the following adaptive laws:

$$\dot{\Theta}^{\mathrm{T}}(t) = -\frac{D \epsilon(t) \zeta^{\mathrm{T}}(t)}{m^2(t)}, \quad (42)$$

$$\dot{\Psi}(t) = -\frac{\Gamma \epsilon(t) \rho^{\mathrm{T}}(t)}{m^2(t)}, \quad (43)$$

$$\dot{\lambda}_i(t) = -\frac{\Gamma_{\lambda i} \epsilon_i(t) \eta_i(t)}{m^2(t)}, \quad i = 2, 3, \ldots, m, \quad (44)$$

where the adaptive gains are selected so that $\Gamma = \Gamma^{\mathrm{T}} > 0$ and $\Gamma_{\lambda i} = \Gamma_{\lambda i}^{\mathrm{T}} > 0$, for $i = 2, 3, \ldots, m$.

Stability analysis: To evaluate the closed-loop stability, define a positive definite function $V(\tilde{\Theta}(t), \tilde{\Psi}(t), \tilde{\lambda}_i(t))$ as

$$V = \mathrm{tr}[\tilde{\Theta}^{\mathrm{T}} S \tilde{\Theta}] + \mathrm{tr}[\tilde{\Psi}^{\mathrm{T}} \Gamma^{-1} \tilde{\Psi}] + \sum_{i=2}^{m} \tilde{\lambda}_i^{\mathrm{T}} \Gamma_{\lambda i}^{-1} \tilde{\lambda}_i, \quad (45)$$

and calculate its time derivative as

$$\begin{aligned} \dot{V} &= 2\left[-\frac{\zeta^{\mathrm{T}}(t) \tilde{\Theta} S D \epsilon(t)}{m^2(t)} - \frac{\rho^{\mathrm{T}}(t) \tilde{\Psi} \epsilon(t)}{m^2(t)} - \sum_{i=2}^{m} \frac{\tilde{\lambda}_i^{\mathrm{T}} \epsilon_i(t) \eta_i(t)}{m^2(t)}\right] \\ &= -\frac{2 \epsilon^{\mathrm{T}}(t) \epsilon(t)}{m^2(t)} \leq 0. \quad (46) \end{aligned}$$

Having $\dot{V} \leq 0$ implies that all the estimation signals are bounded: $\Theta(t) \in L^\infty$, $\Psi(t) \in L^\infty$, and $\lambda_i(t) \in L^\infty$ for $i = 2, \ldots, m$. It also shows that $\epsilon(t)/m(t) \in L^2 \cap L^\infty$, which implies that $\dot{\Theta}(t) \in L^2 \cap L^\infty$, $\dot{\Psi}(t) \in L^2 \cap L^\infty$, and $\dot{\lambda}_i(t) \in L^2 \cap L^\infty$ for $i = 2, \ldots, m$. From these properties, the boundedness of $x_p(t)$ can be shown. The boundedness of $x_p(t)$, $r(t)$, $K_x(t)$, and $K_r(t)$ guarantees that $u(t)$ is bounded, thus all closed-loop control signals are bounded and $\lim_{t\to\infty} e(t)$ approaches zero asymptotically (Hsu, Costa, Imai, and Kokotovic 2001).

4. Quadrotor controller development

This section utilizes established control theory to develop the nominal controller, and then expands to the adaptive control architecture applied to the quadrotor system.

4.1. *Quadrotor nominal controller*

Before embarking on the adaptive controller analysis, it is crucial to understand the characteristics of the vehicle and the structure of the nominal controller. To numerically evaluate the system, consider a quadrotor with the following estimated parameters: $g = 9.8\,\mathrm{m/s^2}$, $m = 0.6\,\mathrm{kg}$, $J_x = J_y = 0.02\,\mathrm{kg\,m^2}$, $J_z = 0.05\,\mathrm{kg\,m^2}$, $J_r = 0.001\,\mathrm{kg\,m^2}$, and $\Omega_r = 0\,\mathrm{rad/s}$, with dimensions $n = 12$ and $m = 4$. The system is a special form where the number of inputs equals the number of outputs. For the nominal controller, the parameter estimates are assumed to be the true values which are used to determine K_x^* and K_r^*.

Using the specified parameters and inserting zeros for all the states yields the following state and input submatrices:

$$A_t = \begin{bmatrix} 0 & -9.8 & 0 \\ 9.8 & 0 & 0 \\ 0 & 0 & 0 \end{bmatrix}, \quad A_r = \begin{bmatrix} 0 & 0 & 0 \\ 0 & 0 & 0 \\ 0 & 0 & 0 \end{bmatrix},$$
$$B_t = \begin{bmatrix} 0 \\ 0 \\ 1.667 \end{bmatrix}, \quad B_r = \begin{bmatrix} 50 & 0 & 0 \\ 0 & 50 & 0 \\ 0 & 0 & 20 \end{bmatrix}, \tag{47}$$

where the system transfer matrix $G(s)$ is described by

$$G(s) = C(sI - A_p)^{-1} B_p = \frac{N(s)}{d(s)}$$
$$= \frac{1}{d(s)} \mathrm{diag}\{1.667s^2, 490, -490, 20s^2\}, \tag{48}$$

where $d(s) = s^4$ is the characteristic equation after pole/zero cancelations. The controllability and observability matrices are both full rank, so the system is fully controllable and observable; and the system transfer matrix $G(s)$ is full rank and strictly proper, which satisfies the design condition.

Finding the poles and zeros of a MIMO system is accomplished with the Smith-McMillan form, described by Goodwin et al. (2001) and Hosoe (1975). Find the greatest common divisors $\chi_i(s)$ of all $i \times i$ minor determinants of $N(s)$. Then $\bar{\epsilon}_i = \chi_i / \chi_{i-1}$ forms $\bar{\epsilon}_i(s)/d(s) = \epsilon_i(s)/\delta_i(s)$, which is used to find the diagonal elements of the Smith-McMillan matrix. Following the procedure, the Smith-McMillan form becomes

$$G_{\mathrm{SM}}(s) = \mathrm{diag}\left\{ \frac{1}{s^2}, \frac{1}{s^4}, \frac{1}{s^4}, \frac{1}{s^2} \right\}, \tag{49}$$

where the system zeros are calculated as

$$p_z(s) = \epsilon_1(s)\epsilon_2(s)\epsilon_3(s)\epsilon_4(s) = 1 \tag{50}$$

and the system poles are found to be

$$p_p(s) = \delta_1(s)\delta_2(s)\delta_3(s)\delta_4(s) = s^{12}, \tag{51}$$

so there are no zeros and 12 poles located at the origin.

The interactor matrix $\xi(s) = \mathrm{diag}\{s^{l_1}, s^{l_2}, s^{l_3}, s^{l_4}\}$ of the quadrotor is generally diagonal, and forms the high-frequency gain matrix

$$K_p = \lim_{s\to\infty} \xi(s) G(s), \tag{52}$$

which must be finite and nonsingular; so the limit of all elements of K_p must not grow to infinity, and the determinant must be nonzero. Selecting $l_2 = l_3 = 4$ and $l_1 = l_4 = 2$ yields $\xi(s) = \mathrm{diag}\{ s^2, s^4, s^4, s^2 \}$, and the high-frequency gain matrix becomes

$$K_p = \mathrm{diag}\{1.667, 490, -490, 20\}, \tag{53}$$

where $|K_p| = 8003333 \neq 0$. The polynomials $d_i(s)$ in the modified interactor matrix $\xi_m(s)$ must have degrees of l_i, and the desired poles can be placed arbitrarily. For this analysis, all poles are located at -2, which gives the following:

$$\partial d_2(s) = l_2 = 4, \quad \partial d_3(s) = l_3 = 4,$$
$$d_2(s) = d_3(s) = (s+2)^4 = s^4 + 8s^3 + 24s^2 + 32s + 16, \tag{54}$$

and

$$\partial d_1(s) = l_1 = 2, \quad \partial d_4(s) = l_4 = 2,$$
$$d_1(s) = d_4(s) = (s+2)^2 = s^2 + 4s + 4. \tag{55}$$

The coefficients d_{ij} from the polynomials

$$d_i(s) = s^4 + d_{i1}s^3 + d_{i2}s^2 + d_{i3}s + d_{i4}, \quad i = 2, 3,$$
$$d_i(s) = s^2 + d_{i1}s + d_{i2}, \quad i = 1, 4, \tag{56}$$

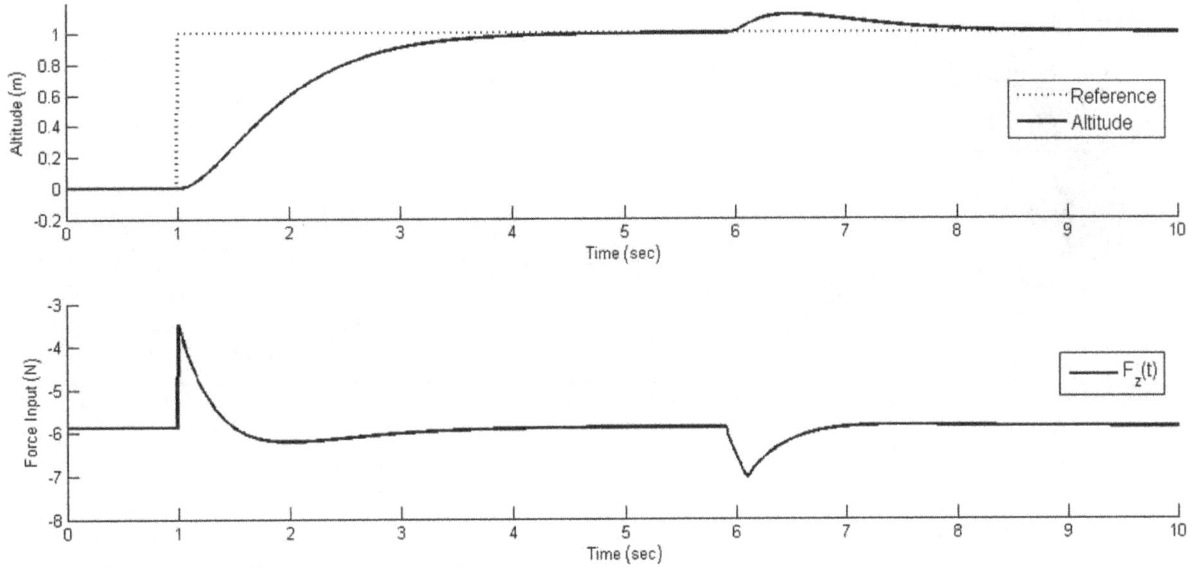

Figure 2. Elevation change: altitude and force input.

are used to populate the K_0 matrix which determines the K_x^* and K_r^* matrices. After pole/zero cancelations, the closed-loop system is described by $G_c(s)$, given as

$$G_c(s) = C(sI - A_p - B_pK_x^*)^{-1}B_pK_r^* \quad (57)$$

$$= C(sI - A_m)^{-1}B_m \quad (58)$$

$$= \begin{bmatrix} \frac{1}{(s+2)^2} & 0 & 0 & 0 \\ 0 & \frac{1}{(s+2)^4} & 0 & 0 \\ 0 & 0 & \frac{1}{(s+2)^4} & 0 \\ 0 & 0 & 0 & \frac{1}{(s+2)^2} \end{bmatrix}, \quad (59)$$

which shows that the transfer matrix fits the desired form, and all of the poles have been successfully placed at $s = -2$.

4.2. *Quadrotor adaptive controller*

When implementing the adaptive controller, the initial parameter estimates are not expected to equal their true values. However, the controller needs to be initialized with some starting values. The initial guesses for the parameters can be used to determine appropriate starting points.

From the nominal controller development, use the initial parameter guesses to find $K_x(0)$ and $K_r(0)$, which provides a starting point to populate $\Theta(0)$. In a similar way, the initial parameter estimates form an approximate value for the high-frequency gain matrix, denoted as $\hat{K}_p$. Because the other two parameter estimates, $\Psi(t)$ and $\lambda_i(t)$, are based on the LDS decomposition of K_p, we can also determine suitable starting values for those matrices.

The structure of $D(t)$ is known to be diagonal, with our choice of γ_i. For simplicity, select $S(0) = I_4$ as a starting point. Ideally, the parameter estimates $\lambda_i(t)$ should be initialized with zero values, and then adapt as needed. Setting $L(0) = I_4$ achieves $\lambda_{ij}(0) = 0$, so selecting values for γ_i is all that remains. The relationship is

$$K_p = \text{LDS} \Leftrightarrow L^{-1}(0)\hat{K}_pS^{-1}(0) = D(0), \quad (60)$$

where

$$\hat{K}_p = \text{diag}[\hat{k}_{p1}, \ldots, \hat{k}_{pm}] \quad \text{and}$$
$$D(0) = \text{diag}[d_1(0), \ldots, d_m(0)] \quad (61)$$

indicate that $D(0)$ should be initialized with $\hat{K}_p$, so the gain is set to $\gamma_i = \text{sign}[K_{pi}]K_{pi} > 0$.

5. Simulation study

Two simulations are presented as part of this research. The first simulation shows the system response to a step change in altitude which is then subjected to a physical disturbance. The second simulation illustrates the vehicle trajectory while following a circular path with an initial position error. When running the adaptive controller simulations, the true parameter values are set so that the mass and the moment of inertias are 110% of their estimated values. Both simulations use the parameter and gain values developed previously.

Elevation change: The quadrotor is initially at rest at the origin. At 1.0 s, the reference signal increases altitude by 1.0 m, and at 6.0 s, the system is disturbed by a 2.0 N force. The controller development arbitrarily sets the poles at −2, which implies the system is overdamped. The elevation change simulation, provided in Figure 2, confirms the

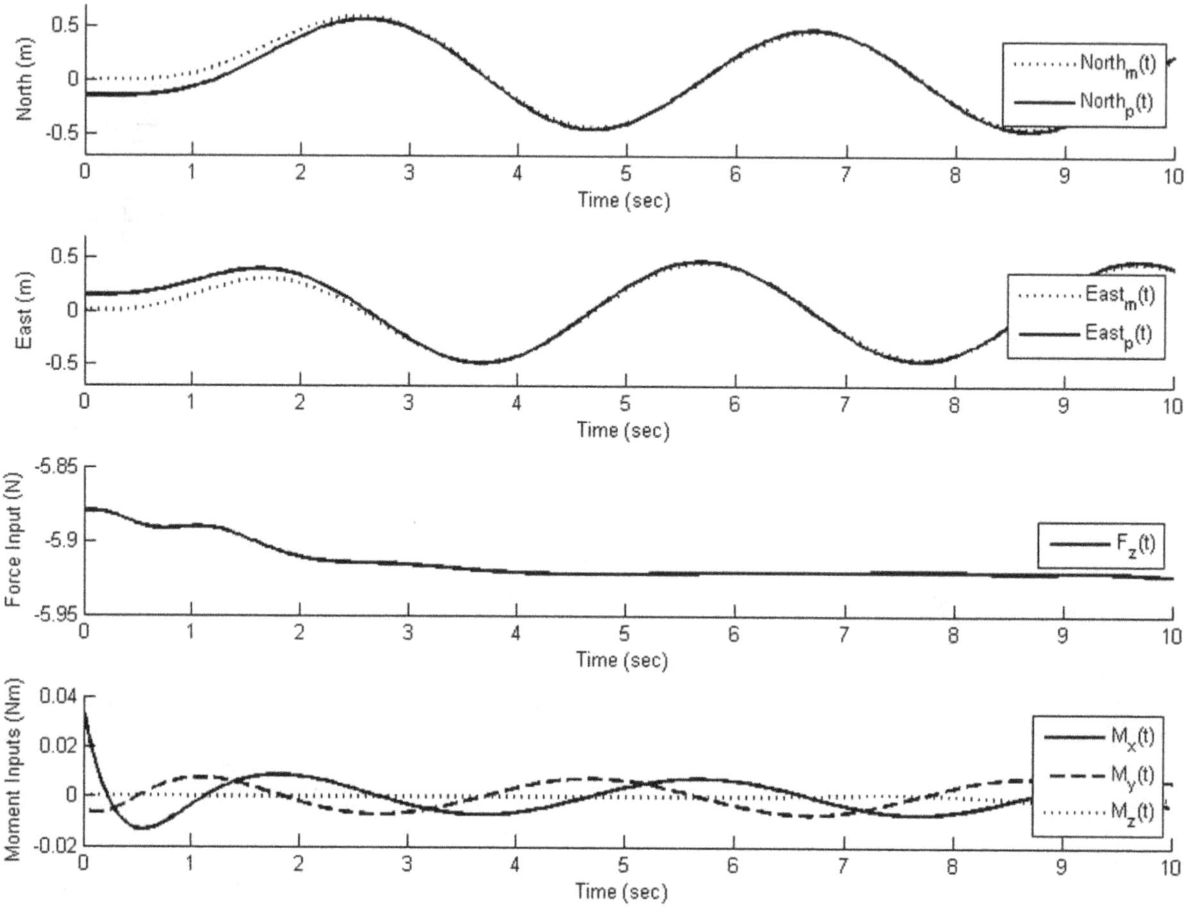

Figure 3. Circular path: north/east positions and force/moment inputs.

system is overdamped, because there is no overshoot and the response has a rise time around 2.5 s. The poles of the characteristic polynomials were arbitrarily selected, so the design can be adjusted to achieve a custom system response. The simulation also confirms that the disturbance is quickly rejected, where the system returns to the reference altitude within 2.0 s.

Circular path: The quadrotor starts at rest with an initial position error of 0.15 m in both the north and east directions. The reference signal trajectory is defined as a loiter where the vehicle circles around the origin following a 0.5 m radius at 4.0 s per cycle. The simulation results are provided in Figure 3, which illustrates the quadrotor is capable of tracking the prescribed reference output. Whereas, a fixed controller will have a steady-state error with a sinusoidal signal, the tracking error with the adaptive controller decays to zero exponentially. For this simulation, the tracking error is eliminated within 3.0 s.

Simulation summary: The two simulations illustrate important characteristics of the adaptive control system. The control system for both simulations uses arbitrarily prescribed characteristics, so the controller can be further adjusted to achieve any particular system response. The first case shows that the quadrotor accurately tracks a given reference signal, and the vehicle adequately rejects disturbances. The second simulation is more complex with two states changing over time. Despite the coupling between inputs and outputs, the system still tracks the reference signal. The controller successfully rejected the initial position error, and the steady-state error decayed to zero exponentially.

6. Conclusions

This research developed the dynamics, equations of motion, and linearization for the quadrotor vehicle which laid the groundwork for the controller development. Existing linear control theory for both fixed and adaptive controllers was reviewed and applied to the quadrotor system. The adaptive controller was based on an LDS decomposition which relaxed the matching condition between the plant and model structure. This approach maintained the simple state feedback controller structure, and avoided the more complicated output feedback structure. Using this decomposition placed the uncertainty of the high-frequency gain matrix K_p within the adaptation process, which allows the controller to adapt to both parameter uncertainty and varying operating points. It was demonstrated that the adaptive control system ensures closed-loop stability and asymptotic output tracking; and a simulation analysis showed that the quadrotor can overcome initial errors, track reference signals, and reject disturbances, despite coupling between states. Expanding the range of operating points for quadrotor UAVs increases their robustness, which may increase their number of applications.

References

Bouabdallah, S., Murrieri, P., & Siegwart, R. (2004, April). *Design and control of an indoor micro quadrotor*. Proc. of the 2004 IEEE international conference on robotics and auto, New Orleans, LA, pp. 4393–4398.

Chen, C. (2013). *Linear system theory and design*. New York: Oxford University Press.

Das, A., Subbarao, K., & Lewis, F. (2009). Dynamic inversion with zero-dynamics stabilization for quadrotor control. *IET Control Theory and Applications*, *3*(3), 303–314.

Diao, C., Xian, B., Yin, Q., Zeng, W., Li, H., & Yang, Y. (2011, May). *A nonlinear adaptive control approach for quadrotor UAVs*. Proc. of the 2011 8th Asian control conference, Kaohsiung, Taiwan.

Dydek, Z., Annaswamy, A., & Lavretsky, E. (2013). Adaptive control of quadrotor UAVs: A design trade study with flight evaluations. *IEEE Transactions on Control Systems Technology*, *21*(4), 1400–1406.

Elliott, H., & Wolovich, W. (1982). A parameter adaptive control structure for linear multivariable systems. *IEEE Transactions on Automatic Control*, *27*(2), 340–352.

Franklin, G., Powell, J., & Emami-Naeini, A. (2009). *Feedback control of dynamic systems*. Upper Saddle River, NJ: Pearson.

Goodwin, G., Graebe, S., & Salgado, M. (2001). *Control system design*. Upper Saddle River, NJ: Prentice Hall.

Goodwin, G., & Sin, K. S. (1984). *Adaptive filtering prediction and control*. Englewood Cliffs, NJ: Prentice Hall.

Guo, J., Liu, Y., & Tao, G. (2009). *Multivariable MRAC with state feedback for output tracking*. Proc. of the 2009 ACC, St. Louis, MO, USA. pp. 592–597.

Guo, J., Tao, G., & Liu, Y. (2011, April). A multivariable MRAC scheme with application to a nonlinear aircraft model. *Automatica*, *47*, 804–812.

Hehn, M., & D'Andrea, R. (2011, May). *A flying inverted pendulum*. IEEE international conference on robotics and automation, Shanghai, China. pp. 763–770.

Hosoe, S. (1975, December). On a time-domain characterization of the numerator polynomials of the Smith-McMillan form. *IEEE Transactions on Automatic Control*, *20*(6), pp. 799–800.

Hsu, L., Costa, R.R., Imai, A.K., & Kokotovic, P. (2001). *Lyapunov-based adaptive control of MIMO systems*. Proc. of the 2001 ACC, Arlington, VA, USA, pp. 4808–4813.

Imai, A.K., Costa, R.R., Hsu, L., Tao, G., & Kokotovic, P. (2001, December). *Multivariable MRAC using high-frequency gain matrix factorization*. Proc. of the 40th IEEE conference on decision and control, Orlando, FL, pp. 1193–1198.

Ioannou, P., & Sun, J. (1996). *Robust adaptive control*. Upper Saddle River, NJ: Prentice Hall.

Isidori, A. (1995). *Nonlinear control systems*. Berlin: Springer-Verlag.

Krstic, M., Kanellakopoulos, I., & Kokotovic, P. V. (1995). *Nonlinear and adaptive control design*. New York: John Wiley & Sons.

Madani, T., & Benallegue, A. (2006, October). *Backstepping control for a quadrotor helicopter*. Proc. of the 2006 IEEE/RSJ, international conference on intelligent robots and systems, Beijing, China, pp. 3255–3260.

Nicol, C., Macnab, C. J. B., & Ramirez-Serrano, A. (2011). Robust adaptive control of a quadrotor helicopter. *Mechatronics*, *21*, 927–938.

Pounds, P., Mahony, R., & Corke, P. (2010). Modelling and control of a large quadrotor robot. *Control Engineering Practice*, *18*, 691–699.

Tao, G. (2003a). *Adaptive control design and analysis*. Hoboken, NJ: John Wiley and Sons.

Tao, G. (2003b). *A unification of multivariable MRAC based on high-frequency gain matrix decompositions*. Proc. of the 2003 ACC, Denver, CO, USA, pp. 945–950.

20

Finite-horizon H_∞ filtering for time-varying delay systems with randomly varying nonlinearities and sensor saturations

Jinling Liang[a]*, Fangbin Sun[a] and Xiaohui Liu[b]

[a]Department of Mathematics, Southeast University, Nanjing 210096, People's Republic of China; [b]School of Computer Science and Technology, Nanjing University of Science and Technology, Nanjing 210094, People's Republic of China

This paper mainly focuses on the H_∞ filtering problem for a class of discrete time-varying systems with delays and randomly varying nonlinearities and sensor saturations. Two sets of binary switching sequences taking values of 1 and 0 are introduced to account for the stochastic phenomena of nonlinearities and sensor saturations which occur and influence the dynamics of the system in a probabilistic way. To further reflect the realities of transmission failure in the measurement, missing observation case is also considered simultaneously. By appropriately constructing a time-varying Lyapunov function and utilizing the stochastic analysis technique, sufficient criteria are presented in terms of a set of recursive linear matrix inequalities (RLMIs) under which the filtering error dynamics achieves the prescribed H_∞ performance over a finite horizon. Moreover, at each time point k, the time-varying filter parameters can be solved iteratively according to the explicit solutions of the RLMIs. Finally, a numerical simulation is exploited to demonstrate the effectiveness of the proposed filter design scheme.

Keywords: H_∞ filtering; time-varying delayed systems; randomly varying sensor saturations; recursive linear matrix inequalities; finite horizon

1. Introduction

In practical engineering fields such as signal processing area, to carry out some specific design tasks, the state information or some combinations of the state information are needed to be known which however, are often unavailable. And this is one of the main backgrounds for investigating the estimation problems. Generally speaking, the aim of the estimation problem is to estimate certain system parameters or state variables by utilizing the accessible measurements, which might be with the existence of stochastic errors. In the literature, much work has been done on various estimation problems, and several filtering methodologies have been proposed (Ahmad & Namerikawa, 2013; Lu, Xie, Zhang, & Wang, 2007; Mohamed, Nahla, & Safya, 2013; Reif & Unbehauen, 1999). Among them, the Kalman filtering (Lu et al., 2007; Reif & Unbehauen, 1999; Xie, Soh, & de Souza, 1994) and the H_∞ filtering (Dong, Wang, Ho, & Gao, 2011; Li, Lam, & Shu, 2010; Zhang, Chen, & Tseng, 2005; Zhang, Feng, & Duan, 2006) are two notable ones. The main idea for the Kalman filtering is to estimate the future values of the signal by utilizing the past/current observations. When employing the H_∞ filtering, criteria are often presented in the form of Riccati difference equations (Gershon, Shaked, & Yaesh, 2001; Hung & Yang, 2003; Xie & de Souza, 1992; Zhang et al., 2006) or linear matrix inequalities (LMIs) (Li et al., 2010; Shen, Wang, & Liu, 2011; Shen, Wang, Shu, & Wei, 2010; Wang, Shen, & Liu, 2012).

It is well known that almost all the realistic systems are intrinsically time-varying, and frequently affected by the nonlinear exogenous disturbances and time delays, which markedly increase the difficulty when analyzing the system due to the complexity. In the past decades, much efforts have been devoted to the filtering and control problems for the discrete time-varying systems (Shen et al., 2011; Shen, Ding, & Wang, 2013). For instance, robust H_∞ filtering problem has been investigated in Dong et al. (2011) for the Markovian jump time-varying systems. When dealing with the time-varying systems in practice, a fundamental issue arises naturally, that is, the state performance constrains are restricted only over a finite horizon instead of the infinite one. Such kind of finite-horizon filtering problem has attracted much attention in recent years, and it is desirable to develop effective and executable algorithms to determine the filter parameters. Motivated by the novel difference linear matrix inequality method proposed in Shaked and Suplin (2001), Gershon and Shaked (2008), Gershon, Shaked, and Yaesh (2005), a new and practical recursive linear matrix inequality (RLMI) approach has been firstly presented in Shen et al. (2010) where the available state estimates have also been utilized which might decrease the conservation of

*Corresponding author. Email: jinlliang@gmail.com

the results obtained. On the other hand, time delays are ubiquitous in systems mainly due to the reasons such as finite capabilities of signal transmission among different parts of the systems. Numerous results pertaining to the filtering problem have been obtained in Lu et al. (2007), Chen and Zheng (2012), Wu and Wang (2009), Wei, Wang, and Shen (2013) for the delayed time-invariant systems. When referring to the time-varying delayed systems, relating results are relatively few (Basin, Shi, Alvarez, & Wang, 2009; Wei, Wang, & Shen, 2010), which might mainly due to the mathematical complexity. For example, in Basin et al. (2009), the central suboptimal H_∞ filters have been defined for the linear continuous-time systems with state or measurement delay. Particularly, a numerically appealing algorithm has been developed in Wei et al. (2010) for the error-constrained filtering problem of the discrete time-varying delay systems with bounded noise, where randomly varying nonlinearities and sensor saturations are not considered.

When the inputs are large enough, the sensors turn to be saturated rather than linear caused by physical constraints. In other words, the sensors possess the nonlinear characteristic when confronting saturations, and it may degrade the filter performance (Liu, Wang, & Yang, 2003; Wang et al., 2012; Yang & Li, 2009) when neglecting the amplitude saturation effect. Therefore, sensor saturation issue is currently an attracting and active research area. As pointed out in Shen et al. (2010), when working circumstances change suddenly or instruments abrade, this will result in the randomly changeable of the nonlinear disturbances in terms of their type and/or intensity and the missing measurement situations. Such kinds of cases are characterized and named with randomly occurring nonlinearities (RONs) and randomly occurring sensor saturations (ROSSs) in Wang et al. (2012) and Shen et al. (2010) and further studied in Shen, Wang, Shu, and Wei (2011), Wang, Wang, and Liu (2010), Wei et al. (2013), Ding, Wang, Shen, and Shu (2012). Specially, an adaptive reliable H_∞ filter method has been developed in Yang and Ye (2007) to against the sensor failure case, and the filter parameter gains are determined based on LMIs by solving two optimization problems. It should be noted that in all these references, time delay effects have not been considered.

Based on the above discussions, in this paper, we will concentrate on the H_∞ filtering problem for a class of discrete time-varying *delayed* systems with *missing measurements* and *randomly varying nonlinearities and sensor saturations over a finite horizon*. Illuminated by the ideas reported in Shen et al. (2010), Dong et al. (2011), Wang, Dong, Shen, and Gao (2013), the H_∞ filtering problem is investigated for the discrete time-varying system with delays by introducing an improved time-varying Lyapunov functional, and sufficient criteria are given which ensure the validity of the H_∞ performance constraint for the filtering error dynamics. Moreover, in the output measurement process, both possible sensor saturations and data-missing phenomena are considered, which are introduced to reflect the intricate working circumstances of the underlying system. In addition, randomly varying nonlinearities between the current and the delayed state nonlinearities are also involved, which together make the H_∞ filtering problem hard to be analyzed, not to mention the design problem for the H_∞ filter. And this is the main aim of this paper to shorten such a gap.

The rest of the paper is organized as follows. In Section 2, the discrete time-varying delayed system with randomly varying nonlinearities and sensor saturations is presented, and the H_∞ filtering problem addressed is formulated. In Section 3, by resorting to the stochastic analysis techniques, sufficient conditions are established in the form of time-varying matrix inequalities under which the output estimation error is assured to meet the constraint of the given H_∞ performance level. Furthermore, the parameters of the H_∞ filter are designed according to the feasible solutions of a set of RLMIs and a recursive filtering algorithm is developed. In Section 4, one illustrative example is given to demonstrate the effectiveness of the results derived. Finally, in Section 5 the conclusion is drawn.

Notations: The notations used here are fairly standard except where otherwise stated. $\mathbb{R}^n$ denotes the n-dimensional Euclidean space. The set of all integers is represented by $\mathbb{Z}$ and $\mathbb{R}$ means the set of all real numbers. The interval $[m, n]$ with $m, n \in \mathbb{Z}$ and $m < n$ denotes the set of integer sequence $\{m, m+1, \ldots, n\}$, and $[a, b]$ with $a, b \in \mathbb{R}$ and $a < b$ represents the set of real numbers between a and b. The notation $X \geq Y$ (respectively, $X > Y$), where X and Y are real symmetric matrices, means that $X - Y$ is positive semi-definite (respectively, positive definite). M^{T} represents the transpose of the matrix M and I is used to denote the identity matrix with compatible dimensions. diag$\{\cdots\}$ stands for a block-diagonal matrix. Moreover, Prob$\{X\}$ means the occurrence probability of the event X and $\mathbb{E}\{x\}$ stands for the expectation of the stochastic variable x with respect to the given probability measure Prob. The asterisk "$*$" in a matrix is used to denote a term that is induced by symmetry. Matrices, if they are not explicitly specified, are assumed to have compatible dimensions.

2. Problem formulation and preliminaries

Consider the following discrete time-varying delayed system defined on the finite horizon $k \in [0, N]$:

$$\begin{aligned} x(k+1) &= A(k)x(k) + A_1(k)x(k-d) + B(k)v(k) \\ &\quad + \alpha(k)f(k, x(k)) + (1-\alpha(k))g(k, x(k-d)), \\ y(k) &= \psi(C(k)x(k)) + D(k)v(k), \\ z(k) &= M(k)x(k), \\ x(s) &= \phi(s), \quad s = -d, -d+1, \ldots, 0, \end{aligned} \tag{1}$$

where $x(k) \in \mathbb{R}^n$ is the state vector, $y(k) \in \mathbb{R}^m$ is the measured output vector, $z(k) \in \mathbb{R}^r$ is the signal to be estimated; $A(k)$, $A_1(k)$, $B(k)$, $C(k)$, $D(k)$ and $M(k)$ are known real time-varying matrices with appropriate dimensions; $d > 0$ is an integer representing the constant delay of the system; $\phi(\cdot) \in \mathbb{R}^n$ is the initial state vector function defined on $[-d, 0]$; $v(k) \in \mathbb{R}^q$ is the exogenous disturbance signal belonging to $l_2[0, N]$ which denotes the space of square summable sequences with the norm

$$\|v\|_{[0,N]}^2 = \mathbb{E}\left\{\sum_{k=0}^{N} \|v(k)\|^2\right\}.$$

The nonlinear functions $f(\cdot,\cdot)$, $g(\cdot,\cdot) : [0, N] \times \mathbb{R}^n \to \mathbb{R}^n$ are assumed to be continuous and satisfy the following sector-bounded conditions (Cao, Lin, & Chen, 2003; Wang, Liu, & Liu, 2008):

$$[f(k,x) - U_1(k)x]^{\mathrm{T}}[f(k,x) - U_2(k)x] \leq 0, \quad (2)$$

$$[g(k,x) - V_1(k)x]^{\mathrm{T}}[g(k,x) - V_2(k)x] \leq 0, \quad (3)$$

where $k \in [0, N]$ and $x \in \mathbb{R}^n$; $U_1(k)$, $U_2(k)$, $V_1(k)$ and $V_2(k)$ are known real matrices with appropriate dimensions, and $U(k) = U_1(k) - U_2(k)$, $V(k) = V_1(k) - V_2(k)$ are symmetric positive-definite matrices.

The random variable $\alpha(k) \in \mathbb{R}$ takes values of 1 and 0 with

$$\mathrm{Prob}\{\alpha(k) = 1\} = \bar{\alpha}, \quad \mathrm{Prob}\{\alpha(k) = 0\} = 1 - \bar{\alpha}, \quad (4)$$

where constant $\bar{\alpha} \in [0, 1]$ is known.

Remark 1 Most of realistic systems are subject to nonlinear disturbances which themselves might change abruptly due mainly to the reasons such as sudden changes in environment, random switching between subsystems, failure connection between part of the nodes of networks as well as asynchronous information transmission within networks. In other words, the nonlinear disturbances might occur in a probabilistic way. Such kind of phenomena has been firstly named RONs in Wang, Wang, and Liang (2009) to account for the probabilistic occurrence of different nonlinear functions. Here, illuminated by the ideas proposed in in Wang et al. (2009), $\alpha(k)$ is used just to account for the phenomena of randomly varying nonlinearities between the current state nonlinearity $f(k, x(k))$ and the delayed state nonlinearity $g(k, x(k - d)))$.

The nonlinear function $\psi(\cdot) : \mathbb{R}^m \to \mathbb{R}^m$ is given as follows:

$$\psi(C(k)x(k)) = \beta(k)\sigma(C(k)x(k)) + (1 - \beta(k))\gamma(k)C(k)x(k), \quad (5)$$

where $\sigma(\cdot) : \mathbb{R}^m \to \mathbb{R}^m$ represents the sensor saturation function with the following form:

$$\sigma(u) = [\sigma_1^{\mathrm{T}}(u_1), \sigma_2^{\mathrm{T}}(u_2), \ldots, \sigma_m^{\mathrm{T}}(u_m)]^{\mathrm{T}}, \quad (6)$$

where $u = (u_1, \ldots, u_m)^{\mathrm{T}} \in \mathbb{R}^m$, $\sigma_i(u_i) = \mathrm{sign}(u_i)\min\{u_{i,\max}, |u_i|\}$ for $i = 1, 2, \ldots, m$ and $u_{i,\max}$ denotes the ith element of the saturation vector $u_{\max}$.

In Equation (5), the random variables $\beta(k) \in \mathbb{R}$ and $\gamma(k) \in \mathbb{R}$ are Bernoulli distributed sequences taking values of 1 and 0 with

$$\begin{aligned} &\mathrm{Prob}\{\beta(k) = 1\} = \bar{\beta}, \quad \mathrm{Prob}\{\beta(k) = 0\} = 1 - \bar{\beta}, \\ &\mathrm{Prob}\{\gamma(k) = 1\} = \bar{\gamma}, \quad \mathrm{Prob}\{\gamma(k) = 0\} = 1 - \bar{\gamma}, \end{aligned} \quad (7)$$

where $\bar{\beta}$, $\bar{\gamma} \in [0, 1]$ are known constants. Here, $\beta(k)$ is introduced to account for the phenomena of randomly varying sensor saturations caused by physical electron device constraints, while $\gamma(k)$ is used to describe the probable data-missing phenomenon caused by mutative working conditions or fluctuant signal transmission channels when the sensors fail to work. Throughout this paper, we further assume that the stochastic variables $\alpha(k)$, $\beta(k)$ and $\gamma(k)$ are mutually independent.

Remark 2 In recent years, the RONs have been extensively studied in Dong et al. (2011), Wang et al. (2010) and Wei, Wang, and Han (2013) for the Markovian jump systems and the complex networks. When such kind of phenomena occur in a sensor network, the notation of ROSSs has been firstly introduced in Wang et al. (2012) and Ding et al. (2012) by further considering the physical limitations of components. It should be noted that such kind of ideas have also been employed in earlier works such as Nahi (1969) and Chau, Qin, Sayed, Wahab, and Yang (2010), where a Markov chain has been proposed to capture the battery recovery (Chau et al., 2010), an the missing measurements (or uncertain observations) have been considered for the optimal estimation problems (Nahi, 1969).

Remark 3 The measurement output model given in Equations (1) and (5) is originated from Wang et al. (2012) where the H_∞ filtering problem has been investigated for the nonlinear sensor networks with time-invariant system matrices. As pointed out in Wang et al. (2012), the main advantage of such kind of measurement output equation is that it is capable of accounting for the phenomena of both ROSSs and missing measurements in a unified form. To be specific, if $\beta(k) = 0$ and $\gamma(k) = 0$, the output observer receives only the noise signal; if $\beta(k) = 0$ and $\gamma(k) = 1$, it means that the output observer works regularly; if $\beta(k) = 1$, whatever the value of $\gamma(k)$ is, only saturated signals are received by the output observer. In practice, the case that the sensor saturation phenomenon and the data-missing phenomenon occur simultaneously does exist. At this time, we only consider the former one since the saturated signals can also be viewed as one special form of the data-missing phenomenon.

Illuminated by the analysis method used in Dong et al. (2011) and Yang and Li (2009), we assume that there exist diagonal matrices $H_1(k)$ and $H_2(k)$ such that $0 \leq H_1(k) < I \leq H_2(k)$, and the saturation function $\sigma(C(k)x(k))$ in

Equation (5) is rewritten as follows:

$$\sigma(C(k)x(k)) = H_1(k)C(k)x(k) + h(C(k)x(k)), \quad (8)$$

where $h(C(k)x(k))$ is a nonlinear vector-valued function satisfying the following inequality:

$$h^{\mathrm{T}}(C(k)x(k))(h(C(k)x(k)) - H(k)C(k)x(k)) \leq 0 \quad (9)$$

with $H(k) = H_2(k) - H_1(k)$.

According to the above discussions, system (1) can be rewritten as follows:

$$\begin{aligned} x(k+1) &= A(k)x(k) + A_1(k)x(k-d) + B(k)v(k) \\ &\quad + \alpha(k)f(k,x(k)) + (1-\alpha(k))g(k,x(k-d)), \\ y(k) &= \beta(k)H_1(k)C(k)x(k) + \beta(k)h(C(k)x(k)) \\ &\quad + (1-\beta(k))\gamma(k)C(k)x(k) + D(k)v(k), \\ z(k) &= M(k)x(k), \\ x(s) &= \phi(s), \quad s = -d, -d+1, \ldots, 0. \end{aligned} \quad (10)$$

In this paper, we will design the following filter for the time-varying system (10):

$$\begin{aligned} \hat{x}(k+1) &= F_f(k)\hat{x}(k) + G_f(k)y(k), \\ \hat{z}(k) &= M_f(k)\hat{x}(k), \end{aligned} \quad (11)$$

where $\hat{x}(k) \in \mathbb{R}^n$ is the state vector of the filter, $\hat{z}(k) \in \mathbb{R}^r$ is the estimate of $z(k)$; $F_f(k)$, $G_f(k)$ and $M_f(k)$ are time-varying filter matrices to be designed. Here, we take $\hat{x}(k) \equiv 0$ for $k \leq 0$, which will be used when designing the filter algorithm in the sequel.

For convenience of expression, we introduce the following notions:

$$\begin{aligned} \tilde{\mathscr{A}}(e(k)) &= (A(k) - \bar{\beta}G_f(k)H_1(k)C(k) \\ &\quad - (1-\bar{\beta})\bar{\gamma}G_f(k)C(k))e(k) \\ &\quad + (B(k) - G_f(k)D(k))v(k) \\ &\quad + (A(k) - \bar{\beta}G_f(k)H_1(k)C(k) \\ &\quad - (1-\bar{\beta})\bar{\gamma}G_f(k)C(k) \\ &\quad - F_f(k))\hat{x}(k) - \bar{\beta}G_f(k)h(C(k)x(k)) \\ &\quad + \bar{\alpha}f(k,x(k)) + (1-\bar{\alpha})g(k,x(k-d)), \\ \tilde{\mathscr{B}}(e(k)) &= -G_f(k)H_1(k)C(k)e(k) - G_f(k)H_1(k)C(k)\hat{x}(k) \\ &\quad - G_f(k)h(C(k)x(k)), \\ \tilde{\mathscr{C}}(e(k)) &= -G_f(k)C(k)e(k) - G_f(k)C(k)\hat{x}(k), \\ \tilde{\mathscr{A}}_1(e(k)) &= A_1(k)e(k-d) + A_1(k)\hat{x}(k-d), \\ \tilde{\mathscr{F}}(e(k)) &= f(k,x(k)) - g(k,x(k-d)), \\ \tilde{\mathscr{M}}(e(k)) &= M(k)e(k) + (M(k) - M_f(k))\hat{x}(k). \end{aligned}$$

By letting $e(k) = x(k) - \hat{x}(k)$ and $\tilde{z}(k) = z(k) - \hat{z}(k)$, the error dynamics can be obtained as follows from Equations (10) and (11):

$$\begin{aligned} e(k+1) &= \tilde{\mathscr{A}}(e(k)) + (\beta(k) - \bar{\beta})\tilde{\mathscr{B}}(e(k)) \\ &\quad + \left((1-\beta(k))\gamma(k) - (1-\bar{\beta})\bar{\gamma}\right)\tilde{\mathscr{C}}(e(k)) \\ &\quad + (\alpha(k) - \bar{\alpha})\tilde{\mathscr{F}}(e(k)) + \tilde{\mathscr{A}}_1(e(k)), \\ \tilde{z}(k) &= \tilde{\mathscr{M}}(e(k)). \end{aligned} \quad (12)$$

The filtering problem to be addressed is as follows: design the filter (11) such that the H_∞ performance constraint (13) is satisfied. More specially, for any nonzero exogenous disturbance $v(k) \in l_2([0,N],\mathbb{R}^q)$, the estimation error $\tilde{z}(k)$ satisfies the following inequality:

$$\|\tilde{z}\|^2_{[0,N]} \leq \gamma^2 \left\{ \|v\|^2_{[0,N]} + \mathbb{E}\left\{ \sum_{k=-d}^{0} e^{\mathrm{T}}(k)S(k)e(k) \right\} \right\} \quad (13)$$

where $\gamma > 0$ is a given disturbance attenuation level and $\{S(k)\}_{-d\leq k\leq 0}$ is a known positive-definite matrix sequence.

Remark 4 In recent years, the finite-horizon filtering problem has attracted much attention for its practicability. For example, the robust H_∞ filtering problem with error variance constraints has been investigated for the discrete linear time-varying systems in Hung and Yang (2003), where the conditions are in the form of forward recursive Riccati equations which are hard to be solved in practice. Recently, novel works have been done in Shen et al. (2010) and Wei et al. (2010), respectively, for the robust H_∞ finite-horizon filtering of systems with RONs and quantization effects and the error-constrained filtering of nonlinear delayed systems with non-Gaussian noises. It should be noted that the delay effects have not been considered in Shen et al. (2010), and in Wei et al. (2010) the phenomena of ROSSs and missing measurements have not been taken into account. By taking the phenomena of time delay, ROSSs and missing measurements together, it will be hard to analyze the H_∞ filtering problem, not to mention the design problem for the time-varying H_∞ filter, which mainly motivates the present work of this article.

3. Main results

In this section, in order to design the filter (11), we first give a sufficient criterion to guarantee that the error system (12) satisfies the H_∞ performance constraint (13) via the RLMI approach, which is given by the following theorem.

THEOREM 1 *Consider the error system* (12) *with known filter parameters* $\{F_f(k)\}_{0\leq k\leq N}$, $\{G_f(k)\}_{0\leq k\leq N}$ *and* $\{M_f(k)\}_{0\leq k\leq N}$. *Let the disturbance attenuation level* $\gamma > 0$ *and the positive-definite matrix sequence* $\{S(k)\}_{-d\leq k\leq 0}$ *be given, the estimation error* $\tilde{z}(k)$ *satisfies the* H_∞ *performance constraint* (13) *if there exist four families of positive scalars* $\{\varepsilon_1(k)\}_{0\le k\le N}$, $\{\varepsilon_2(k)\}_{0\le k\le N}$, $\{\varepsilon_3(k)\}_{0\le k\le N}$, $\{\mu(k)\}_{0\le k\le N+1}$ *and two families of positive-definite matrices* $\{P(k)\}_{0\le k\le N+1}$, $\{Q(k)\}_{-d+1\le k\le N+1}$ *satisfying the following initial condition:*

$$\mathbb{E}\left\{e^{\mathrm{T}}(0)P(0)e(0)+\sum_{k=-d}^{-1}e^{\mathrm{T}}(k)Q(k+1)e(k)\right\}+\mu(0)\le\gamma^2\mathbb{E}\left\{\sum_{k=-d}^{0}e^{\mathrm{T}}(k)S(k)e(k)\right\} \tag{14}$$

and the RLMIs

$$\begin{bmatrix} \Xi_1(k) & \hat{\mathscr{A}}^{\mathrm{T}}(k)P(k+1) & \hat{\mathscr{B}}^{\mathrm{T}}(k)P(k+1) & \hat{\mu}\mathscr{C}^{\mathrm{T}}(k)P(k+1) & \hat{\alpha}\mathscr{F}^{\mathrm{T}}P(k+1) & \mathscr{L}^{\mathrm{T}}(k)\\ * & -P(k+1) & 0 & 0 & 0 & 0\\ * & * & -P(k+1) & 0 & 0 & 0\\ * & * & * & -\hat{\mu}P(k+1) & 0 & 0\\ * & * & * & * & -\hat{\alpha}P(k+1) & 0\\ * & * & * & * & * & -I \end{bmatrix}\le 0 \tag{15}$$

for $0\le k\le N$, *where* $\hat{\alpha}=\bar{\alpha}(1-\bar{\alpha})$, $\hat{\beta}=\bar{\beta}(1-\bar{\beta})$, $\hat{\gamma}=(1-\bar{\beta})\bar{\gamma}-(1-\bar{\beta})^2\bar{\gamma}^2$, $\hat{\mu}=\bar{\gamma}(1-\bar{\gamma})(1-\bar{\beta})$, $\hat{m}=\hat{\beta}^{1/2}$, $\hat{n}=\hat{\beta}^{1/2}\bar{\gamma}$,

$$\Xi_1(k)=\begin{bmatrix} \Gamma_1(k) & 0 & 0 & -\varepsilon_1(k)\tilde{U}_2(k) & 0 & \Gamma_2(k) & \Gamma_3(k)\\ * & \Gamma_4(k) & 0 & 0 & -\varepsilon_2(k)\tilde{V}_2(k) & 0 & \Gamma_5(k)\\ * & * & -\gamma^2 I & 0 & 0 & 0 & 0\\ * & * & * & -\varepsilon_1(k)I & 0 & 0 & \Gamma_6(k)\\ * & * & * & * & -\varepsilon_2(k)I & 0 & \Gamma_7(k)\\ * & * & * & * & * & -\varepsilon_3(k)I & \Gamma_8(k)\\ * & * & * & * & * & * & \Gamma_9(k) \end{bmatrix},$$

$$\Gamma_1(k)=-P(k)+Q(k+1)+M^{\mathrm{T}}(k)M(k)-\varepsilon_1(k)\tilde{U}_1(k),$$

$$\Gamma_2(k)=\frac{\varepsilon_3(k)C^{\mathrm{T}}(k)H^{\mathrm{T}}(k)}{2},$$

$$\Gamma_3(k)=M^{\mathrm{T}}(k)(M(k)-M_f(k))\hat{x}(k)-\varepsilon_1(k)\tilde{U}_1(k)\hat{x}(k),$$

$$\Gamma_4(k)=-Q(k-d+1)-\varepsilon_2(k)\tilde{V}_1(k),$$

$$\Gamma_5(k)=-\varepsilon_2(k)\tilde{V}_1(k)\hat{x}(k-d),$$

$$\Gamma_6(k)=-\varepsilon_1(k)\tilde{U}_2^{\mathrm{T}}(k)\hat{x}(k),$$

$$\Gamma_7(k)=-\varepsilon_2(k)\tilde{V}_2^{\mathrm{T}}(k)\hat{x}(k-d),$$

$$\Gamma_8(k)=\frac{\varepsilon_3(k)H(k)C(k)\hat{x}(k)}{2},$$

$$\Gamma_9(k)=\mu(k+1)-\mu(k)-\varepsilon_1(k)\hat{x}^{\mathrm{T}}(k)\tilde{U}_1(k)\hat{x}(k)-\varepsilon_2(k)\hat{x}^{\mathrm{T}}(k-d)\tilde{V}_1(k)\hat{x}(k-d),$$

$$\hat{\mathscr{A}}(k)=\mathscr{A}(k)+\mathscr{A}_1(k),\quad \hat{\mathscr{B}}(k)=\hat{m}\mathscr{B}(k)-\hat{n}\mathscr{C}(k),$$

$$\mathscr{A}(k)=[\bar{A}(k)\ \ 0\ \ B(k)-G_f(k)D(k)\ \ \bar{\alpha}I(1-\bar{\alpha})I\ -\bar{\beta}G_f(k)\ \ (\bar{A}(k)-F_f(k))\hat{x}(k)],$$

$$\bar{A}(k)=A(k)-\bar{\beta}G_f(k)H_1(k)C(k)-(1-\bar{\beta})\bar{\gamma}G_f(k)C(k),$$

$$\mathscr{A}_1(k)=[0\ A_1(k)\ 0\ 0\ 0\ 0\ A_1(k)\hat{x}(k-d)],$$

$$\mathscr{B}(k)=[\bar{B}(k)\ 0\ 0\ 0\ 0\ -G_f(k)\ \bar{B}(k)\hat{x}(k)],$$

$$\bar{B}(k)=-G_f(k)H_1(k)C(k),\quad \mathscr{F}=[0\ 0\ 0\ I\ -I\ 0\ 0],$$

$$\mathscr{C}(k)=[-G_f(k)C(k)\ 0\ 0\ 0\ 0\ 0\ -G_f(k)C(k)\hat{x}(k)],$$

$$\mathscr{L}(k)=[0\ 0\ 0\ 0\ 0\ 0\ (M(k)-M_f(k))\hat{x}(k)],$$

$$\tilde{U}_1(k)=\frac{(U_1^{\mathrm{T}}(k)U_2(k)+U_2^{\mathrm{T}}(k)U_1(k))}{2},$$

$$\tilde{U}_2(k)=-\frac{(U_1^{\mathrm{T}}(k)+U_2^{\mathrm{T}}(k))}{2},$$

$$\tilde{V}_1(k)=\frac{(V_1^{\mathrm{T}}(k)V_2(k)+V_2^{\mathrm{T}}(k)V_1(k))}{2},$$

$$\tilde{V}_2(k)=-\frac{(V_1^{\mathrm{T}}(k)+V_2^{\mathrm{T}}(k))}{2}.$$

Proof Select a Lyapunov function for the time-varying system (12) as follows:

$$V(k,e(k))=e^{\mathrm{T}}(k)P(k)e(k)+\sum_{s=k-d}^{k-1}e^{\mathrm{T}}(s)Q(s+1)e(s)+\mu(k), \tag{16}$$

where $k=0,1,\ldots,N+1$ and $\{P(k)\}_{0\le k\le N+1}$, $\{Q(k)\}_{-d+1\le k\le N+1}$, $\{\mu(k)\}_{0\le k\le N+1}$ are the solutions of the RLMIs (15) with the initial condition (14).

It follows from Equation (12) that

$$\begin{aligned}&\mathbb{E}\{e^{\mathrm{T}}(k+1)P(k+1)e(k+1)\}\\ &=\mathbb{E}\{\tilde{\mathscr{A}}^{\mathrm{T}}(e(k))P(k+1)\tilde{\mathscr{A}}(e(k))\\ &\quad+(\beta(k)-\bar{\beta})^2\tilde{\mathscr{B}}^{\mathrm{T}}(e(k))P(k+1)\tilde{\mathscr{B}}(e(k))\\ &\quad+((1-\beta(k))\gamma(k)-(1-\bar{\beta})\bar{\gamma})^2\tilde{\mathscr{C}}^{\mathrm{T}}(e(k))\\ &\quad\times P(k+1)\tilde{\mathscr{C}}(e(k))\end{aligned}$$

$$
\begin{aligned}
&+ (\alpha(k) - \bar{\alpha})^2 \tilde{\mathscr{F}}^{\mathrm{T}}(e(k))P(k+1)\tilde{\mathscr{F}}(e(k)) \\
&+ 2\tilde{\mathscr{A}}^{\mathrm{T}}(e(k))P(k+1)[(\beta(k) - \bar{\beta})\tilde{\mathscr{B}}(e(k)) \\
&+ ((1-\beta(k))\gamma(k) - (1-\bar{\beta})\bar{\gamma})\tilde{\mathscr{C}}(e(k)) \\
&+ (\alpha(k) - \bar{\alpha})\tilde{\mathscr{F}}(e(k)) \\
&+ \tilde{\mathscr{A}}_1(e(k))] + 2(\alpha(k) - \bar{\alpha})\tilde{\mathscr{F}}^{\mathrm{T}}(e(k)) \\
&\times P(k+1)\tilde{\mathscr{A}}_1(e(k)) + 2(\beta(k) \\
&- \bar{\beta})\tilde{\mathscr{B}}^{\mathrm{T}}(e(k))P(k+1)[((1-\beta(k))\gamma(k) \\
&- (1-\bar{\beta})\bar{\gamma})\tilde{\mathscr{C}}(e(k)) \\
&+ (\alpha(k) - \bar{\alpha})\tilde{\mathscr{F}}(e(k)) + \tilde{\mathscr{A}}_1(e(k))] \\
&+ \tilde{\mathscr{A}}_1^{\mathrm{T}}(e(k))P(k+1)\tilde{\mathscr{A}}_1(e(k)) \\
&+ 2((1-\beta(k))\gamma(k) - (1-\bar{\beta})\bar{\gamma})\tilde{\mathscr{C}}^{\mathrm{T}}(e(k))P(k+1) \\
&\times [(\alpha(k) - \bar{\alpha})\tilde{\mathscr{F}}(e(k)) + \tilde{\mathscr{A}}_1(e(k))]\} \\
=\ &\mathbb{E}\{\tilde{\mathscr{A}}^{\mathrm{T}}(e(k))P(k+1)\tilde{\mathscr{A}}(e(k)) \\
&+ \bar{\beta}(1-\bar{\beta})\tilde{\mathscr{B}}^{\mathrm{T}}(e(k))P(k+1)\tilde{\mathscr{B}}(e(k)) \\
&+ \tilde{\mathscr{A}}_1^{\mathrm{T}}(e(k))P(k+1)\tilde{\mathscr{A}}_1(e(k)) + ((1-\bar{\beta})\bar{\gamma} \\
&- (1-\bar{\beta})^2\bar{\gamma}^2)\tilde{\mathscr{C}}^{\mathrm{T}}(e(k))P(k+1)\tilde{\mathscr{C}}(e(k)) \\
&+ \bar{\alpha}(1-\bar{\alpha})\tilde{\mathscr{F}}^{\mathrm{T}}(e(k))P(k+1)\tilde{\mathscr{F}}(e(k)) \\
&+ 2\tilde{\mathscr{A}}^{\mathrm{T}}(e(k))P(k+1)\tilde{\mathscr{A}}_1(e(k)) \\
&- 2\bar{\gamma}\bar{\beta}(1-\bar{\beta})\tilde{\mathscr{B}}^{\mathrm{T}}(e(k))P(k+1)\tilde{\mathscr{C}}(e(k))\}, \qquad (17)
\end{aligned}
$$

where the independence properties of $\alpha(k)$, $\beta(k)$ and $\gamma(k)$ in conditions (4) and (7) are utilized. More specifically, to derive the second equality of Equation (17), the following facts have been used:

$$
\begin{aligned}
&\mathbb{E}\{(\alpha(k) - \bar{\alpha})^2\} = \bar{\alpha}(1-\bar{\alpha}), \\
&\mathbb{E}\{(\beta(k) - \bar{\beta})^2\} = \bar{\beta}(1-\bar{\beta}), \\
&\mathbb{E}\{[(1-\beta(k))\gamma(k) - (1-\bar{\beta})\bar{\gamma}]^2\} \\
&\quad = (1-\bar{\beta})\bar{\gamma} - (1-\bar{\beta})^2\bar{\gamma}^2, \\
&\mathbb{E}\{(\beta(k) - \bar{\beta})((1-\beta(k))\gamma(k) - (1-\bar{\beta})\bar{\gamma})\} \\
&\quad = \mathbb{E}\{(\beta(k) - \bar{\beta})(\gamma(k) - \beta(k)\gamma(k) \\
&\qquad + \bar{\beta}\gamma(k) - \bar{\beta}\gamma(k) - \bar{\gamma} + \bar{\beta}\bar{\gamma})\} \\
&\quad = \mathbb{E}\{(\beta(k) - \bar{\beta})((\gamma(k) - \bar{\gamma}) + \gamma(k)(\bar{\beta} - \beta(k)) \\
&\qquad + \bar{\beta}(\bar{\gamma} - \gamma(k)))\} \\
&\quad = \mathbb{E}\{0 - \gamma(k)(\beta(k) - \bar{\beta})^2 + 0\} \\
&\quad = -\bar{\gamma}\bar{\beta}(1-\bar{\beta}).
\end{aligned}
$$

Define $\eta(k) = [e^{\mathrm{T}}(k) \quad e^{\mathrm{T}}(k-d) \quad v^{\mathrm{T}}(k) \quad f^{\mathrm{T}}(k,x(k)) \quad g^{\mathrm{T}}(k,x(k-d)) \quad h^{\mathrm{T}}(C(k)x(k)) \quad 1]^{\mathrm{T}}$, we can easily calculate, in view of Equations (12), (16) and (17) that

$$
\begin{aligned}
&\mathbb{E}\{V(k+1,e(k+1))\} - \mathbb{E}\{V(k,e(k))\} \\
&\quad + \mathbb{E}\{\|\tilde{z}(k)\|^2\} - \gamma^2\mathbb{E}\{\|v(k)\|^2\} \\
&= \mathbb{E}\{\tilde{\mathscr{A}}^{\mathrm{T}}(e(k))P(k+1)\tilde{\mathscr{A}}(e(k)) + \tilde{\mathscr{A}}_1^{\mathrm{T}}(e(k))P(k+1) \\
&\quad \times \tilde{\mathscr{A}}_1(e(k)) + \hat{\alpha}\tilde{\mathscr{F}}^{\mathrm{T}}(e(k))P(k+1)\tilde{\mathscr{F}}(e(k)) \\
&\quad + \hat{\beta}\tilde{\mathscr{B}}^{\mathrm{T}}(e(k))P(k+1)\tilde{\mathscr{B}}(e(k)) \\
&\quad + \hat{\gamma}\tilde{\mathscr{C}}^{\mathrm{T}}(e(k))P(k+1)\tilde{\mathscr{C}}(e(k)) \\
&\quad + 2\tilde{\mathscr{A}}^{\mathrm{T}}(e(k))P(k+1)\tilde{\mathscr{A}}_1(e(k)) \\
&\quad - 2\bar{\gamma}\bar{\beta}(1-\bar{\beta})\tilde{\mathscr{B}}^{\mathrm{T}}(e(k))P(k+1)\tilde{\mathscr{C}}(e(k)) \\
&\quad + e^{\mathrm{T}}(k)Q(k+1)e(k) - e^{\mathrm{T}}(k-d)Q(k-d+1) \\
&\quad \times e(k-d) + \mu(k+1) - e^{\mathrm{T}}(k)P(k)e(k) \\
&\quad - \mu(k) + \tilde{\mathscr{M}}^{\mathrm{T}}(e(k))\tilde{\mathscr{M}}(e(k)) - \gamma^2 v^{\mathrm{T}}(k)v(k)\} \\
&= \mathbb{E}\{\eta^{\mathrm{T}}(k)\mathscr{A}^{\mathrm{T}}(k)P(k+1)\mathscr{A}(k)\eta(k) + \eta^{\mathrm{T}}(k)\mathscr{A}_1^{\mathrm{T}}(k) \\
&\quad \times P(k+1)\mathscr{A}_1(k)\eta(k) + \hat{\alpha}\eta^{\mathrm{T}}(k)\mathscr{F}^{\mathrm{T}}P(k+1)\mathscr{F}\eta(k) \\
&\quad + 2\eta^{\mathrm{T}}(k)\mathscr{A}^{\mathrm{T}}(k)P(k+1)\mathscr{A}_1(k)\eta(k) \\
&\quad - 2\bar{\gamma}\bar{\beta}(1-\bar{\beta})\eta^{\mathrm{T}}(k)\mathscr{B}^{\mathrm{T}}(k)P(k+1)\mathscr{C}(k)\eta(k) \\
&\quad + \hat{\beta}\eta^{\mathrm{T}}(k)\mathscr{B}^{\mathrm{T}}(k)P(k+1)\mathscr{B}(k)\eta(k) + \hat{\gamma}\eta^{\mathrm{T}}(k) \\
&\quad \times \mathscr{C}^{\mathrm{T}}(k)P(k+1)\mathscr{C}(k)\eta(k) + \eta^{\mathrm{T}}(k)\Phi(k)\eta(k)\}, \qquad (18)
\end{aligned}
$$

where

$$
\Phi(k) = \begin{bmatrix}
\Phi_{11}(k) & 0 & 0 & 0 & 0 & 0 & M^{\mathrm{T}}(k)(M(k) - M_f(k))\hat{x}(k) \\
* & -Q(k-d+1) & 0 & 0 & 0 & 0 & 0 \\
* & * & -\gamma^2 I & 0 & 0 & 0 & 0 \\
* & * & * & 0 & 0 & 0 & 0 \\
* & * & * & * & 0 & 0 & 0 \\
* & * & * & * & * & 0 & 0 \\
* & * & * & * & * & * & \hat{\gamma}(k)
\end{bmatrix}
$$

with $\Phi_{11}(k) = -P(k) + Q(k+1) + M^{\mathrm{T}}(k)M(k)$ and $\hat{\gamma}(k) = \mu(k+1) - \mu(k) + \hat{x}^{\mathrm{T}}(k)(M(k) - M_f(k))^{\mathrm{T}} \times (M(k) - M_f(k))\hat{x}(k)$.

On the other hand, utilizing the notations defined in Equation (15), it is straightforward to show that Equations (2) and (3) infer the validity of the inequalities given below:

$$\eta^{\mathrm{T}}(k)\Phi_1(k)\eta(k) \leq 0, \quad \eta^{\mathrm{T}}(k)\Phi_2(k)\eta(k) \leq 0, \tag{19}$$

where

$$\Phi_1(k) = \begin{bmatrix} \tilde{U}_1(k) & 0 & 0 & \tilde{U}_2(k) & 0 & 0 & \tilde{U}_1(k)\hat{x}(k) \\ * & 0 & 0 & 0 & 0 & 0 & 0 \\ * & * & 0 & 0 & 0 & 0 & 0 \\ * & * & * & I & 0 & 0 & \tilde{U}_2^{\mathrm{T}}(k)\hat{x}(k) \\ * & * & * & * & 0 & 0 & 0 \\ * & * & * & * & * & 0 & 0 \\ * & * & * & * & * & * & \hat{x}^{\mathrm{T}}(k)\tilde{U}_1(k)\hat{x}(k) \end{bmatrix},$$

and

$$\Phi_2(k) = \begin{bmatrix} 0 & 0 & 0 & 0 & 0 & 0 & 0 \\ * & \tilde{V}_1(k) & 0 & 0 & \tilde{V}_2(k) & 0 & \tilde{V}_1(k)\hat{x}(k-d) \\ * & * & 0 & 0 & 0 & 0 & 0 \\ * & * & * & 0 & 0 & 0 & 0 \\ * & * & * & * & I & 0 & \tilde{V}_2^{\mathrm{T}}(k)\hat{x}(k-d) \\ * & * & * & * & * & 0 & 0 \\ * & * & * & * & * & * & \hat{x}^{\mathrm{T}}(k-d)\tilde{V}_1(k)\hat{x}(k-d) \end{bmatrix}.$$

Similarly, Equation (9) can be transformed to an equivalent matrix inequality given as follows:

$$\eta^{\mathrm{T}}(k)\Phi_3(k)\eta(k) \leq 0, \tag{20}$$

where

$$\Phi_3(k) = \begin{bmatrix} 0 & 0 & 0 & 0 & 0 & -\dfrac{C^{\mathrm{T}}(k)H^{\mathrm{T}}(k)}{2} & 0 \\ * & 0 & 0 & 0 & 0 & 0 & 0 \\ * & * & 0 & 0 & 0 & 0 & 0 \\ * & * & * & 0 & 0 & 0 & 0 \\ * & * & * & * & 0 & 0 & 0 \\ * & * & * & * & * & I & -\dfrac{H(k)C(k)\hat{x}(k)}{2} \\ * & * & * & * & * & * & 0 \end{bmatrix}.$$

By considering Equations (18)–(20) together, the inequality which guarantees the H_∞ performance of the error system turns to:

$$\begin{aligned} &\mathbb{E}\{V(k+1,e(k+1))\} - \mathbb{E}\{V(k,e(k))\} \\ &\quad + \mathbb{E}\{\|\tilde{z}(k)\|^2\} - \gamma^2\mathbb{E}\{\|v(k)\|^2\} \\ &\leq \mathbb{E}\{\eta^{\mathrm{T}}(k)(\Phi(k) + \mathscr{A}^{\mathrm{T}}(k)P(k+1)\mathscr{A}(k) \\ &\quad + \mathscr{A}_1^{\mathrm{T}}(k)P(k+1)\mathscr{A}_1(k) + 2\mathscr{A}^{\mathrm{T}}(k)P(k+1)\mathscr{A}_1(k) \\ &\quad + \hat{\beta}\mathscr{B}^{\mathrm{T}}(k)P(k+1)\mathscr{B}(k) + \hat{\gamma}\mathscr{C}^{\mathrm{T}}(k)P(k+1)\mathscr{C}(k) \\ &\quad + \hat{\alpha}\mathscr{F}^{\mathrm{T}}P(k+1)\mathscr{F})\eta(k) \\ &\quad - 2\bar{\gamma}\hat{\beta}\mathscr{B}^{\mathrm{T}}(k)P(k+1)\mathscr{C}(k) - \varepsilon_1(k)\eta^{\mathrm{T}}(k)\Phi_1(k)\eta(k) \\ &\quad - \varepsilon_2(k)\eta^{\mathrm{T}}(k)\Phi_2(k)\eta(k) - \varepsilon_3(k)\eta^{\mathrm{T}}(k)\Phi_3(k)\eta(k)\} \\ &= \mathbb{E}\{\eta^{\mathrm{T}}(k)(\Phi(k) + (\mathscr{A}(k) + \mathscr{A}_1(k))^{\mathrm{T}}P(k+1)(\mathscr{A}(k) \\ &\quad + \mathscr{A}_1(k)) + \hat{\alpha}\mathscr{F}^{\mathrm{T}}P(k+1)\mathscr{F} + \hat{\beta}\mathscr{B}^{\mathrm{T}}(k) \\ &\quad \times P(k+1)\mathscr{B}(k) + \hat{\beta}\bar{\gamma}^2\mathscr{C}^{\mathrm{T}}(k)P(k+1)\mathscr{C}(k) \\ &\quad + (\hat{\gamma} - \hat{\beta}\bar{\gamma}^2)\mathscr{C}^{\mathrm{T}}(k)P(k+1)\mathscr{C}(k) \\ &\quad - 2\bar{\gamma}\hat{\beta}\mathscr{B}^{\mathrm{T}}(k)P(k+1)\mathscr{C}(k))\eta(k) \\ &\quad - \varepsilon_1(k)\eta^{\mathrm{T}}(k)\Phi_1(k)\eta(k) \\ &\quad - \varepsilon_2(k)\eta^{\mathrm{T}}(k)\Phi_2(k)\eta(k) - \varepsilon_3(k)\eta^{\mathrm{T}}(k)\Phi_3(k)\eta(k)\} \\ &= \mathbb{E}\{\eta^{\mathrm{T}}(k)(\Xi_1(k) + (\mathscr{A}(k) + \mathscr{A}_1(k))^{\mathrm{T}}P(k+1)(\mathscr{A}(k) \\ &\quad + \mathscr{A}_1(k)) + \hat{\alpha}\mathscr{F}^{\mathrm{T}}P(k+1)\mathscr{F} + \mathscr{L}^{\mathrm{T}}(k)\mathscr{L}(k) \\ &\quad + (\hat{m}\mathscr{B}(k) - \hat{n}\mathscr{C}(k))^{\mathrm{T}}P(k+1)(\hat{m}\mathscr{B}(k) - \hat{n}\mathscr{C}(k)) \\ &\quad + \hat{\mu}\mathscr{C}^{\mathrm{T}}(k)P(k+1)\mathscr{C}(k))\eta(k)\}, \end{aligned} \tag{21}$$

where $\Xi_1(k) = \Phi(k) - \varepsilon_1(k)\Phi_1(k) - \varepsilon_2(k)\Phi_2(k) - \varepsilon_3(k)\Phi_3(k) - \mathscr{L}^{\mathrm{T}}(k)\mathscr{L}(k)$. Applying the Schur complement formula (Boyd, EI Ghaoui, Feron, & Balakrishnan, 1994) and noticing Equation (15), we can easily obtain from Equation (21) that

$$\begin{aligned} &\mathbb{E}\{V(k+1,e(k+1))\} - \mathbb{E}\{V(k,e(k))\} + \mathbb{E}\{\|\tilde{z}(k)\|^2\} \\ &\quad - \gamma^2\mathbb{E}\{\|v(k)\|^2\} \leq 0. \end{aligned} \tag{22}$$

Summing up both sides of Equation (22) with k varying from 0 to N yields

$$\begin{aligned} \|\tilde{z}\|^2_{[0,N]} &\leq \gamma^2\|v\|^2_{[0,N]} + \mathbb{E}\left\{e^{\mathrm{T}}(0)P(0)e(0) \right. \\ &\quad \left. + \sum_{k=-d}^{-1} e^{\mathrm{T}}(k)Q(k+1)e(k)\right\} + \mu(0), \end{aligned} \tag{23}$$

and the H_∞ performance constraint (13) is assured by also considering the initial condition (14). The proof of this theorem is complete. ■

According to the H_∞ performance analysis established in Theorem 1, the design problem of the H_∞ filter for the stochastic system (1) is reduced to the one for finding feasible solutions to a set of RLMIs.

THEOREM 2 *Let the disturbance attenuation level $\gamma > 0$ and the matrix sequence $\{S(k)\}_{-d\leq k\leq 0}$ be given. The H_∞ filtering problem is solvable for the discrete time-varying system (1) if there exist two families of positive-definite matrices $\{P(k)\}_{0\leq k\leq N+1}$, $\{Q(k)\}_{-d+1\leq k\leq N+1}$, two families of matrices $\{X(k)\}_{0\leq k\leq N}$, $\{Y(k)\}_{0\leq k\leq N}$ and four families of positive scalars $\{\varepsilon_1(k)\}_{0\leq k\leq N}$, $\{\varepsilon_2(k)\}_{0\leq k\leq N}$, $\{\varepsilon_3(k)\}_{0\leq k\leq N}$, $\{\mu(k)\}_{0\leq k\leq N+1}$ satisfying the initial condition (14) and the*

RLMIs

$$\begin{bmatrix} \Xi_1(k) & \Xi_2(k) \\ * & \Xi_3(k) \end{bmatrix} \le 0 \qquad (24)$$

for $0 \le k \le N$, *where* $\Xi_1(k)$ *is the same as defined in Equation* (15) *and*

$$\Xi_2(k) = \begin{bmatrix} \tilde{\Gamma}_1(k) & \tilde{\Gamma}_2(k) & -\hat{\mu}C^T(k)Y^T(k) & 0 & 0 \\ A_1^T(k)P(k+1) & 0 & 0 & 0 & 0 \\ \tilde{\Gamma}_3(k) & 0 & 0 & 0 & 0 \\ \bar{\alpha}P(k+1) & 0 & 0 & \hat{\alpha}P(k+1) & 0 \\ (1-\bar{\alpha})P(k+1) & 0 & 0 & -\hat{\alpha}P(k+1) & 0 \\ -\bar{\beta}Y^T(k) & -\hat{m}Y^T(k) & 0 & 0 & 0 \\ \tilde{\Gamma}_4(k) & \tilde{\Gamma}_5(k) & \tilde{\Gamma}_6(k) & 0 & \hat{x}^T(k)\left(M(k)-M_f(k)\right)^T \end{bmatrix},$$

$$\Xi_3(k) = diag\{-P(k+1), -P(k+1), -\hat{\mu}P(k+1), -\hat{\alpha}P(k+1), -I\},$$

in which

$$\tilde{\Gamma}_1(k) = A^T(k)P(k+1) - \bar{\beta}C^T(k)H_1^T(k)Y^T(k) - (1-\bar{\beta})\bar{\gamma}C^T(k)Y^T(k),$$

$$\tilde{\Gamma}_2(k) = -\hat{m}C^T(k)H_1^T(k)Y^T(k) + \hat{n}C^T(k)Y^T(k),$$

$$\tilde{\Gamma}_3(k) = B^T(k)P(k+1) - D^T(k)Y^T(k),$$

$$\begin{aligned}\tilde{\Gamma}_4(k) = {} & \hat{x}^T(k)A^T(k)P(k+1) - \bar{\beta}\hat{x}^T(k)C^T(k)H_1^T(k)Y^T(k) \\ & - (1-\bar{\beta})\bar{\gamma}\hat{x}^T(k)C^T(k)Y^T(k) \\ & - \hat{x}^T(k)X^T(k) + \hat{x}^T(k-d)A_1^T(k)P(k+1),\end{aligned}$$

$$\tilde{\Gamma}_5(k) = -\hat{m}\hat{x}^T(k)C^T(k)H_1^T(k)Y^T(k) + \hat{n}\hat{x}^T(k)C^T(k)Y^T(k),$$

$$\tilde{\Gamma}_6(k) = -\hat{\mu}\hat{x}^T(k)C^T(k)Y^T(k)$$

and the other symbols are the same as defined in Theorem 1. *Moreover, for each* $0 \le k \le N$, *if inequalities* (14) *and* (24) *are feasible, the desired filter is given by Equation* (11) *with the parameters as*

$$F_f(k) = P^{-1}(k+1)X(k), \quad G_f(k) = P^{-1}(k+1)Y(k), \qquad (25)$$

and the filter matrix $M_f(k)$ can be obtained by solving the corresponding LMI at time point k.

Proof By substituting Equations (25) into (24) and applying Theorem 1, we can easily conclude the validity of the result. ■

According to Theorem 2, the following RLMIs algorithm is given for the H_∞ filtering problem which is illuminated by the design ideas in Shen et al. (2010) and Dong et al. (2011). The H_∞ filtering problem can be implemented recursively as follows:

Step 1. Let the H_∞ performance index γ, the final time N, the initial matrix sequence $\{S(k)\}_{-d\le k\le 0}$ and the initial states $\{\phi(k)\}_{-d\le k\le 0}$ be given. Select appropriate positive-definite matrix $P(0)$, positive-definite matrix sequence $\{Q(s)\}_{-d+1\le s\le 0}$ and positive scalar $\mu(0)$ satisfying the initial condition (14) and set $k=0$;

Step 2. Solve the RLMIs (24) to obtain the positive-definite matrices $P(k+1)$, $Q(k+1)$, the positive scalars $\mu(k+1)$, $\varepsilon_1(k)$, $\varepsilon_2(k)$, $\varepsilon_3(k)$, the matrices $X(k)$, $Y(k)$ and the filter parameter matrix $M_f(k)$ by utilizing the known parameters $P(k)$, $\{Q(s)\}_{-d+1\le s\le k}$, $\mu(k)$ and $\hat{x}(k)$;

Step 3. From Equation (25), derive the other two filter parameter matrices $F_f(k)$ and $G_f(k)$, then get the state estimate $\hat{x}(k+1)$ according to Equation (11);

Step 4. If $k<N$, set $k=k+1$ and go to *Step* 2, else go to *Step* 5;

Step 5. Exit.

4. Numerical example

In this section, we provide a numerical example to test the proposed design algorithm. Consider a discrete time-varying delayed system described by model (1) with the following time-varying parameters:

$$A(k) = \begin{bmatrix} 0 & 0.1\sin(k) \\ \sin(6k) & 0.2 \end{bmatrix},$$

$$A_1(k) = \begin{bmatrix} 0 & 0.12 \\ -0.12 & 0.1\sin(6k) \end{bmatrix}, \quad D(k) = 1,$$

$$B(k) = \begin{bmatrix} 0.15 \\ 0.3 \end{bmatrix}, \quad C(k) = [0.12\sin(6k) \quad 0.1],$$

$$M(k) = [0.1 \quad 0.1],$$

and the nonlinear functions $f(\cdot,\cdot)$, $g(\cdot,\cdot)$ are the same as those in Shen et al. (2011) represented as follows:

$$f(k,x(k)) = \frac{1}{4}\begin{bmatrix} 0.1x_1(k) + 0.1x_2(k) + 0.25x_2(k)\sin(x_1(k)) \\ \dfrac{0.5(x_1(k) + (1/3)x_2(k))}{1+x_2^2(k)} - 0.1x_1(k) + 0.3x_2(k) \end{bmatrix},$$

$$g(k,x(k-d)) = \frac{1}{4}\begin{bmatrix} \dfrac{0.3(x_1(k-d) + x_2(k-d))}{1+x_2^2(k-d)} + x_1^2(k-d) + 0.1x_1(k-d) + 0.1x_2(k-d) \\ 0.3x_1(k-d) + 0.3x_2(k-d) \end{bmatrix}.$$

It has been shown in Shen et al. (2011) that the above nonlinearities satisfying conditions (2) and (3) with

$$U_1(k) = V_1(k) = \begin{bmatrix} 0.1 & 0.05 \\ 0 & 0.1 \end{bmatrix},$$

$$U_2(k) = V_2(k) = \begin{bmatrix} -0.05 & 0 \\ -0.05 & 0.05 \end{bmatrix}.$$

It is assumed that the saturation function is

$$\sigma(u(k)) = \begin{cases} u(k) & \text{if} - u_{\max} \leq u(k) \leq u_{\max}, \\ u_{\max} & \text{if} u(k) > u_{\max}, \\ -u_{\max} & \text{if} u(k) < -u_{\max}, \end{cases}$$

where $u(k)$ is the position value of target. In this example, the saturation value $u_{\max}$ is taken as 0.08, $H_1(k)$ and $H_2(k)$ are set as $H_1(k) = 0.3(1 + |\tanh(k)|)$, $H_2(k) \equiv 1$.

In this simulation, we choose the exogenous disturbance input $v(k)$ to be $v(k) = \exp(-k/20) \times n(k)$, where $n(k)$ is uniformly distributed over $[-0.05, 0.05]$. The random variables $\alpha(k)$, $\beta(k)$ and $\gamma(k)$ are with the probabilities as $\bar{\alpha} = 0.8$, $\bar{\beta} = 0.5$, $\bar{\gamma} = 0.6$ and the constant time delay is set with $d = 1$. For illustration purposes, set $N = 20$, the H_∞ performance level $\gamma = 0.5$, $S(k) = \text{diag}\{20, 1\}$ and the initial value $x(-1) = [0.3 \quad -0.2]^{\mathrm{T}}$, $x(0) = [0.2 \quad 0]^{\mathrm{T}}$. The initial condition (14) is satisfied with $P(0) = I, Q(0) = \text{diag}\{2, 2\}, \mu(0) = 0.3$. According to the given RLMI algorithm, the time-varying LMIs in Equation (24) can be solved recursively and the desired filter matrices $F_f(k)$, $G_f(k)$ and $M_f(k)$ from time $k = 0$ to $k = 6$ are given in Table 1. It follows from Theorem 2 that the H_∞ filtering problem is solvable for the discrete time-varying system (1).

Table 1. The desired filter parameters.

k	0	1	2	3
$F_f(k)$	$\begin{bmatrix} 0 & 0 \\ 0 & 0 \end{bmatrix}$	$\begin{bmatrix} 0.0283 & 0.0654 \\ 0.0692 & 0.0195 \end{bmatrix}$	$\begin{bmatrix} 0.1794 & 0.1440 \\ 0.1702 & -0.3478 \end{bmatrix}$	$\begin{bmatrix} -0.0105 & -0.0629 \\ -0.0859 & 0.2099 \end{bmatrix}$
$G_f(k)$	$\begin{bmatrix} 0.0998 \\ 0.2057 \end{bmatrix}$	$\begin{bmatrix} 0.1029 \\ 0.2105 \end{bmatrix}$	$\begin{bmatrix} 0.1032 \\ 0.2209 \end{bmatrix}$	$\begin{bmatrix} 0.1235 \\ 0.2834 \end{bmatrix}$
$M_f(k)$	$\begin{bmatrix} 0 & 0 \end{bmatrix}$	$\begin{bmatrix} 0.0524 & 0.1066 \end{bmatrix}$	$\begin{bmatrix} 0.0597 & 0.1066 \end{bmatrix}$	$\begin{bmatrix} 0.0073 & 0.0676 \end{bmatrix}$
k	4	5	6	...
$F_f(k)$	$\begin{bmatrix} 0.0186 & 0.2248 \\ 0.3614 & -0.1823 \end{bmatrix}$	$\begin{bmatrix} -0.0021 & -0.0063 \\ -0.0105 & 0.0618 \end{bmatrix}$	$\begin{bmatrix} -0.2377 & 0.0155 \\ 0.0572 & -0.1260 \end{bmatrix}$	...
$G_f(k)$	$\begin{bmatrix} 0.1103 \\ 0.2924 \end{bmatrix}$	$\begin{bmatrix} 0.1308 \\ 0.3489 \end{bmatrix}$	$\begin{bmatrix} 0.1405 \\ 0.3104 \end{bmatrix}$	...
$M_f(k)$	$\begin{bmatrix} 0.0077 & 0.0681 \end{bmatrix}$	$\begin{bmatrix} -0.0176 & -0.0417 \end{bmatrix}$	$\begin{bmatrix} -0.1006 & 0.0598 \end{bmatrix}$	...

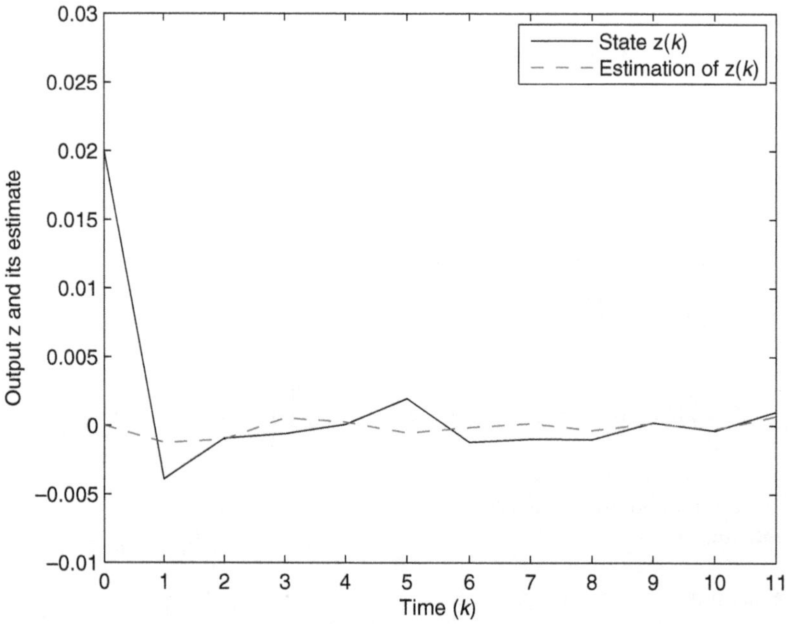

Figure 1. Trajectories of the output z and its estimate $\hat{z}$.

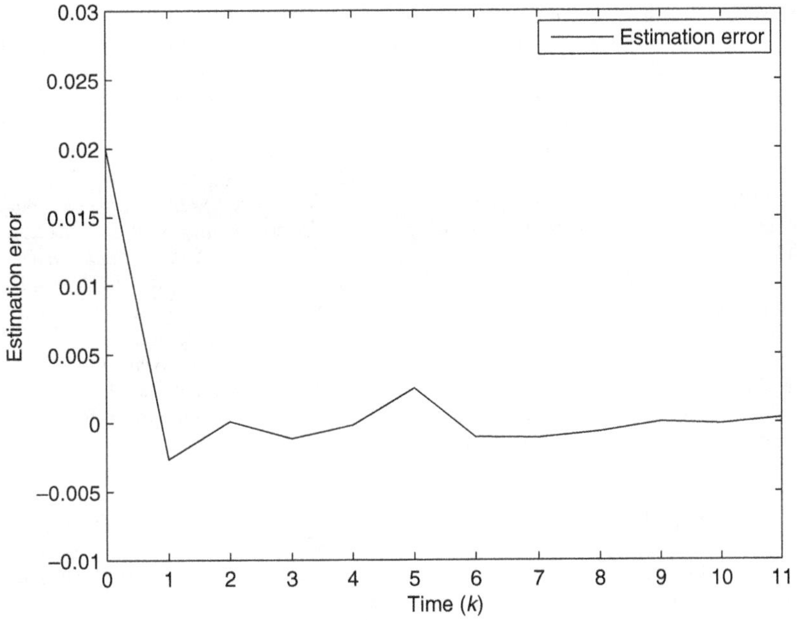

Figure 2. Trajectories of the estimation error $\tilde{z}$.

By employing the filter (11) with the parameters as given in Table 1, the estimation trajectories are shown in Figures 1 and 2. More specially, Figure 1 presents the output $z(k)$ and its estimate $\hat{z}(k)$ and Figure 2 draws the estimation error output $\tilde{z}(k)$. The simulation results further demonstrate the effectiveness of the filter design scheme.

5. Conclusions

In this paper, we have investigated the finite horizon H_∞ filtering problem for the time-varying delayed system with incomplete information such as RONs, ROSSs as well as missing measurements. A time-varying filter has been designed for the system under consideration such that the filtering error system satisfies the H_∞ performance constraints on the finite horizon. By resorting to the stochastic analysis and matrix inequality techniques, sufficient conditions have been derived in the form of RLMIs which not only guarantee the error system to preserve the H_∞ performance but also give the explicit expressions of the desired filtering parameters. Simulation results further demonstrate the feasibility of the proposed filtering methods.

Acknowledgements

This work was supported in part by the National Natural Science Foundation of China under Grants 61174136, the Natural Science Foundation of Jiangsu Province of China under Grant BK2011598 and BK20130017, the Programme for New Century Excellent Talents in University under Grant NCET-12-0117 and the Fundamental Research Funds for the Central Universities under Grant 2242012155.

References

Ahmad, H., & Namerikawa, T. (2013). Extended Kalman filter-based mobile robot localization with intermittent measurements. *Systems Science & Control Engineering: An Open Access Journal, 1*, 113–126.

Basin, M. V., Shi, P., Alvarez, D. C., & Wang, J. (2009). Central suboptimal H_∞ filter design for linear time-varying systems with state or measurement delay. *Circuits, Systems and Signal Processing, 28*(2), 305–330.

Boyd, S., El Ghaoui, L., Feron, E., & Balakrishnan, V. (1994). *Linear matrix inequalities in system and control theory*. Philadelphia, PA: SIAM.

Cao, Y-Y., Lin, Z., & Chen, B. M. (2003). An output feedback H_∞ controller design for linear systems subject to sensor nonlinearities. *IEEE Transactions on Circuits and Systems – I, 50*(7), 914–921.

Chau, C-K., Qin, F., Sayed, S., Wahab, M. H., & Yang, Y. (2010). Harnessing battery recovery effect in wireless sensor networks: Experiments and analysis. *IEEE Journal on Selected Areas in Communications, 28*(7), 1222–1232.

Chen, Y., & Zheng, W. (2012). Stochastic state estimation for neural networks with distributed delays and Markovian jump. *Neural Networks, 25*, 14–20.

Ding, D., Wang, Z., Shen, B., & Shu, H. (2012). H_∞ state estimation for discrete-time complex networks with randomly occurring sensor saturations and randomly varying sensor delays. *IEEE Transactions on Neural Networks and Learning Systems, 23*(5), 725–736.

Dong, H., Wang, Z., Ho, D. W. C., & Gao, H. (2011). Robust H_∞ filtering for Markovian jump systems with randomly occurring nonlinearities and sensor saturation: The finite-horizon case. *IEEE Transactions on Signal Processing, 59*(7), 3048–3057.

Gershon, E., & Shaked, U. (2008). H_∞ output-feedback control of discrete-time systems with state-multiplicative noise. *Automatica, 44*(2), 574–579.

Gershon, E., Shaked, U., & Yaesh, I. (2001). H_∞ control and filtering of discrete-time stochastic systems with multiplicative noise. *Automatica, 37*(3), 409–417.

Gershon, E., Shaked, U., & Yaesh, I. (2005). *H_∞ control and estimation of state-multiplicative linear systems*. New York, NY: Springer-Verlag.

Hung, Y. S., & Yang, F. (2003). Robust H_∞ filtering with error variance constraints for discrete time-varying systems with uncertainty. *Automatica, 39*(7), 1185–1194.

Li, P., Lam, J., & Shu, Z. (2010). H_∞ positive filtering for positive linear discrete-time systems: An augmentation approach. *IEEE Transactions on Automatic Control, 55*(10), 2337–2342.

Liu, J., Wang, J., & Yang, G. (2003). Reliable guaranteed variance filtering against sensor failures. *IEEE Transactions on Signal Processing, 51*(5), 1403–1411.

Lu, X., Xie, L., Zhang, H., & Wang, W. (2007). Robust Kalman filtering for discrete-time systems with measurement delay. *IEEE Transactions on Circuits and Systems – II, 54*(6), 522–526.

Mohamed, M., Nahla, K., & Safya, B. (2013). Design of a nonlinear observer for mechanical systems with unknown inputs. *Systems Science & Control Engineering: An Open Access Journal, 1*, 105–112.

Nahi, N. E. (1969). Optimal recursive estimation with uncertain observations. *IEEE Transactions on Information Theory, IT-15*(4), 457–462.

Reif, K., & Unbehauen, R. (1999). The extended Kalman filter as an exponential observer for nonlinear systems. *IEEE Transactions on Signal Processing, 47*(8), 2324–2328.

Shaked, U., & Suplin, V. (2001). A new bounded real lemma representation for the continuous-time case. *IEEE Transactions on Automatic Control, 46*(9), 1420–1426.

Shen, B., Ding, S. X., & Wang, Z. (2013). Finite-horizon H_∞ fault estimation for linear discrete time-varying systems with delayed measurements. *Automatica, 49*(1), 293–296.

Shen, B., Wang, Z., & Liu, X. (2011). Bounded H_∞ synchronization and state estimation for discrete time-varying stochastic complex networks over a finite horizon. *IEEE Transactions on Neural Networks, 22*(1), 145–157.

Shen, B., Wang, Z., Shu, H., & Wei, G. (2010). Robust H_∞ finite-horizon filtering with randomly occurred nonlinearities and quantization effects. *Automatica, 46*(11), 1743–1751.

Shen, B., Wang, Z., Shu, H., & Wei, G. (2011). H_∞ filtering for uncertain time-varying systems with multiple randomly occurred nonlinearities and successive packet dropouts. *International Journal of Robust and Nonlinear Control, 21*(14), 1693–1709.

Wang, Z., Dong, H., Shen, B., & Gao, H. (2013). Finite-horizon H_∞ filtering with missing measurements and quantization effects. *IEEE Transactions on Automatic Control, 58*(7), 1707–1718.

Wang, Z., Liu, Y., & Liu, X. (2008). H_∞ filtering for uncertain stochastic time-delay systems with sector-bounded nonlinearities. *Automatica, 44*(5), 1268–1277.

Wang, Z., Shen, B., & Liu, X. (2012). H_∞ filtering with randomly occurring sensor saturations and missing measurements. *Automatica, 48*(3), 556–562.

Wang, Y., Wang, Z., & Liang, J. (2009). Global synchronization for delayed complex networks with randomly occurring nonlinearities and multiple stochastic disturbances. *Journal of Physics A: Mathematical and Theoretical, 42*(13), 135101.

Wang, Z., Wang, Y., & Liu, Y. (2010). Global synchronization for discrete-time stochastic complex networks with randomly occurred nonlinearities and mixed time delays. *IEEE Transactions on Neural Networks, 21*(1), 11–25.

Wei, G., Wang, L., & Han, F. (2013). A gain-scheduled approach to fault-tolerant control for discrete-time stochastic delayed systems with randomly occurring actuator faults. *Systems Science & Control Engineering: An Open Access Journal, 1*, 82–90.

Wei, G., Wang, Z., & Shen, B. (2010). Error-constrained filtering for a class of nonlinear time-varying delay systems with non-gaussian noises. *IEEE Transactions on Automatic Control, 55*(12), 2876–2882.

Wei, G., Wang, Z., & Shen, B. (2013). Probability-dependent gain-scheduled control for discrete stochastic delayed systems with randomly occurring nonlinearities. *International Journal of Robust and Nonlinear Control, 23*(7), 815–826.

Wu, L., & Wang, Z. (2009). Fuzzy filtering of nonlinear fuzzy stochastic systems with time-varying delay. *Signal Processing, 89*(9), 1739–1753.

Xie, L., Soh, Y. C., & de Souza, C. E. (1994). Robust Kalman filtering for uncertain discrete-time systems. *IEEE Transactions on Automatic Control, 39*(6), 1310–1314.

Xie, L., & de Souza, C. E. (1992). Robust H_∞ control for linear systems with norm-bounded time-vayring uncertainty. *IEEE Transactions on Automatic Control, 37*(8), 1188–1191.

Yang, F., & Li, Y. (2009). Set-membership filtering for systems with sensor saturation. *Automatica, 45*(8), 1896–1902.

Yang, G., & Ye, D. (2007). Adaptive reliable H_∞ filtering against sensor failures. *IEEE Transactions on Signal Processing, 55*(7), 3161–3171.

Zhang, H., Feng, G., & Duan, G. (2006). H_∞ filtering for multiple-time-delay measurements. *IEEE Transactions on Signal Processing, 54*(5), 1681–1688.

Zhang, W., Chen, B., & Tseng, C. (2005). Robust H_∞ filtering for nonlinear stochastic systems. *IEEE Transactions on Signal Processing, 53*(2), 589–598.

21

Energy-efficient gas pipeline transportation

R. Whalley and A. Abdul-Ameer*

Faculty of Engineering, The British University in Dubai, PO Box 345015, Dubai, UAE

Gas transportation via long pipelines is considered. Distributed parameter, dynamic modelling with series and shunt energy dissipation and gas stream, equivalent capacitance and inductance effects are proposed. Hybrid analysis techniques, wherein both the distributed and the lumped, concentrated elements of the pipeline system are included in the overall model, are advocated. Constrained optimisation procedures, with the introduction of the Hamiltonian function to minimise the pipeline, inflow–outflow difference, are invoked thereby promoting impedance matching and the energy-efficient transportation of the specified, gas volume flow rate. Illustrative application studies are outlined thereby validating the analytical methods employed and the determination of the optimum, pipeline exit resistance.

Keywords: gas; pipeline; transportation; modelling; optimisation

Nomenclature

C_j	capacitance of the jth airway per unit length	scalar
L_j	inductance of the jth airway per unit length	scalar
r_j	series resistance of the jth airway per unit length	scalar
g_j	shunt conductance (admittance) of the jth airway per unit length	scalar
$p_j(s)$	pressure at inlet to the jth airway per unit length	function
$p_{j+1}(s)$	pressure at outlet from the jth airway per unit length	function
$q_j(s)$	volume flow at inlet to the jth airway per unit length	function
$q_{j+1}(s)$	volume flow at outlet from the jth airway per unit length	function
$\zeta_j(s)$	characteristics impedance of the jth airway	function
$\Gamma_j(s)$	propagation function for the jth airway	function
T	phase lead time constant	scalar
τ	phase lag time constant	scalar
R	frictional resistance at pipeline exit	scalar
ρ	gas density	scalar

1. Introduction

The transportation of gas over long distances by pipelines will be considered in this contribution. This method of supply is a relatively safe, reliable and a cost-effective form of conveying natural gas and oil which is universally employed.

Constructing and the installation of gas or oil pipelines is an expensive, labour-intensive and politically sensitive operation. These networks often span remote regions, cross-national boundaries and ecologically protected areas resulting in international agreements following delicate, protracted negotiations.

Beyond this, the running cost associated with gas pipelines is substantial. Owing to the frictional energy dissipation arising from the internal pipeline roughness, welds, joints, bends and discontinuities, there is a continuous reduction in the gas stream pressure and, hence, volume gas steam flow rate. To counter this effect, compressors and heat exchangers are installed, at strategic locations along the pipeline thereby restoring the pressure loss and gas volume flow rate.

Due to the length of gas pipelines and the proportional pressure loss, the compressors and heat exchangers employed operate continuously. Consequently, the running cost, maintenance and refit charges associated with this requirement are substantial.

This problem is exacerbated by the remote locations, monitoring and operation of the compressor drive systems and gas coolers. These active devices must also respond to varying load demands with the requirement for constant delivery pressure and supply rates.

As with all continuously operated, long-duration, duty cycle systems, any operational economy is translated into significant savings owing to the accumulating reduction in running, maintenance and delivery costs. To assess

*Corresponding author. Email: alaa.ameer@buid.ac.ae

demands under which maximum economy can be achieved whilst satisfying specified delivery conditions, an accurate model for long pipeline systems is required. Once this has been established, optimum operating conditions can be investigated.

Unfortunately, the classical theory for spatially dispersed pipeline systems results in irrational, multivariable, input–output models which are incomplete in the Laplace transform variable; see, for example, Schwartz & Friedland (1965) and Takahashi (1970). Technically, it is possible to obtain the predicted system responses from these models. However, the procedures involved do not provide simple, usable results which can be incorporated into design, analysis or optimisation investigations.

Finite element techniques may be used to assess the pipeline dynamics. With this procedure, large matrix models arise from the modelling procedure thereby attracting computational errors. Considerable speculation surrounding the computed pipeline performance may also be encountered in that the number and composition of the elements employed are unspecified in finite element modelling, as discussed in Section 2.

In view of this, the focus of this contribution will be on deriving a pipeline modelling method which includes all of the salient features of classical analysis whilst avoiding the above complications arising from multiple, lumped parameter methods. Once established, this provision invites the use of constrained optimisation and the employment of impedance matching to confine operational energy consumption.

Ultimately, the delivery of specified, regulated gas supplies is the principal attraction. This, when considering pipelines with varying, distributed frictional characteristics, will be investigated using the calculus of variations following the derivation of an accurate, unambiguous model, for the pipeline system.

2. Modelling methods

In the transportation of gases via long pipelines and compressor networks, geographical dispersion is a significant feature. Owing to this, obtaining estimates for the volume flow of gas using conventional, lumped parameter theory is inappropriate.

It may be considered that using multiple lumping with the application of finite element techniques would be sufficient. However, this is not so, since the dynamics of spatially dispersed systems comprise a combination of travelling, stationary and reflected pressure and volume flow waveforms. Hence, there is no equivalent lumped, cascaded model counterpart since travelling and reflected transient components cannot materialise in lumped representations, in that all spatial dispersion effects are absent from these realisations.

In any case when considering pipelines of $>10\,\text{km}$ in length, the matrix models, derived from finite element methods, would be dimensionally very large (Esahanian & Behbani-Nejad, 2002). Consequently, in addition to the analytical disadvantages cited above, numerical computational errors which would further contaminate the results would be encountered (Bradie, 2006).

Alternatively, with the employment of hybrid, distributed–lumped modelling an accurate modelling method is available (Whalley, 1988; Whalley & Abdul-Ameer, 2009). This procedure allows pipeline elements, which are clearly distributed, to be modelled using distributed parameter methods. Otherwise, relatively point-wise components and sub-assemblies such as valves, compressors, bends and restrictions may be represented using lumped analysis methods without too much loss of accuracy. This allows engineering judgement to be exercised in selecting the appropriate modelling method for each sub-system element.

Importantly, the many boundary conditions and complexities arising from the use of distributed parameter methods universally are avoided with this approach, whilst including simple point-wise representations for components which form the pipeline connecting elements and series or parallel branches.

Following the modelling of the individual elements of pipelines, distributed–lumped analysis allows the construction of an overall hybrid matrix configuration for the system. This final model provides a general, component-identifiable, accurate, impedance admittance realisation enabling dynamic simulation, analysis and regulator design.

3. Series and parallel representations

Elements comprising an overall pipeline system may be assembled series, parallel or in series–parallel form where in each case the steady-state volume flow would be inversely proportional to the pipeline input impedance.

In the analysis herein, lumped parameter components are represented by simple passive impedances, $g_{j11}(s)$. However, any analogous two-port network representation could be employed to model these units, as shown in Whalley (1990).

4. The distributed parameter modelling of pipelines

Although the flow of gas in pipeline systems is usually turbulent, three-dimensional and nonlinear, there are compelling reasons for the formulation of simple, usable models for perturbed flow changes, relative to the given steady-state conditions. Essentially, this type of model would enable the analysis of complex, interconnected applications. Ideally this would be via general solutions for the spatially distributed pressure–flow relationships following input or disturbance changes. Moreover, simulation studies could be easily accommodated using this form of representation with the advantage that regulator design exercises could now be embarked upon using existing theoretical techniques and algorithms.

In fact, pioneering work, as detailed in Iberall (1960), Nichols (1961) and Brown (1962), showed that theoretically derived, first-order, perturbed, acoustic, one-dimensional approximations for the Navier–Stokes equations were available. These modelling restrictions included zero bulk modulus and radiant heat transfer effects. A continuous, homogeneous medium was also assumed with no radial or axial heat transfer effects.

Further work within this framework was undertaken in Bartlett and Whalley (1998) where a general discrete, distributed–lumped parameter representation for linear systems was presented. Low-temperature application studies using this approach were proposed in Bartlett and Whalley (1995) where all of these representations related to the perturbed pressure-variation dynamics relative to steady-state equilibrium conditions.

Extending this work, this contribution focuses on the distributed parameter system model element shown in Appendix 1, Figure A1 where L_j, C_j, r_j and g_j are the pipeline system distributed inductance, capacitance and series (longitudinal) and shunt (radial) flow resistance and conductance effects per metre length of pipeline, respectively. The governing equations for this type of element for the jth pipeline section are

$$\frac{\partial p_j}{\partial x}(t,x) = -L_j\frac{\partial q_j}{\partial t} - r_j q_j(t,x) \tag{1}$$

and

$$\frac{\partial q_j}{\partial x}(t,x) = -C_j\frac{\partial p_j}{\partial t} - g_j p_j(t,x) \text{ as shown in Figure 1.} \tag{2}$$

Following Laplace transformation with zero initial conditions, Equations (1) and (2) yield the solution for the jth distributed parameter model of a system of m elements, as shown in Appendix 1:

$$\begin{bmatrix} P_j(s,l_j) \\ P_{j+1}(s,l_{j+1}) \end{bmatrix} = \begin{bmatrix} \zeta_j(s)(s)w_j(s) & -\zeta_j(s)\left(w_j^2(s)-1\right)^{1/2} \\ \zeta_j(s)(w_j^2(s)-1)^{1/2} & -\zeta_j(s)w_j(s) \end{bmatrix} \times \begin{bmatrix} Q_j(s,l_j) \\ Q_{j+1}(s,l_{j+1}) \end{bmatrix}, \tag{3}$$

where $j = 2k+1$, $k = 0, 1, \ldots, m-1$ (m is the number of distributed parameter elements),

$$\zeta_j(s) = \left[\frac{(L_j s + r_j)}{(C_j s + g_j)}\right]^{1/2},$$

$$w_j(s) = \frac{(e^{2\Gamma_j(s)l_j}+1)}{(e^{2\Gamma_j(s)l_j}-1)}$$

and $$\Gamma_j(s) = \left[\left(\frac{L_j s + r_j}{C_j s + g_j}\right)\right]^{1/2}.$$

Consequently, even with all of the constraints mentioned earlier the input–output relationship for a typical distributed parameter gas pipeline network model is multivariable, irrational and is incomplete in the Laplace variable s. This difficulty effectively masks any correspondence between the actual system performance and the governing equations so that extracting dynamic information from this representation is markedly impaired.

Although in Equation (3), for example, $\Gamma_j(s)$ and $\zeta_j(s)$ have branch points in the complex frequency plane, inversion with the use of Bromwich's contour and the Laplace error function, $\mathrm{Erf}(t) = (2/\pi)\int_0^c e^{-u^2}\,du$, see, for example, Spiegel (1965), is possible yielding the time–response characteristics following input pressure changes.

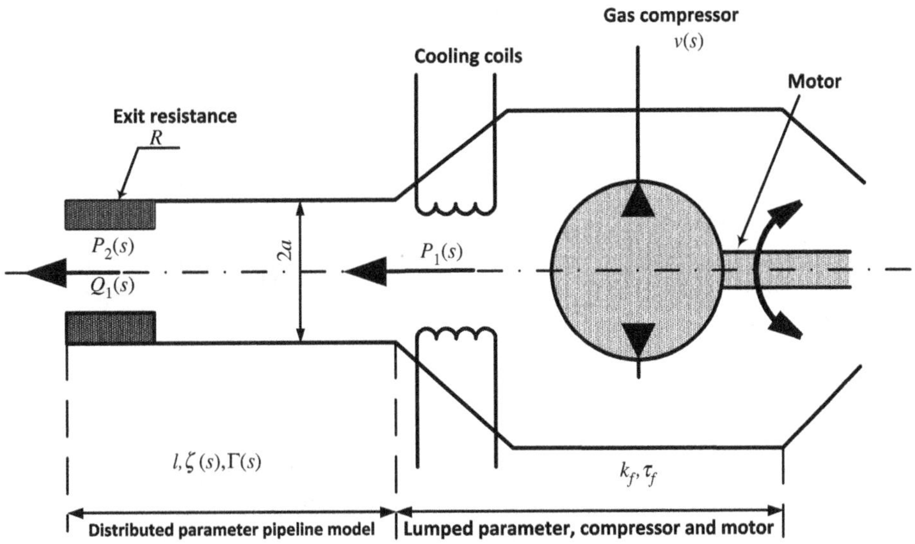

Figure 1. Distributed–lumped parameter model of pipeline, compressor and motor.

However, this does little to aid the recovery of the original aim of producing a "user friendly" distributed parameter description for large-scale, pipeline system modelling, analysis, simulation and regulator design.

In this regard, of interest here is the nature of the series impedance and shunt admittance of the infinitesimal pipeline element, shown in Figure A1. The series impedance frictional drag r_j, for example, represents the effect of the gas flow on the pressure gradient arising from shear action at the pipeline wall boundary layer. Contrasting this, the shunt g, admittance or conductance arises from compressibility effects, as shown in Robertson and Crowe (1990) where the frictional drag arises from varying gas path compliance, owing to cross-flow turbulence and molecular heat transfer.

In pipeline systems, the flow impedance is principally due the entrance/exit losses and the r_j and g_j, distributed pipeline frictional resistance effects, where in general:

$$\frac{1}{g_j} \neq r_j. \tag{4}$$

In dimensionally "long" pipelines both the series r_j and the shunt g_j frictional factors contribute to the overall pressure drop and diminishing volume flow characteristics. The analysis herein also confirms that the inclusion of both these dissipation mechanisms is mandatory.

Moreover, the per unit length energy storage parameters, as shown in Eckman (1958) and Palm (2005), for a circular pipeline, diameter $2a_j$, are the gas path capacitance and inductance of

$$C_j = \frac{\pi a_j^2}{\gamma R_g \theta_j} \tag{5}$$

and

$$L_j = \frac{1}{\pi a_j^2}, \tag{6}$$

respectively, where

$$L_j \gg C_j \tag{7}$$

for engineering applications.

In view of the inequalities of Equations (4) and (7) rationality may be recovered by equating

$$\left(\frac{C_j s}{g_j} + 1\right) \frac{\prod_{k=1}^{N} (T_{jk}s + 1)^2}{\prod_{k=1}^{N} (\tau_{jk}s + 1)^2} \cong \left(\frac{L_j s}{r_j} + 1\right). \tag{8}$$

It should be noted that with an appropriate choice of T_{jk} and τ_{jk} the approximation of Equation (8), with $s = \mathrm{i}\omega$, is accurate at

$$\text{low frequencies} \quad \frac{\omega < r_j}{L_j},$$

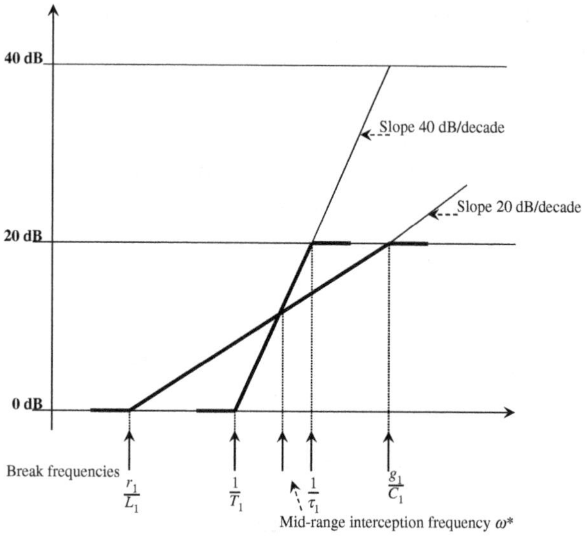

Figure 2. Bode diagram asymptotic approximation.

$$\text{high frequencies} \quad \frac{\omega > g_j}{C_j}$$

and at $(2N - 1)$ intermediate frequencies ω^* where

$$\frac{r_j}{L_j} < \omega^* < \frac{g_j}{C_j}$$

as shown by the Bode plot of Figure 2, where $N = 1$.

Then from Equation (A16) since

$$\Gamma_j(s) = \left[r_j \left(\frac{L_j s}{r_j} + 1\right) \left(\frac{C_j s}{g_j} + 1\right) g_j \right]^{1/2}. \tag{9}$$

Equation (9) becomes, following the substitution shown in Equation (8):

$$\Gamma_j(s) = \alpha_j \frac{(T_{j1}s + 1)\,(T_{j2}s + 1)}{(\tau_{j1}s + 1)\,(\tau_{j2}s + 1)} \left(\frac{C_j s}{g_j} + 1\right), \tag{10}$$

where $\alpha_j = \sqrt{r_j g_j}$.

Equally, following Equations (A13) and (A14) since

$$\zeta_j(s) = \sqrt{\frac{(L_j s + r_j)}{(C_j s + g_j)}} \tag{11}$$

then Equation (11), with the substitution of Equation (8), continuing with $N = 2$ for illustration purposes, is

$$\zeta_j(s) = \bar{\alpha}_j \frac{(T_{j1}s + 1)\,(T_{j2}s + 1)}{(\tau_{j1}s + 1)\,(\tau_{j2}s + 1)} \quad \text{where } \bar{\alpha}_j = \sqrt{\frac{r_j}{g_j}}. \tag{12}$$

The remaining important function of Equation (2) is $w_j(s)$. Since

$$w_j(s) = \frac{(e^{2l_j\Gamma_j(s)} + 1)}{(e^{2l_j\Gamma_j(s)} - 1)} \tag{13}$$

then with $\Gamma_j(s)$ from Equation (10):

$$w_j(s) = \frac{(e^{2l_j\sqrt{r_jg_j}\chi_j(s)} + 1)}{(e^{2l_j\sqrt{r_jg_j}\chi_j(s)} - 1)}. \tag{14}$$

Consequently, upon substituting for Equation (14):

$$(w_j^2(s) - 1)^{1/2} = \frac{2e^{l_j\sqrt{r_jg_j}\chi_j(s)}}{(e^{2l_j\sqrt{r_jg_j}\chi_j(s)} - 1)} \tag{15}$$

where in Equation (15):

$$\chi_j(s) = \frac{(T_{j1}s+1)}{(\tau_{j1}s+1)}\frac{(T_{j2}s+1)}{(\tau_{j2}s+1)}\left(\frac{C_js}{g_j}+1\right). \tag{16}$$

From Equation (16), following division, the truncated series evaluation is always a simple P+D term, given by

$$\chi_j \cong a_js + b_j, \tag{17}$$

where from Equation (17), expanding $\chi(s)$ for high frequencies gives

$$a_j = \left[\frac{C_j}{g_j} + (T_{j1} + T_{j2}) - (\tau_{j1} + \tau_{j2})\right] \quad \text{and} \quad b_j = 1.$$

It is evident from Equations (13)–(17) that the distributed parameter model is now in an attractive form. The functions comprising Equation (3) are free from origin branch point problems with each component $\zeta_j(s)$, $w_j(s)$ and $(w_j^2(s) - 1)^{1/2}$ being single valued and complete, in the Laplace variable s with simple steady-state values of

$$\zeta_j(0) = \bar{\alpha}_j,\ w_j(0) = \frac{e^{2l_j\alpha_j\chi_j(0)}+1}{e^{2l_j\alpha_j\chi_j(0)}-1} \quad \text{and}$$

$$(w_j^2(0) - 1)^{1/2} = \frac{2e^{l_j\alpha_j\chi_j(0)}}{e^{2l_j\alpha_j\chi_j(0)}-1}.$$

Although values for the equivalent of the energy storage via the distributed capacitance and inductance effects per unit length in gas pipelines can be obtained with accuracy from Equations (5) and (6), the distributed, per unit length, series and shunt resistance values are more difficult to ascertain. From the theory of fluid dynamics, the pressure drop due to friction is inversely proportional to the Reynolds number, $Re < 2500$, for a given gas flow velocity, gas density, ducting length and cross-sectional area. For higher Reynolds numbers, owing to the elevated velocities in gas pipelines the empirical law of Blasius could be used to obtain estimates of the frictional flow coefficients, as discussed in Rogers and Mayhew (1970). These estimates and those obtained from Moody diagrams (Gautam, 2009) however are based on steady flow conditions, in a pipeline of constant cross-sectional area. Consequently, as in all engineering system problems, the frictional coefficient values are known with least confidence. Empirically based results, derived from direct measurement, may be used if the system exists. Otherwise, upper and lower bounded values for r_j and g_j may be employed, with the system response characteristics reflecting these estimates.

5. Lumped parameter elements

The assumption here is that pipeline restrictions can be modelled using elements having the ladder network structure shown in Langill (1965). In the overall configuration, presented in Equation (18), $g_{j+1,1}(s), g_{j+1,2}(s)$ and $g_{j+1,3}(s)$ are the lumped impedances for the jth lumped element where $j = 2k, k = 1, \ldots, m$. Models of greater sophistication could be employed if so desired. However, for many applications the arrangement shown provides an adequate, equivalent analogue representation for sub-assemblies such as compressors, bends and valves for analysis purposes. The generic multivariable representation given in Rosenbrock (1974) may be adopted here, if required, to replicate impedance models of greater complexity.

6. Integrated hybrid model

An overall model structure must now be derived enabling the assembly of the system matrix. For purposes of illustration, if a distributed–lumped configuration is assumed, then an appropriate system matrix can be constructed by adding consecutive distributed or lumped system descriptions and in so doing eliminate all intermediate variables.

In this case, the system model for a total of m distributed–lumped interconnected sections results in the system equation:

$$(P_1(s), 0, 0, \ldots, 0)^{\mathrm{T}} = \Omega(s)(Q_1(s), Q_2(s), \ldots, Q_m(s))^{\mathrm{T}}, \tag{18}$$

where

$$\Omega(s) = \begin{bmatrix} \zeta_1(s)w_1(s) & -\zeta_1(s)(w_1^2(s)-1)^{1/2} & 0 & 0 & \cdots \\ \zeta_1(s)(w_1^2(s)-1)^{1/2} & -\zeta_1(s)w_1(s) - \bar{g}_{1,11}(s) & -g_{1,12}(s) & 0 & \cdots \\ 0 & -g_{1,12}(s) & -\bar{g}_{1,22}(s) + \zeta_2(s)w_2(s) & \zeta_2(s)(w_2^2(s)-1)^{1/2} & \cdots \\ 0 & 0 & \zeta_2(s)(w_2^2(s)-1)^{1/2} & -\zeta_2(s)w_2(s) - \bar{g}_{3,11}(s) & \cdots \\ \vdots & \vdots & 0 & 0 & 0 \\ 0 & 0 & 0 & 0 & 0 \end{bmatrix}.$$

This impedance matrix representation is in respect of the distributed–lumped–distributed–lumped system topology with $\bar{g}_{1,11}(s), \bar{g}_{1,22}$, etc. being the diagonal elements of the termination matrix. However, there is no restriction on assembling alternative matrix descriptions for distributed–distributed–distributed realisations, for example, when modelling a series of purely distributed parameter pipeline elements of varying dimensions.

As Equation (18) shows, $\Omega(s)$ is a skew, symmetric tri-diagonal matrix enabling simple recursive procedures to be employed in the computation of the determinant, as shown in Barnett (1992).

The theoretical basis outlined here is sufficient for the analysis procedures required. With these methods the dynamic models for complex series–parallel pipeline systems can be constructed in a simple, systematic and scientific manner. Moreover, the admittance functions, following the inversion of Equation (18), are easily realisable in terms of rational and irrational Laplace functions which can be used for the analysis or simulation purposes for the complete hybrid distributed–lumped system model of Equation (18), as illustrated in the following application study.

7. Pipeline and compressor model

In this application a model representing a single, long pipeline and a compressor will be considered. The arrangement for analysis purposes is shown in Figure 1.

From the theory of Section 6, the system Equation (18) is relevant, since there is only a single distributed parameter section and a single termination, lumped resistance element.

Hence

$$\begin{bmatrix} P_1(s) \\ P_2(s) \end{bmatrix} = \begin{bmatrix} \zeta(s)w(s) & -\zeta(s)(w^2(s)-1)^{1/2} \\ \zeta(s)(w^2(s)-1)^{1/2} & -\zeta(s)w(s) \end{bmatrix} \times \begin{bmatrix} Q_1(s) \\ Q_2(s) \end{bmatrix}, \quad (19)$$

where in Equation (19), the termination relationship between the transformed pressure change $P_2(s)$ and the transformed airflow change $Q_2(s)$ is simply

$$P_2(s) = RQ_2(s).$$

Consequently, following inversion Equation (19) becomes

$$\frac{\begin{bmatrix} \zeta(s)w(s)+R & -\zeta(s)(w^2(s)-1)^{1/2} \\ \zeta(s)(w^2(s)-1)^{1/2} & -\zeta(s)w(s) \end{bmatrix}}{\zeta(s)(w(s)R+\zeta(s))} \begin{bmatrix} P_1(s) \\ 0 \end{bmatrix} = \begin{bmatrix} Q_1(s) \\ Q_2(s) \end{bmatrix}. \quad (20)$$

If the compressor unit is assumed to be relatively lumped, in comparison with the pipeline, comprising rotors and bearings, see, for example, Hodder (2008) then

$$P_1(s) = \frac{k_f U(s)}{(\tau_f(s)+1)},$$

where k_f is the gain and τ_f is the fan time constant and $U(s)$ is the applied voltage for the electrical drive. For this particular application, the parameters are as follows:

a	pipeline radius	0.5 m
θ	absolute gas temperature	313 K
R	exit resistance	$1.0\,\mathrm{N\,s/m^5}$
R_g	characteristic gas constant	287 J/kg K
l	pipeline length	1000, 5000 and 10,000 m
τ_c	compressor time constant	5.0 s
k_c	compressor gain	$10\,\mathrm{kg\,s/m^3v}$
γ	adiabatic index	1.31
g	shunt conductance per metre	$10^{-4}\,\mathrm{m^5/N\,s}$
r	series resistance per metre	$0.6\times10^{-4}\,\mathrm{N\,s/m^5}$

From Equations (5) and (6), the gas capacitance and inductance effects per metre length of pipeline are

$$C = \frac{\pi a^2}{\gamma R_g \theta} = 0.643\times10^{-4}\,\mathrm{m}^2$$

and

$$L = \frac{1}{\pi a^2} = 1.2835\,\mathrm{m}^2,$$

respectively.

From Equations (10) and (12), and (11), respectively, we have:

$$\bar{\alpha} = \sqrt{\left(\frac{r}{g}\right)} = 0.7745,$$

$$\alpha = \sqrt{rg} = 0.7745\times10^{-4}$$

$$\text{and } \zeta(s) = \bar{\alpha}\left(\frac{T_1 s+1}{\tau_1 s+1}\right)\left(\frac{T_2 s+1}{\tau_2 s+1}\right).$$

Also Equation (8) requires that for a single pipeline section where for illustration purposes, with $N = 2$ and where the double subscripts have been dropped for purposes of clarity:

$$\frac{(T_1 s+1)^2}{(\tau_1 s+1)^2}\frac{(T_2 s+1)}{(\tau_2 s+1)} \cong \frac{(L/r)s+1}{(C/g)s+1},$$

where $L/r = 2.1392\times10^4$ s, $C/g = 0.0643$ s, $T_1 = 3750$, $T_2 = 3.0$, $\tau_1 = 110.1$ and $\tau_2 = 0.175$ s.

The break frequencies selected for the Bode characteristics for $1/(L/r), 1/(C/g), 1/T_1, 1/T_2, 1/\tau_1$ and $1/\tau_2$ are $0.4674\times10^{-4}, 15.552, 6\times10^{-4}, 0.4, 0.05$ and 6.02 rad/s, respectively.

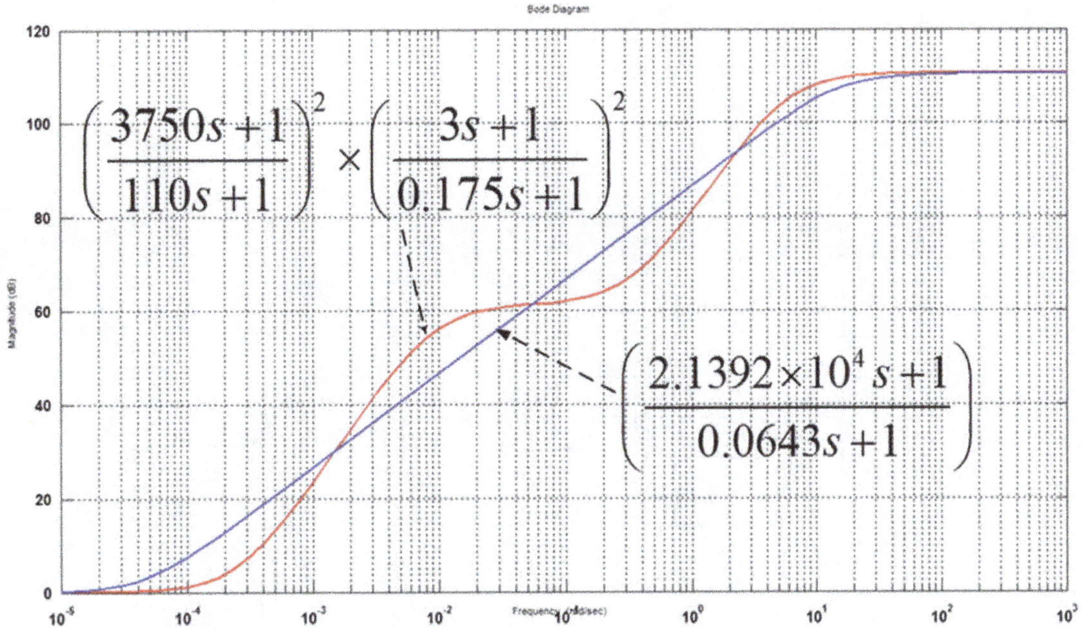

Figure 3. Frequency response diagrams for $N = 2$ with $\left|\left(\frac{3750i\omega+1}{110i\omega+1}\right)^2 \times \left(\frac{3i\omega+1}{0.175i\omega+1}\right)^2\right|$ and $\left|\frac{2.1392\times10^4 i\omega+1}{0.0643i\omega+1}\right|$.

It is evident from Figure 3 that the approximations have three mid-range frequency intersections and exact low- and high-frequency correspondence, as stated earlier.

If now, as in Equation (16)

$$\chi(s) = \frac{(T_1 s+1)(T_2 s+1)}{(\tau_1 s+1)(\tau_2 s+1)}\left(\frac{C}{g}s+1\right) \tag{21}$$

is evaluated for the values given that $w(s)$ and $\zeta(s)$, in Equation (19), are fully defined and the outputs may now be computed from

$$\frac{Q_1(s)}{P_1(s)} = \frac{(\zeta(s)w(s)+R)}{(\zeta(s)(w(s)R+\zeta(s)))} \tag{22}$$

and

$$\frac{Q_2(s)}{P_1(s)} = \frac{(w^2(s)-1)^{1/2}}{(w(s)R+\zeta(s))}, \tag{23}$$

where in delay form $w(s) = (1+e^{-2l\alpha\chi(s)})/(1-e^{-2l\alpha\chi(s)})$ and

$$\hat{w}(s) = (w^2(s)-1)^{1/2} = \frac{2e^{-l\alpha x(s)}}{(1-e^{-2l\alpha x(s)})}.$$

Alternatively, commensurate with the pipeline system topology, Equations (22) and (23) may be written as

$$\frac{Q_1(s)}{P_1(s)} = \frac{(\zeta(s)w(s)+R)}{\zeta(s)(w(s)R+\zeta(s))} \tag{24}$$

and

$$\frac{Q_2(s)}{Q_1(s)} = \frac{\zeta(s)(w^2(s)-1)^{1/2}}{(\zeta(s)w(s)+R)}. \tag{25}$$

The block diagram for the representation given by Equations (24) and (25) is in a series form, as shown in Figure 4, and whereas Equations (22) and (23) provide the parallel, equivalent realisation, for this system model.

To simplify the simulation process it would be prudent to construct sub-system blocks for $w(s)$ and $\hat{w}(s)$. From Equation (17)

$$\chi(s) = as+b,$$

where the low-frequency approximation is

$$a = \frac{C}{g} + (T_1+T_2) - (\tau_1+\tau_2) \quad \text{and} \quad b = 1 \tag{26}$$

so that

$$w(s) = \frac{(1+e^{-2l\alpha(as+b)})}{(1-e^{-2l\alpha(as+b)})}. \tag{27}$$

Following a step input change on the $w(s)$ sub-system, it is easy to show that

$$\frac{w(s)}{s} = \frac{1}{s}\left(1+2\sum_{n=1}^{\infty} e^{-2nl\alpha(as+b)}\right)$$

so that the output following any arbitrary finite input change would be stable since

$$b > 0.$$

From the geometry of the approximation given by Equation (8), given in Figure 1, evidently

$$(T_1+T_2) > (\tau_1+\tau_2)$$

so that

$$a > 0,$$

resulting in the finite time delay $e^{-2l\alpha a}$.

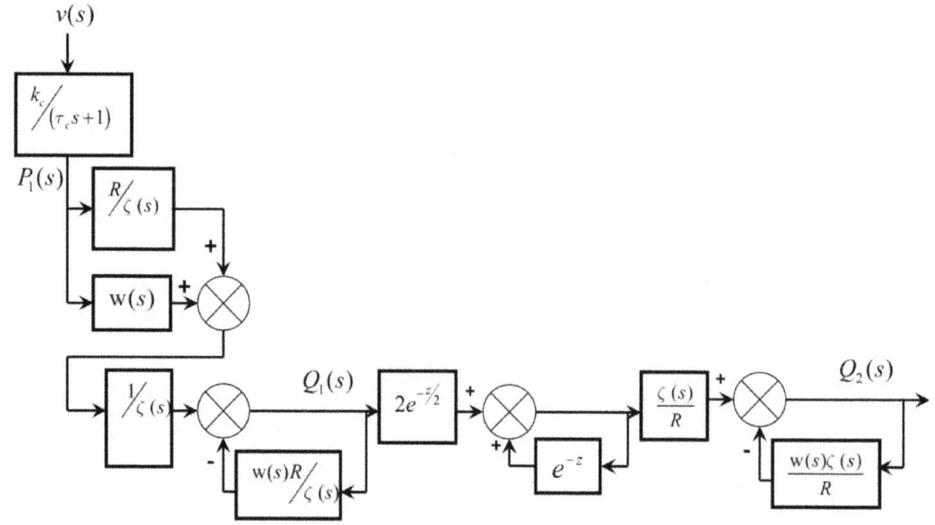

$$w(s)=\frac{\left(1+e^{-z(s)}\right)}{\left(1-e^{-z(s)}\right)},\ z(s)=-2l\alpha\,x(s),\ \alpha=\sqrt{(rg)},\ \bar{\alpha}=\sqrt{\frac{r}{g}},\ \zeta(s)=\bar{\alpha}\sqrt{\left[\frac{(Ts+1)}{(\tau s+1)}\right]}$$

Figure 4. Distributed–lumped parameter series representation block diagram for a pipeline and compressor, l = 1000, 5000 and 10,000 m.

Also, $\hat{w}(s)$ in delay form is

$$\hat{w}(s)=\frac{2\mathrm{e}^{-l\alpha(as+b)}}{(1-\mathrm{e}^{-2l\alpha(as+b)})} \tag{28}$$

then following a unit step change on this sub-system

$$\frac{\hat{w}(s)}{s}=\frac{1}{s}\sum_{n=1}^{\infty}\mathrm{e}^{-nl\alpha(as+b)}$$

and this produces a stable output response since again $b>0$ and $a>0$ gives a finite time delay $\mathrm{e}^{-nl\alpha a}$ and attenuation $\mathrm{e}^{-nl\alpha b}$, repectively.

In this case, when

$$r=0.6\times10^{-4}\,\mathrm{Nsec/m^5}$$

then following division of the low-frequency approximation from Equation (26) is

$$\chi(s)=1+1648.9s \quad \text{and}$$

$$\zeta(s)=\bar{\alpha}\left(\frac{1666.6s+1}{20s+1}\right)\left(\frac{2.5s+1}{0.166s+1}\right).$$

As the pipeline length varies, the characteristic impedance $\zeta(s)$, the exit resistance R and the compressor model remain invariant. Consequently, only $w(s)$ and $\hat{w}(s)$ need to be adjusted in the simulation model to obtain the gas flow characteristics for any length of pipeline with the same diameter and per unit length resistance values. In this regard, substituting for $\chi(s)$ in the equations for $w(s)$ and $\hat{w}(s)$ are given by Equations (27) and (28).

Hence for the 1000 m pipeline

$$w(s)=\frac{(1+0.8555\mathrm{e}^{-77.0627s})}{(1-0.8555\mathrm{e}^{-77.0627s})},$$

$$\hat{w}(s)=\frac{1.8508\mathrm{e}^{-38.5813s}}{(1-0.8555\mathrm{e}^{-77.0627s})},$$

for the 5000 m pipeline

$$w(s)=\frac{(1+0.4609\mathrm{e}^{-385.3s})}{(1-0.4609\mathrm{e}^{-385.3s})},$$

$$\hat{w}(s)=\frac{1.3578\mathrm{e}^{-192.5794s}}{(1-0.4609\mathrm{e}^{-385.3s})}$$

and for the 10,000 m pipeline

$$w(s)=\frac{(1+0.2125\mathrm{e}^{-770.62s})}{(1-0.2125\mathrm{e}^{-770.62s})},$$

$$\hat{w}(s)=\frac{0.9218\mathrm{e}^{-385.15s}}{(1-0.2125\mathrm{e}^{-770.62s})}.$$

The distributed–lumped parameter block representation for the series configuration including the compressor unit can be given by

$$\frac{P_1(s)}{v(s)}=\frac{k_c}{\tau_c s+1} \text{ is shown in Figure 4.}$$

Following unit step changes on the compressor motor voltage input, the changes in the volume flow at $Q_1(t)$ and $Q_2(t)$ are shown in Figure 5, in dotted and bold lines, respectively. This figure is initially for a 1000 m long, 1.0 m diameter, distributed parameter pipeline model, with $r=0.6\times10^{-4}\,\mathrm{N\,s/m^5}$ and $g=10^{-4}\,\mathrm{m^5/N\,s}$, with all the

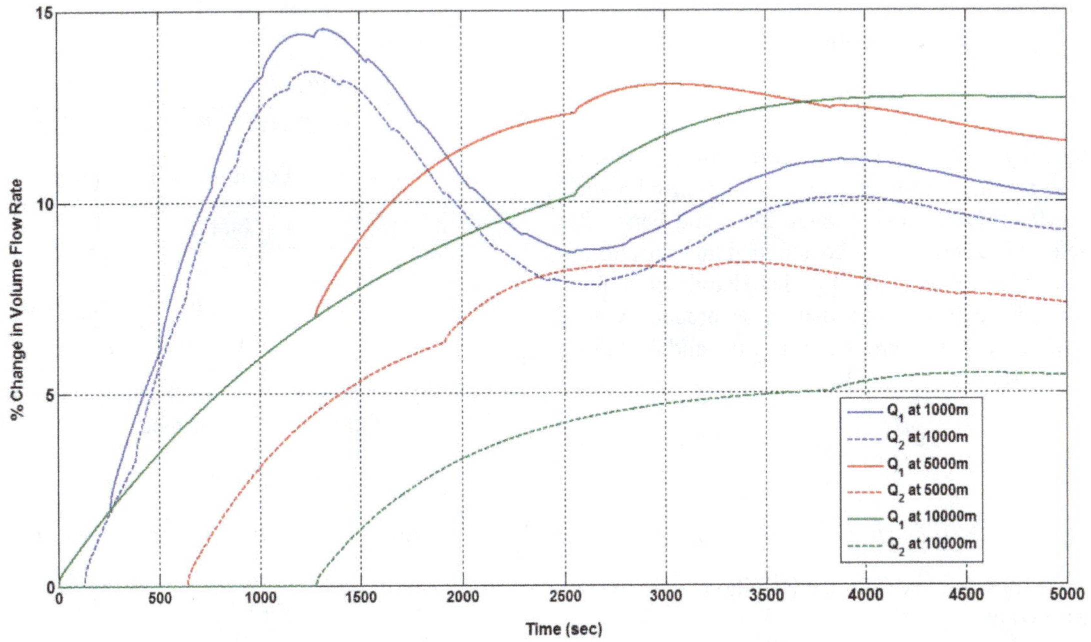

Figure 5. Percentage changes in inlet and exit pipeline volume flow rates for 1 m diameter gas pipeline with an exit resistance of 1 N s/m^5.

remaining parameters given earlier. Upon increasing the pipeline length to 5000 m and then 10,000 m the responses, following a 1% change in the compressor motor voltage, are also shown in Figure 5. Once again dotted and bold traces for the inlet and exit volume flow transients, respectively, are employed.

8. Gas flow transportation and energy requirement

The steady-state energy required to transport the input gas flow to the pipeline exit is proportional to $Q_1^2(0) - Q_2^2(0)$. This energy difference between the input and output gas flow rate increases with pipeline roughness and discontinuities in the form of connections, bends and branches and with the gas, volume flow delivery rate. These aspects are also instrumental in increasing turbulence, noise, uneven gas flow and, hence, pipeline erosion.

These damaging prospects are exacerbated in that a larger capacity compressor and electrical drives are then required to maintain constant gas flows dissipating thereby increasing electrical drive and energy consumption. Consequently, frictional losses are minimised by the use of smooth bore pipelines which may be lined.

To further reduce the pipeline energy losses, of interest here is the relationship between the pipeline characteristic impedance and the exit resistance R. With appropriate "matching" between R and $\zeta(0)$ the possibility of minimising the input–output energy dissipation difference remains a tempting proposition.

Since the direct application of minimisation analysis would result in negative and/or infinite exit resistances, the problem requires the employment of constrained minimisation (Barnett, 1992). To achieve this, the use of Hamiltonian function and Lagrange multiplier for the pipeline model is proposed.

9. Constrained steady-state optimisation

From Equation (3) the squared, steady-state difference in the volume gas flow is given by

$$\frac{(\zeta(0)w(0)+R)^2-\zeta^2(0)\hat{w}^2(0)}{(\zeta(0)+w(0)R)^2\zeta^2(0)} = Q_1^2(0) - Q_2^2(0) = J, \tag{29}$$

where in Equation (29)

$$w(0) = \frac{(e^{2l\alpha\chi(0)}+1)}{(e^{2l\alpha\chi(0)}-1)} \quad \text{and} \quad \hat{w}(0) = \frac{2e^{l\alpha\chi(0)}}{(e^{2l\alpha\chi(0)}-1)},$$

and the performance index $J > 0$, since $\zeta(0), R, w(0)$ and $\hat{w}(0) > 0$ and $w(0) > \hat{w}(0)$.

Minimising the squared, steady-state volume gas flow difference would also minimise the squared, steady-state input and exit gas velocity difference which is proportional to the energy dissipation in the delivery process. Minimising this difference also minimises the continuous, input–exit volume gas stream expansion or compression effects. Consequently, these changes would be instrumental in minimising erosion, vibration, noise and turbulence within the pipeline system.

Essentially, the principal energy dissipation effects in this horizontal, constant-diameter gas pipeline transportation system arise from the pipeline series and shunt frictional resistance and conductance, r and g, respectively, and from the exit resistance R. Since r and g arise from the pipeline characteristics, the constraint adopted here relates to the pipeline roughness parameter $\zeta(0)$ and exit resistance R, via the simple constraint equation

$$R = k\zeta(0). \tag{30}$$

Consequently, if the squared volume flow difference between the pipeline input and exit is minimised with respect to $\zeta(0)$, then a suitable value for k and hence R could be selected to achieve this constrained minimisation.

From Equations (29) and (30) the Hamiltonian for this system can be formed, so that in accordance with (Hodder, 2008) the supremum value for $\zeta(0)$ and R could be determined from

$$\frac{\partial H}{\partial \zeta(0)} = 0 \tag{31}$$

$$\text{and } \frac{\partial H}{\partial R} = 0, \tag{32}$$

where in Equations (31) and (32), the Hamiltonian function H is represented by

$$H = \frac{(\zeta(0)w(0) + R)^2 - \zeta^2(0)\hat{w}^2(0)}{(\zeta(0) + w(0)R)^2\zeta^2(0)} + (R - k\zeta(0))\lambda,$$

where λ is the Lagrange multiplier associated with the constraint, given by Equation (30). Upon evaluating Equations (31) and (32)

$$\begin{aligned}\frac{\partial H}{\partial \zeta(0)} &= \frac{2[(\zeta(0)w(0) + R)w(0) - \zeta(0)\hat{w}^2(0)]}{(\zeta(0) + w(0)R)^2\zeta^2(0)} \\ &- \frac{2[(\zeta(0)w(0) + R)^2 - \zeta^2(0)\hat{w}^2(0)]}{(\zeta(0) + w(0)R)^3\zeta^2(0)} \\ &- \frac{2[(\zeta(0)w(0) + R)^2 w(0) - \zeta(0)\hat{w}^2(0)]}{(\zeta(0) + w(0)R)^2\zeta^3(0)} \\ &- \lambda k = 0,\end{aligned} \tag{33}$$

$$\begin{aligned}\frac{\partial H}{\partial R} &= \frac{2[(\zeta(0)w(0) + R)]}{(\zeta(0) + w(0)R)^2\zeta^2(0)} \\ &- \frac{2w(0)[(\zeta(0)w(0) + R)^2 - \zeta^2(0)\hat{w}^2(0)]}{(\zeta(0) + w(0)R)^3\zeta^2(0)} \\ &+ \lambda = 0.\end{aligned} \tag{34}$$

In Equations (33) and (34) if

$$x = \frac{2[(\zeta(0)w(0) + R)]}{(\zeta(0) + w(0)R)^2\zeta^2(0)}$$

$$\text{and } y = \frac{2[(\zeta(0)w(0) + R)^2 - \zeta^2(0)\hat{w}^2(0)]}{(\zeta(0) + w(0)R)^3\zeta^2(0)}$$

then Equations (33) and (34) are simply

$$xw(0) - y - y\frac{(\zeta(0) + w(0)R)}{\zeta(0)} = k\lambda, \tag{35}$$

$$\text{and } x - yw(0) = -\lambda. \tag{36}$$

Hence, from Equations (35) and (36) following the elimination of λ

$$\frac{x}{y} = \frac{[(\zeta(0)w(0) + R)(\zeta(0) + w(0)R)]}{(\zeta(0)w(0) + R)^2 - \zeta^2(0)\hat{w}^2(0)}. \tag{37}$$

Since, $R = k\zeta(0)$ Equation (37) becomes

$$\begin{aligned}&\frac{[\zeta^2(0)(w(0) + k)(1 + w(0)k)]}{(w(0) + k)^2\zeta^2(0) - \zeta^2(0)\hat{w}^2(0)} \\ &= \frac{2\zeta(0) + w(0)\zeta(0)k + w(0)\zeta(0)k}{\zeta(0)(w(0) + k)}.\end{aligned} \tag{38}$$

Finally Equation (38) yields the simple optimum, least volume flow difference relationships of

$$k = \sqrt{2}\hat{w}(0) - w(0). \tag{39}$$

From Equation (39), the optimum value of R is given by

$$R = (\sqrt{2}\hat{w}(0) - w(0))\zeta(0) \tag{40}$$

guaranteeing $R > 0$.

10. Steady-state volume flow difference

If from Equations (22) and (23), the final values of $Q_1(0)$ and $Q_2(0)$, for a step change on $P_1(s)$, $(P_1(s) = 1/s, s \to 0)$, are given by

$$Q_1(0) = \frac{(\zeta(0)w(0) + R)}{\zeta(0)\,(\zeta(0) + w(0)R)} \tag{41}$$

$$\text{and } Q_2(0) = \frac{\hat{w}(0)}{(\zeta(0) + w(0)R)} \tag{42}$$

then the steady-state volume flow difference constrained minimum for $(Q_1^2(0) - Q_2^2(0))$ would occur with

$$R = (\sqrt{2}\hat{w}(0) - w(0))\zeta(0)$$

and from Equations (41) and (42)

$$Q_1(0) = \frac{\sqrt{2}\hat{w}(0)}{[(1 - w^2(0)) + \sqrt{2}w(0)\hat{w}(0)]\zeta(0)} \tag{43}$$

$$\text{and } Q_2(0) = \frac{\hat{w}(0)}{[(1 - w^2(0)) + \sqrt{2}w(0)\hat{w}(0)]\zeta(0)}. \tag{44}$$

Consequently

$$Q_1(0) - Q_2(0) = \frac{(\sqrt{2} - 1)\hat{w}(0)}{[(1 - w^2(0)) + \sqrt{2}w(0)\hat{w}(0)]\zeta(0)}. \tag{45}$$

Then, from Equations (43) and (44)

$$Q_2(0) = \frac{1}{\sqrt{2}}Q_1(0). \tag{46}$$

Hence, with this output resistance, $R = (\sqrt{2}\hat{w}(0) - w(0))\zeta(0)$, the steady-state volume output flow, following input changes on $P_1(s)$, always increases to 0.7071 of the input volume flow, irrespective of the pipeline roughness characteristics, r and g. This is an important result.

From Equations (43) and (44) it is also evident that

$$Q_1^2(0) - Q_2^2(0) = \frac{\hat{w}(0)}{\zeta^2(0)[(1 - w^2(0)) + \sqrt{2}w(0)\hat{w}(0)]}$$

so that $Q_1^2(0) = 2Q_2^2(0)$.

Hence, the efficiency of the delivery with $Q_2(0) = \frac{1}{\sqrt{2}}Q_1(0)$, and $R = (\sqrt{2}\hat{w}(0) - w(0))\zeta(0)$, with the constraint that $R = k\zeta(0), k > 0$, rises to 50%.

Higher delivery efficiencies than these are achievable with increasing values of R. However, lower steady-state rates of volume flow, for $Q_2(0)$, would be experienced with $Q_2(0)$ falling to zero, as R is increased to large values.

11. Application study

For the 1000 m long, 1.0 m diameter pipeline considered earlier, determine the optimum value of the pipeline exit resistance R which would deliver a volume flow of $Q_2(0) = 0.7071Q_1(0)$, minimising the input–output volume flow difference, with the constraint $R = k\zeta(0), k > 0$.

The pipeline series and shunt resistance and conductance values here, as stated earlier, are $r = 0.6 \times 10^{-4}$ N s/m^5 and $g = 10^{-4}$ m^5/N s, respectively. Consequently

$$\bar{\alpha} = \zeta(0) = \sqrt{\frac{r}{g}} = 0.7745$$

$$\text{and } \alpha = \sqrt{rg} = 0.7745 \times 10^{-4},$$

$$w(0) = \frac{(1 + e^{-2l\alpha})}{(1 - e^{-2l\alpha})} = 12.9374$$

$$\text{and } \hat{w}(0) = \frac{2e^{-l\alpha}}{(1 - e^{-2l\alpha})} = 12.8987.$$

Hence, from Equation (39) the optimum output resistance is

$$R = (\sqrt{2}\hat{w}(0) - w(0))\zeta(0) = 4.0817$$

and from the constraint, given by Equation (30)

$$k = \frac{R}{\zeta(0)} = 5.2701 \quad \text{and} \quad \tan^{-1} k \cong 80°.$$

This result can be validated graphically by plotting the performance index J against $\zeta(0)$. The least value of the J contours which contact the constraint line, angle 80°, gives the minimum input–output volume flow which in this case is $J = 0.0147$. Figure 6 shows the constant J contours occupy the first quadrant of the $R - \zeta$ plane, $(R, \zeta) > 0$, with the $J = \infty$ contour mapping into the origin. The 80° constraint line is tangential to $J = 0.0147$, as predicted and this is the least value of J commensurate with the constraint requirement, implied by Equation (30).

12. Energy consumption efficiency

The block diagram representation for this system is shown in Figure 4. The transient response for the system following a 1% change on the compressor motor voltage is as shown earlier, for the 1000 m pipeline with $R = 1$, with $Q_1(t)$ and $Q_2(t)$ increasing to their maximum steady-state values in approximately 6000 s.

It is very easy to obtain the proportional steady-state energy dissipation curves from this simulation, for a range

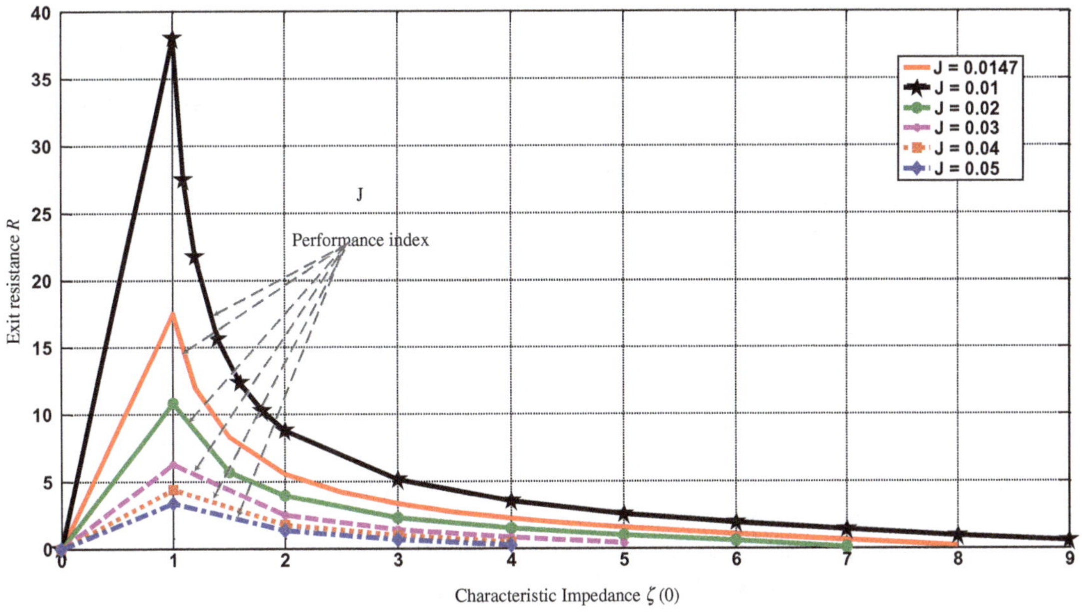

Figure 6. Constant performance index J contours against characteristic impedance $\zeta(0)$ and exit resistance R.

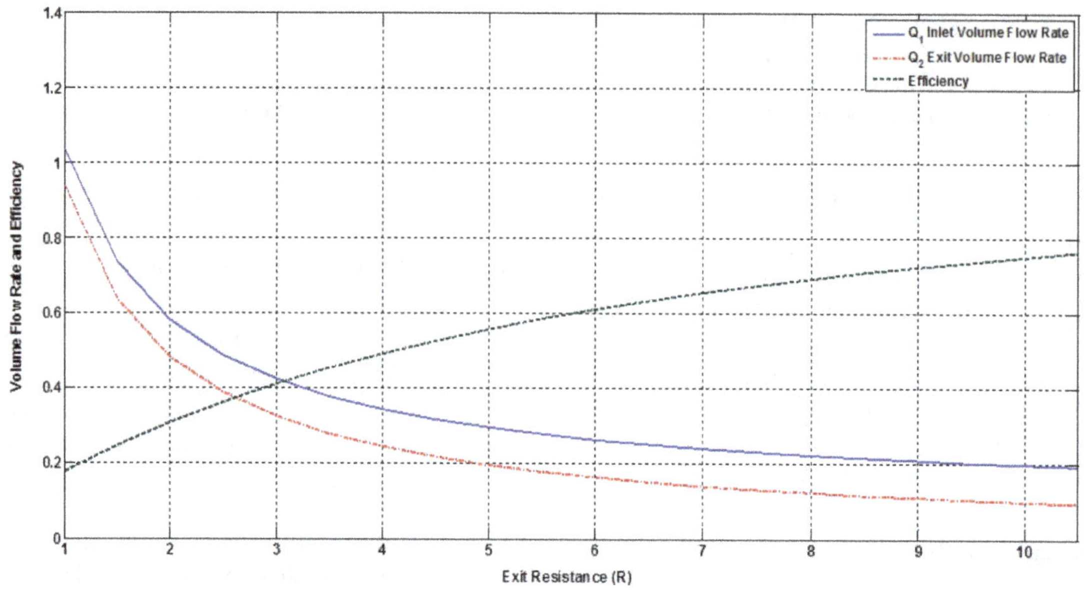

Figure 7. Inlet and exit proportional volume flow rates and transportation energy efficiency against exit resistance R.

of exit resistances R. Figure 7 shows the steady-state performance $Q_1(t), Q_2(t)$ and the steady-state delivery curves for this application.

The steady-state energy consumption efficiency is proportional to $[(Q_1^2(0) - Q_2^2(0))/Q_1^2(0)]$. This equation indicates that either a large volume gas flow, $Q_2(0)$, can be delivered with low energy efficiency or small volume gas flow can be delivered with high energy efficiency. From Section 10, the squared flow difference $(Q_1^2(0) - Q_2^2(0))$ with the constraint $R = k\zeta(0)$ ensures that this can always be achieved with an efficiency of 50% whilst guaranteeing a high delivery load of $Q_2(0) = 0.7071Q_1(0)$.

This delivery load, volume flow, is independent of the pipeline roughness.

13. Conclusion

In this contribution, the theory for modelling the gas flow in long pipelines was presented. It was shown that accurate, unambiguous, simple models could be easily constructed for pipeline–compressor configurations with the model-simulation block diagram requiring no more than four basic sub-system models, which would include the compressor model and for the pipeline functions of $w(s)$, $(w^2(s) - 1)^{1/2}$ and $\zeta(s)$ for the complete system representation.

The distributed parameter theory involved was also uncomplicated with the incorporation of the continuous series and shunt resistances and the gas stream equivalent, capacitance and inductance, energy storage effects. Consequently, the perturbed transient volume flow responses following any input variation are readily available from the theoretical development and simulation process established herein.

The compact volume flow response functions derived invite the optimisation analysis of Sections 8–11. This advancement considered the constrained minimum energy dissipation problem via impedance matching. By selecting the constraint $R = k\zeta(0)$, the analysis restricts the identification of the output valve exit resistance R to practical realisable values.

To do this, the Hamiltonian function for the system and a Lagrange multiplier were employed. This led to the sample elegant solution relating the exit resistance to the pipeline characteristics with the final result showing that

$$Q_2(0) = 0.7071Q_1(0)$$

with the inclusion of the optimum exit resistance of

$$R = (\sqrt{2}\hat{w}(0) - w(0))\zeta(0).$$

Importantly, the absolute pipeline roughness values would not be required to evaluate this resistance.

This exit resistance for the 1000 m pipeline under consideration would give the smoothest exit flow with the least energy dissipation for this steady-state exit volume flow rate.

References

Barnett, S. (1992). *Matrices, methods and applications*. Oxford: Clarendon Press.

Bartlett, H., & Whalley, R. (1995). Gas flow in pipes and tunnels. *Proceedings of the Institution of Mechanical Engineers, Part I, 209* (6), 41–52.

Bartlett, H., & Whalley, R. (1998). Analogue solution to the modelling and simulation of distributed-lumped parameter systems. *Proceedings of the Institution of Mechanical Engineers, Part I, 212* (12), 99–114.

Bradie, B. (2006). *Numerical analysis*. New Jersey: Pearson Int.

Brown, F. T. (1962). The transient response of fluid lines. *ASME Journal of Basic Engineering, 15*, 547–553.

Eckman, D. P. (1958). *Automatic process control*. London: J. Wiley.

Esahanian, Y., & Behbani-Nejad, M. (2002). Reduced order modelling of unsteady flows about complex configurations using boundary element method. *Journal of Fluid Engineering, Transactions of ASME, 124*, 988–993.

Gautam, R. (2009). *Intricacies of design of a gas pipeline*. 6th Progress on Oil and Gas Transportation (PETROFED), New Delhi.

Hodder, R. (2008, February). *Screw compressors*. Compressor and Optimisation Conference, paper 15, Aberdeen.

Iberall, A. S. (1960). Attenuation of oscillatory pressures in instrument lines. *Journal of Research, National Bureau of Standards, 4*, pp 2115.

Langill, A. W. (1965). *Automatic control system engineering*. New Jersey: Prentice-Hall.

Nichols, N. B. (1961, May). *The linear properties of pneumatic transmission lines*. Instrument Society of America Joint Automatic Control Conference, Boulder.

Palm, W. J. (2005). *System dynamics*. New York, NY: McGraw-Hill.

Robertson, J. A., & Crowe, C. T. (1990). *Engineering fluid mechanics*. Boston, MA: Houghton Miffin.

Rogers, G. F. C., & Mayhew, Y. R. (1970). *Engineering thermodynamics, work and heat transfer*. London: Longmans Green.

Rosenbrock, H. H. (1974). *Computer aided control system design*. London: Academic Press.

Schwartz, R., & Friedland, B. (1965). *Linear systems*. New York, NY: McGraw-Hill.

Spiegel, M. R. (1965). *Laplace transforms*. New York: Schaum Pub.

Takahashi, Y. (1970). *Control and dynamics systems*. New York: Addison-Wesley.

Whalley, R. (1988). The response of distributed-lumped parameter systems. *Proceedings of the Institution of Mechanical Engineers, Part C, 202* (66), 421–429.

Whalley, R. (1990). Interconnected spatially distributed systems. *Transactions of the Institute of Measurement and Control, 12* (5), 260–271.

Whalley, R., & Abdul-Ameer, A. (2009). The computation of torsional, dynamic stresses. *Proceedings of the Institution of Mechanical Engineers, Journal of Mechanical Engineering Science, Part C, 223*, 1799–1814.

Appendix 1. Derivation of the input–output matrix for a distributed parameter element

Acoustic approximations for first-order, small-amplitude Navier–Stokes equations with zero bulk viscosity and neglecting radiant heat transfer were derived in Iberall (1960) and Nichols (1961).

If in addition to these constraints axial heat transfer is negligible and the pressure over any cross-section is assumed to be uniform, then in a continuous medium for small-amplitude disturbances with rigid, constant temperature walls the longitudinal dynamics for long pipelines may be considered via the elemental, infinitesimal, per unit length model shown in Figure A1.

These constraints are adopted and employed in Brown (1962), Bartlett and Whalley (1998), and Bartlett and Whalley (1995). This enables analysis procedures based on partial differential calculus

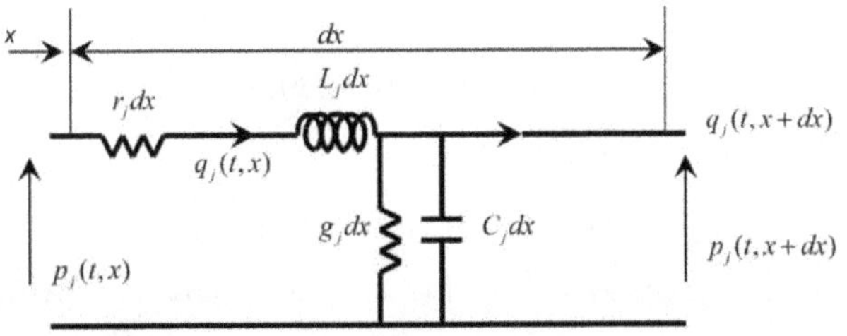

Figure A1. Incremental, distributed gas glow element.

where for the element shown in Figure A1.

$$p(t,x+\mathrm{d}x) - p(t,x) = -\left(L\frac{\partial q}{\partial t}(t,x) + rq(t,x)\right)\mathrm{d}x. \quad \text{(A1)}$$

Also

$$q(t,x+\mathrm{d}x) - q(t,x) = -\left(C\frac{\partial p}{\partial t}(t,x) + gp(t,x)\right)\mathrm{d}x. \quad \text{(A2)}$$

By dividing by dx and taking the limit as $\mathrm{d}x \to 0$ Equations (A1) and (A2) become

$$\frac{\partial p}{\partial x}(t,x) = -L\frac{\partial q}{\partial t}(t,x) - rq(t,x) \quad \text{(A3)}$$

and

$$\frac{\partial q}{\partial x}(t,x) = -C\frac{\partial p}{\partial t}(t,x) - gp(t,x). \quad \text{(A4)}$$

Taking Laplace transformations with respect to time with zero initial conditions allows Equations (A3) and (A4) to be written as

$$\frac{\mathrm{d}P}{\mathrm{d}x} = -(Ls+r)Q, \quad \text{(A5)}$$

$$\frac{\mathrm{d}Q}{\mathrm{d}x} = -(Cs+g)P, \quad \text{(A6)}$$

where $Q = Q(s,x)$ and $P = P(s,x)$. Differentiating Equations (A5) and (A6) with respect to x gives

$$\frac{\mathrm{d}^2Q}{\mathrm{d}x^2} = -(Cs+g)\frac{\mathrm{d}P}{\mathrm{d}x} \quad \text{and} \quad \frac{\mathrm{d}^2P}{\mathrm{d}x^2} = -(Ls+r)\frac{\mathrm{d}Q}{\mathrm{d}x}.$$

Hence

$$\frac{\mathrm{d}^2Q}{\mathrm{d}x^2} - (Ls+r)(Cs+g)Q = 0 \quad \text{(A7)}$$

and

$$\frac{\mathrm{d}^2P}{\mathrm{d}x^2} = -(Ls+r)\frac{\mathrm{d}Q}{\mathrm{d}x} = (Ls+r)(Cs+g)P.$$

Hence

$$\frac{\mathrm{d}^2P}{\mathrm{d}x^2} - (Ls+r)(Cs+g)P = 0. \quad \text{(A8)}$$

Consequently, as Equations (A7) and (A8) show, the homogeneous output and input equations are of the same form and have identical, general solutions. If now a propagation function is defined as

$$\Gamma(s) = [(Ls + r)(Cs + g)]^{1/2}$$

then the solutions required are

$$P(s,x) = A\cosh\Gamma(s)x + B\sinh\Gamma(s)x, \quad \text{(A9)}$$

$$Q(s,x) = \bar{C}\sinh\Gamma(s)x + D\cosh\Gamma(s)x \quad \text{(A10)}$$

then when $x = 0$

$$A = p(s,0),$$
$$D = q(s,0).$$

Differentiating Equations (A9) and (A10) with respect to x and equating to (A5) and (A6) gives

$$-(Ls + r)Q(s,x) = A\Gamma(s)\sinh\Gamma(s)x + B\Gamma(s)\cosh\Gamma(s)x, \quad \text{(A11)}$$

$$-(Cs + g)P(s,x) = \bar{C}\Gamma(s)\cosh\Gamma(s)x + D\Gamma(s)\sinh\Gamma(s)x. \quad \text{(A12)}$$

Again by setting $x = 0$ from Equation (A11)

$$B = -\frac{(Ls + r)}{\Gamma(s)}Q(s,0) = -\sqrt{\frac{(Ls + r)}{(Cs + g)}}Q(s,0) \quad \text{(A13)}$$

and from Equation (A12)

$$\bar{C} = -\frac{(Cs + g)}{\Gamma(s0)}P(s,0) = -\sqrt{\frac{(Cs + g)}{(Ls + r)}}P(s,0) \quad \text{(A14)}$$

or, from Equations (A13) and (A14):

$$B = -\zeta(s)Q(s,0) \quad \text{and} \quad \bar{C} = -\zeta^{-1}(s)P(s,0),$$

where $\zeta(s) = \sqrt{\frac{(Ls+r)}{(Cs+g)}}$ is the characteristic impedance.

With this notation Equations (A9) and (A10) become

$$P(s,x) = \cosh\Gamma(s)x\,P(s,0) - \zeta(s)\sinh\Gamma(s)x\,Q(s,0),$$

$$Q(s,x) = -\zeta^{-1}(s)\sinh\Gamma(s)x\,P(s,0) + \cosh\Gamma(s)x\,Q(s,0).$$

Consequently, the output at l is given by

$$\begin{bmatrix} P(s,l) \\ Q(s,l) \end{bmatrix} = \begin{bmatrix} \cosh\Gamma(s)l & -\zeta(s)\sinh\Gamma(s)l \\ -\zeta^{-1}(s)\sinh\Gamma(s)l & \cosh\Gamma(s)l \end{bmatrix} \begin{bmatrix} P(s,0) \\ Q(s,0) \end{bmatrix}. \quad \text{(A15)}$$

Equation (A15) can be arranged in impedance form with:

$$\text{ctnh}\Gamma(s)l = \frac{(e^{2\Gamma(s)l} + 1)}{(e^{2\Gamma(s)l} - 1)} = w(s)$$

$$\text{and } \text{csch}\Gamma(s)l = (w^2(s) - 1)^{1/2}.$$

With this notation Equation (A15) becomes for the jth distributed section of pipeline

$$\begin{bmatrix} P_j(s,l_j) \\ P_{j+1}(s,l_{j+1}) \end{bmatrix} = \begin{bmatrix} \zeta_j(s)w_j(s) & -\zeta_j(s)(w_j^2(s) - 1)^{1/2} \\ \zeta_j(s)(w_j^2(s) - 1)^{1/2} & -\zeta_j(s)w_j(s) \end{bmatrix} \times \begin{bmatrix} Q_j(s,l_j) \\ Q_{j+1}(s,l_{j+1}) \end{bmatrix} \quad \text{(A16)}$$

which completes the analysis.

22

Online suboptimal control of linearized models

V. Costanza* and P.S. Rivadeneira

"Grupo de Sistemas No Lineales", INTEC-Facultad de Ingeniería Química (UNL-CONICET), Güemes 3450, 3000 Santa Fe, Argentina

A novel approach to approximately solving the restricted-control linear quadratic regulator problem online is substantiated and applied in two case studies. The first example is a one-dimensional system whose exact solution is known. The other one refers to the temperature control of a metallic strip at the exit of a multi-stand rolling mill. The new (online-feedback) strategy employs a convenient version of the gradient method, where partial derivatives of the cost are taken with respect to the final penalization matrix coefficients and to the switching times where the control (de)saturates. The calculations are based on exact algebraic formula, which do not involve trajectory simulations, and so reducing in principle the computational effort associated with receding horizon or nonlinear programming methods.

Keywords: optimal control; restricted controls; linear quadratic regulator problem; online optimization

1. Introduction

Linearized models are frequently employed to treat nonlinear systems evolving near an equilibrium point, or when tracking a reference trajectory. These approximate models are accepted provided both deviations (of the state from the given target and of the manipulated variable from the reference control) are 'small'. Therefore, restrictions on the control values appear naturally when working with linearizations. In most cases it is implicitly assumed that the linear approximation is BIBO stable and controllable as to guarantee that if the control moves between the imposed bounds, then the states will depart within tolerances from the target (and tend asymptotically to it). When the problem concerning an n-dimensional system and an additive cost objective is regular, i.e. when the Hamiltonian of the problem can be uniquely optimized by a control value u^0 depending continuously on the remaining variables (t, x, λ), then a set of $2n$ ordinary differential equations (ODEs) with two-point boundary-value conditions, known as Hamilton (or Hamiltonian) canonical equations (HCEs), has to be solved to obtain the optimal solution. For the linear-quadratic regulator (LQR) with a finite horizon, there exist well-known methods (see for instance Costanza & Neuman, 2009; Sontag, 1998) to transform the boundary-value problem into an initial-value one. In the infinite-horizon, bilinear-quadratic regulator, and change of set-point servo problems, there also exists an attempt to find the missing initial condition for the co-state variable from the data of each particular problem, which allows to integrate the Hamiltonian equations online with the underlying control process (Costanza & Neuman, 2006). For nonlinear systems this line of work is in its beginnings (Costanza & Rivadeneira, 2008, Costanza, Rivadeneira, & Spies, 2009).

Whenever an optimal performance is desired, the bounded-control context may lead to non-regular optimal control problems, for whose solution there are not standard recipes (Athans & Falb, 2006; Qin & Badgwell, 2003; Sontag, 1998; Speyer & Jacobson, 2010). Since the early 1960s, the Pontryagin's maximum principle (PMP) has been at the core of the development of modern optimal control theory (Pontryagin, Boltyanskii, Gamkrelidze, & Mishchenko, 1964) to treat non-regular situations. This paper takes advantage of the relationships between PMP and the Hamilton–Jacobi approaches to the LQR problem. The decisive theoretical finding may be phrased as follows: the optimal solution to a given restricted LQR problem can be generated by saturating the solution of another unrestricted LQR problem, with the same dynamics and cost objective as the original one, but starting at a different initial condition and subject to a quadratic final penalization with a different matrix coefficient. Offline and online schemes were developed to detect this new initial condition and final penalization matrix (Costanza, Rivadeneira, & González, 2014). An online algorithm in this direction is the main contribution of this article from the practical point of view. Since the strategy is intended to work in feedback form when the control is between bounds, then the precise knowledge of the initial condition of the subjacent unrestricted LQR process is not substantial. In fact, in such a context what becomes a priority is the location of

*Corresponding author. Email: tsinoli@santafe-conicet.gov.ar

the time instants where the control saturates/desaturates. Therefore, the proposed scheme updates the final penalization matrix and the saturation/desaturation times referred above, while the total cost is being reduced *via* a straightforward version of the gradient method (Pardalos & Pytlak, 2008). The resulting control will in general be suboptimal with respect to the given initial state, since: (i) the eventual occurrence of state perturbations which deviate the system from the optimal trajectory of the original problem and also that (ii) the number of online updating of the parameters, each consuming some computer time, may not be sufficient to reach their optimal values during the available time horizon. The numerical scheme takes advantage of the online availability of the Riccati matrices that correspond to a range of final penalty parameter values, generated from the solutions to a pair of first-order partial differential equations (PDE) (Costanza & Rivadeneira, 2013). For its simplicity and small computational effort, the online algorithm can be considered as a potential tool to be used in combination with the receding or shrinking horizon policies (Camacho & Bordons, 2004), in an enlarged MPC context that would contemplate strict finite-horizon problems, and may be considered as an alternative to nonlinear programming approaches (Cannon, Liao, & Kouvaritakis, 2008, Rao et al., 2008), which depend on the time and space discretization adopted.

The article has the following structure: after the Introduction, the regular LQR results and the auxiliary matrices that will be used in the sequel are presented. Then the bounded-control version of the problem is described and the main theoretical fact used here is stated. Afterwards the algebraic formulas to be employed in the numerical updating of the parameters are explicitly given. Two applications of the numerical scheme to linearized systems are then illustrated, one of them arising from an industrial problem. The usual Conclusions are given at the end.

2. Equations for regular LQR optimal control problems

The finite-horizon, time-constant formulation of the LQR problem with free final states and unconstrained controls attempts to minimize the (quadratic) cost

$$\mathcal{J}(u) = \int_0^{t_f} [x'(\tau)Qx(\tau) + u'(\tau)Ru(\tau)]\,d\tau + x'(t_f)Sx(t_f), \quad (1)$$

with respect to all the admissible (here piecewise-continuous) control trajectories $u : [0, t_f] \to \mathbb{R}^m$ of duration t_f, applied to some fixed, finite-dimensional, deterministic plant. Then control strategies affect the $\mathbb{R}^n$-valued states x through some initialized, autonomous, dynamical constraint

$$\dot{x} = Ax + Bu := f(x, u), \quad x(0) = x_0 \neq 0. \quad (2)$$

This will be called a $(A, B, Q, R, S, t_f, \mathbb{R}^m, x_0)$-problem.

The (real, time-constant) matrices in Equations (1) and (2) will be assumed to have the following properties: Q and S are positive-semidefinite $n \times n$ matrices, R is $m \times m$ and positive-definite, A is $n \times n$, B is $n \times m$, and the pair (A, B) is controllable. The expression under the integral is usually known as the 'Lagrangian' L of the cost, namely

$$L(x, u) := x'Qx + u'Ru. \quad (3)$$

Under these conditions the Hamiltonian of the problem, namely the $\mathbb{R}^n \times \mathbb{R}^n \times \mathbb{R}^m \to \mathbb{R}$ function defined by

$$H(x, \lambda, u) := L(x, u) + \lambda' f(x, u) \quad (4)$$

is known to be regular, i.e. that H is uniquely minimized with respect to u, and this occurs when u takes the explicit control value

$$u^0(x, \lambda) = -\tfrac{1}{2}R^{-1}B'\lambda, \quad (5)$$

(in this case, independently of x), which is usually called 'the H-minimal control'. The 'Hamiltonian' form of the problem (see, for instance, Sontag, 1998) requires then to solve the two-point boundary-value problem for the HCEs

$$\dot{x} = H^0_\lambda(x, \lambda), \quad x(0) = x_0, \quad (6)$$

$$\dot{\lambda} = -H^0_x(x, \lambda), \quad \lambda(t_f) = 2Sx(t_f), \quad (7)$$

where $H^0(x, \lambda)$, usually called the minimized (or control) Hamiltonian, stands for

$$H^0(x, \lambda) := H(x, \lambda, u^0(x, \lambda)), \quad (8)$$

and H^0_λ and H^0_x for the column vectors with i-components $\partial H^0/\partial \lambda_i$, $\partial H^0/\partial x_i$ respectively, i.e. Equations (6) and (7) here take the form

$$\begin{aligned} \dot{x} &= Ax - \tfrac{1}{2}W\lambda, \\ \dot{\lambda} &= -2Qx - A'\lambda, \end{aligned} \quad (9)$$

respectively, with $W := BR^{-1}B'$. It is well known that the solution to the unrestricted regular problem, as posed above, relies in turn on the solution $P(\cdot)$ to the Riccati differential equation (RDE)

$$\dot{P} = PWP - PA - A'P - Q, P(t_f) = S, \quad (10)$$

which establishes a useful relationship between the optimal state $x^*(\cdot)$ and the costate $\lambda^*(\cdot)$ trajectories, namely

$$\lambda^*(t) = 2P(t)x^*(t), \quad (11)$$

and, based on Equation (5), leads to the optimal control trajectory

$$u^*(t) = u^0(x^*(t), \lambda^*(t)) = -R^{-1}B'P(t)x^*(t), \quad (12)$$

or equivalently to the optimal feedback law

$$u_f(t, x) = -R^{-1}B'P(t)x. \quad (13)$$

When the control values are restricted, the global regularity of the Hamiltonian cannot be assured, and therefore the search for the optimal control strategy becomes more involved, as may be observed in the following sections. Additional relevant objects from the LQR theory will be used in the sequel, for instance, the matrices $\alpha(T,S), \beta(T,S)$, solutions to the following pair of first-order, quasilinear PDE (see Costanza & Neuman, 2009 for details):

$$\alpha_T - \alpha_S M = -\alpha N, \quad \alpha(0,S) = I, \tag{14}$$

$$\beta_T - \beta_S M = -\beta N, \quad \beta(0,S) = 2S, \tag{15}$$

where the subindices denote partial derivatives and the matrix coefficients result

$$M := A'S + SA + Q - SWS, \tag{16}$$

$$N := A - WS. \tag{17}$$

These matrices allow us to calculate, for any unbounded LQR problem, the solution $P(\cdot, t_f, S)$ to its RDE through the formula

$$P(t,t_f,S) = \tfrac{1}{2}\beta(t_f - t, S)[\alpha(t_f - t, S)]^{-1} \quad \forall\, t \in [0,t_f], \tag{18}$$

and in such a case the matrices α and β are also related to the boundary conditions by the following relations:

$$x(0) = \alpha(t_f,S)x(t_f), \lambda(0) = \beta(t_f,S)x(t_f). \tag{19}$$

3. The bounded-control case

The manipulated variable in most of the control systems appearing in practical applications can assume only a bounded set of values. The term 'manipulated' indicates that a person or an instrument assigns a value to a signal generated by physical means, and therefore this value cannot take more than a physically realizable amount. Commonly, the manipulated variable can move inside and on the boundary of some bounded subset of a metric space, then it is natural to assume that the admissible set of control values is a compact subset of $\mathbb{R}$. The qualitative features of optimal control solutions to bounded problems are significantly different from those of unbounded ones (Pontryagin et al., 1964). But questions about how much they actually differ, which classes of problems lead to bang–bang controls, and whether their solutions are just saturations of the optimal trajectories of unbounded problems, are still open. Linearizations are accepted provided fluctuations are small. This principle affects state and control deviations. Particularly, the speed variations cannot be allowed to trespass appropriate bounds, i.e.

$$u(t) \in \mathbb{U} := [u_{\min}, u_{\max}]. \tag{20}$$

This restriction determines the structure of the optimal control problem. When $\mathbb{U}$ is bounded (and closed, as in Equation (20)), then the problem tends to lose regularity; derivatives of the Hamiltonian and the value function are not even guaranteed to exist. The search for solutions to restricted problems most frequently falls in the domains of the Pontryagin's maximum principle (PMP; Pontryagin et al., 1964). However, even when solved, PMP is not flexible enough to treat state perturbations: no optimal feedback laws arise from the application of PMP equations, but only open-loop control strategies.

In this paper the following result (Costanza & Rivadeneira, 2013) will be exploited:

Let us assume that there exists a time instant $t \in (0, t_f)$ where $u^*_{x_0}(t) \in (u_{\min}, u_{\max})$. Then there exists a time interval $I \subset (0, t_f)$ containing t such that the optimal phase trajectory $\{x^*_{x_0}, \lambda^*_{x_0}\}$ of the original $(A, B, Q, R, S, t_f, \mathbb{U}, x_0)$-problem coincides with the optimal phase trajectory $\{\hat{x}, \hat{\lambda}\}$ corresponding to a $(A, B, Q, R, \hat{S}, t_f, \mathbb{R}, \hat{x}_0)$-problem.

In what follows, it will be assumed that there exists just one maximal 'regular' interval $(\tau_1, \tau_2) \subset (0, t_f)$ where the control takes values in $(u_{\min}, u_{\max})$. The generation of a suboptimal control strategy by approximating the unknown parameters $(\hat{S}, \hat{x}_0)$ has already been published in the minimal-control-energy situation (Costanza et al., 2014), and the extension to the general LQR problem was announced in Costanza and Rivadeneira (2013), although there the updating of the unknown parameters still required the simulation of state trajectories. In this paper, explicit algebraic formulas are given for parameters updating, avoiding ODE integrations, and thus reducing the computer time in the process of decreasing the total cost.

3.1. *Algebraic formulas used in the online procedure*

3.1.1. *Auxiliary objects*

The following type of feedback control laws will be frequently used in the sequel leading online to a suboptimal control:

$$\tilde{u}(t) := \begin{cases} u_{\min}, & \forall\, t \in [0, \tau_1), \\ -R^{-1}B'P(t,\tilde{S})x(t), & \forall\, t \in [\tau_1, \tau_2), \\ u_{\max}, & \forall\, t \in [\tau_2, t_f], \end{cases} \tag{21}$$

where $\tilde{u}(t)$ is a short notation for $\tilde{u}_{\tilde{S},\tau_1,\tau_2}(t)$, which will be used when it is necessary to indicate that the feedback law is associated to the parameters $(\tilde{S}, \tau_1, \tau_2)$. The 'seed' strategy will be adopted to start the online iterative method below, and has the same structure, namely

$$u_{\text{seed}}(t) := \begin{cases} u_{\min} & \text{if } -R^{-1}B'P(t,S)x(t) \le u_{\min}, \\ u_{\max} & \text{if } -R^{-1}B'P(t,S)x(t) \ge u_{\max}, \\ -R^{-1}B'P(t,S)x(t) & \text{otherwise.} \end{cases} \tag{22}$$

The state trajectory corresponding to the control u_{seed} and starting at x_0, i.e. $x_{u_{\text{seed}}}$, will be denoted as x_{seed}. Note that through the seed control and state trajectories, simulated offline for the nominal final matrix S, the first values for the saturation times, denoted $\tau_{1,0} \le \tau_{2,0}$, are detected if they exist. The initial approximation to the unknown penalization matrix $\hat{S}$ will be denoted $\tilde{S}_0 := S$, and consistently for the control, $\tilde{u}_0 := u_{\text{seed}}$. The Hamiltonian matrix of the original problem,

$$\mathbf{H} := \begin{pmatrix} A & -\frac{W}{2} \\ -2Q & -A' \end{pmatrix}, \tag{23}$$

and the associated fundamental matrix

$$\mathbf{V}(t) := \mathrm{e}^{\mathbf{H}t}, \tag{24}$$

will also be employed in devising algebraic formula for the partial derivatives of the cost. The matrix $\mathbf{V}(t)$ is $2n \times 2n$. For convenience, its $n \times n$ partition is denoted as

$$\begin{pmatrix} \mathbf{V}_1(t) & \mathbf{V}_2(t) \\ \mathbf{V}_3(t) & \mathbf{V}_4(t) \end{pmatrix} := \mathbf{V}(t), \tag{25}$$

which can also be expressed in terms of the auxiliary matrices α, β, and their derivatives since, Costanza and Neuman (2009)

$$\mathbf{V} = (\mathrm{e}^{\mathbf{H}T})^{-1} = \begin{pmatrix} \alpha - \alpha_S S & \alpha_S/2 \\ \beta - \beta_S S & \beta_S/2 \end{pmatrix}^{-1}. \tag{26}$$

In what follows $P(t, \tilde{S})$ will denote the solution to the RDE (10) for π, with final condition $\pi(t_f) = \tilde{S}$. When the value of $\tilde{S}$ is clear in the text, the notation may simplify from $P(t, \tilde{S})$ to $P(t)$. The following identity will also be used:

$$\begin{aligned} \frac{\partial P(t,\tilde{S})}{\partial \tilde{S}} &= \frac{\partial[(1/2)\beta(t_f - t, \tilde{S})[\alpha(t_f - t, \tilde{S})]^{-1}]}{\partial \tilde{S}} \\ &= \tfrac{1}{2}[\beta_S \alpha^{-1} - \beta\alpha^{-1}\alpha_S\alpha^{-1}](t_f - t, \tilde{S}) \\ &= \tfrac{1}{2}[\beta_S - 2P(t,\tilde{S})\alpha_S]\alpha^{-1}. \end{aligned} \tag{27}$$

The 'saturated' fundamental matrix:

$$\Psi(t, \tau) := \int_\tau^t \mathrm{e}^{A(t-\sigma)}\,\mathrm{d}\sigma = \mathrm{e}^{At}\int_\tau^t \mathrm{e}^{-A\sigma}\,\mathrm{d}\sigma, \tag{28}$$

and the related matrices

$$\check{\Psi}(t, \tau) := \int_\tau^t \mathrm{e}^{A'(\sigma-\tau)} Q\, \mathrm{e}^{A(\sigma-\tau)}\,\mathrm{d}\sigma, \tag{29}$$

$$\hat{\Psi}(t, \tau) := \int_\tau^t \Psi'(\sigma, \tau) Q\, \mathrm{e}^{A(\sigma-\tau)}\,\mathrm{d}\sigma \tag{30}$$

will also be needed in the sequel. These matrices are calculated and interpolated offline. So it is assumed that, in real-time applications, they will be available as functions of their two variables (t, τ), in the range $[0, t_f] \times [0, t_f]$.

Since $u(t) \equiv u_{\min}$ in $[0, \tau_1]$, then the state at time τ_1 results

$$x(\tau_1) = \mathrm{e}^{At}x_0 + \Psi(\tau_1, 0)Bu_{\min}. \tag{31}$$

Here it is important to note that the state at τ_2 may be calculated in two equivalent ways: (i) either from the Hamiltonian flow, since for each $t \in [\tau_1, \tau_2)$ the control is $-R^{-1}B'P(t,\tilde{S})x(t)$ and the costate (corresponding to this piece of a regular trajectory), denoted $\tilde{\lambda}(t)$, is $\tilde{\lambda}(t) = 2P(t,\tilde{S})x(t)$, and therefore

$$\begin{pmatrix} x(t) \\ \tilde{\lambda}(t) \end{pmatrix} = \mathbf{V}(t-\tau_1)\begin{pmatrix} x(\tau_1) \\ \tilde{\lambda}(\tau_1) \end{pmatrix}, \tag{32}$$

implying that

$$\begin{aligned} x(\tau_2) &= \mathbf{V}_1(\tau_2 - \tau_1)x(\tau_1) + \mathbf{V}_2(\tau_2 - \tau_1)\tilde{\lambda}(\tau_1) \\ &= (\mathbf{V}_1(\tau_2 - \tau_1) + 2\mathbf{V}_2(\tau_2 - \tau_1)P(\tau_1, \tilde{S}))x(\tau_1), \end{aligned} \tag{33}$$

or (ii) $x(t)$ should also coincide with the state of some process having the same final penalization $\tilde{S}$ and starting at some initial state $\tilde{x}(0)$, i.e.

$$\begin{pmatrix} x(t) \\ \tilde{\lambda}(t) \end{pmatrix} = \mathbf{V}(t)\begin{pmatrix} \tilde{x}(0) \\ 2P(0,\tilde{S})\tilde{x}(0) \end{pmatrix}, \tag{34}$$

$$x(\tau_2) = (\mathbf{V}_1(\tau_2) + 2\mathbf{V}_2(\tau_2)P(0,\tilde{S}))\tilde{x}(0). \tag{35}$$

In what follows, the arguments t of $\mathbf{V}(t)$ will also be omitted when the context is clear. The state at the final time t_f is

$$x(t_f) = \mathrm{e}^{A(t_f-\tau_2)}x(\tau_2) + \Psi(t_f, \tau_2)Bu_{\max}. \tag{36}$$

3.1.2. *The partial derivatives of the cost*

It is known (Dhamo & Tröltzsch, 2011) that the total cost $\mathcal{J}(\tilde{u})$ is differentiable as a function of the variables $(\tilde{S}, \tau_1, \tau_2)$. The total cost is time-partitioned here for convenience:

$$J(\tilde{S}, \tau_1, \tau_2) := \mathcal{J}(\tilde{u}) := J_1 + J_2 + J_3 + J_4, \tag{37}$$

$$J_1 := \int_0^{\tau_1} L(x_{\tilde{u}}(t), \tilde{u}(t))\,\mathrm{d}t = Ru_{\min}^2\tau_1 + \int_0^{\tau_1} x'(t)Qx(t)\,\mathrm{d}t, \tag{38}$$

$$J_2 := \int_{\tau_1}^{\tau_2} L\,\mathrm{d}t = x'(\tau_1)P(\tau_1,\tilde{S})x(\tau_1) - x'(\tau_2)P(\tau_2,\tilde{S})x(\tau_2), \tag{39}$$

$$J_3 := \int_{\tau_2}^{t_f} L\,\mathrm{d}t = Ru_{\max}^2(t_f - \tau_2) + \int_{\tau_2}^{t_f} x'(t)Qx(t)\,\mathrm{d}t, \tag{40}$$

$$J_4 := x'(t_f)Sx(t_f). \tag{41}$$

Based on the preliminary formulas, the partial derivatives of each partial cost can be expressed as

$$D_{\tau_1}J_1 = Ru_{\min}^2 + x'(\tau_1)Qx(\tau_1), \tag{42}$$

where $x(\tau_1)$ and $x'(\tau_1)$ should be replaced by their corresponding expressions from Equation (31), as in all succeeding partial derivatives. Next,

$$D_{\tau_1}J_2 = 2x'(\tau_1)P(\tau_1,\tilde{S})\dot{x}(\tau_1) + x'(\tau_1)\dot{P}(\tau_1,\tilde{S})x(\tau_1) - \cdots - \frac{\partial}{\partial\tau_1}(x'(\tau_2)P(\tau_2,\tilde{S})x(\tau_2)) \quad (43)$$

$$= 2x'(\tau_1)P(\tau_1)[Ax(\tau_1) + Bu_{\min}] + \cdots + x'(\tau_1)[P(\tau_1)WP(\tau_1) - P(\tau_1)A - A'P(\tau_1) - Q]x(\tau_1), \quad (44)$$

since $(\partial/\partial\tau_1)[x'(\tau_2)P(\tau_2)x(\tau_2)] = 2x'(\tau_2)P(\tau_2)(\partial x(\tau_2)/\partial\tau_1)$, and, from Equation (35), $\partial x(\tau_2)/\partial\tau_1 = 0$.

$$D_{\tau_1}J_3 = D_{\tau_1}\int_{\tau_2}^{t_f} x'(t)Qx(t)\,dt = \int_{\tau_2}^{t_f} 2x'(t)Q\frac{\partial x(t)}{\partial\tau_1}\,dt = 0, \quad (45)$$

since inside the integral of the state $x(t) = e^{A(t-\tau_2)}x(\tau_2) + \Psi(t_f,\tau_2)Bu_{\max}$, $\partial x(t)/\partial\tau_1 = e^{A(t-\tau_2)}\ \partial x(\tau_2)/\partial\tau_1$, and $\partial x(\tau_2)/\partial\tau_1 = 0$.

$$D_{\tau_1}J_4 = 2x'(t_f)S\frac{\partial x(t_f)}{\partial\tau_1} = 2x'(t_f)S\,e^{A(t_f-\tau_2)}\frac{\partial x(\tau_2)}{\partial\tau_1} = 0. \quad (46)$$

Similarly, the derivatives with respect to τ_2 are

$$D_{\tau_2}J_1 = 0, \quad (47)$$

$$D_{\tau_2}J_2 = -[2x'(\tau_2)P(\tau_2,\tilde{S})\dot{x}(\tau_2) + x'(\tau_2)\dot{P}(\tau_2)x(\tau_2)] = -\{2x'(\tau_2)P(\tau_2)(Ax(\tau_2) + Bu_{\max}) + \cdots + x'(\tau_2)P(\tau_2)WP(\tau_2) - P(\tau_2)A - A'P(\tau_2) - Q)x(\tau_2)\}, \quad (48)$$

where Equations (33) and (31) should be used to replace $x(\tau_2)$, as in the next partial derivative:

$$D_{\tau_2}J_3 = -[Ru_{\max}^2 + x'(\tau_2)Qx(\tau_2)], \quad (49)$$

$$D_{\tau_2}J_4 = 2x'(t_f)S\frac{\partial x(t_f)}{\partial\tau_2} = -2x'(t_f)S\,e^{A(t_f-\tau_2)}[WP(\tau_2,\tilde{S})x(\tau_2) + Bu_{\max}], \quad (50)$$

since, from Equation (36) $\partial x(t_f)/\partial\tau_2 = -A\,e^{A(t_f-\tau_2)}x(\tau_2) + e^{A(t_f-\tau_2)}\dot{x}(\tau_2) - e^{A(t_f-\tau_2)}Bu_{\max}$, and $\dot{x}(\tau_2) = (A - WP(\tau_2,\tilde{S}))x(\tau_2)$. The derivatives with respect to the elements $\tilde{S}_{ij}$ of $\tilde{S}$, globally denoted as $D_{\tilde{S}}$, are

$$D_{\tilde{S}}J_1 = 0, \quad (51)$$

$$D_{\tilde{S}}J_2 = x'(\tau_1)\left[\frac{\partial P(\tau_1)}{\partial\tilde{S}} - 4(\mathbf{V}_1 + 2\mathbf{V}_2P(\tau_1))'P(\tau_2)\mathbf{V}_2\frac{\partial P(\tau_1)}{\partial\tilde{S}} - \cdots - (\mathbf{V}_1 + 2\mathbf{V}_2P(\tau_1))'\frac{\partial P(\tau_2)}{\partial\tilde{S}}(\mathbf{V}_1 + 2\mathbf{V}_2P(\tau_1))\right]x(\tau_1), \quad (52)$$

$$D_{\tilde{S}}J_3 = \int_{\tau_2}^{t_f} 2x'(t)Q\frac{\partial x(t)}{\partial\tilde{S}}\,dt = 4[x'(\tau_2)\check{\Psi}(t_f,\tau_2) + u_{\max}B'\hat{\Psi}(t_f,\tau_2)]\mathbf{V}_2(\tau_2-\tau_1)\frac{\partial P(\tau_1,\tilde{S})}{\partial\tilde{S}}x(\tau_1), \quad (53)$$

after replacing $x'(t) = x'(\tau_2)\,e^{A'(t-\tau_2)} + B'\Psi'(t,\tau_2)u_{\max}$, and expanding $\partial x(t)/\partial\tilde{S} = e^{A(t-\tau_2)}(\partial x(\tau_2)/\partial\tilde{S}) = e^{A(t-\tau_2)}2\mathbf{V}_2(\tau_2-\tau_1)(\partial P(\tau_1,\tilde{S})/\partial\tilde{S})x(\tau_1)$. Finally, after similar manipulations,

$$D_{\tilde{S}}J_4 = 2x'(t_f)S\frac{\partial x(t_f)}{\partial\tilde{S}} = 4x'(t_f)S\,e^{A(t_f-\tau_2)}\mathbf{V}_2(\tau_2-\tau_1)\frac{\partial P(\tau_1,\tilde{S})}{\partial\tilde{S}}x(\tau_1). \quad (54)$$

3.1.3. *Updating the parameters*

First approximations $\tau_{1,0}$ and $\tau_{2,0}$ to the optimal saturation points τ_1 and τ_2 become available after (offline) simulating the state trajectory x_{seed}. A subdivision of the time-horizon in 'sampling times' of the form $t_0 = 0 < t_1 < t_2 < \cdots < t_N = T$ is adopted to make possible intermediate calculations, updating parameters, and deciding changes in the control strategy. For $t \in [0, t_1]$ the control is set to

$$u(t) \equiv u_{\text{seed}}(t). \quad (55)$$

Then, during this initial sampling interval (associated with $k = 0$) and through, the parameters $(\tilde{S}, \tau_1, \tau_2)$ are updated to construct successive control strategies $\tilde{u}_{k,j}$, $j = 1, 2, \ldots$that decrease the value of the total cost:

$$\mathcal{J}(\tilde{u}_{k,j+1}) \leq \mathcal{J}(\tilde{u}_{k,j}) \leq \cdots \mathcal{J}(u_{\text{seed}}), \quad j = 1, 2, \ldots \quad (56)$$

according to the prescriptions of the simplest gradient method:

$$\tilde{S}_{k,j} := \tilde{S}_{k,j-1} - \gamma_S\frac{\partial J}{\partial\tilde{S}}(\tilde{S}_{k,j-1}, \tau_{1,k}, \tau_{2,k}), \quad j = 1, 2, \ldots. \quad (57)$$

The last updating of $\tilde{S}_k$ that can be computed during the sampling interval is denoted

$$\tilde{S}_{k+1} \approx \lim_j \tilde{S}_{k,j}, \quad (58)$$

and on the same lines,

$$\tau_{i,k+1} \approx \lim_j\left(\tau_{i,k} - \gamma_{\tau_i}\frac{\partial J}{\partial\tau_i}\right)_j, \quad i = 1, 2, \quad (59)$$

where $\gamma_S, \gamma_{\tau_1}, \gamma_{\tau_2}$ are appropriate constants, tuned by the user for each experiment. During the next sampling interval $(t_{k+1}, t_{k+2}]$, the control is set to

$$u(t) \equiv \tilde{u}_{\tilde{S}_{k+1},\tau_{1,k+1},\tau_{2,k+1}}(t). \quad (60)$$

4. Applications and numerical results

4.1. *A one-dimensional example*

The first case study of this paper is the optimal control problem defined by the following objects:

$$\dot{x}(t) = u(t), \quad 0 \leq t \leq 1,$$
$$x(0) = 1 - e,$$
$$u(t) \in [1.44, 2] \subset \mathbb{R}, \tag{61}$$
$$J(u) = \int_0^1 (x^2 + u^2)\mathrm{d}t + 13\,[x(1)]^2 .$$

The optimal solution is given in Costanza and Rivadeneira (2013). A slightly different version of this problem, aimed to minimize only the control energy, was presented in Troutman (1996). For any real constant C, this system can also be considered as a linearized version of the dynamics

$$\dot{z}(t) = U(t) + C, \tag{62}$$

with state $z(t)$, control $U(t)$, steady-state $z_{SS}(t) \equiv 0$, equilibrium control $U_{SS}(t) = -C$, and deviations $x(t) := z(t) - z_{SS}(t), u(t) := U(t) - U_{SS}(t)$.

The gradient method was first tried offline (without discretizing the time-horizon into sampling periods), to obtain: $\tilde{S} = 9.60305$, $\tau_1 = 0.09289$, $\tau_2 = 0.62095$, $J_{\text{off}} = 3.73091 \approx J^* = 3.7309$ (Costanza & Rivadeneira, 2013). It was also implemented online, as described above, by adopting a fixed sampling period $\Delta t_k = t_{k+1} - t_k = 0.025$ and allowing for a maximum of 30 iterations inside each Δt_k. The total cost obtained was $J_{\text{on}} = 3.73123$, slightly higher than $J_{\text{off}} \approx J^*$. The resulting online control trajectory, and the evolution of the required parameters $\tau_1, \tau_2, \tilde{S}$, updated at each sampling-time, are depicted in Figures 1 and 2, respectively.

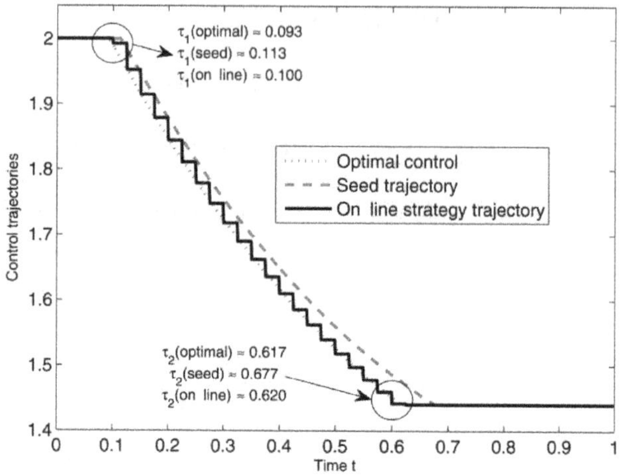

Figure 1. Control strategy resulting from the online application of the gradient method to the one-dimensional example.

4.2. *A typical linearized model situation: the rolling mill*

4.2.1. *The nonlinear first-order PDE setup*

The second case study models a rolling mill described in Hearns and Grimble (2010), whose (infinite dimensional) dynamics (from a standard energy balance) obeys the following first-order PDE

$$\frac{\partial \theta}{\partial t} = -V\frac{\partial \theta}{\partial z} + a(\theta_{\text{a}} - \theta) + b(\theta_{\text{a}}^4 - \theta^4), \tag{63}$$

where $\theta(t, z)$ is the temperature of the metallic strip at time t and location z in the trend, $V(t)$ is the linear speed of the strip, and θ_{a} is the ambient temperature (assumed constant in this set up). The coefficients a, b weigh the rate of heating due to conduction and radiation, respectively. The system

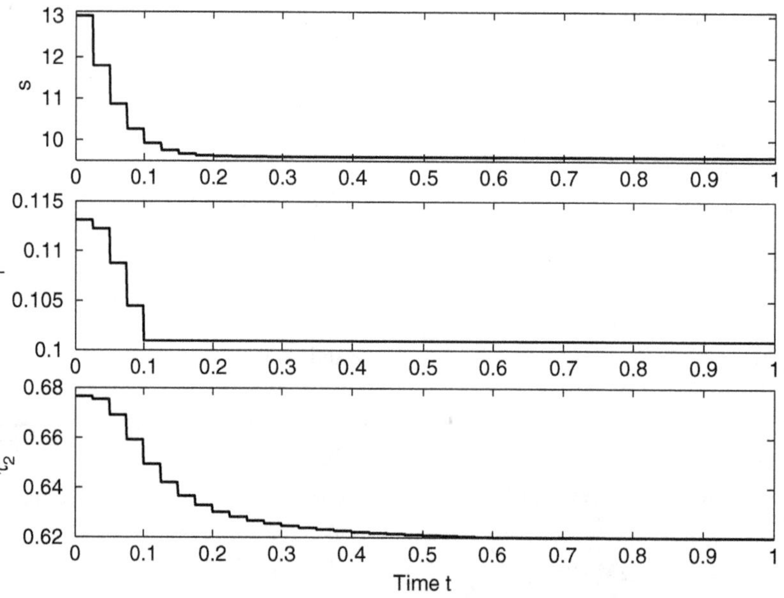

Figure 2. Evolution of the parameters $\tilde{S}, \tau_1, \tau_2$, during the online application of the gradient method.

is simplified by neglecting radiation (small b), and by supposing that the temperature will stay around the equilibrium profile

$$\theta_{SS}(z) = \theta_a + (\theta_0 - \theta_a)\exp\left(-\frac{az}{V_0}\right), \tag{64}$$

which is the solution to Equation (63) with $b = 0$, $\partial\theta/\partial t = 0$, $V(t) \equiv V_0$, and $\theta_0 := \theta_{SS}(0)$, some appropriate constant characterizing each physical set up. The following definitions

$$\Delta\theta(t,z) := \theta(t,z) - \theta_{SS}(z), \quad u(t) := V(t) - V_0 \tag{65}$$

allow one to approximately express the dynamics of the fluctuations through the 'linearized' version of Equation (63), namely

$$\frac{\partial\Delta\theta}{\partial t} = -V_0\frac{\partial\Delta\theta}{\partial z} - a\Delta\theta + \left[\frac{a}{V_0}(\theta_0 - \theta_a)\exp\left(-\frac{az}{V_0}\right)\right]u, \tag{66}$$

after neglecting the term $u\partial\Delta\theta/\partial z$, on the argument that it is the product of two 'small' quantities.

Remark 1 The linear approximation of the original problem implicitly supposes not only that deviations $\Delta\theta(t,z)$ from the steady-state temperature $\theta_{SS}(z)$ are small, but also that the manipulated variable $V(t)$ will also be near the steady-state velocity V_0. These assumptions imply, especially, that the control variable $u(t)$ should not be allowed to take unbounded values, even if their physical realization are possible, because the linearized model will risk to depart too much from the original dynamics.

4.2.2. *The z-discretization approach leading to a finite-dimensional linear control system*

From the control theory perspective, the state in Equation (66) is at each time t the z-function $\Delta\theta(t,\cdot)$. This, in principle, makes the system under study infinite-dimensional, whose treatment is out of the scope of this paper. An n-dimensional approximation has then been constructed by discretizing the z-variable in the form:

$$z_i := (i-1)h, \quad i = 1,\ldots,n, \tag{67}$$

next by defining n state variables x_i (or equivalently a vector state variable $x(\cdot)$ with values $x(t)$ in $\mathbb{R}^n$),

$$x_i(t) := \Delta\theta(t,z_i), \quad i = 1,\ldots,n, \tag{68}$$

$$x(t) := (x_1(t), x_2(t), \ldots, x_n(t))', \tag{69}$$

and finally by approximating the z-partial derivative by some appropriate linear combination of the function $\Delta\theta(t,\cdot)$ evaluated at the discretized values z_i, for instance,

$$\frac{\partial\Delta\theta}{\partial z}(t,z_i) \approx \frac{x_{i+1}(t) - x_i(t)}{h}, \quad i = 1,\ldots,n-1, \tag{70}$$

$$\frac{\partial\Delta\theta}{\partial z}(t,z_n) \approx \frac{x_n(t) - x_{n-1}(t)}{h}. \tag{71}$$

After such manipulations the following structure of a linear control system is obtained

$$\dot{x} = Ax + Bu, \tag{72}$$

where the $n \times n$ matrix A and the column n-vector B take the form:

$$A = (a_{ij}) : \begin{cases} a_{ii} = \dfrac{V_0}{h} - a, & a_{i,i+1} = -\dfrac{V_0}{h}, \\ & i = 1,\ldots,n-1, \\ a_{n,n-1} = \dfrac{V_0}{h}, & a_{nn} = -\left(a + \dfrac{V_0}{h}\right), \\ \text{all remaining} & \\ \text{elements equal to 0,} & \end{cases} \tag{73}$$

$$B = (b_i) = \frac{a}{V_0}(\theta_0 - \theta_a)\exp\left(-\frac{az_i}{V_0}\right), \quad i = 1,\ldots,n. \tag{74}$$

The eigenvalues of the matrix A are dominated by the relation between the heat gained at each position by convection versus the heat extracted at that point by the environment, implicit in the term $V_0/h - a$ which appears in the main diagonal, except in its last element. Now, from one side, the free evolution has to be stable to keep any physical meaning in the equations (the temperature cannot grow forever). But, if control has to be relevant to increase stability, it is appropriate to explore those situations near where the system might lose stability (for instance, due to environmental perturbations). With this contradictory objectives in mind, the following values for the parameters were investigated

$$V_0 = h = 1, \quad a = 1.001. \tag{75}$$

The discretized, ODE version (72) of Equation (66) was numerically confirmed to be an acceptable approximation.

4.2.3. *Numerical simulation of the online strategy*

The initial state $x_0 = x(0)$ used for simulation of the system defined by Equations (72)–(74) was

$$x_i(0) = 100\sin\left(\frac{2\pi z_i}{10}\right), \quad i = 1,\ldots,n, \tag{76}$$

with the following values for the reference temperatures (in °C)

$$\theta_a = 20, \quad \theta_0 = 700. \tag{77}$$

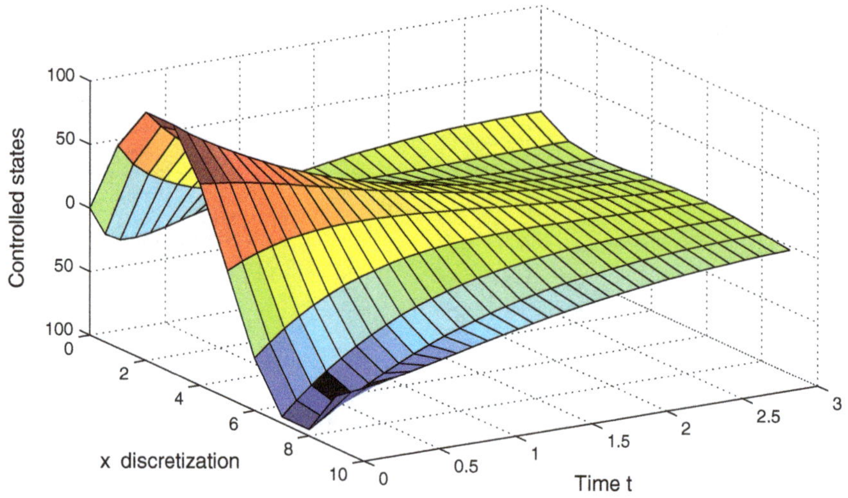

Figure 3. Evolution of the states under the online control strategy, after a sinusoidal initial profile.

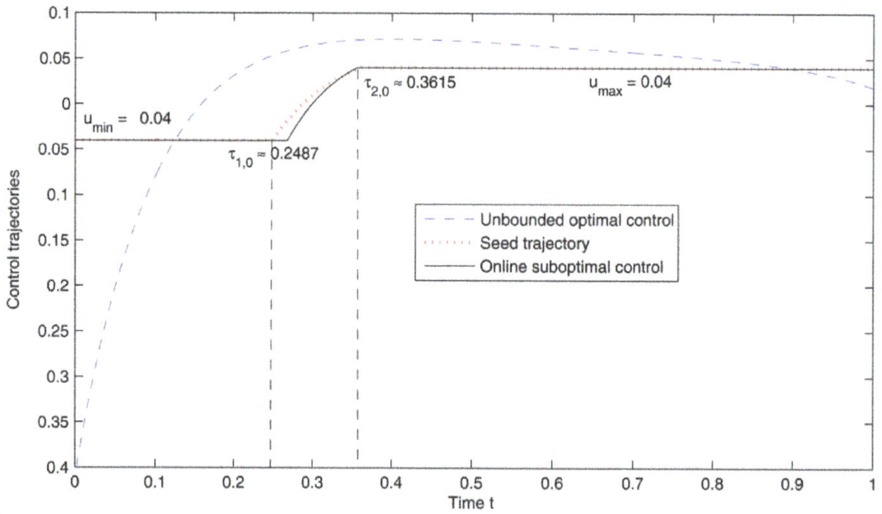

Figure 4. Unbounded optimal control, restricted seed control, and the resulting suboptimal control after applying the gradient method.

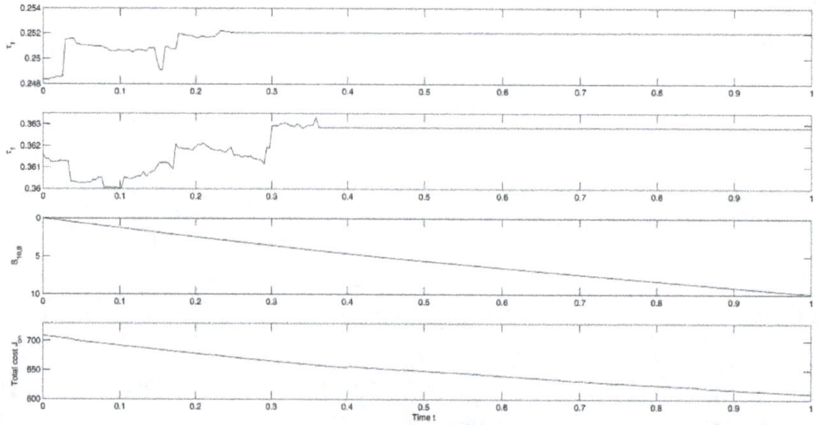

Figure 5. Evolution of some updated parameters and their corresponding total costs.

The cost objective of the LQR problem was given the following parameters:

$$t_f = 1, \quad Q = 0.1 I_{10}, \quad R = 500, \quad S = 50 I_{10}, \qquad (78)$$

and the bounds imposed on control values were

$$[u_{min}, u_{max}] = [-0.04, 0.04]. \qquad (79)$$

After simulating the seed control and state trajectories, it was found that

$$\tau_{1,0} = 0.2487, \quad \tau_{2,0} = 0.3615, \quad J_{seed} = 721. \qquad (80)$$

Results are shown in Figures 3–5. In Figure 3 the evolution of the states under control is shown. All the components of the state tend to equilibrium, as expected. The unbounded optimal control is shown in Figure 4, together with the seed control trajectory and the suboptimal bounded control resulting from applying the online strategy with fixed sampling intervals $\Delta t = 0.1$. The updating of the parameters τ_1 and τ_2, together with $\tilde{S}_{10,8} = \tilde{S}_{8,10}$ (just one non-diagonal coefficient of $\tilde{S}$ for illustration), and the evolution of the total cost associated with those parameters, are depicted in Figure 5. The relevant final values for the parameters were

$$\begin{aligned} \tau_1 &= 0.2590, \quad \tau_2 = 0.3623, \quad J_{on} = 601, \\ \mathrm{diag}(\tilde{S}) &= (50.0, \ldots, 49.9944, 53.5244, 68.3213), \\ \mathrm{subdiag}(\tilde{S}) &= (0, \ldots, -0.7454, 0.1887), \\ \mathrm{subsubdiag}(\tilde{S}) &= (0, \ldots, -10.5406). \end{aligned} \qquad (81)$$

The reduction in the total cost J_{on} with respect to the cost of the seed strategy J_{seed} was 16.78%, mostly due to the high values assigned to R and S in the original cost objective formulation.

5. Conclusions

An efficient online scheme to calculate suboptimal control strategies for linearized models has been presented. Linearized models are thought as approximate versions of real-life nonlinear systems. Linearizations are accurate provided deviations from equilibrium values are small. This context imposes restrictions on the control values, and then normally the Pontryagin approach is required to solve the bounded LQR problem, instead of the more agreeable Hamilton–Jacobi theory. Although feedback laws may be preferred in practice, when perturbations are expected to appear, a closed-loop control is in general suboptimal when there exist constraints in the manipulated variable. With these limitations in mind, an efficient online algorithm is devised to approximate the open-loop optimal control *via* feedback, based on recent theoretical results, and its features are illustrated when applied to two case-studies. The resulting strategies are quite different from the saturated form of the optimal control corresponding to the unrestricted problem with same parameters and initial condition, which is used here just as a first approximation, and so labelled as a 'seed' strategy. Such a seed feedback is often naively adopted in Engineering practice during the whole optimization period, although it has been shown that it is far from optimal. Reductions in total cost, in both examples, validate this assertion. It should be acknowledged that the alternative procedure proposed here will also be suboptimal. This is because the application of PMP to obtain the optimal solution is essentially an offline calculation leading to an open-loop recipe, and if any deviation from the optimal solution occurs (by mistake or by ignorance), then optimality will immediately be lost, no matter the subsequent effort. However, when the PMP solution was not previously found, or when only the 'seed' strategy is available, or when state perturbations appear in a real process-control situation; then no more than a suboptimal performance better than the seed's one can be expected. The online updating of the parameter $\tilde{S}$ and the saturation times τ_i, as long as the total cost is reduced (guaranteed by the gradient method), will clearly improve the seed strategy as time evolves. This new scheme will result in the optimal strategy only when: (i) the right (optimal) $\tilde{S}$ value is reached before the Riccati gain $P(\tilde{S})$ has to be applied, and (ii) no state perturbations occur. As a consequence, the stability of the method is guaranteed since the cost is not allowed to increase, and it is bounded from below. Some positive features of the new online proposed strategy are as follows:

- The method is based on theoretical results ensuring that the hidden final penalization $\hat{S}$ and the appropriate (two at the most) saturation times τ_1 and τ_2 are the critical objects to be ascertained.
- It takes advantage of the availability of α and β as functions of $(T - t, S)$, and consequently on the possibility of generating Riccati matrices $P(t, T, \tilde{S})$ online by simple algebraic manipulations, as $\tilde{S}$ is updated; i.e. the RDE does not need to be solved for any value of $\tilde{S}$, not even offline.
- The control in Equation (21) is given in feedback form, and therefore the algorithm is unaffected by state perturbations due to fluctuations in environmental conditions.
- The updating of parameters $(\tilde{S}, \tau_1, \tau_2)$ is performed via the gradient of the cost of the process, and this cost is calculated by simple algebraic formulas instead of by predicting state, control, and cost trajectories by ODE integrations, as in most 'predictive control' techniques. This reduces the computational effort and allows for updating in shorter sampling intervals.
- Another conceptual difference with currently available approaches is that there exists a unique matrix $\hat{S}$ to look for in treating each LQR problem. This allows for further reduction on the computing effort,

since there is no need for updating Riccati equations through receding horizon schemes.

- It is under exploration the online generation of the matrices α and β involved in the calculation of the optimal feedback gain at each sampling time. This step will improve the applicability of the algorithm to large-dimensional processes, especially to those governed by PDE.

Acknowledgements

We thank Prof. Michael Grimble, at the Industrial Control Centre, Strahclyde University, for introducing us to the problem of temperature control in a metallic strip leaving a rolling mill.

References

Athans, M., & Falb, P.L. (2006). *Optimal control: An introduction to the theory and its applications*. New York: Dover.

Camacho, E. F., & Bordons, C. (2004). *Model predictive control* (2nd ed.). London: Springer.

Cannon, M., Liao, W., & Kouvaritakis, B. (2008). Efficient MPC optimization using Pontryagin's minimum principle. *International Journal of Robust and Nonlinear Control, 18*, 831–844.

Costanza, V., & Neuman C. E. (2006). Optimal control of nonlinear chemical reactors via an initial-value Hamiltonian problem. *Optimal Control Applications & Methods, 27*, 41–60.

Costanza, V., & Neuman, C. E. (2009). Partial differential equations for missing boundary conditions in the linear-quadratic optimal control problem. *Latin American Applied Research, 39*, 207–212.

Costanza, V., & Rivadeneira, P. S. (2008). Finite-horizon dynamic optimization of nonlinear systems in real time. *Automatica, 44*, 2427–2434.

Costanza, V., & Rivadeneira, P. S. (2013). Optimal saturated feedback laws for LQR problems with bounded controls. *Computational and Applied Mathematics, 32*, 355–371.

Costanza, V., Rivadeneira, P. S., & González, A.H. (2014). Minimizing control-energy in a class of bounded-control LQR problems. *Optimal Control Applications & Methods, 35*, 361–382.

Costanza, V., Rivadeneira, P. S., & Spies, R. D. (2009). Equations for the missing boundary values in the Hamiltonian formulation of optimal control problems. *Journal of Optimization Theory and Applications, 149*, 26–46.

Dhamo, V., & Tröltzsch, F. (2011). Some aspects of reachability for parabolic boundary control problems with control constraints. *Computational Optimization and Applications, 50*, 75–110.

Hearns, G., & Grimble, M. J. (2010, June 30–July 02). *Temperature control in transport delay systems*. Paper presented at the 2010 American control conference, Baltimore, MD, USA.

Pardalos P., & Pytlak, R. (2008). *Conjugate gradient algorithms in nonconvex optimization*. New York: Springer.

Pontryagin, L.S., Boltyanskii, V. G., Gamkrelidze, R. V., & Mishchenko, E. F. (1964). *The mathematical theory of optimal processes*. New York: Macmillan.

Qin, S. J., & Badgwell, T. A. (2003). A survey of industrial model predictive control technology. *Control Engineering Practice, 11*, 733–764.

Rao, A. V., Benson, D. A., Huntington, G. T., Francolin, C., Darby, C. L., & Patterson, M. A., *User's manual for GPOPS: A MATLAB package for dynamic optimization using the gauss pseudospectral method*, University of Florida Report, August 2008.

Sontag, E. D. (1998). *Mathematical control theory* (2nd ed.). New York: Springer.

Speyer, J. L., & Jacobson, D. H. (2010). *Primer on optimal control theory*. Philadelphia, PA: SIAM Books.

Troutman, J. L. (1996). *Variational calculus and optimal control*. New York: Springer.

23

Estimation of tire–road friction coefficient and its application in chassis control systems

Kanwar Bharat Singh[a]* and Saied Taheri[a,b]

[a]*Department of Mechanical Engineering, Virginia Tech, Randolph Hall (MC0238), Blacksburg, VA 24061, USA;* [b]*NSF I/UCRC Center for Tire Research (CenTiRe), Virginia Tech, Blacksburg, VA, USA*

Knowledge of tire–road friction conditions is indispensable for many vehicle control systems. In particular, friction information can be used to enhance the performance of wheel slip control systems, for example, knowledge of the current maximum coefficient of friction would allow an anti-lock brake system (ABS) controller to start braking with the optimal brake pressure, meaning the early cycles of operation are more efficient, resulting in shorter stopping distances. Also, from a passive safety perspective, it may be useful to present the driver with friction information so they can adjust their driving style to the road conditions. Hence, it is highly desirable to estimate friction using existing onboard vehicle sensor information. Many approaches for estimating tire–road friction estimation have been proposed in the literature with different sensor requirements and relative excitation levels. This paper aims at estimating the tire–road friction coefficient by using a well-defined model of the tire behavior. The model adopted for this purpose is the physically based brush tire model. In its simplest formulation, the brush model describes the relationship between the tire force and the slip as a function of two parameters, namely, tire stiffness and the tire–road friction coefficient. Knowledge of the shape of the force–slip characteristics of the tire, possibly obtained through the estimation of both friction and tire stiffness using the brush model, provides information about the slip values at which maximum friction is obtained. This information could be used to generate a target slip set point value for controllers, such as an ABS or a traction control system. It is also important to realize that a model-based approach is inherently limited to providing road surface friction information when the tire is exposed to an excitation with high utilization levels (i.e. under high-slip conditions). To be of greatest use to active safety control systems, an estimation method needs to offer earlier knowledge of the limits. In order to achieve the aforementioned objective, an integrated approach using an intelligent tire-based friction estimator and the brush tire model-based estimator is presented. An integrated approach gives us the capability to reliably estimate friction for a wider range of excitations (both low-slip and high-slip conditions).

Keywords: Brush model; friction estimation; Levenberg–Marquardt; nonlinear least squares; intelligent tire

1. Introduction

Tire friction forces, as the primary forces affecting planar vehicle motions, are physically limited by the road surface coefficient of friction (μ) and the instantaneous tire normal forces (Figure 1). Therefore, the ability to reliably estimate the tire–road friction coefficient is important for maximizing the performance of vehicle control systems, which work well only when the tire force command computed by the safety systems is within the friction limit.

Instantaneous knowledge of the friction potential will result in improved performance by several of the active chassis control systems. Examples of vehicle control systems that can benefit from the knowledge of tire–road friction include anti-lock braking systems (ABS), electronic stability control (ESC), adaptive cruise control (ACC), and collision warning or collision avoidance systems (Braghin et al., 2009; Cheli, Leo, Melzi, & Sabbioni, 2010; Cheli et al., 2011a, 2011b; Erdogan, Hong, Borrelli, & Hedrick, 2011; Sabbioni, Kakalis, & Cheli, 2010; Singh, Arat, & Taheri, 2012, 2013). The quality of traffic management and road maintenance work (e.g. salt application and snow plowing) can also be improved if the estimated friction value is communicated to the traffic and highway authorities.

The importance of friction estimation is reflected by the considerable amount of work that has been done in this field (Google Scholar) (Table 1). In normal driving conditions, the frictional force is not fully utilized, and the developed tire forces will be somewhere in the interior of the friction circle. When inputs are imposed on a tire, a relative motion between the tire structure and the road surface will arise. This relative motion is referred to as tire slip. The relation between the resulting tire forces and slip depends on many factors, namely, tire inflation pressure, vertical load, tire wear state, temperature, etc., and contains information about the available friction. When the

*Corresponding author. Email: kbsingh@vt.edu

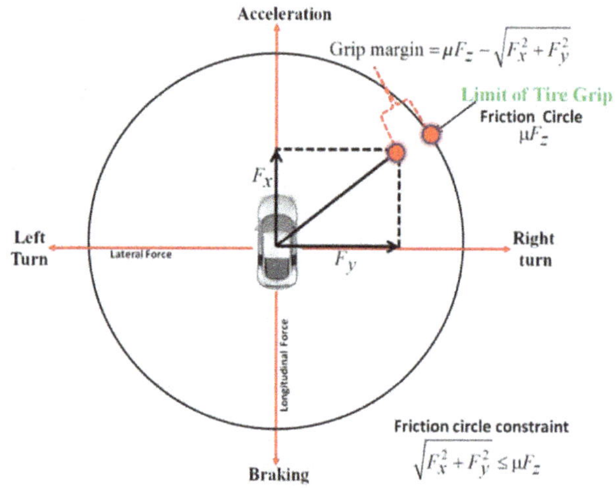

Figure 1. Friction circle of a tire.

Table 1. State-of-the-art literature review.

Tire road friction estimation	
Lateral dynamics based	**Longitudinal dynamics based**
Pasterkamp: JVSD 1997 [11]	Germann: IEEE 1994 [1]
Ray: Automatica 1997 [2]	Ray: Automatica 1997 [2]
Yasui: SAE 2004 [12]	Gustafsson: Automatica 1997 [3]
Klomp: AVEC 2006 [13]	Muller: ASME 2003 [4]
Hsu: PhD Thesis 2009 [14]	Lee: IEEE 2004 [5]
Erdogan:PhD Thesis 2009 [15]	Li: IMechE 2007 [6]
Andersson: IVSS 2010 [9]	Svendenius: PhD Thesis 2007 [7]
Ahn: PhD Thesis 2011 [10]	Rajamani: IEEE 2010 [8]
Nishihara: ASME 2011 [16]	Andersson: IVSS 2010 [9]
Matilainen: IEEE 2011 [17]	AhnPhD Thesis 2011 [10]
Luque: JVSD 2013 [33]	Rajamani: IEEE/ASME 2012 [31]
	Li: IEEE/ASME 2013 [32]

tire is exposed to excitation with high utilization, beyond the point corresponding to the maximum available friction force, the tire starts sliding and the resulting tire force directly corresponds to the friction coefficient.

Hence, determination of friction coefficient is straightforward in cases where tire forces are saturated, such as under hard braking conditions. The difficulty lies in obtaining a friction estimate under more normal driving circumstances, in which the tire slip is smaller (lower utilization conditions). In these cases, a model-based approach can be advantageous (Andersson et al., 2010). By fitting tire force and moment data to a model of the tire, the model parameters, including friction coefficient (μ), may be estimated. This approach may allow the estimation of friction without requiring tire force saturation. This study investigates the use of a model-based approach to estimate the tire–road friction coefficient and, more importantly, determines the level of tire force excitation necessary to allow the estimate to converge within a specified range of accuracy.

2. Estimation of vehicle states and tire forces

A model-based friction estimation method is based on the assumption that the lateral force, traction/braking force, aligning torque, vertical load and the two tire kinematic variables, slip angle and wheel slip, can be estimated indirectly using observers developed based on vehicle dynamics measurements (acceleration, yaw and roll rates, suspension deflections, etc.), or measured directly using certain sensor-based advanced tire concepts. In this study, we propose using an integrated vehicle state estimator, comprising a series of model-based and kinematics-based observers and an effectively designed merging scheme that ensures robust estimation performance even during the vehicle maneuvers which show highly nonlinear tire characteristics and in the existence of road inclination or bank angle. It is assumed that measurements from a six-axis inertial measurement unit (three axes of rotation rate measurement and three axes of acceleration measurement), wheel speed sensors and steering wheel angle sensor are available.

The block diagram in Figure 2 explicitly shows the estimation process in its entirety.

The entire process is separated into five blocks: the first block serves to identify the road bank and grade angles (using a kinematics-based observer) and vehicle chassis roll (using a Kalman filter) and pitch angles (with vehicle mass adaptation); the second block contains a bias compensation algorithm (gravity compensation in accelerometer measurements), a vehicle longitudinal speed estimation algorithm (based on the measurements of the four wheel rotational speeds and the gravity-compensated longitudinal vehicle acceleration) and a tire load estimation algorithm (using gravity-compensated acceleration information and roll/pitch states); the third block contains a tire longitudinal/lateral force estimation observer (sliding-mode observer based), while the fourth block contains a nonlinear vehicle longitudinal and lateral velocity observer (based on an unscented Kalman filter), designed for the purpose of vehicle sideslip estimation. Finally, the fifth block makes use of the estimations provided by the third and fourth blocks to estimate the tire slip ratio and slip angle (Luenberger observer based).

It is worth stressing the fact that the present work is concerned with examining the feasibility of estimating the tire–road friction coefficient (μ) in real time. Hence, a complete description of the integrated vehicle state estimator as shown in Figure 2 is beyond the scope of this work. A thorough coverage of this topic with specific details about the different schemes used for tire/vehicle state and parameter estimation is included in our previous work (Arat, Singh, & Taheri, 2013, 2014; Singh, 2012; Singh et al., 2012; Singh and Taheri, 2013).

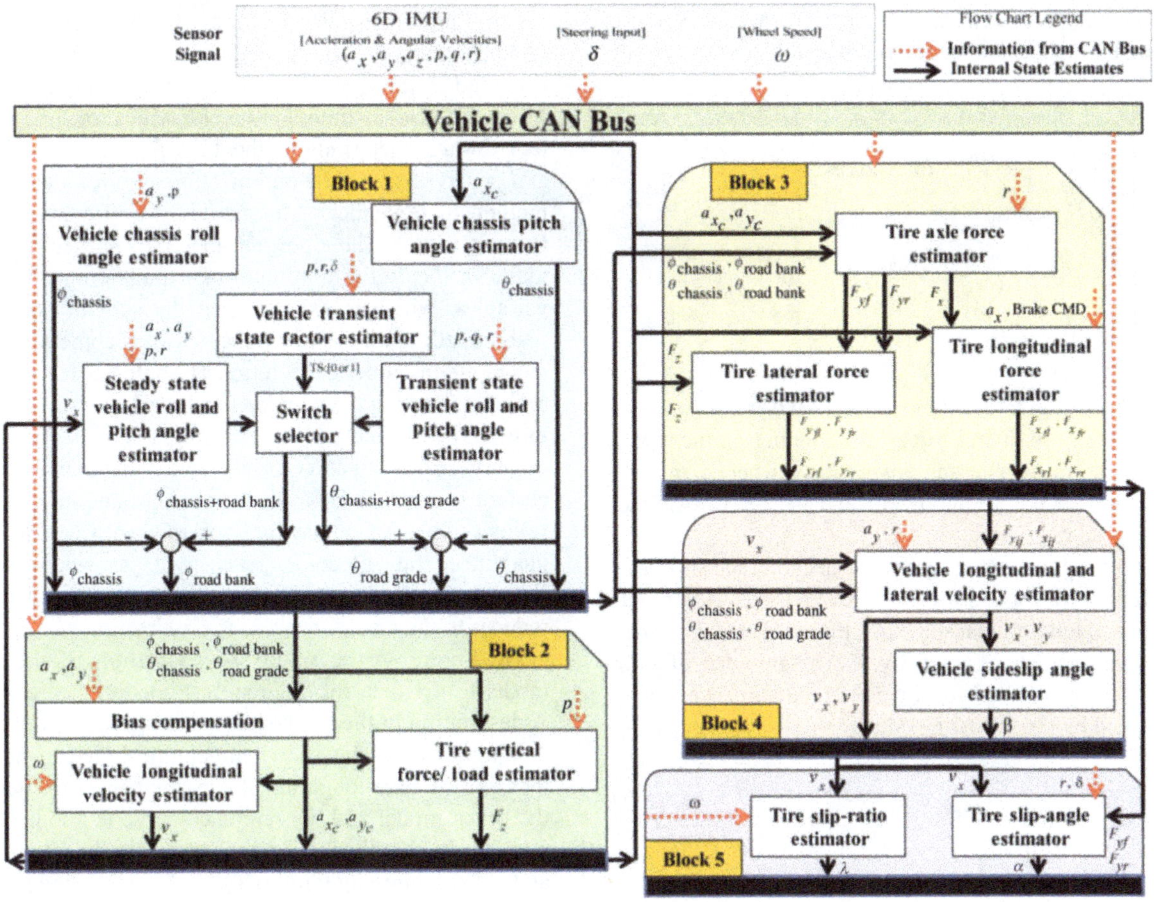

Figure 2. Functional diagram of the estimation process.

3. Tire model selection

Critical to the success of a model-based approach is the choice of model structure. Li, Fei-Yue, and Qunzhi (2006) provide a comprehensive summary of various models that have been developed to describe the complex nonlinear behavior of a tire. As this study focuses on parameter estimation, it is desirable to choose a model with a small number of parameters. The brush model (Pacejka, 2005) is well suited to these requirements, containing only stiffness

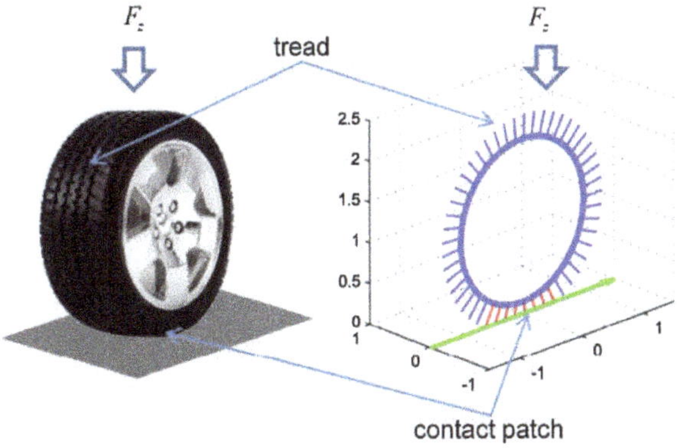

Figure 3. Tire and brush model (Ahn, 2011).

and friction parameters. The basic concept of the brush model is to represent the tire as a row of elastic bristles which touch the road plane and can deflect in a direction that is parallel to the road surface (Figure 3).

As a result, a tire can be modeled as a thin disk with brushes along the circumference that represent the tire treads. Treads in the contact patch are compressed and experience vertical stresses. The distribution of vertical stress is assumed to be parabolic. The generated forces or moment can be computed by integrating the stress of all brushes in the contact patch. A thorough coverage of the brush model is included in Pacejka (2005).

In a purely longitudinal slip case, the tire longitudinal force can be represented as follows:

$$F_x = \begin{cases} C_x\left(\frac{\lambda}{\lambda+1}\right) - \left(\frac{1}{3}\frac{C_x^2|\lambda/(\lambda+1)|(\lambda/(\lambda+1))}{\mu F_z}\right) \\ \qquad - \left(\frac{1}{27}\frac{C_x^3(\lambda/\lambda+1)^3}{(\mu F_z)^2}\right) \quad \text{for} \quad |\lambda| \le |\lambda_{sl}|, \\ \mu F_z \ \mathrm{sign}(\lambda) \quad \text{for} \quad |\lambda| > |\lambda_{sl}|, \end{cases} \tag{1}$$

where F_x, F_z, C_x, μ, λ and λ_{sl} stand for tire longitudinal force, tire normal force, tire longitudinal stiffness, tire–road friction coefficient, slip ratio and slip ration where transition from partial to full sliding occurs, respectively.

In a purely lateral slip case, the tire lateral force and tire aligning moment can be represented as follows:

$$F_y = \begin{cases} -C_y \tan(\alpha) + \left(\frac{1}{3}\frac{C_y^2|\tan(\alpha)|\tan(\alpha)}{\mu F_z}\right) \\ -\left(\frac{1}{27}\frac{C_y^3\tan^3(\alpha)}{(\mu F_z)^2}\right) \quad \text{for} \quad |\alpha| \le |\alpha_{sl}|, \\ -\mu F_z \text{sign}(\alpha) \quad \text{for} \quad |\alpha| > |\alpha_{sl}|, \end{cases} \tag{2}$$

$$\tau_a = \begin{cases} \frac{C_y \tan(\alpha) a_{cpl}}{3}\left(1 - \left|\frac{C_y \tan(\alpha)}{3\mu F_z}\right|\right)^3 & \text{for} \quad |\alpha| \le |\alpha_{sl}|, \\ 0 & \text{for} \quad |\alpha| > |\alpha_{sl}|, \end{cases} \tag{3}$$

where in addition to the above terms F_y, τ_a, C_y, α, α_{sl} and a_{cpl} stand for tire lateral force, tire aligning moment, tire cornering stiffness, slip angle, slip angle where transition from partial to full sliding occur and half of tire contact patch length, respectively.

The force and moment equations in combined slip cases are similar to the equations for pure slip cases. If both lateral slip and longitudinal slip exist, the treads are deformed in the direction determined by the magnitudes of both slips. The brush model for the combined slip case can be represented by the following equation:

$$F_x = F\frac{\sigma_x}{\sigma}, \quad F_y = F\frac{\sigma_y}{\sigma}, \quad M_z = -t(\sigma) \times F_y, \tag{4}$$

where

$$F(\lambda, \alpha, \mu) = \begin{cases} \mu F_z(1-\rho^3) & \text{for} \quad |\sigma| \le |\sigma_{sl}|, \\ \mu F_z \,\text{sgn}(\alpha) & \text{for} \quad |\sigma| > |\sigma_{sl}|, \end{cases}$$

$$\sigma_x = \frac{\lambda}{\lambda+1}, \quad \sigma_y = \frac{\tan(\alpha)}{\lambda+1}, \quad \sigma = \sqrt{\sigma_x^2 + \sigma_y^2},$$

$$a_{cpl} = a_{cpl_0}\sqrt{\frac{F_z}{F_{z_0}}}, \quad C = 2c_p a_{cpl}^2$$

$$\theta = \frac{C}{3\mu F_z}, \quad \sigma_{sl} = \frac{1}{\theta}, \quad \rho = 1 - \theta\sigma,$$

$$t(\sigma) = \frac{l(1-|\theta\sigma|)^3}{(3 - 3|\theta\sigma| + |\theta\sigma|^2)}.$$

4. Brush model adaptation

In order to use the brush model as a basis for friction estimation, it is desirable to validate the model. For validation purposes, tire force and moment data were created using "magic formula" tire model coefficients available in the literature (Pacejka, 2005). To account for the influence of road friction on the tire force characteristics, "magic formula" scaling factors previously published in the literature (Arosio, Braghin, Cheli, & Sabbioni, 2005; Braghin, Cheli, & Sabbioni, 2006) were used. To approximate the measurement/estimation uncertainty, the simulation data were corrupted with zero mean white noise. The model fitting algorithm is based on storing data points in the force–slip/moment–slip plane and using those points to compute the tire longitudinal/cornering stiffness and the tire–road friction coefficient by optimization. In other words, it is a method for the identification of parameters through determining the best fit between modeled and observed data. The optimization algorithm used is the method of Levenberg–Marquardt (LM) (Lourakis, 2005; Roweis, 1996). The algorithm of LM is an iterative technique to locate the minimum of a function with several variables, which is expressed as the sum of squares of real-valued nonlinear functions. It has become a standard technique in numerical solution of nonlinear least-squares problems and widely adopted in a broad spectrum of disciplines. The LM method can be thought of as a combination of gradient descent and Gauss–Newton methods. When the current solution is far from the correct one, the algorithm follows the gradient descent scheme, with a slower but guaranteed rate of convergence, whereas when the current solution is close to the correct solution, the algorithm reduces to the Gauss–Newton approximation.

In Figure 4, the results of adaptation of the brush model to tire data for different test conditions are shown. A description of the test conditions are given in Table 2.

From the results shown in Figure 4, it can be seen that, for the pure longitudinal slip, the coherence between the brush model and the reference curve is good. For the pure lateral slip, there are discrepancies in the lateral force and the self-aligning torque (SAT). As mentioned in previous research (Svendenius, 2003), the main reason for this discrepancy is the assumption of a stiff carcass.

5. Real-time implementation

The real-time parameter estimation algorithm used in this study is similar to the one presented in Hsu (2009). The complete real-time estimation algorithm is outlined below:

- Iteratively perform nonlinear least squares (NLLS) to the brush model on the batch of force–slip/moment–slip data, starting with initial estimates of braking/cornering stiffness and friction coefficient.
- To ensure that there is enough data for the NLLS fit to be meaningful, first initialize the process by placing a tire slip level threshold. The tire slip must exceed the threshold value before parameter estimation begins.
- The next step is to determine whether the tire force/moment has saturated sufficiently enough to estimate μ. In parallel to the NLLS fit, apply the method of least squares to the data points to find the slope of the line through the origin. Calculate the incremental mean-squared error of both fits from the most recent vector of data points of length N. If

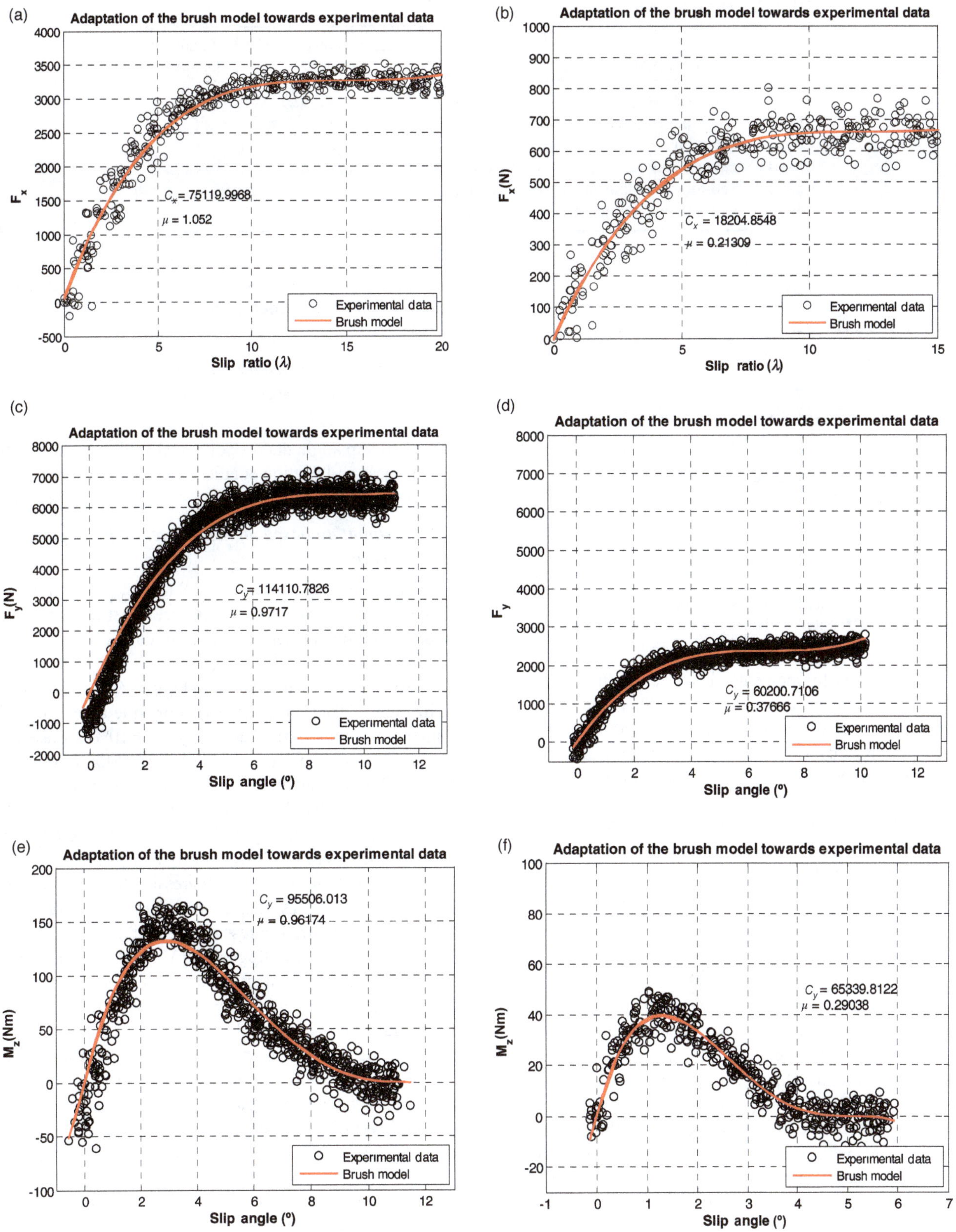

Figure 4. (a)–(f) Adaptation of the brush–tire model to tire measurement data (ref. Table 2 for a description of the test conditions). (a) Case 1, (b) case 2, (c) case 3, (d) case 4, (e) case 5 and (f) case 6.

Table 2. Tire test conditions.

	Test description	Surface condition	Measured/estimated signals	Estimated parameters
Case 1	Longitudinal force test	High μ	F_x, F_z, λ	C_x, μ
Case 2	Longitudinal force test	Low μ	F_x, F_z, λ	C_x, μ
Case 3	Lateral force test	High μ	F_y, F_z, α	C_y, μ
Case 4	Lateral force test	Low μ	F_y, F_z, α	C_y, μ
Case 5	Self-aligning moment test	High μ	τ_a, F_z, α	C_y, μ
Case 6	Self-aligning moment test	Low μ	τ_a, F_z, α	C_y, μ

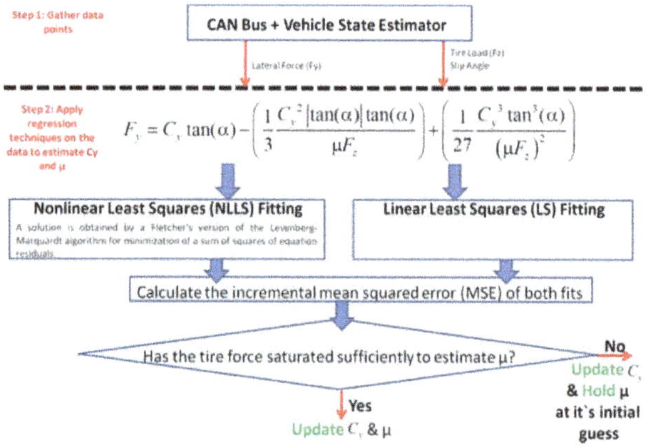

Figure 5. Tire parameter estimation algorithm – Lateral Force–Slip Regression Method.

the force–slip/moment–slip data are sufficiently nonlinear, update the value of μ. Otherwise, hold the coefficient of friction estimate at its previous value.

- As new data are collected and appended to the existing batch, repeat the above algorithm.

A schematic representation of the "Lateral Force–Slip Regression Method" is shown in Figure 5.

Similar estimation schemes were used for the "Longitudinal Force–Slip Regression Method" and the "Moment–Slip Regression Method."

6. Parameter estimation results

Friction estimation using the real-time estimation technique presented in Section 4 has been tested on tire measurement data. This section investigates how the value of the highest available slip (tire force utilization level) affects the accuracy of the friction estimation. The following subsections present the results for each of the estimation methods, namely, "Longitudinal Force–Slip Regression Method," "Lateral Force–Slip Regression Method" and "Moment–Slip Regression Method."

6.1. Estimation based on the Longitudinal Force–Slip Regression Method

Using the estimation algorithm described previously with $\lambda_{\text{thres}} = 0.1$, $\mu_{\text{initial}} = 0.8$ and $C_{x_{\text{initial}}} = 110000$, force and slip ratio data are post-processed to yield longitudinal stiffness and friction coefficient estimates (Figure 6).

6.2. Estimation based on the Lateral Force–Slip Regression Method

With $\alpha_{\text{thres}} = 2°$, $\mu_{\text{initial}} = 0.8$ and $C_{y_{\text{initial}}} = 110000$, force and slip angle data is post-processed to yield cornering stiffness and friction coefficient estimates (Figure 7).

6.3. Estimation based on the Moment–Slip Regression Method

With $\alpha_{\text{thres}} = 0.5°$, $\mu_{\text{initial}} = 0.8$ and $C_{y_{\text{initial}}} = 110000$, moment and slip angle data are post-processed to yield cornering stiffness and friction coefficient estimates (Figure 8).

Based on the results shown in Figures 6–8, the required utilization of friction necessary to provide a friction estimate within the specified accuracy of $\pm 10\%$ is presented in Table 3. In the case of the "Force–Slip Regression Method," more than 75–80% of the available friction force must be generated before an accurate estimate can be derived. However, it is possible to estimate the tire road friction coefficient for lower levels of utilization ($\sim$30–40%) if SAT ("Moment–Slip Regression Method") is used as a basis for the estimator instead of the lateral force ("Lateral Force–Slip Regression Method").

The "Moment–Slip Regression Method" presents a better opportunity of estimating the friction coefficient for lower levels of utilization, since the SAT saturates before the lateral force is saturated (Figure 9).

It is also important to realize that the force-based approach is inherently limited to operate during either longitudinal or lateral excitation, that is, either during acceleration/braking or cornering. Since they are active during different instants, it is advisable to combine two or more methods (Figure 10) as also suggested in previous work (Ahn, 2011) and hence provide a continuous estimate of friction.

To be of greatest use to active safety control systems, an estimation method needs to offer earlier knowledge of the limits. The next section presents an implementation strategy for estimating the tire–road friction coefficient under low-slip conditions.

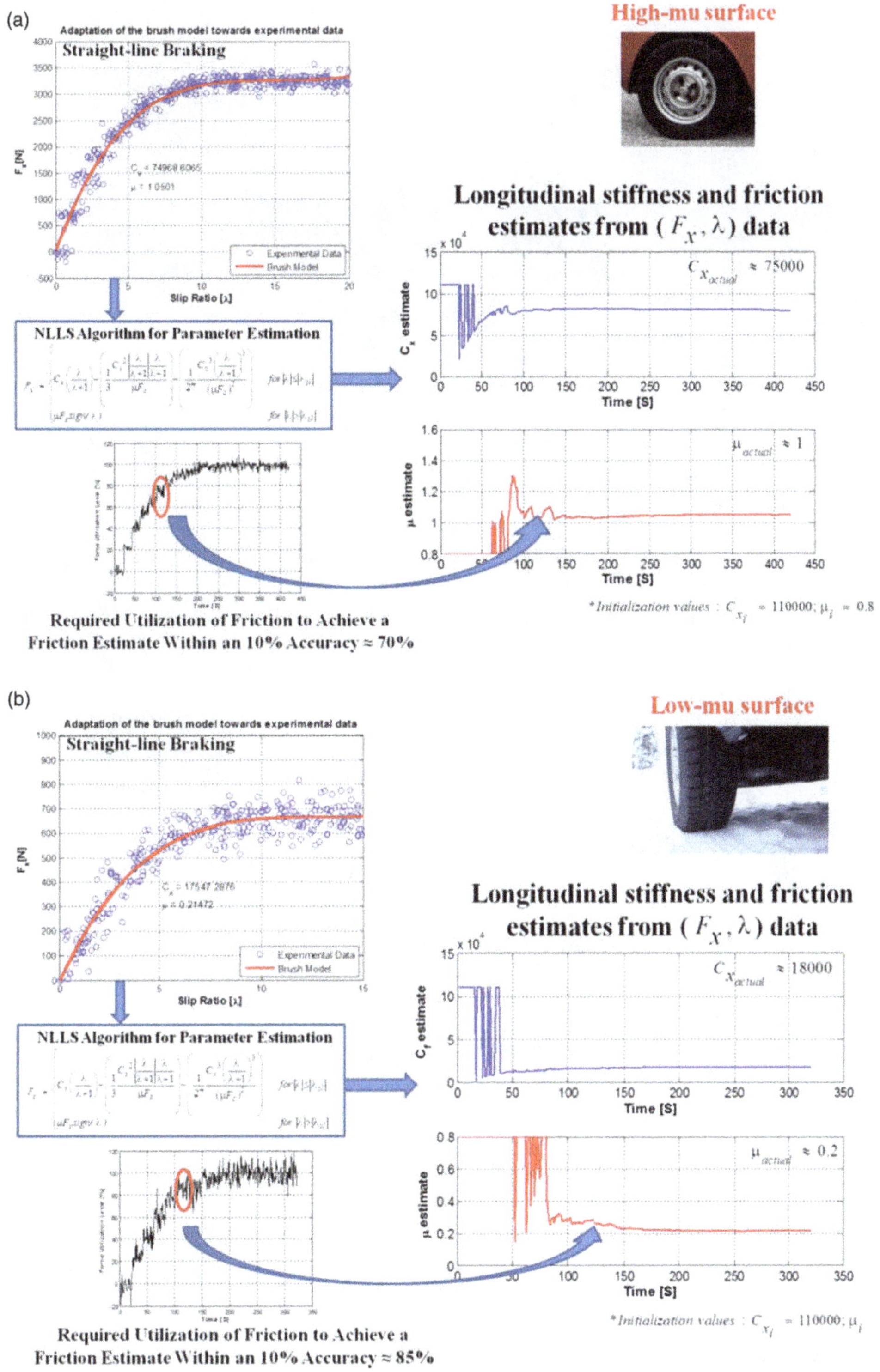

Figure 6. Longitudinal stiffness and friction coefficient estimates under (a) high μ conditions (case 1) and (b) low μ conditions (case 2).

7. Integrated tire–road friction estimation scheme

Availability of certain new technologies, popularly known as "intelligent tires" or "smart tires" (Morinaga; Yasushi Hanatsuka and Morinaga, 2013), hold the potential of providing real-time road surface condition information under low-slip rolling conditions. The implementation strategy for one such algorithm that utilizes sensor signals from an instrumented tire is presented in this section. The instrumented tire system was developed by placing accelerometers on the inner liner of a tire (Figure 11(a)). Figure 11(b) shows the final assembly of the instrumented tire with a high-speed slip ring attached to the wheel. Extensive

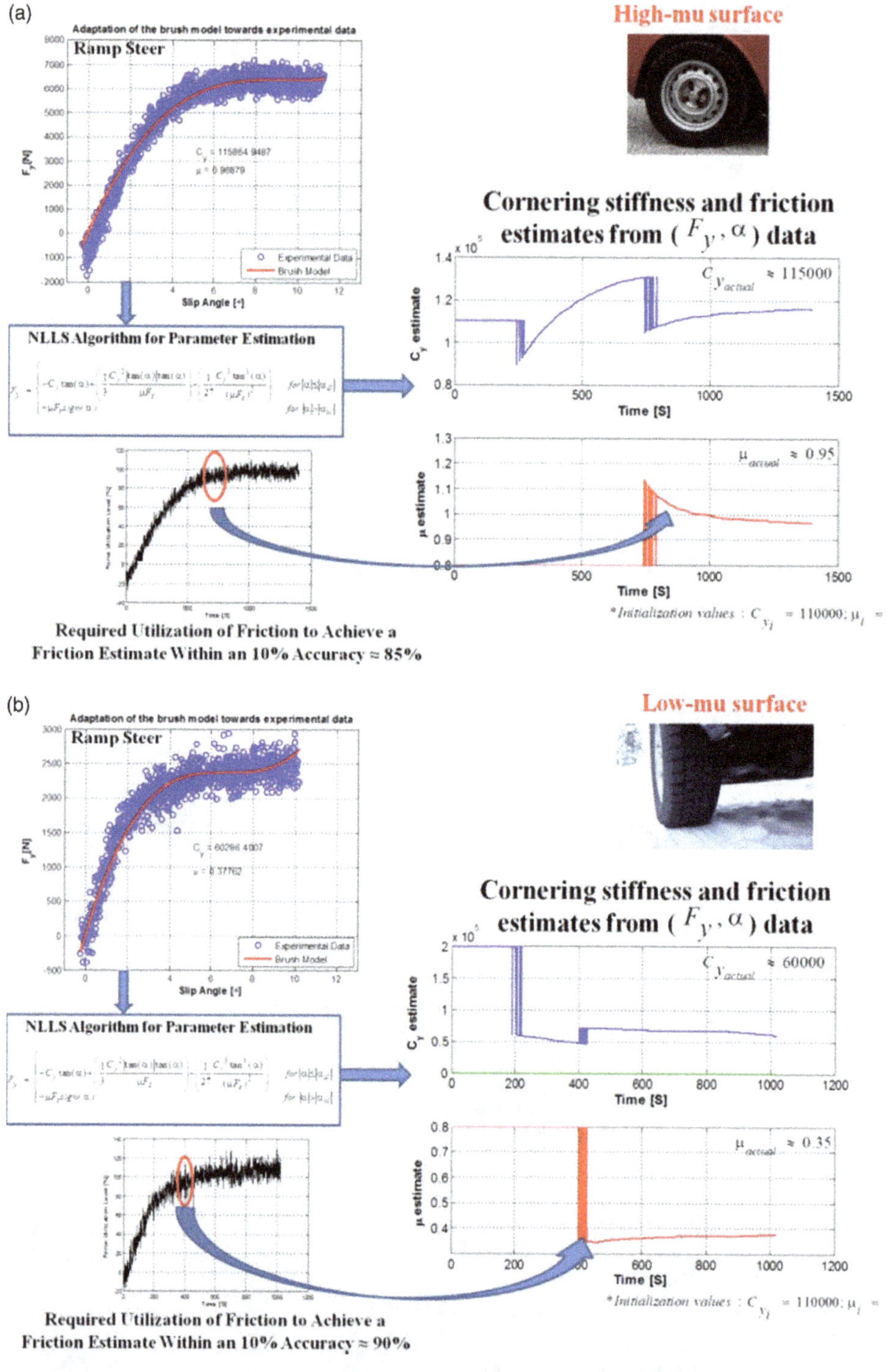

Figure 7. Lateral stiffness and friction coefficient estimates under (a) high μ conditions (case 1) and (b) low μ conditions (case 2).

dynamic tests of the tire were conducted using the in-house mobile tire test rig shown in Figure 11(c) and 11(d). An example of the acceleration signal is shown in Figure 12.

The effect of tire load, translational speed, varying pressure conditions and road surface roughness on the tire vibration spectra were studied by varying each of these parameters by carrying out extensive outdoor testing of the instrumented tire under free-rolling, traction/braking and steering conditions.

The power spectrum road of each accelerometer signal from these tests was computed using Welch's averaged modified periodogram method for spectral estimation (Figure 13). Analyzing the dynamic test results, it was

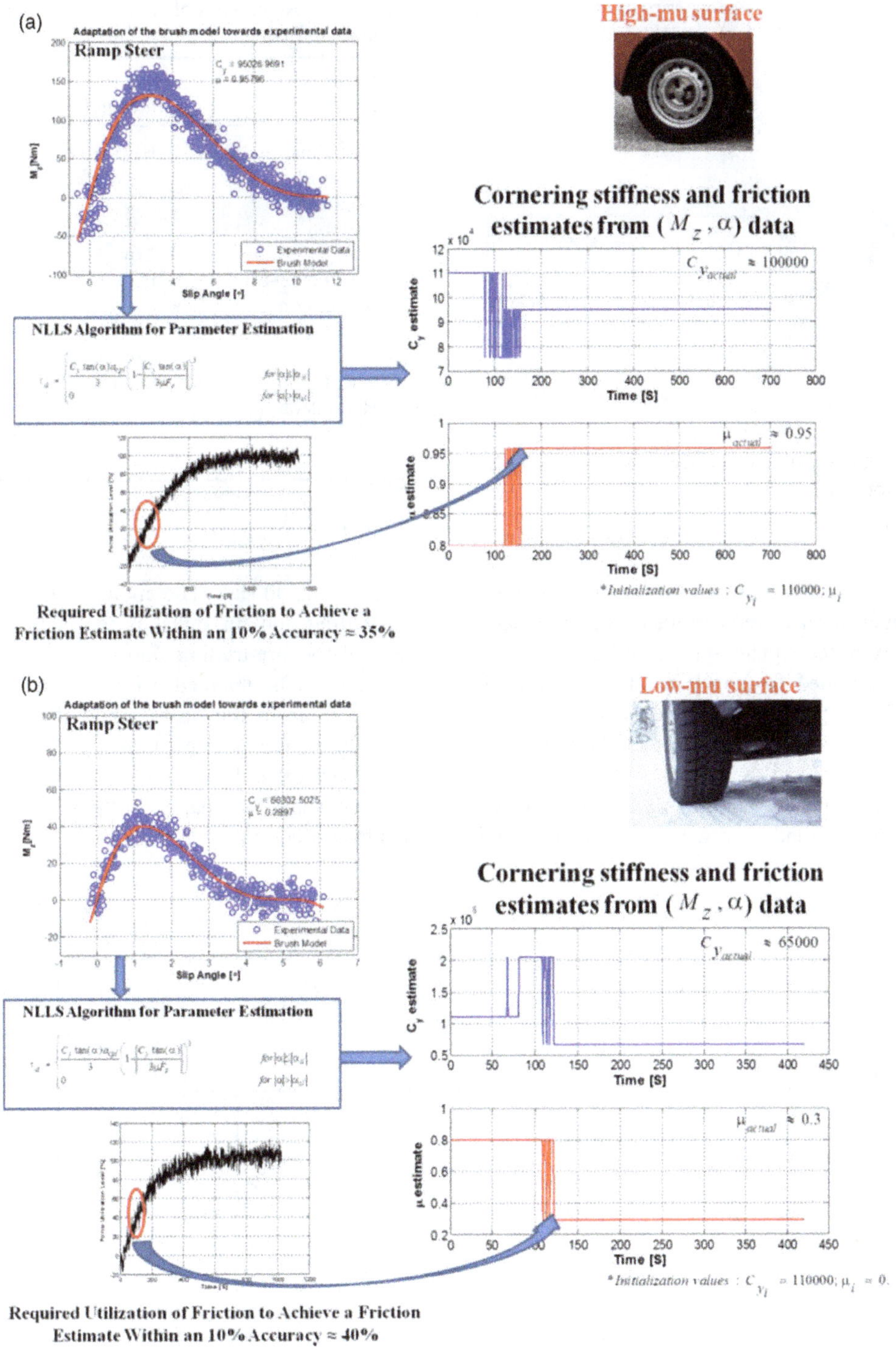

Figure 8. Lateral stiffness and friction coefficient estimates under (a) high μ conditions (case 1) and (b) low μ conditions (case 2).

concluded that, a marked difference was noticed in the concentration of the higher frequencies on the spectrum of the circumferential acceleration signal of the tire tested on different surface conditions (Figure 13(d)). This variation in the circumferential acceleration signal power spectral density (PSD) on different road surface conditions presented an opportunity to characterize the road condition using the tire vibration pattern information.

The proposed intelligent tire-based surface condition estimating algorithm consists of detecting the circumferential vibration of a tire of a running vehicle; dividing the detected tire vibration into vibration in a pre-trailing domain, the domain existing before a trailing edge position; and vibration in a post-trailing domain, the domain existing after a trailing edge position. Thereafter extracting signals of tire vibration only from the pre-trailing domain;

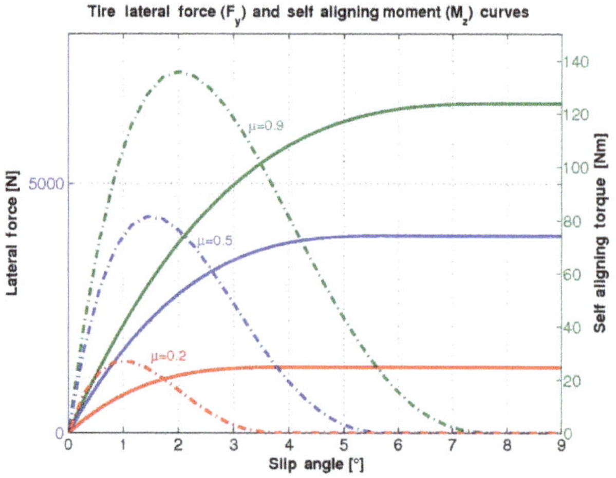

Figure 9. Lateral force and aligning torque versus slip angle. General behavior for a pneumatic tire.

Table 3. Required utilization of friction (in percent) to achieve a friction estimate within an accuracy of ± 10%.

Estimation methodology	Friction coefficient (μ)	Required utilization of friction (%)
Longitudinal Force–Slip Regression Method	High μ	70
	Low μ	85
Lateral Force–Slip Regression Method	High μ	85
	Low μ	90
Moment–Slip Regression Method	High μ	35
	Low μ	40

obtaining a time-series waveform of tire vibration including only the frequencies in a predetermined frequency band by passing the extracted signals through a band-pass filter of the predetermined frequency band; calculating a vibration level in the predetermined frequency band and estimating a road surface condition based on the calculated vibration level.

The predefined frequency bands being a low-frequency band (e.g. 10–500 Hz band) and a high-frequency band (e.g. 600–2500 Hz). The motivation for only using the pre-trailing domain signal is the larger difference in the PSD of the pre-trailing domain signal, when compared with the PSDs obtained using the entire signal or using the signal from the post-trailing domain. To determine these differences, the instrumented tire was first driven on a dry surface and then on a wet road surface at different speeds and the change in the vibration level ratio (R) was measured, where R is the ratio of the aforementioned

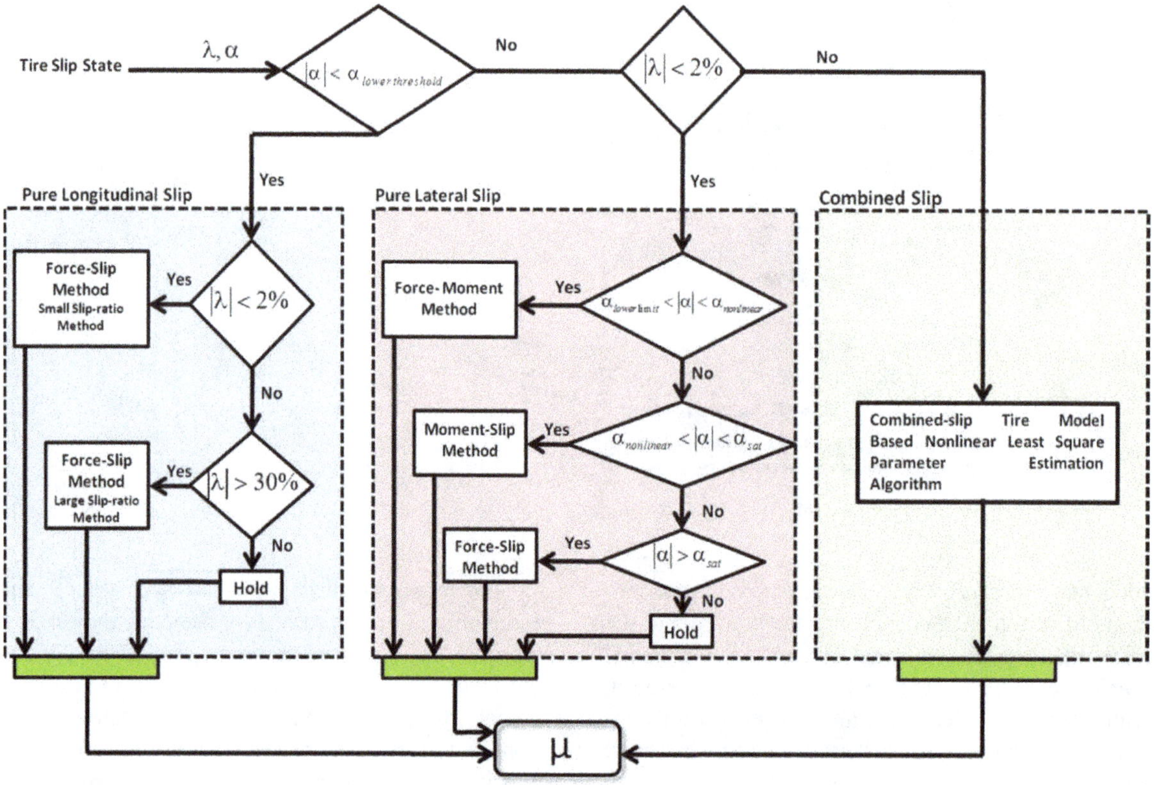

Figure 10. Integrated friction estimation algorithm – flow diagram.

Figure 11. Intelligent tire application: (a) sensor mounting location, (b) instrumented tire assembly, (c) mobile tire test rig and (d) test rig attached to the towing vehicle.

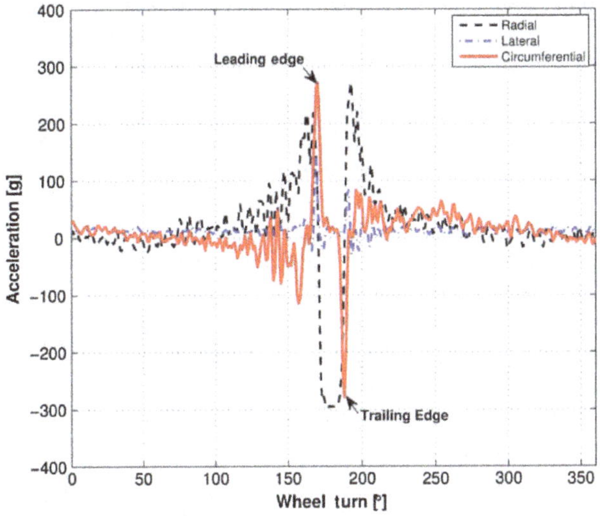

Figure 12. Measured acceleration signal for one rotation.

high-frequency band and the low-frequency band vibration level. It is evident from Figure 14 that the vibration level ratio R increased as the tire was tested on the wet road surface. This change in the vibration level ratio can be attributed to the increased slippage of the tire, and thus, it has been confirmed that the slipperiness of a road surface can be decided by setting a proper threshold value.

For this purpose, a fuzzy logic classification approach was developed for the real-time implementation of the proposed algorithm. The application of fuzzy logic to solve the classification problem is motivated by its noise tolerance to the vibration data retrieved from sensors, and its ability for real-time implementation while ensuring robustness with respect to imprecise or uncertain signal interpretation. Figure 15 shows the fuzzy controller architecture.

Based on the interdependence of all the inputs for a given road surface condition and the way they effect the vibration spectra of a tire, a set of linguistic rules were developed to identify the road surface condition. The classifier performance was validated on smooth asphalt, regular asphalt, rough asphalt and wet asphalt (Figure 16).

Two different tests were performed to study the classifier performance. The first test involved testing the tire under free-rolling and low-slip conditions (low force utilization). The second test involved testing the tire under high-slip conditions (high force utilization). For the first test (free-rolling and low-slip conditions), the classifier was successfully able to distinguish between the different road surface conditions as shown in Figure 17. However, for the second test (high-slip conditions), classifier performance was unsatisfactory (Figure 18). Higher misclassification rates under high-slip conditions were attributed to the increased vibration levels in the circumferential

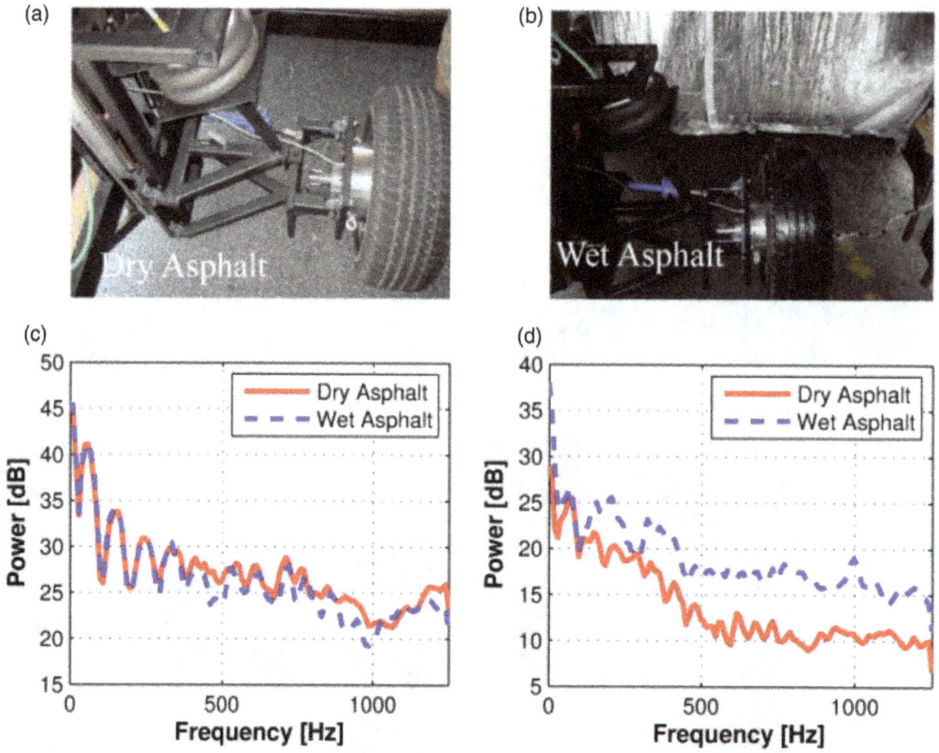

Figure 13. Tire tested on different road surface conditions: (a) dry surface testing and (b) wet surface testing; roughness dependence study: (c) radial signal PSD and (d) circumferential signal PSD.

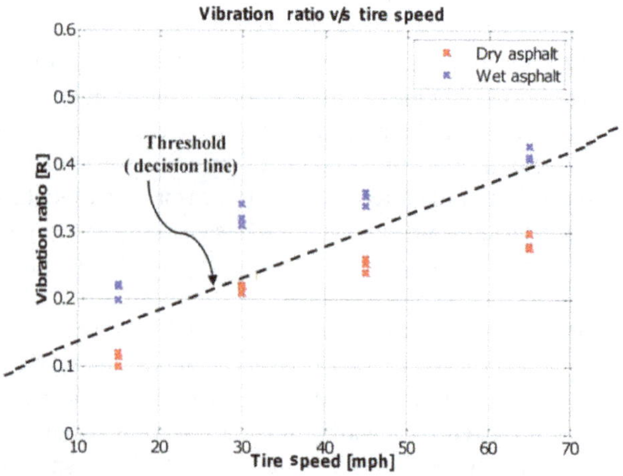

Figure 14. Vibration ratio on dry and wet surface conditions for a range of tire speeds.

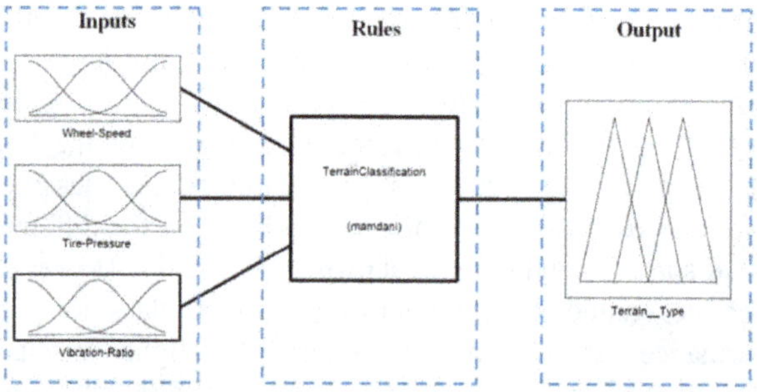

Figure 15. Fuzzy logic-based controller architecture.

acceleration signal due to the stick/slip phenomenon linked to the tread block vibration modes (Figure 19).

The requirement of a more complex event detection algorithm for distinguishing the high-frequency content of a signal due to the tread block mobility effects, under high-slip conditions, makes the proposed fuzzy logic classifier unsuitable for friction estimation under high-slip conditions.

Hence, it was proposed to use a model-based approach as presented in the previous section to estimate road surface friction under high-slip conditions. Finally, it was proposed to use an integrated approach (Figure 20) using the intelligent tire-based friction estimator and the model-based estimator which would reliably estimate friction for a wider range of excitations (both low-slip and high-slip conditions). Using an integrated approach, the road surface friction classifier was successfully able to distinguish between the different road surface conditions as shown in Figures 21 and 22.

8. Application of road friction information in vehicle control systems – development of new control strategies

A change in the peak grip potential of the tire (Figure 23(a)) not only affects the "limit" handling behavior of the vehicle, but is also known to affect the vehicle

Figure 16. Tire tested on different road surface conditions: (a) rough asphalt, (b) regular asphalt, (c) smooth asphalt and (d) wet asphalt.

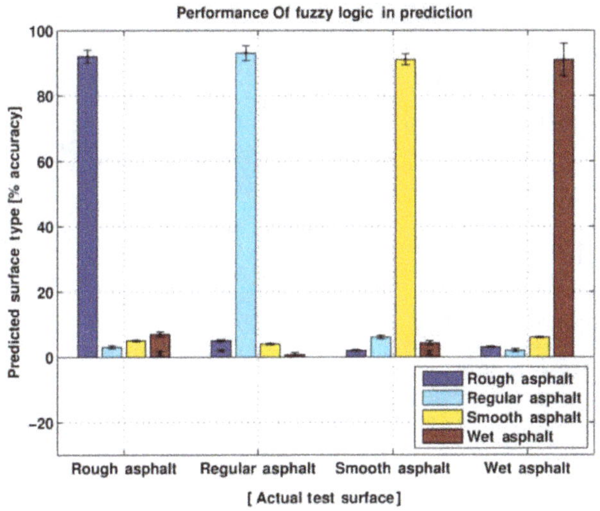

Figure 17. Performance of the fuzzy logic classifier – low-slip conditions.

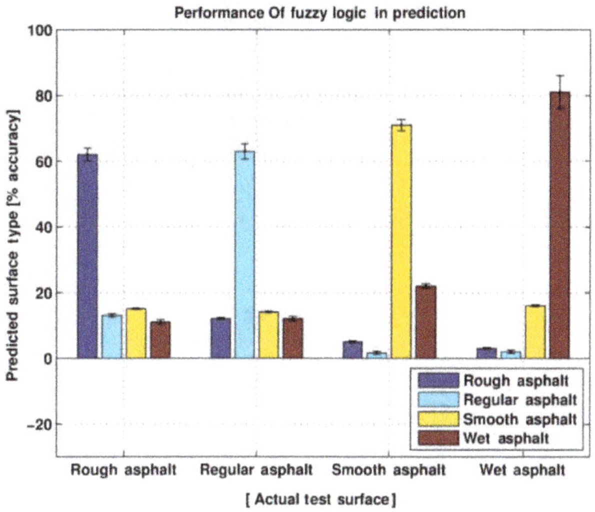

Figure 18. Performance of the fuzzy logic classifier – high-slip conditions.

"linear range" handling behavior. This change in the vehicle "linear range" behavior is due to the influence of road friction on the tire stiffness in the low-slip region (Figure 23(b)). Moreover, the peak slip ratio position of the maximum coefficient of friction varies for different road conditions (Figure 24). Hence, in the context of an anti-lock brake system (ABS) based on a fixed thresholding rule-based algorithm, it cannot be expected that an ABS controller that is optimized for dry asphalt performs as reliably and efficiently on wet or icy surfaces.

To quantify the performance benefits for an ABS controller using road friction condition information, a modified ABS algorithm has been developed, as shown in Figure 25. The modified controller leverages friction information

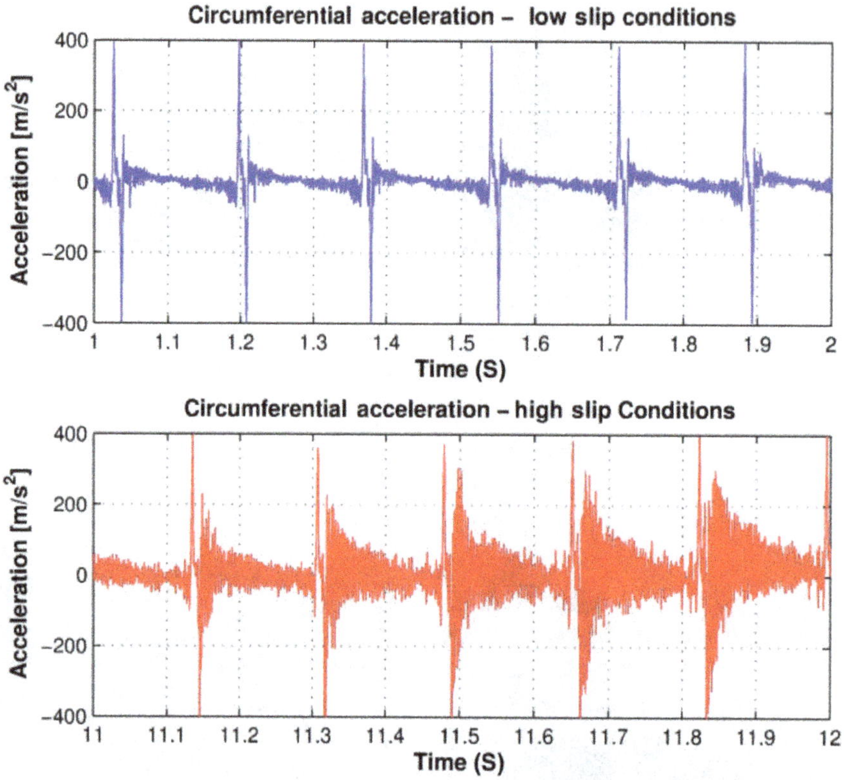

Figure 19. Circumferential acceleration signal under low-slip conditions (top), and increased vibration levels in the circumferential acceleration signal under high-slip conditions (bottom).

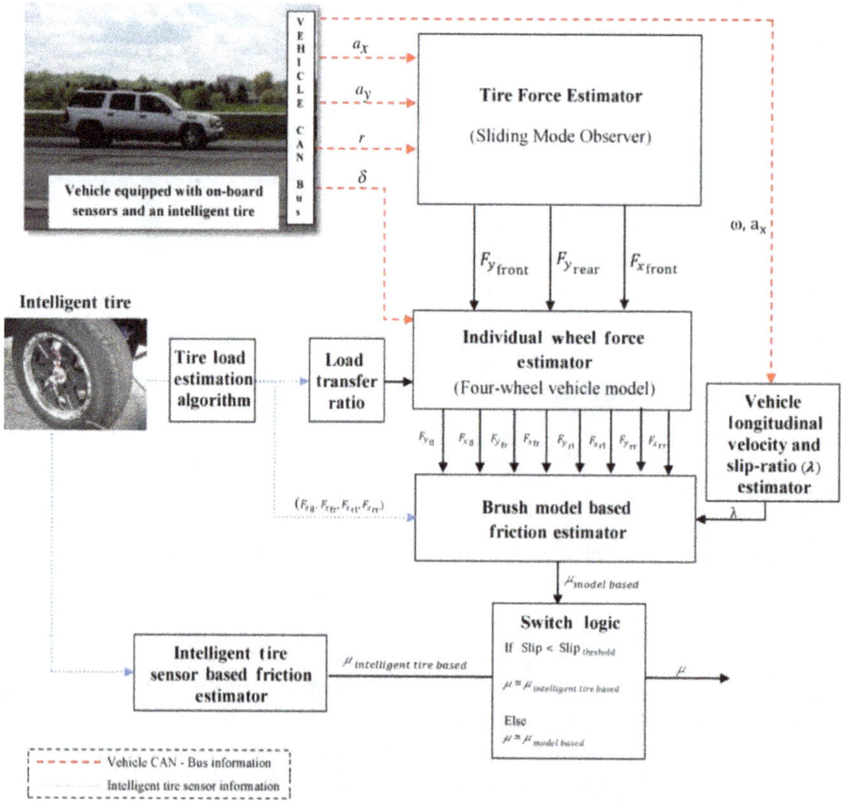

Figure 20. Architecture of the proposed integrated approach using an intelligent tire-based friction estimator and the model-based estimator.

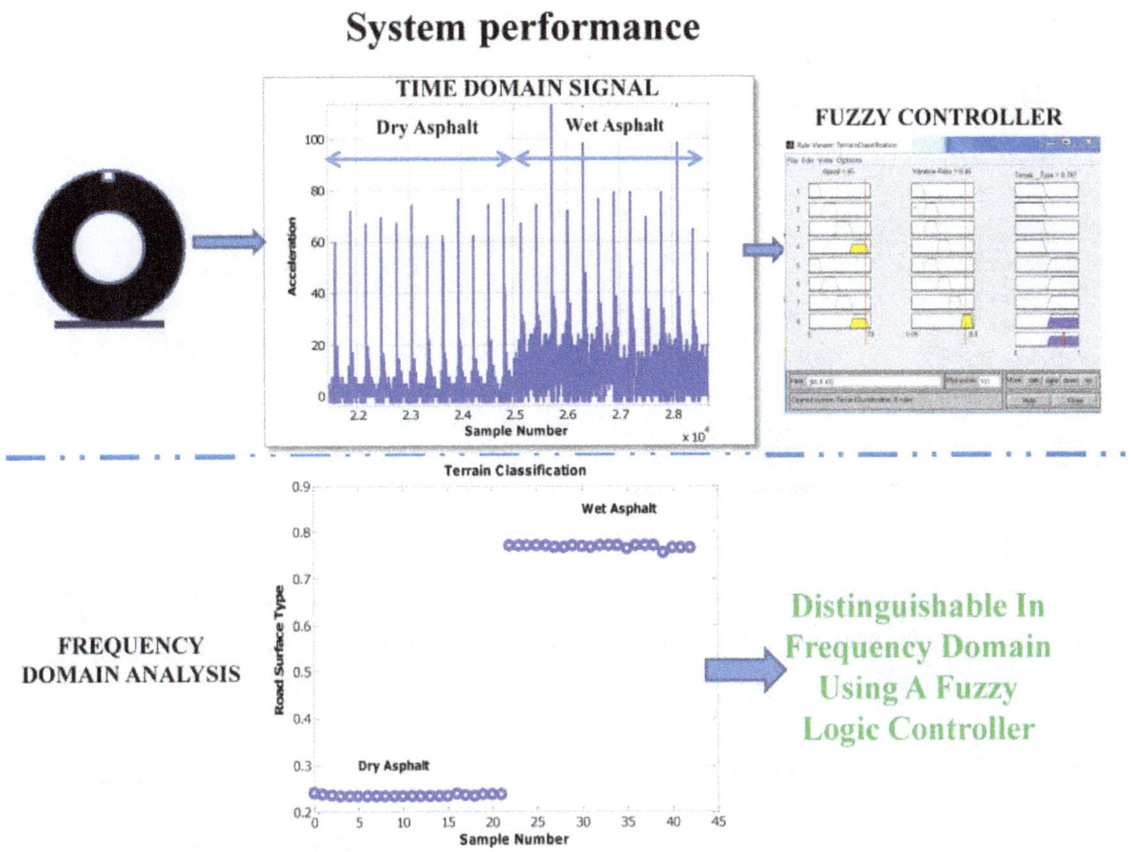

Figure 21. Classification performance on dry and wet asphalt.

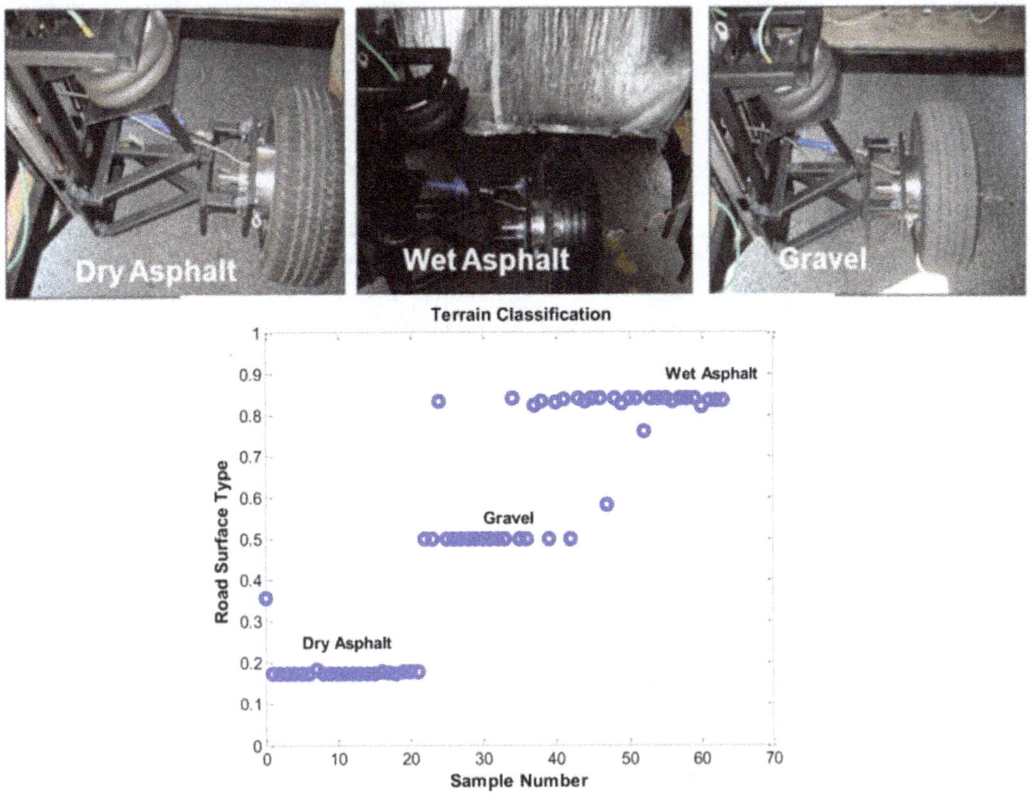

Figure 22. Classification performance on dry asphalt, gravel and wet asphalt.

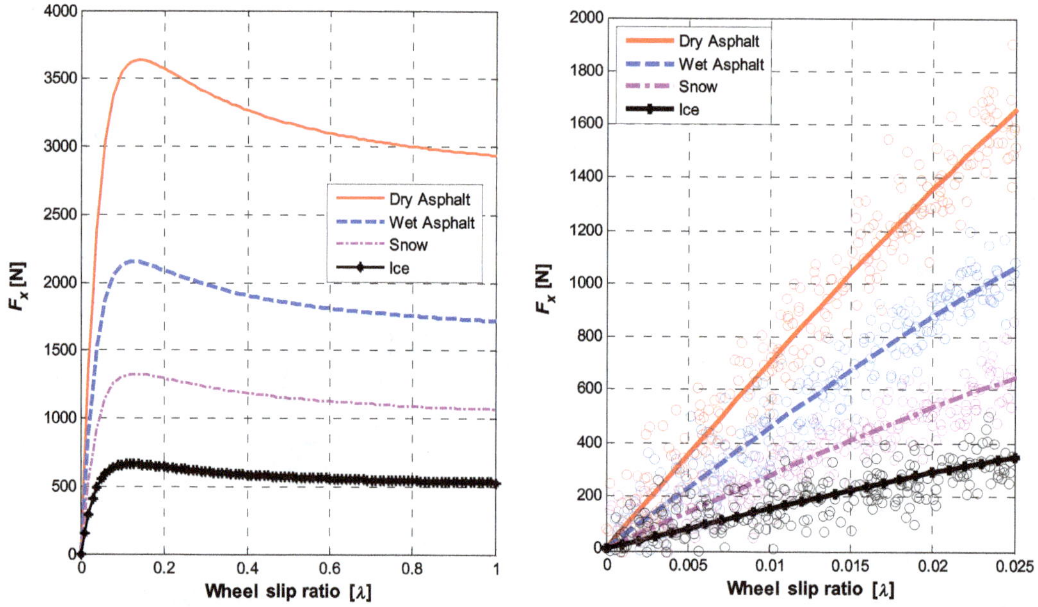

Figure 23. (a) Longitudinal tire force under different road surface conditions, and (b) longitudinal tire force in the small-slip region under different road conditions.

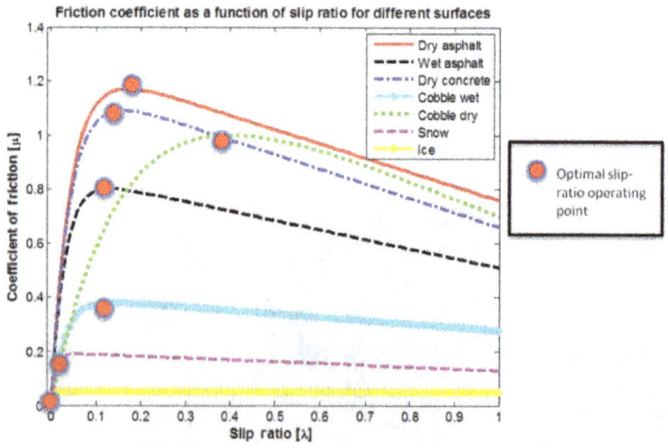

Figure 24. Position of slip values at which maximum friction (braking force) is obtained.

to change the optimal slip point thresholds, thus maximizing the braking force. Moreover, the controller uses a brake preconditioning algorithm with a friction adaptation strategy to start braking with the optimal brake pressure, and thus ensuring that the early cycles of braking are more efficient. Simulations were carried out on a series of braking maneuvers to examine the possible improvements in the ABS system performance. The results reveal that the presence of road condition information allows for a considerable decrease in the stopping distance (Singh et al., 2013). Most impressive improvements are obtained for the jump-μ tests (Figure 26). In light of these results, we conclude that the knowledge of road surface condition can be quite favorable for enhancing the current ABS algorithms.

In the context of a classical ESC system based on a model reference approach, the desired values of yaw rate and body sideslip angle are generated from a reference model, which takes into account the vehicle velocity, the driver input, tire/axle load and cornering stiffness (understeer/oversteer behavior). The weighting factor, which determines the balance between the yaw rate tracking and sideslip regulation, depends primarily on a term combining the estimated rear axle sideslip angle and its derivative. Therefore, an accurate online estimate of the vehicle sideslip angle is critical for the effective operation of an ESC system. In production cars, the vehicle sideslip angle is not measured because this measurement requires expensive equipment such as optical correlation sensors. Most production vehicles rely on observers based on vehicle dynamics models for indirectly estimating the vehicle sideslip angle. These observers perform reasonably well in normal driving situations, when the steering characteristics specify a tight connection between the steering wheel angle, yaw rate, lateral acceleration and vehicle sideslip angle. When a vehicle is near or at the limit of adhesion, tire forces and consequently the yaw dynamics strongly depend on the surface coefficient of friction. For example, limit tire forces on ice can be about 10 times smaller than on dry surface. The vehicle model used within the observer should therefore be adapted to the changing surface friction. The coefficient of friction, however, is unknown and has to be estimated. Thus, the estimation of sideslip angle depends on another estimate, which increases the potential for errors. Uncertainties in sideslip angle estimation lead the ESC system to be conservatively calibrated in order to balance robustness concerns with performance. This inherently sacrifices some level of performance.

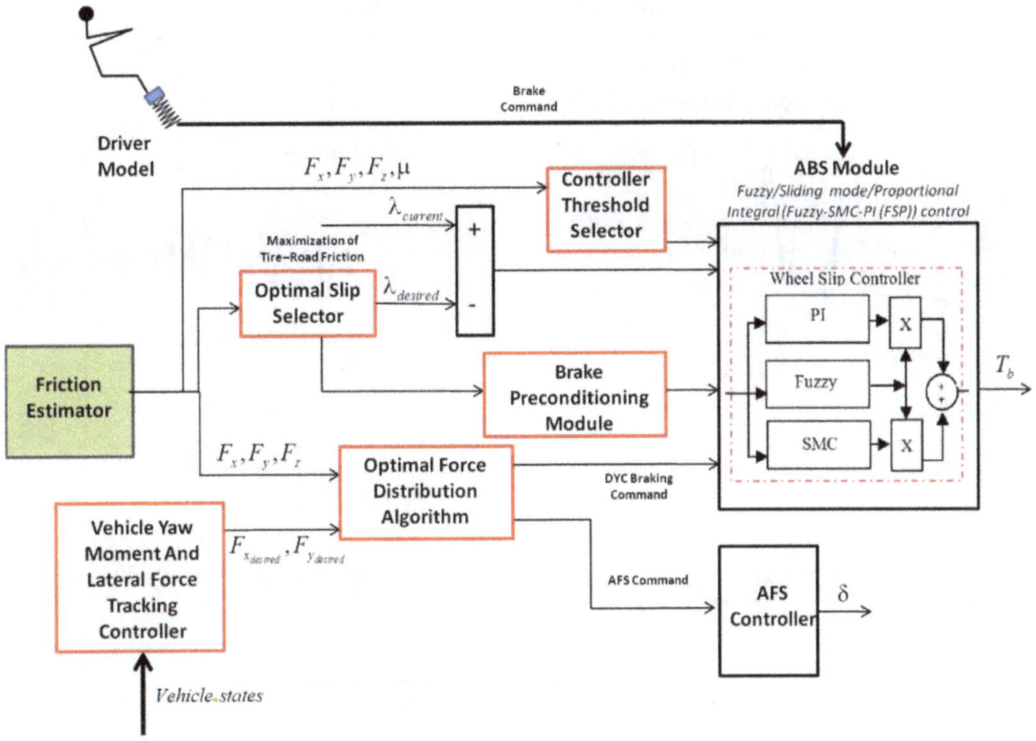

Figure 25. Modified ABS algorithm designed to use road friction information.

From the foregoing discussion, it is apparent that the knowledge of road surface condition would be beneficial in improving the accuracy of sideslip angle observers, which eventually would enhance the performance of ESC systems. To quantify the performance benefits, an enhanced ESC controller based on active front steering (AFS) and direct yaw moment control (DYC) has been developed (Figure 27). The proposed controller consists of a full vehicle state estimator with friction adaptation (Singh, 2012). The vehicle sideslip is estimated using an extended Kalman filter (EKF)-based observer. More details pertaining to the modified ESC algorithm are given in Table 4.

Simulation results show that the new control strategy aiming to use all of the information available from the vehicle state estimator can significantly enhance vehicle stability during emergency evasive maneuvers on various road conditions ranging from dry asphalt to very slippery packed snow road surfaces (Figure 28).

Another vehicle safety system that is becoming more prevalent in the vehicle industry is the advanced driver assistance system (ADAS). Typically, ADAS features three technologies: collision mitigation braking system (CMBS), lane keeping assist system and ACC. CMBS is an active safety system that helps the driver to avoid or mitigate rear-end collisions. It uses forward-looking sensors to detect obstacles ahead of the vehicle. The systems use relative distance, relative velocity and vehicle velocity information to warn the driver or control the vehicle. Specifically, a warning critical distance is defined as a function of vehicle velocity and relative velocity.

From Figure 29, we can see that if friction information would be available, the critical warning and critical braking distances could be calculated more precisely (since the deceleration rate for the vehicles (α_1, α_2) depends on the maximum tire–road friction available). Using a high default value for the friction coefficient causes the systems to lose some of their safety potential on low-friction surfaces. Using a low or medium default value would on the other hand cause the safety systems to activate too early in high-friction conditions, taking the driver "out of the loop" possibly unnecessarily.

The modified algorithm used in this study assumes to have full knowledge of the road conditions (Figure 30). Consequently, the collision mitigation algorithm adapts its critical distance (warning/braking distance) definitions when the road conditions change (Figure 31).

A parametric analysis aimed at evaluating the benefits induced by the introduction of friction information has been carried out. These simulations will be used to show the benefit of using friction estimation in conjunction with a collision mitigation brake system algorithm. In the test case, the host and the lead car are both traveling at 27.8 m/s with a separation of 50 m. The lead car suddenly applies the brakes and decelerates. The host vehicle maintains its velocity, which simulates a driver

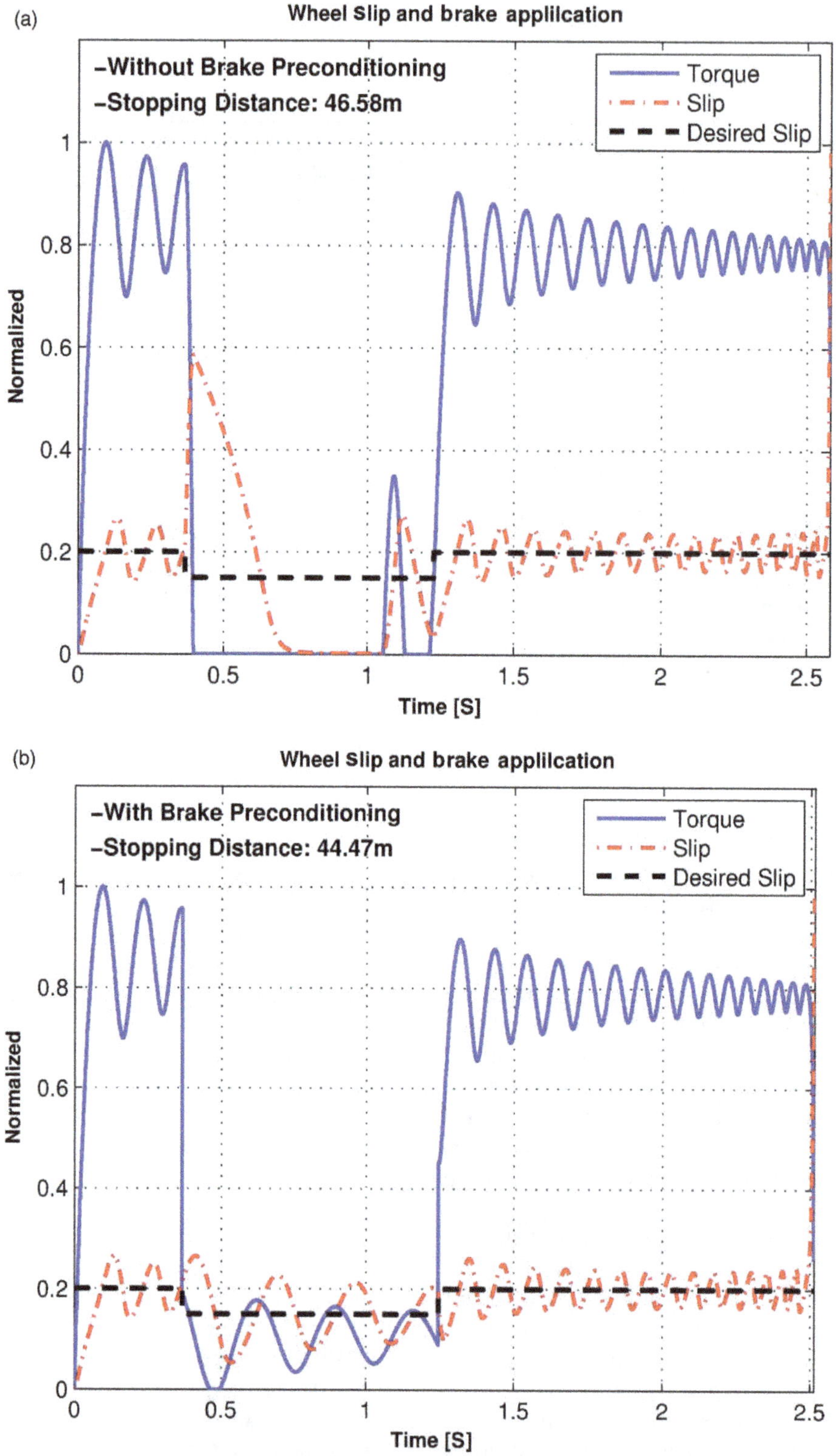

Figure 26. ABS performance – (a) without knowledge of road friction conditions and (b) with knowledge of road friction conditions.

who is unaware of the critical nature of the situation. Figure 32 shows the vehicle response when the collision mitigation brake system algorithm without the friction adaptation strategy is used (i.e. friction information is unavailable). The relative velocity at impact in this case is 14.5 m/s.

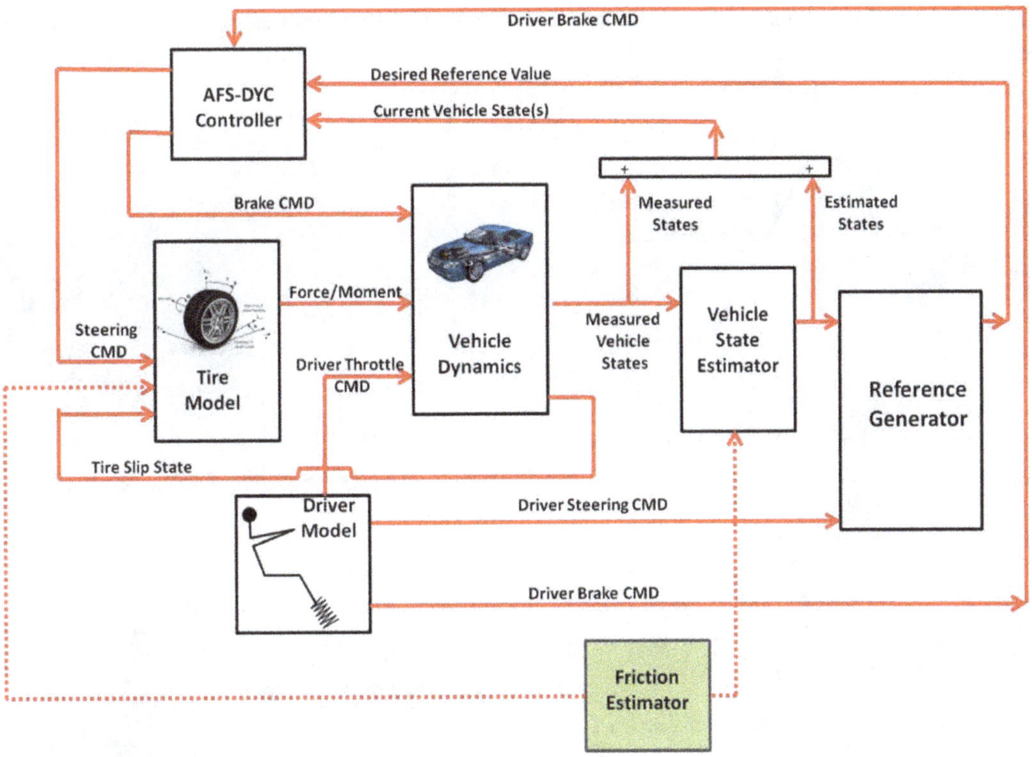

Figure 27. Modified ESC algorithm designed to use road friction information.

Table 4. Enhanced ESC system model details.

Reference generator	Sliding mode control (SMC) strategy	The model responses of vehicle sideslip angle and yaw rate are described. Total lateral force and the total yaw moment required for the controlled vehicle to follow the model responses are estimated using an SMC strategy
AFS controller	SMC strategy	Difference between vehicle measured yaw rate and desired yaw rate is considered as the sliding surface. Control law is based on tire force feedback and is obtained from a nonlinear eight-degree-of-freedom vehicle model
DYC controller	Rule based	Implemented through braking one of the four wheels based on detection of understeer or oversteer driving situations
Tire model	Combined slip model	Tire forces are modeled using magic formula tire model *with friction adaptation*
Vehicle state estimator (Singh, 2012)	Full state estimation using a nonlinear vehicle model	Vehicle sideslip is estimated using an EKF-based observer *with friction adaptation*

Figure 33 shows the vehicle response when the collision mitigation brake system algorithm with the friction adaptation strategy is used (i.e. friction information is available). In this simulation, the driver is completely out of the loop, so the collision mitigation brake system brings the vehicle to a rest. Notice that the plot shows that the vehicles collide at $\sim$7.7 s. The relative velocity at impact is 9.5 m/s. Clearly, the modified algorithm with friction information (Figure 21) applies the brakes sooner during degraded road conditions, which gives the vehicle more time to slow down. As a result, the impact speed and the impact energy are reduced. These results can be improved even further by increasing the friction adaption scaling factors.

From the foregoing discussion, we can thus conclude that road friction condition information would enable slip control systems (ABS, traction control system, ESC, etc.) to be started with the optimal initial parameters for the friction situation at hand. Moreover, accurate friction information would also enable the CMBS to start intervention from a more optimal distance in every road condition.

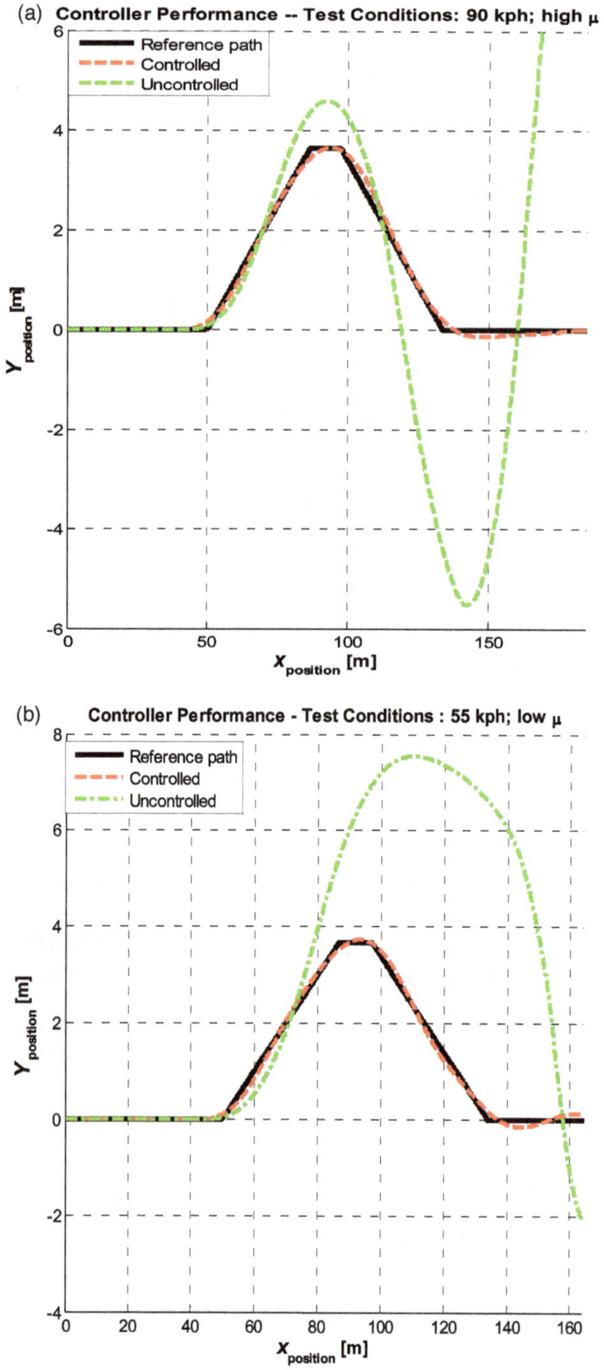

Figure 28. ESC performance: (a) high μ conditions and (b) low μ conditions.

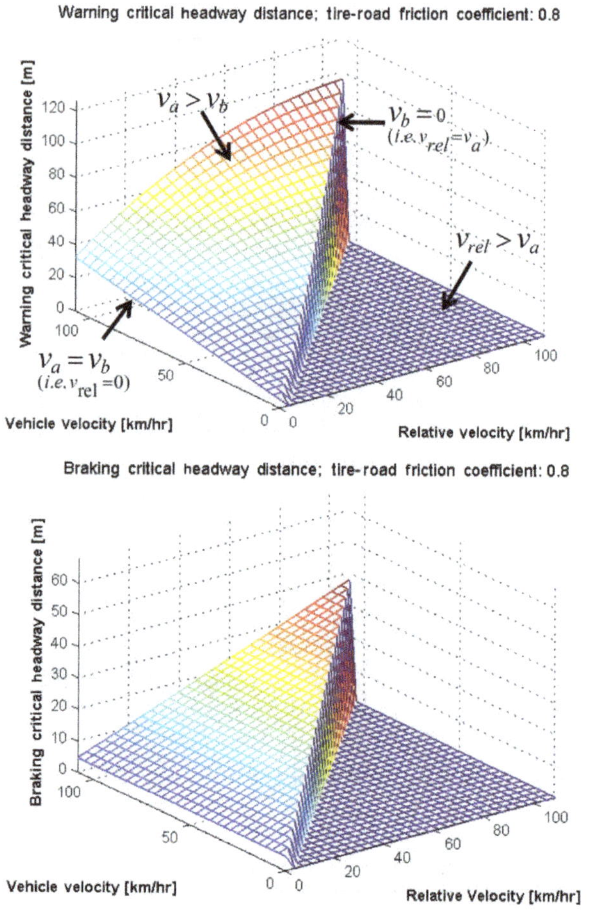

Figure 29. (a) Critical warning distance and (b) critical braking distance.

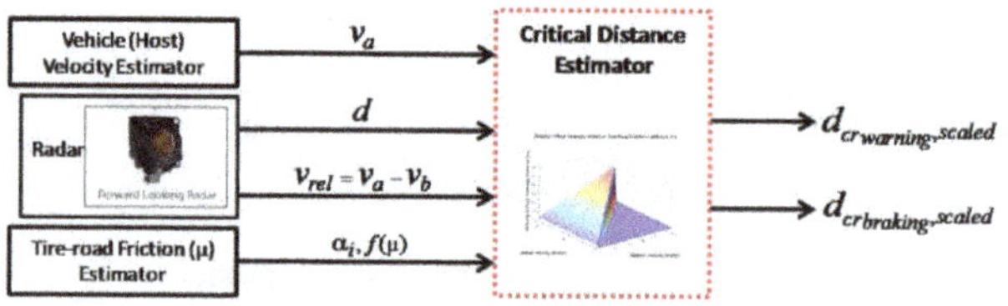

Figure 30. Modified algorithm designed use tire–road friction information to adapt its critical distance definitions.

9. Conclusion

Road friction is an important parameter for vehicle safety applications, but difficult to estimate accurately in all driving situations and weather conditions. Friction estimation methods based on the applied slip angle and slip ratio have been proposed earlier, but in the case of a freely rolling tire, the friction estimation is still an unsolved topic. This study uses a three-axis accelerometer on the inner liner of a tire to detect the tire road friction potential. More specifically, a two pronged approach was adopted to estimate tire–road friction coefficient. Firstly, a "force–slip" and "moment–slip" model-based approach is proposed. The primary shortcoming of the "Force–Slip Regression Method" identified is the requirement that the vehicle must enter the nonlinear region of handling before friction can be estimated. Thus, in the case of the "Force–Slip Regression Method," an estimation algorithm based on the brush model will work, but high friction utilization ($\sim$75–80%) is required for an accurate friction estimate. The "Moment–Slip Regression Method" based on total aligning moment has the benefit of knowing the coefficient of friction earlier, that is, under lower levels of utilization ($\sim$30–40%).

(a) **The scaled warning distance is defined as follows:**

$$d_{cr\ warning,\ scaled} = \left(\frac{v_a^2}{\mu g} - \frac{(v_a - v_{rel})^2}{\mu g} \right) + v_a \cdot \tau_1 + (v_a - v_b) \cdot \tau_2 + d_o$$

$$\alpha_1 = \alpha_2 \approx \mu g \quad \right] \quad \text{Maximum deceleration condition – depends on tire-road friction coefficient}$$

$$*ma = \sum F_x => ma = \mu(F_{z1} + F_{z2} + F_{z3} + F_{z4})$$
$$or\ ma = \mu mg => a = \mu g$$

(b) **The scaled braking distance is defined as follows:**

$$d_{cr\ braking,\ scaled} = \left(v_{rel} \cdot (\tau_1 + \tau_2) + 0.5 \cdot \alpha_2 \cdot (\tau_1 + \tau_2)^2 \right) \cdot f(\mu)$$

$$where:$$

$$f(\mu) = \begin{cases} f(\mu_{min}) & if\ \mu \leq \mu_{min} \\ f(\mu_{min}) + \dfrac{f(\mu_{norm}) - f(\mu_{min})}{\mu_{norm} - \mu_{min}} \cdot (\mu - \mu_{min}) & if\ \mu_{min} < \mu < \mu_{norm} \\ f(\mu_{norm}) & if\ \mu \geq \mu_{norm} \end{cases}$$

(piece-wise linear scaling function)

*Parameters :
$f(\mu_{norm}) = 1$
$f(\mu_{min}) = 2.4$
$\mu_{norm} = 0.8$
$\mu_{min} = 0.15$

*Parameters :
$\tau_1 = 0.9s$
$\tau_2 = 0.2s$
$d_o = 5m$
$\alpha_2 = 7.5 m/s^2$

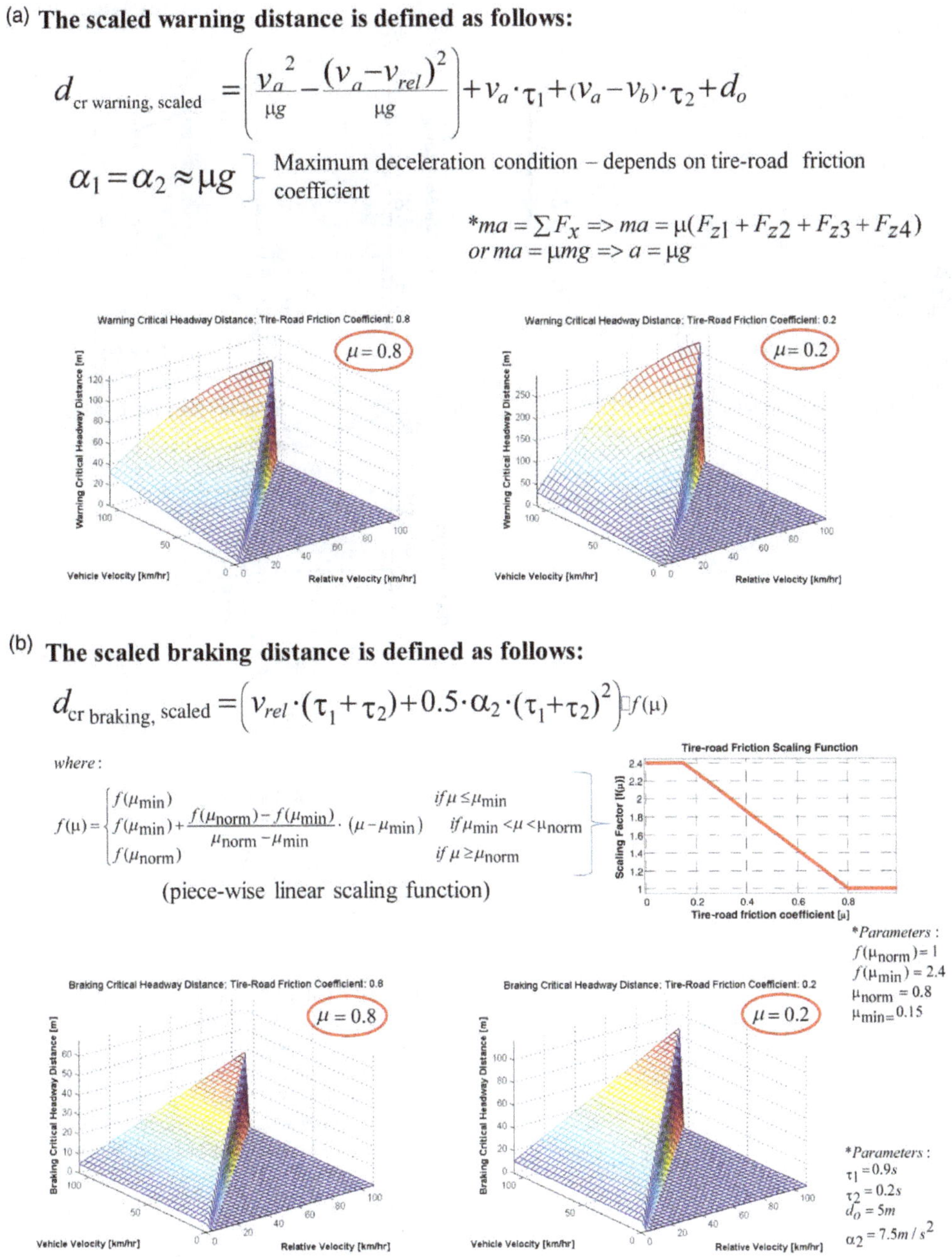

Figure 31.   (a) Scaled warning distance and (b) scaled braking distance.

To be of greatest use to active safety control systems, an estimation method needs to offer even earlier knowledge of the limits, that is, ideally offer knowledge on peak friction level under free-rolling conditions. To achieve this, an integrated approach using the intelligent tire-based friction estimator and the model-based estimator is proposed. This would give us the capability to reliably estimate friction for a wider range of excitations. The proposed intelligent tire-based method characterizes the road surface friction level using the measured frequency response of the tire vibrations and provides the capability to estimate the tire road friction coefficient under extremely lower levels of force utilization, that is, under free-rolling to low-slip excitation conditions. The ability to reliably estimate tire–road friction coefficient is important for maximizing the performance of vehicle control systems, which work well only when the tire force command computed by the safety systems is within the friction limit. The development of a sensorized smart/intelligent tire system is expected to eliminate some of the vehicle sensors and provide accurate, reliable and real-time information about magnitudes, directions and limits of force for each tire. Benefits of application of knowledge of friction potential have been demonstrated for an ABS, ESC system and CMBS.

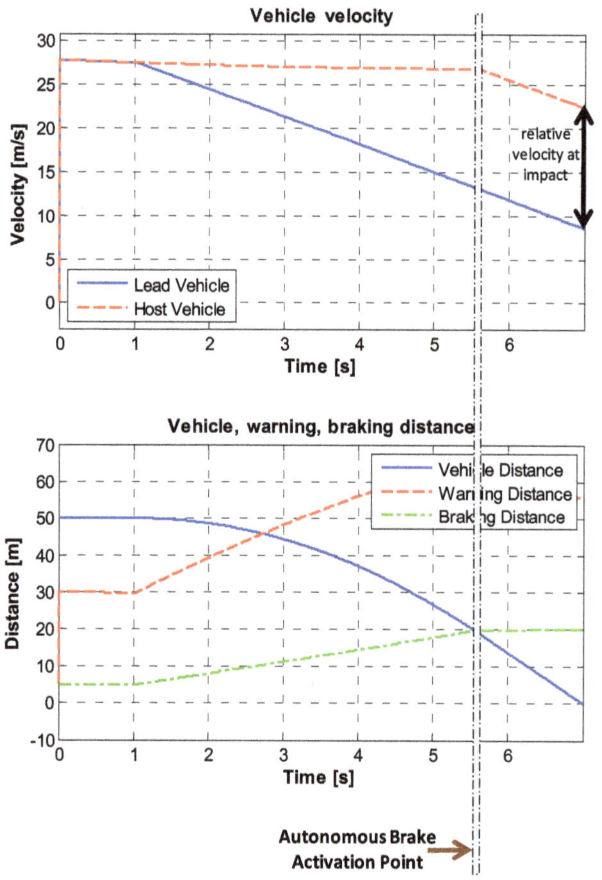

Figure 32. Friction condition unknown (assumed $\mu = 0.8$; actual $\mu = 0.25$).

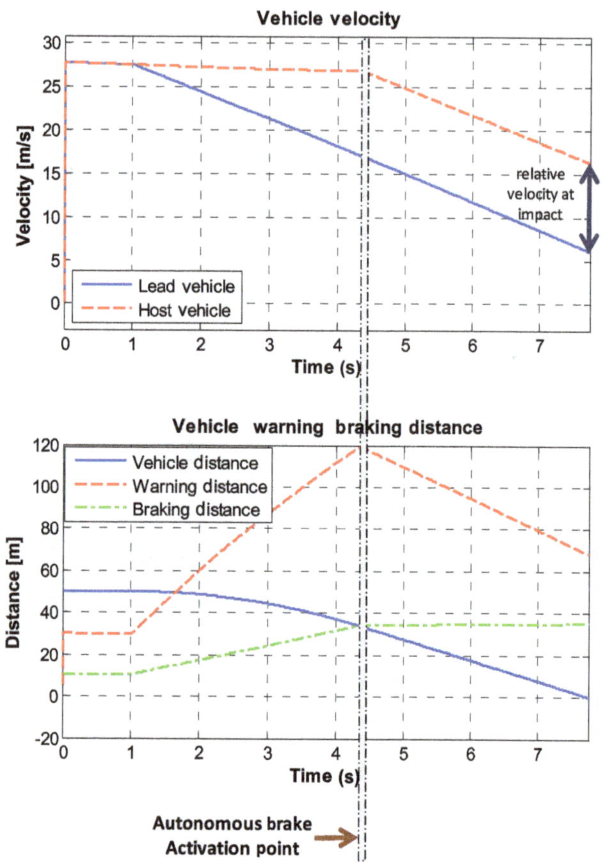

Figure 33. Friction condition known (assumed $\mu = 0.25$; actual $\mu = 0.25$).

Disclosure statement

No potential conflict of interest was reported by the author(s).

ORCID

Kanwar Bharat Singh http://orcid.org/0000-0001-7795-007X

References

Ahn, C. S. (2011). *Robust estimation of road friction coefficient for vehicle active safety systems*. Ann Arbor: Department of Mechanical Engineering, The University of Michigan.

Andersson, M., Bruzelius, F., Casselgren, J., Hjort, M., Löfving, S., Olsson, G., ... Yngve, S. (2010). *Road friction estimation, Part II – IVSS project report*. Retrieved from http://www.ivss.se/upload/ivss_refll_slutrapport.pdf

Arat, M. A., Singh, K. B., & Taheri, S. (2013). Optimal tire force allocation by means of smart tire technology. *SAE International Journal of Passenger Cars–Mechanical Systems*, *6*, 163–176.

Arat, M. A., Singh, K. B., & Taheri, S. (2014). An intelligent tire based adaptive vehicle stability controller. *International Journal of Vehicle Design*, *65*(2/3), 118–143.

Arosio, D., Braghin, F., Cheli, F., & Sabbioni, E. (2005). Identification of Pacejka's scaling factors from full-scale experimental tests. *Vehicle System Dynamics*, *43*, 457–474.

Braghin, F., Cheli, F., & Sabbioni, E. (2006). Environmental effects on Pacejka's scaling factors. *Vehicle System Dynamics*, *44*, 547–568.

Braghin, F., Cheli, F., Melzi, S., Sabbioni, E., Mancosu, F., & Brusarosco, M. (2009). *Development of a cyber tire to enhance performances of active control systems*. Presented at the 7th EUROMECH Solid Mechanics Conference, Lisbon, Portugal.

Chankyu, L., Hedrick, K., & Kyongsu, Y. (2004). Real-time slip-based estimation of maximum tire–road friction coefficient. *IEEE/ASME Transactions on Mechatronics*, *9*, 454–458.

Cheli, F., audisio, G., brusarosco, M., mancosu, F., Cavaglieri, D., & melzi, S. (2011a). *Cyber tyre: A novel sensor to improve vehicle's safety*. Presented at the SAE 2011 World Congress & Exhibition, Detroit, MI.

Cheli, F., Sabbioni, E., Sbrosi, M., Brusarosco, M., Melzi, S., & d'alessandro, V. (2011b). *Enhancement of ABS performance through on-board estimation of the tires response by means of smart tires*. Presented at the SAE 2011 World Congress & Exhibition, Detroit, MI.

Cheli, F., Leo, E., Melzi, S., & Sabbioni, E. (2010). On the impact of "smart tyres" on existing ABS/EBD control systems. *Vehicle System Dynamics*, *48*, 255–270.

Erdogan, G. (2009). *New sensors and estimation systems for the measurement of tire–road friction coefficient and tire slip variables*. Minneapolis, MN: Mechanical Engineering, University of Minnesota.

Erdogan, G., Hong, S., Borrelli, F., & Hedrick, K. (2011). *Tire sensors for the measurement of slip angle and friction coefficient and their use in stability control systems*. Presented at the SAE 2011 World Congress & Exhibition, Detroit, MI.

Germann, S., Wurtenberger, M., & Daiss, A. (1994). Monitoring of the friction coefficient between tyre and road surface. In *Control applications*. Proceedings of the Third IEEE Conference on, 1, 613–618.

Google Scholar search articles on tire road friction estimation. Retrieved from http://scholar.google.com/scholar?q = tire + road + friction + estimation&hl = en&as_sdt = 0&as_vis = 1&oi = scholart&sa = X&ei = eSU8VJbFLcf3yQTyy YLIAg&ved = 0CBsQgQMwAA

Gustafsson, F. (1997). Slip-based estimation of tire–road friction. *Automatica, 33*, 1087–1099.

Hsu, Y.-H. J. (2009). *Estimation and control of lateral tire forces using steering torque*. Stanford, CA: Department of Mechanical Engineering, Stanford University.

Klomp, M., & Lidberg, M. (2006). *Safety margin estimation in steady state maneuvers*. Presented at the Proceedings of AVEC '06 – The 8th International Symposium on Advanced Vehicle Control, Taipei, Taiwan.

Li, B., Du, H., & Li, W. (2013). *A novel cost effective method for vehicle tire–road friction coefficient estimation*. Advanced Intelligent Mechatronics (AIM), 2013 IEEE/ASME International Conference on, 1528–1533.

Li, K., Misener, J. A., & Hedrick, K. (March 1, 2007). On-board road condition monitoring system using slip-based tyre–road friction estimation and wheel speed signal analysis. *Proceedings of the Institution of Mechanical Engineers, Part K: Journal of Multi-body Dynamics, 221*, 129–146.

Li, L., W. Fei-Yue, & Qunzhi, Z. (2006). Integrated longitudinal and lateral tire/road friction modeling and monitoring for vehicle motion control. *IEEE Transactions on Intelligent Transportation Systems, 7*, 1–19.

Lourakis, M. I. A. (2005). A brief description of the Levenberg-Marquardt algorithm implemented by levmar. *Foundation of Research and Technology, 4*, 1–6.

Luque, P., Mántaras, D. A., Fidalgo, E., Álvarez, J., Riva, P., Girón, P., . . . Ferran, J. (2013). Tyre–road grip coefficient assessment – Part II: online estimation using instrumented vehicle, extended Kalman filter, and neural network. *Vehicle System Dynamics, 51*(12), 1872–1893.

Matilainen, M. J., & Tuononen, A. J. (2011). *Tire friction potential estimation from measured tie rod forces*. Intelligent Vehicles Symposium (IV), 2011 IEEE, 320–325.

Morinaga, H., *CAIS technology for detailed classification of road surface condition*. Presented at the Tire Technology Expo, Cologne, Germany.

Muller, S., Uchanski, M., & Hedrick, K. (2003). Estimation of the maximum tire–road friction coefficient. *Journal of Dynamic Systems, Measurement, and Control, 125*, 607–617.

Nishihara, O., & Masahiko, K. (2011). Estimation of road friction coefficient based on the brush model. *Journal of Dynamic Systems, Measurement, and Control, 133*, 041006–9.

Pacejka, H. B. (2005). Tyre brush model. In *Tyre and vehicle dynamics* (pp. 93–134). 2nd ed. Oxford: Elsevier.

Pasterkamp, W. R., & Pacejka, H. B. (1997). The tyre as a sensor to estimate friction. *Vehicle System Dynamics, 27*, 409–422.

Rajamani, R., Phanomchoeng, G., Piyabongkarn, D., & Lew, J. Y. (2012). Algorithms for real-time estimation of individual wheel tire–road friction coefficients. *IEEE/ASME Transactions on Mechatronics, 17*, 1183–1195.

Rajamani, R., Piyabongkarn, N., Lew, J., Yi, K., & Phanomchoeng, G. (2010). Tire–road friction-coefficient estimation. *IEEE Control Systems, 30*, 54–69.

Ray, L. R. (1997). Nonlinear tire force estimation and road friction identification: Simulation and experiments. *Automatica, 33*, 1819–1833.

Roweis, S. (1996). Levenberg–Marquardt optimization. Lecture notes, Department of Computer Science, University of Toronto. Retrieved from https://www.cs.nyu.edu/ ~ roweis/notes/lm.pdf

Sabbioni, E., Kakalis, L., & Cheli, F. (2010). *On the impact of the maximum available tire–road friction coefficient awareness in a brake-based torque vectoring system*. Presented at the SAE 2010 World Congress & Exhibition, Detroit, MI.

Singh, K. B. (2012). *Development of an intelligent tire based tire – vehicle state estimator for application to global chassis control* (Master's thesis). Dept of Mechanical Engineering, Virginia Tech.

Singh, K. B., Arat, M., & Taheri, S. (2012). Enhancement of collision mitigation braking system performance through real-time estimation of tire–road friction coefficient by means of smart tires. *SAE International Journal of Passenger Cars – Electronic and Electrical Systems, 5*(2), 607–624.

Singh, K. B., Arat, M., & Taheri, S. (2013). An intelligent tire based tire–road friction estimation technique and adaptive wheel slip controller for antilock brake system. *Journal of Dynamic Systems, Measurement, and Control, 135*, 031002–031002.

Singh, K. B., Bedekar, V., Taheri, S., & Priya, S. (2012). Piezoelectric vibration energy harvesting system with an adaptive frequency tuning mechanism for intelligent tires. *Mechatronics, 22*, 970–988.

Singh, K. B., & Taheri, S. (October–December, 2013). Piezoelectric vibration-based energy harvesters for next-generation intelligent tires. *Tire Science and Technology, 41*(4), 262–293.

Svendenius, J. (2003). *Tire models for use in braking applications*. Masters, Lund, Sweden: Department of Automatic Control, Lund Institute of Technology.

Svendenius, J. (2007). *Tire modeling and friction estimation*. Lund: Department of Automatic Control, Lund University.

Yasushi Hanatsuka, Y. W., & Morinaga, Hiroshi. (2013). *Method for estimating condition of road surface* (US 20130116972 A1). USA Patent.

Yasui, Y., Tanaka, W., Muragishi, Y., Ono, E., Momiyama, M., Katoh, H., . . . Imoto, Y. (2004). *Estimation of lateral grip margin based on self-aligning torque for vehicle dynamics enhancement*. Presented at the SAE 2004 World Congress & Exhibition, Detroit, MI.

24

Analysis of a delayed hand–foot–mouth disease epidemic model with pulse vaccination

G.P. Samanta*†

Institute of Mathematics, National Autonomous University of Mexico, Mexico D.F., C.P. 04510, Mexico

In this paper, we have considered a dynamical model of hand–foot–mouth disease (HFMD) with varying total population size, saturation incidence rate and discrete time delay to become infectious. It is assumed that there is a time lag (τ) to account for the fact that an individual infected with virus is not infectious until after some time after exposure. The probability that an individual remains in the latency period (exposed class) at least t time units before becoming infectious is given by a step function with value 1 for $0 \leq t < \tau$ and value zero for $t > \tau$. The probability that an individual in the latency period has survived is given by $e^{-\mu\tau}$, where μ denotes the natural mortality rate in all epidemiological classes. It is reported that the first vaccine to protect children against enterovirus 71, or EV71 has been discovered [Zhu, F. C., Meng, F. Y., Li, J. X., Li, X. L., Mao, A. Y., Tao, H., …, Shen, X. L. (2013, May 29). Efficacy, safety, and immunology of an inactivated alum-adjuvant enterovirus 71 vaccine in children in China: A multicentre, randomised, double-blind, placebo-controlled, phase 3 trial. *The Lancet, 381*, 2024–2032. doi:10.1016/S0140-6736(13)61049-1]. Pulse vaccination is an effective and important strategy for the elimination of infectious diseases and so we have analyzed this model with pulse vaccination. We have defined two positive numbers R_1 and R_2. It is proved that there exists an infection-free periodic solution which is globally attractive if $R_1 < 1$ and the disease is permanent if $R_2 > 1$. The important mathematical findings for the dynamical behavior of the HFMD model are also numerically verified using MATLAB. Finally epidemiological implications of our analytical findings are addressed critically.

Keywords: hand–foot–mouth disease; pulse vaccination; permanence; extinction; global stability

1. Introduction

Infectious diseases have tremendous influence on human life and are usually caused by pathogenic microorganisms, such as bacteria, viruses, parasites, or fungi. The diseases can be spread directly or indirectly. Hand–foot–mouth disease (HFMD) is a contagious (transmitted by bodily contact with an infected individual) disease of early childhood caused by viruses that belong to the enterovirus (EV) genus (group) which includes polioviruses, coxsackieviruses, echoviruses, and EVs. EVs are among the most common human viruses infecting around one billion persons worldwide annually and is divided into 10 species. Most EV infections are asymptomatic (Bracho, González-Candelas, Valero, Córdoba, & Salazar, 2011). The most common viruses causing the spread of HFMD are coxsackievirus A16 (COX A16) and enterovirus 71 (EV71) (Yang, Chen, & Zhang, 2013; Zhu, Hao, Ma, Yu, & Wang, 2011). HFMD may also be caused by other EVs. It is common in children (<10 years of age) but can also occur in adults and most patients with fatal complications are infected by EV71 and coxsackievirus A16 (Bracho et al., 2011). It usually takes 3–7 days for a person to get symptoms of HFMD disease after being exposed to the virus. This is called the incubation period of HFMD. Although many HFMD-infected people remain asymptomatic, the symptoms of HFMD include sores in or on the mouth and on the hands, feet, and sometimes the buttocks and legs. The sores may be painful. The virus spreads easily through coughing and sneezing. It can also spread through infected stool.

Although HFMD is classically known as a mild disease, outbreaks in Asia have been associated with a high incidence of fatal cardiopulmonary and neurologic complications (Bracho et al., 2011; Wong, Yip, Lau, & Yuen, 2010). It not only causes health problems but also has great social and economical impacts throughout the world. Because of its global spread and the associated morbidity and mortality it inflicts, much attention has been focused on devising methods for controlling the spread of HFMD based on appropriate preventive measures. These measures include quarantine mechanisms (a strict isolation imposed to prevent the spread of the disease) and personal protection against exposure to infected persons (Liu, 2011). In the past, there was no specific treatment for HFMD. On 29 May 2013, it is reported that Chinese scientists have developed the first vaccine to protect children against enterovirus 71, or EV71, that causes the common and sometimes deadly HFMD (Zhu et al., 2013).

*Email: gpsamanta@math.becs.ac.in

†Present address: Department of Mathematics, Bengal Engineering and Science University, Shibpur, Howrah-711103, India.

The pulse vaccination strategy (PVS) consists of repeated application of vaccine at discrete time with equal interval in a population in contrast to the traditional constant vaccination (Gakkhar & Negi, 2008; Zhou & Liu, 2003). Compared to the proportional vaccination models, the study of pulse vaccination models is in its infancy (Zhou & Liu, 2003). At each vaccination time a constant fraction of the susceptible population is vaccinated successfully. Since 1993, attempts have been made to develop mathematical theory to control infectious diseases using pulse vaccination (Agur, Cojocaru, Mazor, Anderson, & Danon, 1993; Gakkhar & Negi, 2008). Nokes & Swinton, (1995) discussed the control of childhood viral infections by PVS. Stone, Shulgin, & Agur, (2000) presented a theoretical examination of the PVS in the susceptible-infected-recovered (SIR) epidemic model and d'Onofrio (2002a, 2002b) analyzed the use of pulse vaccination policy to eradicate infectious disease for SIR and susceptible-exposed-infected-recovered (SEIR) epidemic models. Different types of vaccination policies and strategies combining pulse vaccination policy, treatment, pre-outbreak vaccination or isolation have already been introduced by several researchers (Babiuk, Babiuk, & Baca-Estrada, 2002; d'Onofrio, 2005; Gao, Chen, Nieto, & Torres, 2006; Gao, Chen, & Teng, 2007; Gjorrgjieva et al., 2005; Tang, Xiao, & Clancy, 2005; Wei & Chen, 2008).

Mathematical epidemiology is the study of the spread of diseases, in space and time, with the objective to identify factors that are responsible for or contributing to their occurrence. Mathematical models are becoming important tools in analyzing the spread and control of infectious diseases. Epidemic models of ordinary differential equations have been studied by a number of researchers (Anderson & May, 1992; Brauer & Castillo-Chavez, 2001; Cai, Li, Ghosh, & Guo, 2009; Capasso, 1993; Diekmann & Heesterbeek, (2000); Kermack & Mckendrick, 1927; Ma, Song, & Takeuchi, 2004; Mena-Lorca & Hethcote, 1992; Meng, Chen, & Cheng, 2007; Naresh, Tripathi, & Omar, 2006; Thieme, 2003). The basic and important objectives for these models are the existence of the threshold values which distinguish whether the infectious disease will be going to extinct, the local and global stability of the disease-free equilibrium and the endemic equilibrium, the existence of periodic solutions and the persistence of the disease. Stability, persistence and permanence in population biology have been studied by many researchers (Takeuchi, Cui, Rinko, & Saito, 2006a, 2006b). Hence, as a part of population biology, permanence of disease plays an important role in mathematical epidemiology.

Although HFMD is a disease of significant public health importance, the transmission dynamics of the HMFD has not yet received adequate research attention in the mathematical modeling of epidemiology literature. It is noted here that very little attention has been paid to the mathematical modeling and analysis of HMFD to gain insight into its transmission dynamics at population level. Urashima, Shindo, & Okabe, (2003) and Wang & Sung, (2004) attempted to find the relationship between the outbreaks of HFMD with the weather patterns in Tokyo and Taiwan respectively. Chuo, Tiing, & Labadin, (2008) used a deterministic SIR model to predict the number of infected and the duration of an outbreak of HMFD when it occurs in Sarawak. Then Roy & Halder (2010) proposed a deterministic SEIR model of HFMD and did only numerical simulations. Recently, Liu (2011) and Yang et al. (2013) used the SEIQRS model to take into account of the quarantine measure. Motivated by the above works and the recent development of the first vaccine to children against enterovirus 71, or EV71 (Zhu et al., 2013), in this paper, we are concerned with the effect of pulse vaccination and saturation incidence on the dynamic of a delayed SEI_AI_SQRS epidemic model of HFMD. Here we have used the Kermack–McKendrick compartmental modeling framework, which entails sub-dividing the entire high-risk human population into mutually exclusive epidemiological compartments (based on disease status), to gain insights into the qualitative features of HFMD in a human population (with the aim of finding effective ways to control its spread). The main feature of this paper is to introduce time delay, saturation incidence rate with valid PVS. We have introduced two threshold values R_1 and R_2 and further obtained that the disease will be going to extinct when $R_1 < 1$ and the disease will be permanent when $R_2 > 1$. The important mathematical findings for the dynamical behavior of the HFMD model are numerically verified using MATLAB and also epidemiological implications of our analytical findings are addressed critically in the Section 5. The aim of the analysis of this model is to trace the parameters of interest for further study, with a view to informing and assisting policy-maker in targeting prevention and treatment resources for maximum effectiveness.

2. Model derivation and preliminaries

In the following, we consider a dynamical model of HFMD caused by EVs with discrete time delay and PVS which satisfies the following assumptions:

The underlying high-risk human population is split up into six mutually exclusive classes (compartments), namely, susceptible (S), exposed (infected but not yet infectious) (E), infective in asymptomatic phase (showing no symptoms of HFMD) (I_A), infective in symptomatic phase (showing symptoms of HFMD) (I_S), infective in symptomatic phase who follow quarantine (a strict isolation imposed to prevent the spread of the disease) mechanisms and personal protection against infecting others (Q) and recovered (infectious people who have cleared (or recovered from) HFMD infection) (R).

The susceptible population increases through birth (a constant influx Λ of susceptible is assumed) and from recovered hosts and decreases due to direct contact with an infectious individual (in I_A or I_S compartments), natural death and PVS.

Standard epidemiological models use a bilinear incidence rate βSI based on the law of mass action (Anderson & May, 1979; 1992) and it is reasonable when the mixing of susceptible with infective is considered to be homogeneous. If the population is saturated with infective, there are three types of incidence forms used in epidemiological model: the proportionate mixing incidence $\beta(SI/N)$ (Anderson & May, 1992; Cooke & van Den Driessche, 1996; Wang, 2002), nonlinear incidence $\beta S^p I^q$ (Hethcote & van Den Driessche, 1991; Hui & Chen, 2004) and saturation incidence $\beta(SI/(1+\sigma S))$ (Anderson & May, 1992; May & Anderson, 1978) or $\beta(SI^p/(1+\sigma I^q))$ (Ruan & Wang, 2003). Here incidence rates $\beta_1(SI_A/(1+\sigma_1 S))$ and $\beta_2(SI_S/(1+\sigma_2 S))$ have been considered.

The infected classes are increased by infection of susceptible. A fraction of the exposed individuals will start to show symptoms of HFMD (and move to the class I_S), while the remaining fraction will not (but still remain capable of infecting others and move to the class I_A). Also, a fraction of the infective in symptomatic phase takes appropriate preventive measures and move to the quarantined class Q. It is assumed that there is a time lag to account for the fact that an individual infected with HFMD is not infectious until after some time (typically 3–7 days (Yang et al., 2013)) after exposure. A fraction of the asymptomatically infectious individuals eventually show disease symptoms (and move to the class I_S) and a fraction recover (and move to the class R). The infected classes are decreased through recovery from infection, by disease-related death and by natural death. Motivated by the recent development of the first vaccine to protect children against enterovirus 71, or EV71 (Zhu et al. 2013), we incorporate a PVS in which a fraction p of the susceptible population is vaccinated successfully at discrete time $t = T, 2T, 3T, \ldots$.

Thus, the following dynamical model of HFMD with discrete time delay and PVS is formulated:

$$
\begin{aligned}
\frac{dS(t)}{dt} &= \Lambda - \beta_1 \frac{S(t)I_A(t)}{1+\sigma_1 S(t)} - \beta_2 \frac{S(t)I_S(t)}{1+\sigma_2 S(t)} \\
&\quad - \mu S(t) + \alpha R(t), \quad t \neq nT, \\
\frac{dE(t)}{dt} &= \beta_1 \frac{S(t)I_A(t)}{1+\sigma_1 S(t)} + \beta_2 \frac{S(t)I_S(t)}{1+\sigma_2 S(t)} \\
&\quad - \beta_1 e^{-\mu\tau} \frac{S(t-\tau)I_A(t-\tau)}{1+\sigma_1 S(t-\tau)} \\
&\quad - \beta_2 e^{-\mu\tau} \frac{S(t-\tau)I_S(t-\tau)}{1+\sigma_2 S(t-\tau)} - \mu E(t), \quad t \neq nT, \\
\frac{dI_A(t)}{dt} &= \rho e^{-\mu\tau} S(t-\tau) \\
&\quad \times \left\{ \frac{\beta_1 I_A(t-\tau)}{1+\sigma_1 S(t-\tau)} + \frac{\beta_2 I_S(t-\tau)}{1+\sigma_2 S(t-\tau)} \right\} \\
&\quad - (r_1 + d_1 + \mu) I_A(t), \quad t \neq nT, \\
\frac{dI_S(t)}{dt} &= (1-\rho) e^{-\mu\tau} S(t-\tau) \\
&\quad \times \left\{ \frac{\beta_1 I_A(t-\tau)}{1+\sigma_1 S(t-\tau)} + \frac{\beta_2 I_S(t-\tau)}{1+\sigma_2 S(t-\tau)} \right\} \\
&\quad + (1-k) r_1 I_A(t) - (q + r_2 + d_2 \\
&\quad + \mu) I_S(t), \quad t \neq nT, \\
\frac{dQ(t)}{dt} &= q I_S(t) - (r_3 + d_3 + \mu) Q(t), \quad t \neq nT, \\
\frac{dR(t)}{dt} &= k r_1 I_A(t) + r_2 I_S(t) + r_3 Q(t) \\
&\quad - \mu R(t) - \alpha R(t), \quad t \neq nT, \\
S(t^+) &= (1-p) S(t), \quad t = nT, \ n = 1, 2, \ldots \\
E(t^+) &= E(t), \quad t = nT, \ n = 1, 2, \ldots \\
I_A(t^+) &= I_A(t), \quad t = nT, \ n = 1, 2, \ldots \\
I_S(t^+) &= I_S(t), \quad t = nT, \ n = 1, 2, \ldots \\
Q(t^+) &= Q(t), \quad t = nT, \ n = 1, 2, \ldots \\
R(t^+) &= R(t) + p S(t), \quad t = nT, \ n = 1, 2, \ldots,
\end{aligned} \tag{1}
$$

where all coefficients are positive constants. Here $S(t)$ denotes the number of susceptible, $E(t)$ denotes the number of exposed, $I_A(t)$ denotes the number of infective in asymptomatic compartment, $I_S(t)$ denotes the number of infective in symptomatically infected compartment, $Q(t)$ denotes the number of symptomatically infective in quarantined compartment and $R(t)$ denotes the number of recovered individuals. The pulse vaccination does not give life-long immunity, there is an immunity waning for the vaccination with the per capita immunity waning rate α, and return to the susceptible class. The influx of susceptible comes from two sources: a constant recruitment Λ and from recovered hosts (αR). The parameters $\beta_1, \beta_2, \mu, \rho, d_1, d_2, d_3, r_1, r_2, r_3, \tau, p$ are:

β_1: The coefficient of transmission rate from infective in asymptomatic compartment to susceptible humans (and become exposed) and the rate of transmission of infection is of the form:

$$\beta_1 \frac{S(t)I_A(t)}{1+\sigma_1 S(t)}.$$

β_2: The coefficient of transmission rate from infective in symptomatically infected compartment to susceptible humans (and become exposed) and the rate of transmission of infection is of the form:

$$\beta_2 \frac{S(t)I_S(t)}{1+\sigma_2 S(t)}.$$

μ: The coefficient of natural death rate of all epidemiological human classes.

d_1: The coefficient of additional disease-related death rate of infective in asymptomatic compartment (I_A).

d_2: The coefficient of additional disease-related death rate of infective in symptomatically infected compartment (I_S).

d_3: The coefficient of additional disease-related death rate of infective in quarantined compartment (Q).

$(1-\rho)$: The fraction of the exposed individuals will start to show disease symptoms and move to the class I_S. The remaining fraction ρ $(0<\rho<1)$ will not start to show disease symptoms (but still remain capable of infecting others) and move to the class I_A.

$(1-k)r_1$: The rate at which the asymptomatically infectious individuals eventually show disease symptoms (move to the class I_S) and recover at the rate kr_1 $(0<k<1)$ (move to the class R).

r_2: The rate at which the infectious individuals showing symptoms of HFMD (in symptomatically infected compartment I_S) clear infections and move to the class R.

r_3: The rate at which symptomatically infected individuals (infective in quarantined compartment Q) clear infections and move to the class R.

q: The quarantine rate.

τ: The constant latency period from the time of being infected (exposed) to the time of being infectious (capable of infecting others). The probability that an individual remains in the latency period (exposed class) at least t time units before becoming infectious is given by a step function with value 1 for $0 \le t < \tau$ and value zero for $t > \tau$. The probability that an individual in the latency period has survived is given by $e^{-\mu\tau}$. The time interval $[t-\tau, t]$ is typically 3–7 days (Yang et al. 2013).

$p(0<p<1)$: The fraction of susceptible who are vaccinated successfully at discrete time $t = T, 2T, 3T, \ldots$, which is called impulsive vaccination rate.

The total high-risk human population size $N(t) = S(t)+E(t)+I_A(t)+I_S(t)+Q(t)+R(t)$ can be determined by the following differential equation:

$$\frac{dN(t)}{dt} = \Lambda - \mu N(t) - d_1 I_A(t) - d_2 I_S(t) - d_3 Q(t), \quad (2)$$

which is derived by adding first six equations of system (1). Therefore,

$$\begin{aligned}
&\Lambda - (\mu + d_1 + d_2 + d_3)N(t) \le \frac{dN(t)}{dt} \le \Lambda - \mu N(t) \\
&\Rightarrow \frac{\Lambda}{\mu + d_1 + d_2 + d_3} \le \liminf_{t\to\infty} N(t) \\
&\le \limsup_{t\to\infty} N(t) \le \frac{\Lambda}{\mu}.
\end{aligned} \quad (3)$$

Let us simplify the model (1) as follows:

$$\begin{aligned}
\frac{dS(t)}{dt} &= \Lambda - \beta_1 \frac{S(t)I_A(t)}{1+\sigma_1 S(t)} - \beta_2 \frac{S(t)I_S(t)}{1+\sigma_2 S(t)} \\
&\quad - \mu S(t) + \alpha R(t), \quad t \ne nT, \\
\frac{dI_A(t)}{dt} &= \rho\, e^{-\mu\tau} S(t-\tau) \\
&\quad \times \left\{ \frac{\beta_1 I_A(t-\tau)}{1+\sigma_1 S(t-\tau)} + \frac{\beta_2 I_S(t-\tau)}{1+\sigma_2 S(t-\tau)} \right\} \\
&\quad - (r_1 + d_1 + \mu) I_A(t), \quad t \ne nT, \\
\frac{dI_S(t)}{dt} &= (1-\rho)\, e^{-\mu\tau} S(t-\tau) \\
&\quad \times \left\{ \frac{\beta_1 I_A(t-\tau)}{1+\sigma_1 S(t-\tau)} + \frac{\beta_2 I_S(t-\tau)}{1+\sigma_2 S(t-\tau)} \right\} \\
&\quad + (1-k) r_1 I_A(t) - (q + r_2 + d_2 \\
&\quad + \mu) I_S(t), \quad t \ne nT, \\
\frac{dQ(t)}{dt} &= q I_S(t) - (r_3 + d_3 + \mu) Q(t), \quad t \ne nT, \\
\frac{dR(t)}{dt} &= k r_1 I_A(t) + r_2 I_S(t) + r_3 Q(t) - \mu R(t) \\
&\quad - \alpha R(t), \quad t \ne nT, \\
\frac{dN(t)}{dt} &= \Lambda - \mu N(t) - d_1 I_A(t) - d_2 I_S(t) \\
&\quad - d_3 Q(t), \quad t \ne nT, \\
S(t^+) &= (1-p)S(t), t = nT, n = 1, 2, \ldots \\
I_A(t^+) &= I_A(t), t = nT, n = 1, 2, \ldots \\
I_S(t^+) &= I_S(t), t = nT, n = 1, 2, \ldots \\
Q(t^+) &= Q(t), t = nT, n = 1, 2, \ldots \\
R(t^+) &= R(t) + pS(t), t = nT, n = 1, 2, \ldots \\
N(t^+) &= N(t), t = nT, n = 1, 2, \ldots,
\end{aligned} \quad (4)$$

with initial conditions

$$\begin{aligned}
&S(\vartheta) = \varphi_1(\vartheta), \quad I_A(\vartheta) = \varphi_2(\vartheta), \quad I_S(\vartheta) = \varphi_3(\vartheta), \\
&Q(\vartheta) = \varphi_4(\vartheta), \quad R(\vartheta) = \varphi_5(\vartheta), \\
&N(\vartheta) = \varphi_6(\vartheta), \quad \text{such that } \varphi_i(\vartheta) \\
&\qquad \ge 0 (i = 1, 2, 3, 4, 5, 6), \quad \forall \vartheta \in [-\tau, 0],
\end{aligned} \quad (5)$$

where $\varphi_i(\vartheta) \ge 0$ $(i = 1, 2, 3, 4, 5, 6)$ are nonnegative continuous functions on $\vartheta \in [-\tau, 0]$. For a biological meaning, we further assume that $\varphi_i(0) > 0$ $(i = 1, 2, 3, 4, 5, 6)$. There exists a unique solution of Equation (4) with initial conditions (5) since the right-hand sides of Equation (4) and the pulse are smooth functions (Bainov & Simeonov, 1993; 1995; Lakshmikantham, Bainov, & Simeonov, 1989).

From biological considerations, we analyze system (4) and (5) in the closed set:

$$\begin{aligned}
G = \Big\{ &(S(t), I_A(t), I_S(t), Q(t), R(t), N(t)) \in \mathbb{R}_+^6 : 0 \le S \\
&+ I_A + I_S + Q + R, N \le \frac{\Lambda}{\mu} \Big\},
\end{aligned} \quad (6)$$

where $\mathbb{R}_+^6$ represents the nonnegative cone of $\mathbb{R}^6$ including its lower dimensional faces. It can be verified that G is positively invariant with respect to Equations (4) and (5).

Before starting our main results, we give the following two lemmas which will be essential for study.

LEMMA 2.1 (Song & Chen, 2001) *Consider the following equation*:

$$\frac{\mathrm{d}x(t)}{\mathrm{d}t} = ax(t-\tau) - bx(t) - cx^2(t), \tag{7}$$

where $a, b, c, \tau > 0$; $x(t) > 0$, *for* $-\tau \leq t \leq 0$. *We have*

$$(I)\ \text{if } a > b, \text{ then } \lim_{t\to\infty} x(t) = \frac{a-b}{c};$$

$$(II)\ \text{if } a < b, \text{ and } c \geq 0, \text{ then } \lim_{t\to\infty} x(t) = 0.$$

LEMMA 2.2 *Consider the following impulsive differential equation*:

$$\begin{aligned} &\frac{\mathrm{d}u(t)}{\mathrm{d}t} = a - bu(t), \quad t \neq kT, \\ &u(t^+) = (1-p)u(t), \quad t = kT, \ k = 1, 2, \ldots \end{aligned} \tag{8}$$

where $a > 0$, $b > 0$, $0 < p < 1$. *Then there exists a unique positive periodic solution of system (8)*:

$$\tilde{u}_e(t) = \frac{a}{b} + \left(u^* - \frac{a}{b}\right)\mathrm{e}^{-b(t-kT)}, \quad kT < t \leq (k+1)T,$$

$$\text{where } u^* = \frac{a(1-p)(1-\mathrm{e}^{-bT})}{b\{1-(1-p)\,\mathrm{e}^{-bT}\}},$$

and $\tilde{u}_e(t)$ *is globally asymptotically stable.*

Proof From the first equation of system (8) we get,

$\frac{\mathrm{d}}{\mathrm{d}t}(\mathrm{e}^{bt}u(t)) = a\,\mathrm{e}^{bt}$. Integrating between pulses:

$$\times \int_{kT}^{t} \mathrm{d}(\mathrm{e}^{bt}u(t)) = \int_{kT}^{t} a\,\mathrm{e}^{bt}\,\mathrm{d}t$$

$$\Rightarrow u(t) = \frac{a}{b} + \left\{u(kT) - \frac{a}{b}\right\}\mathrm{e}^{-b(t-kT)},$$

$$kT < t \leq (k+1)T,$$

where $u(kT)$ is the initial value at time kT. Using the second equation of system (8) we have the following stroboscopic map:

$$\begin{aligned} u((k+1)T) &= (1-p)\left[\frac{a}{b} + \left\{u(kT) - \frac{a}{b}\right\}\mathrm{e}^{-bT}\right] \\ &= f(u(kT)), \end{aligned} \tag{9}$$

where $f(u) = (1-p)\left\{\frac{a}{b} + \left(u - \frac{a}{b}\right)\mathrm{e}^{-bT}\right\}$.

Solving the following equation:

$u = (1-p)\left\{\frac{a}{b} + \left(u - \frac{a}{b}\right)\mathrm{e}^{-bT}\right\}$, we get,

$$u^* = \frac{a(1-p)(1-\mathrm{e}^{-bT})}{b\{1-(1-p)\,\mathrm{e}^{-bT}\}}.$$

Since $|f'(u)| = (1-p)\,\mathrm{e}^{-bT} < 1$, as $0 < p < 1$ and $b > 0$, the system (9) has a unique positive equilibrium $u^* = a(1-p)(1-\mathrm{e}^{-bT})/b\{1-(1-p)\,\mathrm{e}^{-bT}\}$ which is globally asymptotically stable. Hence the corresponding periodic solution of system (8)

$$\tilde{u}_e(t) = \frac{a}{b} + \left(u^* - \frac{a}{b}\right)\mathrm{e}^{-b(t-kT)}, \quad kT < t \leq (k+1)T,$$

$$\text{where } u^* = \frac{a(1-p)(1-\mathrm{e}^{-bT})}{b\{1-(1-p)\,\mathrm{e}^{-bT}\}}$$

is globally asymptotically stable. This completes the proof. ■

3. Global stability of the disease-free periodic solution

In this section, we discuss the existence of the disease-free periodic solution of system (4), in which infectious individuals (in I_A, I_S, Q compartments) are completely absent, that is, $I_A(t) = 0, \forall t \geq 0$, $I_S(t) = 0, \forall t \geq 0$ and $Q(t) = 0, \forall t \geq 0$. Under this circumstances, system (4) reduces to the following impulsive system without delay:

$$\begin{aligned} &\frac{\mathrm{d}S(t)}{\mathrm{d}t} = \Lambda - \mu S(t) + \alpha R(t), \quad t \neq nT, \\ &\frac{\mathrm{d}R(t)}{\mathrm{d}t} = -\mu R(t) - \alpha R(t), \quad t \neq nT, \\ &\frac{\mathrm{d}N(t)}{\mathrm{d}t} = \Lambda - \mu N(t), \quad t \neq nT, \\ &S(t^+) = (1-p)S(t), \quad t = nT, \ n = 1, 2, \ldots \\ &R(t^+) = R(t) + pS(t), \quad t = nT, \ n = 1, 2, \ldots \\ &N(t^+) = N(t), \quad t = nT, \ n = 1, 2, \ldots. \end{aligned} \tag{10}$$

From the third and sixth equations of system (10), we have $\lim_{t\to\infty} N(t) = \Lambda/\mu$.

Further, from the second and eighth equations of system (1) it follows that

$$\lim_{t\to\infty} E(t) = 0 \text{ as } I_A(t) = I_S(t) = Q(t) = 0, \quad \forall\, t \geq 0.$$

In the following, we shall show that the susceptible population $S(t)$ and recovered population $R(t)$ oscillate with period T, in synchronization with the periodic impulsive vaccination strategy under some condition. Consider the following limit system of system (10) as per the previous discussions:

$$R(t) = \frac{\Lambda}{\mu} - S(t),$$
$$\frac{dS(t)}{dt} = (\mu+\alpha)\left\{\frac{\Lambda}{\mu} - S(t)\right\}, \quad t \neq nT, \tag{11}$$
$$S(t^+) = (1-p)S(t), \quad t = nT, \ n = 1,2,\ldots.$$

Using Lemma 2.2, the periodic solution of system (11) is given below:

$$\tilde{S}_e(t) = \frac{\Lambda}{\mu} + \left(S^* - \frac{\Lambda}{\mu}\right) e^{-(\mu+\alpha)(t-nT)},$$
$$nT < t \leq (n+1)T, \quad \text{where}$$
$$S^* = \frac{\Lambda(1-p)(1-e^{-(\mu+\alpha)T})}{\mu\{1-(1-p)e^{-(\mu+\alpha)T}\}} \tag{12}$$

and $\tilde{S}_e(t)$ is globally asymptotically stable.

Denote $R_1 = \dfrac{\beta e^{-\mu\tau} A}{\{(1+\sigma A)\theta\}}$, where $\beta = \max\{\beta_1, \beta_2\}$,

$$\sigma = \min\{\sigma_1, \sigma_2\}, \quad A = \frac{\Lambda(1-e^{-(\mu+\alpha)T})}{\mu\{1-(1-p)e^{-(\mu+\alpha)T}\}}$$
$$\text{and } \theta = \min\{kr_1 + d_1 + \mu, r_2 + d_2 + \mu, r_3 + d_3 + \mu\} > 0. \tag{13}$$

THEOREM 3.1 *If $R_1 < 1$, then the disease-free periodic solution $(\tilde{S}_e(t), 0, 0, 0, \Lambda/\mu - \tilde{S}_e(t), \Lambda/\mu)$ of system (4) with initial conditions (5) is globally asymptotically stable.*

Proof Since $R_1 < 1$, we can choose $\epsilon > 0$ small enough such that

$$\frac{\beta e^{-\mu\tau}(A+\epsilon)}{1+\sigma(A+\epsilon)} < \theta, \quad \text{where } \beta = \max\{\beta_1, \beta_2\},$$
$$\sigma = \min\{\sigma_1, \sigma_2\}, \quad A = \frac{\Lambda(1-e^{-(\mu+\alpha)T})}{\mu\{1-(1-p)e^{-(\mu+\alpha)T}\}}$$
$$\text{and } \theta = \min\{kr_1 + d_1 + \mu, r_2 + d_2 + \mu, r_3 + d_3 + \mu\} > 0. \tag{14}$$

From the first and seventh equations of (4), it follows that

$$\frac{dS(t)}{dt} \leq (\mu+\alpha)\left\{\frac{\Lambda}{\mu} - S(t)\right\}, \quad t \neq nT,$$
$$S(t^+) = (1-p)S(t), \quad t = nT, \ n = 1,2,\ldots.$$

So, we consider the following comparison impulsive differential system:

$$\frac{dz(t)}{dt} = (\mu+\alpha)\left\{\frac{\Lambda}{\mu} - z(t)\right\}, \quad t \neq nT,$$
$$z(t^+) = (1-p)z(t), \quad t = nT, \ n = 1,2,\ldots. \tag{15}$$

By Equations (11) and (12), we know that the periodic solution of system (15),

$$\tilde{z}_e(t) = \tilde{S}_e(t) = \frac{\Lambda}{\mu} + \left(S^* - \frac{\Lambda}{\mu}\right) e^{-(\mu+\alpha)(t-nT)},$$
$$nT < t \leq (n+1)T, \tag{16}$$
$$\text{where } S^* = \frac{\Lambda(1-p)(1-e^{-(\mu+\alpha)T})}{\mu\{1-(1-p)e^{-(\mu+\alpha)T}\}}$$

is globally asymptotically stable. Let $(S(t), I_A(t), I_S(t), Q(t), R(t), N(t))$ be the solution of system (4) with initial conditions (5) and $S(0^+) = S_0 > 0$. If $z(t)$ be the solution of system (15) with initial value $z(0^+) = S_0 > 0$, then by the comparison theorem for impulsive differential equation (Lakshmikantham et al. 1989) there exists an integer $n_1 > 0$ such that

$$S(t) < z(t) < \tilde{z}_e(t) + \epsilon, \quad nT < t \leq (n+1)T, \ n > n_1$$
$$\Rightarrow S(t) < \tilde{z}_e(t) + \epsilon \leq \frac{\Lambda(1-e^{-(\mu+\alpha)T})}{\mu\{1-(1-p)e^{-(\mu+\alpha)T}\}}$$
$$+ \epsilon = \xi \text{ (say)}. \tag{17}$$

Further, from the second, third and fourth equations of system (4), we have $\forall t > nT + \tau$ and $\forall n > n_1$,

$$\frac{d}{dt}\{I_A(t) + I_S(t) + Q(t)\} \leq \frac{\beta\xi e^{-\mu\tau}}{1+\sigma\xi}\{I_A(t-\tau) + I_S(t-\tau)$$
$$+ Q(t-\tau)\} - \theta\{I_A(t) + I_S(t) + Q(t)\}. \tag{18}$$

Consider the following comparison equation:

$$\frac{dy(t)}{dt} = \frac{\beta\xi e^{-\mu\tau}}{1+\sigma\xi} y(t-\tau) - \theta y(t). \tag{19}$$

From Equation (14), we have

$$\frac{\beta\xi e^{-\mu\tau}}{1+\sigma\xi} < \theta \Rightarrow \lim_{t\to\infty} y(t) = 0, \text{ by Lemma 2.1.} \tag{20}$$

Set $(S(t), I_A(t), I_S(t), Q(t), R(t), N(t))$ be the solution of system (4) with initial conditions (5) and $I_A(\vartheta) = \varphi_2(\vartheta) \geq 0$, $I_S(\vartheta) = \varphi_3(\vartheta) \geq 0$, $Q(\vartheta) = \varphi_4(\vartheta) \geq 0, \forall\vartheta \in [-\tau, 0]$ where $\varphi_i(0) > 0 (i = 2,3,4)$, $y(t)$ be the solution of Equation (19) with initial condition $y(\vartheta) = \varphi_2(\vartheta) + \varphi_3(\vartheta) + \varphi_4(\vartheta) \geq 0, \forall\vartheta \in [-\tau, 0]$ where $\varphi_2(0) + \varphi_3(0) + \varphi_4(0) > 0$. By the comparison theorem of differential equation and the positivity of solution (with $I_A(t) \geq 0$, $I_S(t) \geq 0$, $Q(t) \geq 0$), we have

$$\lim_{t\to\infty}\{I_A(t) + I_S(t) + Q(t)\} = 0 \Rightarrow \lim_{t\to\infty} I_A(t) = \lim_{t\to\infty}$$
$$I_S(t) = \lim_{t\to\infty} Q(t) = 0. \tag{21}$$

Hence for any $\epsilon_1 > 0$ (sufficiently small), there exists a positive integer n_2, where $n_2T > n_1T + \tau$, such that $0 <$

$I_A(t), I_S(t), Q(t) < \epsilon_1$, $\forall t > n_2T$. Using the sixth equation of system (4), we get

$$\frac{dN(t)}{dt} > \Lambda - \mu N(t) - (d_1 + d_2 + d_3)\epsilon_1, \quad \forall t > n_2T. \tag{22}$$

Now,

$$\frac{dz_1(t)}{dt} = \{\Lambda - (d_1 + d_2 + d_3)\epsilon_1\} - \mu z_1(t) \Rightarrow \lim_{t\to\infty} z_1(t) = \frac{\Lambda - (d_1 + d_2 + d_3)\epsilon_1}{\mu}.$$

So, by the comparison theorem, there exists an integer $n_3 > n_2$ such that

$$N(t) \geq \frac{\Lambda - (d_1 + d_2 + d_3)\epsilon_1}{\mu} - \epsilon_1, \quad \forall t > n_3T$$
$$\Rightarrow \lim_{t\to\infty} N(t) = \frac{\Lambda}{\mu} \text{ (as } \epsilon_1 > 0 \text{ is arbitrarily small).} \tag{23}$$

It follows from Equations (21) and (23) that there exists an integer $n_4 > n_3$ such that

$$0 < I_A(t), I_S(t), Q(t) < \epsilon_1, \quad N(t) > \frac{\Lambda}{\mu} - \epsilon_1, \quad \forall t > n_4T. \tag{24}$$

Therefore, from the second equation of system (1), we have

$$\frac{dE(t)}{dt} \leq \frac{2\Lambda\beta\epsilon_1}{\mu + \sigma\Lambda} - \mu E(t), \quad \forall t > n_4T. \tag{25}$$

It is clear that there exists an integer $n_5 > n_4$ such that

$$E(t) < A_1 + \epsilon_1, \forall t > n_5T, \quad \text{where } A_1 = \frac{2\Lambda\beta\epsilon_1}{\mu(\mu + \sigma\Lambda)}. \tag{26}$$

So, from the first and seventh equations of system (4) we get

$$\begin{aligned} \frac{dS(t)}{dt} &\geq \left(\Lambda + \frac{\alpha\Lambda}{\mu} - \alpha A_1 - 5\alpha\epsilon_1\right) \\ &\quad - (2\beta\epsilon_1 + \mu + \alpha)S(t), t \neq nT, \\ S(t^+) &= (1-p)S(t), \quad t = nT, \ n = 1, 2, \ldots. \end{aligned} \tag{27}$$

Let us consider the following comparison impulsive differential system $\forall t > n_5T$ and $\forall n > n_5$:

$$\begin{aligned} \frac{dz_2(t)}{dt} &= \left(\Lambda + \frac{\alpha\Lambda}{\mu} - \alpha A_1 - 5\alpha\epsilon_1\right) \\ &\quad - (2\beta\epsilon_1 + \mu + \alpha)z_2(t), \quad t \neq nT, \\ z_2(t^+) &= (1-p)z_2(t), \quad t = nT, \ n = 1, 2, \ldots. \end{aligned} \tag{28}$$

By Lemma 2.2, we know that the periodic solution of system (28) is

$$\begin{aligned} \tilde{z}_{2e}(t) &= \Phi + \left(z_2^* - \Phi\right) e^{-(2\beta\epsilon_1+\mu+\alpha)(t-nT)}, \\ &\quad nT < t \leq (n+1)T, \\ \text{where } \Phi &= \frac{\Lambda + \alpha\Lambda/\mu - \alpha A_1 - 5\alpha\epsilon_1}{2\beta\epsilon_1 + \mu + \alpha} \\ \text{and } z_2^* &= \Phi\frac{(1-p)(1 - e^{-(2\beta\epsilon_1+\mu+\alpha)T})}{\{1 - (1-p)\, e^{-(2\beta\epsilon_1+\mu+\alpha)T}\}}, \end{aligned} \tag{29}$$

which is globally asymptotically stable.

By the comparison theorem for impulsive differential equation (Lakshmikantham et al. 1989), there exists an integer $n_6 > n_5$ such that

$$S(t) > \tilde{z}_{2e}(t) - \epsilon_1, \quad nT < t \leq (n+1)T, \quad n > n_6. \tag{30}$$

Making $\epsilon_1 \to 0$, it follows from Equations (17) and (30) that

$$\begin{aligned} \tilde{S}_e(t) &= \frac{\Lambda}{\mu}\left\{1 - \frac{p e^{-(\mu+\alpha)(t-nT)}}{1 - (1-p)\, e^{-(\mu+\alpha)T}}\right\}, \\ &\quad nT < t \leq (n+1)T, \end{aligned} \tag{31}$$

is globally attractive and so

$$\lim_{t\to\infty} S(t) = \tilde{S}_e(t). \tag{32}$$

By the positivity of $E(t)$ and making $\epsilon_1 \to 0$, it follows from Equation (26) that

$$\lim_{t\to\infty} E(t) = 0. \tag{33}$$

Using Equations (21), (23), (32), (33) and from the restriction $N(t) = S(t) + E(t) + I_A(t) + I_S(t) + Q(t) + R(t)$, we have

$$\lim_{t\to\infty} R(t) = \frac{\Lambda}{\mu} - \tilde{S}_e(t). \tag{34}$$

Therefore, we conclude that if $R_1 < 1$, then the disease-free periodic solution $(\tilde{S}_e(t), 0, 0, 0, \Lambda/\mu - \tilde{S}_e(t), \Lambda/\mu)$ of system (4) with initial conditions (5) is globally asymptotically stable. This completes the proof. ■

4. Permanence

In this section, we wish to discuss the permanence of the system (4), this means that the long-term survival (i.e. will not vanish in time) of all components of the system (4), with initial conditions (5). It demonstrates how the disease will be permanent (i.e. will not vanish in time) under some conditions.

DEFINITION *The system (4) is said to be permanent, i.e. the long-term survival (will not vanish in time) of all components of the system (4), if there are positive constants* $m_i (i = 1, 2, 3, 4)$ *such that:*

$$\liminf_{t\to\infty} S(t) \geq m_1, \quad \liminf_{t\to\infty}\{I_A(t) + I_S(t) + Q(t)\} \geq m_2,$$
$$\liminf_{t\to\infty} R(t) \geq m_3, \quad \liminf_{t\to\infty} N(t) \geq m_4,$$

hold for any solution $(S(t), I_A(t), I_S(t), Q(t), R(t), N(t))$ *of Equation (4) with initial conditions (5). Here* m_i $(i = 1, 2, 3, 4)$ *are independent of Equation (5).*

THEOREM 4.1 *If* $R_2 > 1$, *then there exists a positive constant m such that each positive solution* $(S(t), I_A(t), I_S(t), Q(t), R(t), N(t))$ *of the system (4) with initial conditions (5) satisfies* $(I_A(t) + I_S(t) + Q(t)) \geq m$ *for sufficiently large time t, where*

$$\begin{aligned} R_2 &= \left(\frac{\beta' e^{-\mu\tau}}{\theta'} - \sigma'\right)\frac{\Lambda(1-p)(1-e^{-\mu T})}{\mu\{1-(1-p)e^{-\mu T}\}},\\ \beta' &= \min\{\beta_1, \beta_2\},\\ \sigma' &= \max\{\sigma_1, \sigma_2\},\\ \theta' &= \max\{kr_1 + d_1 + \mu, r_2 + d_2 + \mu, r_3 + d_3 + \mu\}. \end{aligned} \tag{35}$$

Proof From the second, third and fourth equations of system (4), we have

$$\begin{aligned} D'(t) &= \frac{dD(t)}{dt} \geq \beta' e^{-\mu\tau}\frac{S(t-\tau)D(t-\tau)}{1+\sigma' S(t-\tau)} - \theta' D(t)\\ &= D(t)\left\{\beta' e^{-\mu\tau}\frac{S(t)}{1+\sigma' S(t)} - \theta'\right\}\\ &\quad - \beta' e^{-\mu\tau}\frac{d}{dt}\int_{t-\tau}^{t}\frac{S(u)D(u)}{1+\sigma' S(u)}du, \end{aligned}$$

where $D(t) = I_A(t) + I_S(t) + Q(t)$. (36)

$$\text{Define, } V(t) = D(t) + \beta' e^{-\mu\tau}\int_{t-\tau}^{t}\frac{S(u)D(u)}{1+\sigma' S(u)}du$$
$$\Rightarrow V'(t) = \frac{dV(t)}{dt} \geq D(t)\left\{\beta' e^{-\mu\tau}\frac{S(t)}{1+\sigma' S(t)} - \theta'\right\},$$

(using Equation (36))

$$= \theta' D(t)\left\{\frac{\beta' e^{-\mu\tau} S(t)}{\theta'(1+\sigma' S(t))} - 1\right\}. \tag{37}$$

Define, $D^* = \dfrac{\mu}{\beta}(R_2 - 1) > 0$, (since $R_2 > 1$)

$$\Rightarrow D^* \to 0^+ \text{ as } \epsilon = (R_2-1) \to 0^+ \Rightarrow \frac{\beta' e^{-\mu\tau}\xi'}{\theta'(1+\sigma'\xi')} > 1,$$

$$\begin{aligned} &\text{where } \xi' = \frac{\Lambda(1-p)(1-e^{-(\beta D^*+\mu)T})}{(\beta D^*+\mu)\{1-(1-p)e^{-(\beta D^*+\mu)T}\}} - \epsilon > 0,\\ &\beta = \max\{\beta_1, \beta_2\}, \end{aligned} \tag{38}$$

for a sufficiently small $\epsilon > 0$.

If possible, let there exists a $t_1 > 0$ such that $D(t) < D^*, \forall t \geq t_1$. It follows from the first and sixth equations of (4):

$$\begin{aligned} &\frac{dS(t)}{dt} > \Lambda - (\beta D^* + \mu)S(t), \quad \beta = \max\{\beta_1, \beta_2\}, \quad t \neq nT,\\ &S(t^+) = (1-p)S(t), \quad t = nT, \quad n = 1, 2, \ldots. \end{aligned} \tag{39}$$

Let us consider the following comparison impulsive differential system $\forall t \geq t_1$:

$$\begin{aligned} &\frac{dz_3(t)}{dt} = \Lambda - (\beta D^* + \mu)z_3(t), \quad t \neq nT,\\ &z_3(t^+) = (1-p)z_3(t), \quad t = nT, \quad n = 1, 2, \ldots. \end{aligned} \tag{40}$$

By Lemma 2.2, we know that the periodic solution of system (40) is

$$\begin{aligned} &\tilde{z}_{3e}(t) = \frac{\Lambda}{\beta D^* + \mu} + \left\{z_3^* - \frac{\Lambda}{\beta D^* + \mu}\right\}e^{-(\beta D^*+\mu)(t-nT)},\\ &nT < t \leq (n+1)T,\\ &\text{where } z_3^* = \frac{\Lambda(1-p)(1-e^{-(\beta D^*+\mu)T})}{(\beta D^*+\mu)\{1-(1-p)e^{-(\beta D^*+\mu)T}\}}, \end{aligned} \tag{41}$$

which is globally asymptotically stable.

By the comparison theorem for impulsive differential equation (Lakshmikantham et al. 1989), there exists $t_2 > t_1 + \tau$ such that the followings hold:

$$S(t) > \tilde{z}_{3e}(t) - \epsilon \Rightarrow S(t) > z_3^* - \epsilon = \xi', \quad \forall t \geq t_2. \tag{42}$$

$$\text{Next, let } D_1 = \min_{t\in[t_2, t_2+\tau]} D(t) \Rightarrow D(t) \geq D_1, \quad \forall t \geq t_2. \tag{43}$$

Otherwise, there exists a $T_0 > 0$ such that $D(t) \geq D_1, \forall t \in [t_2, t_2 + \tau + T_0]$, where $D(t_2 + \tau + T_0) = D_1$ and $D'(t_2 + \tau + T_0) \leq 0$. However, from Equations (36), (38) and (42), we get

$$D'(t_2 + \tau + T_0) > \theta' D_1\left\{\frac{\beta' e^{-\mu\tau}\xi'}{\theta'(1+\sigma'\xi')} - 1\right\} > 0. \tag{44}$$

So, we have got a contradiction and hence $D(t) \geq D_1, \forall t \geq t_2$. As a consequence of Equations (37), (38), (42) and (43), we get

$$V'(t) > \theta' D_1 \left\{ \frac{\beta' \mathrm{e}^{-\mu\tau} \xi'}{\theta'(1+\sigma'\xi')} - 1 \right\} > 0,$$

$$\forall t \geq t_2 \Rightarrow V(t) \to \infty \text{ as } t \to \infty. \quad (45)$$

This is a contradiction because

$$V(t) = D(t) + \beta' \mathrm{e}^{-\mu\tau} \int_{t-\tau}^{t} \frac{S(u)D(u)}{1+\sigma' S(u)} \mathrm{d}u$$
$$\leq D(t) + \beta' \mathrm{e}^{-\mu\tau} \int_{t-\tau}^{t} S(u)D(u)\, \mathrm{d}u$$
$$\leq \frac{\Lambda}{\mu} + \beta' \mathrm{e}^{-\mu\tau} \int_{t-\tau}^{t} \left(\frac{\Lambda}{\mu}\right)^2 \mathrm{d}u = \frac{\Lambda}{\mu} \left\{ 1 + \frac{\Lambda\tau\beta' \mathrm{e}^{-\mu\tau}}{\mu} \right\}. \quad (46)$$

Table 1. Parameter values for Figure 1.

Parameter	Values
Λ	0.1
β_1	0.09
β_2	0.1
σ_1	0.3
σ_2	0.4
μ	0.01
α	0.2
ρ	0.4
r_1	0.4
r_2	0.2
r_3	0.3
d_1	0.2
d_2	0.15
d_3	0.05
k	0.1
p	0.8
q	0.1
τ	1
T	5

Therefore, we conclude for any $t_1 > 0$, the inequality $D(t) < D^*$ cannot hold for all $t \geq t_1$. Thus we are left to consider the following two cases:

(i) $D(t) \geq D^*$ for sufficiently large t;
(ii) $D(t)$ oscillates about D^* for sufficiently large t.

It is clear that if $D(t) \geq D^*$ for sufficiently large t, then our desired result is obtained. So, we only need to consider the case (ii). Let

$$m = \min \left\{ \frac{D^*}{2}, D^* \mathrm{e}^{-\theta'\tau} \right\},$$

where $\theta' = \max\{kr_1 + d_1 + \mu, r_2 + d_2 + \mu, r_3 + d_3 + \mu\}$. (47)

Now, we will show that $D(t) \geq m$ for sufficiently large t. Let $t^* > 0$ and $t_0 > 0$ satisfy $D(t^*) = D(t^* + t_0) = D^*$ and $D(t) < D^*$ for $t^* < t < t^* + t_0$, where t^* is sufficiently large such that $S(t) > \xi'$ for $t^* < t < t^* + t_0$. It is clear that $D(t)$ is uniformly continuous since the positive solution of Equation (4) is ultimately bounded and $D(t) = (I_A(t) + I_S(t) + Q(t))$ is not affected by impulsive effects. Hence there exists a constants T_1, where $0 < T_1 < \tau$ and T_1 is independent of t^*, such that $D(t) > D^*/2$ for $t^* \leq t \leq t^* + T_1$. If $t_0 \leq T_1$, the required result is obtained. If $T_1 < t_0 \leq \tau$, since $D'(t) > -\theta' D(t)$ and $D(t^*) = D^*$, it follows that $D(t) \geq D^* \mathrm{e}^{-\theta'\tau}$ for $t^* < t < t^* + t_0$. If $t_0 > \tau$; we have $D(t) \geq D^* \mathrm{e}^{-\theta'\tau}$ for $t^* < t < t^* + \tau$ and by using the same arguments we can obtain $D(t) \geq D^* \mathrm{e}^{-\theta'\tau}$ for $t^* + \tau < t < t^* + t_0$ as the interval $[t^*, t^* + t_0]$ can be chosen arbitrarily. So, we can conclude that $D(t) \geq m$ for sufficiently large t. On the basis of the previous discussions, the choice of m is independent of the positive solution of (4) and hence any positive solution of (4) satisfies $D(t) \geq m$ for t large enough. This completes the proof. ■

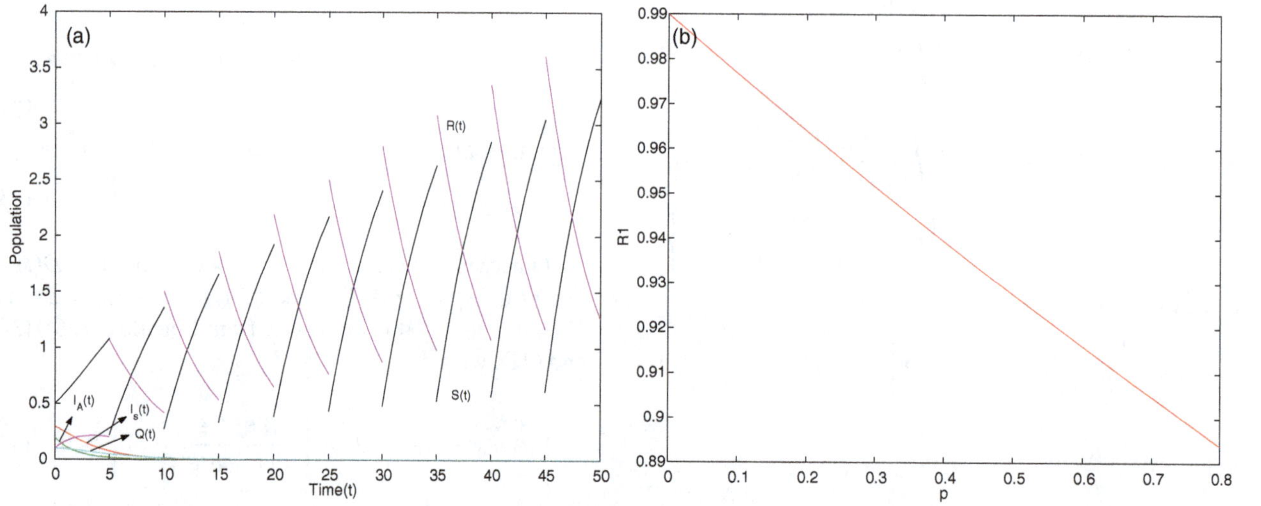

Figure 1. (a) Movement paths of $S(t), I_A(t), I_S(t), Q(t)$ and $R(t)$ for $R_1 = 0.8938 < 1$, (b) the effects of pulse vaccination (p) on the threshold value R_1, with parameter values given in Table 1.

THEOREM 4.2 *If $R_2 > 1$, then the system (4) with initial conditions (5) is permanent.*

Proof Suppose $(S(t), I_A(t), I_S(t), Q(t), R(t), N(t))$ be any solution of system (4) with initial conditions (5). From the first and sixth equations of system (4), we have

$$\begin{aligned} \frac{\mathrm{d}S(t)}{\mathrm{d}t} &\geq \Lambda - \beta_1 S(t) I_A(t) - \beta_2 S(t) I_S(t) - \mu S(t) \\ &\geq \Lambda - \left(\frac{\beta\Lambda}{\mu} + \mu\right) S(t), \\ &\beta = \max\{\beta_1, \beta_2\} \text{ and } t \neq nT, \\ S(t^+) &= (1-p)S(t), \quad t = nT, \quad n = 1, 2, \ldots. \end{aligned} \tag{48}$$

Table 2. Parameter values for Figure 2.

Parameter	Values
Λ	0.1
β_1	0.8
β_2	0.9
σ_1	0.2
σ_2	0.3
μ	0.01
α	0.2
ρ	0.4
r_1	0.04
r_2	0.02
r_3	0.03
d_1	0.02
d_2	0.015
d_3	0.005
k	0.1
p	0.8
q	0.1
τ	1
T	5

Let us consider the following comparison impulsive differential system:

$$\begin{aligned} \frac{\mathrm{d}z_4(t)}{\mathrm{d}t} &= \Lambda - \left(\frac{\beta\Lambda}{\mu} + \mu\right) z_4(t), \quad t \neq nT, \\ z_4(t^+) &= (1-p) z_4(t), \quad t = nT, \quad n = 1, 2, \ldots. \end{aligned} \tag{49}$$

By Lemma 2.2, we know that the periodic solution of system (49) is

$$\begin{aligned} \tilde{z}_{4e}(t) &= \frac{\mu\Lambda}{\beta\Lambda + \mu^2} + \left\{z_4^* - \frac{\mu\Lambda}{\beta\Lambda + \mu^2}\right\} \\ &\quad \times \mathrm{e}^{-(\beta\Lambda/\mu + \mu)(t - nT)}, \\ &nT < t \leq (n+1)T, \end{aligned} \tag{50}$$

$$\text{where } z_4^* = \frac{\mu\Lambda(1-p)(1 - \mathrm{e}^{-(\beta\Lambda/\mu+\mu)T})}{(\beta\Lambda + \mu^2)\{1 - (1-p)\,\mathrm{e}^{-(\beta\Lambda/\mu+\mu)T}\}},$$

which is globally asymptotically stable.

By the comparison theorem for impulsive differential equation, there exists sufficiently small $\epsilon_1 > 0$ such that the following holds:

$$\lim_{t\to\infty} S(t) \geq \frac{\mu\Lambda(1-p)(1 - \mathrm{e}^{-(\beta\Lambda/\mu+\mu)T})}{(\beta\Lambda + \mu^2)\{1 - (1-p)\,\mathrm{e}^{-(\beta\Lambda/\mu+\mu)T}\}} - \epsilon_1 > 0. \tag{51}$$

From the fifth equation of system (4) and using Theorem 4.1, we have

$$\begin{aligned} &\frac{\mathrm{d}R(t)}{\mathrm{d}t} \geq rm - (\mu + \alpha) R(t) \\ &\Rightarrow \lim_{t\to\infty} R(t) \geq \frac{rm}{\mu + \alpha} - \epsilon_2 > 0, \text{ where} \\ &r = \min\{kr_1, r_2, r_3\}, \end{aligned} \tag{52}$$

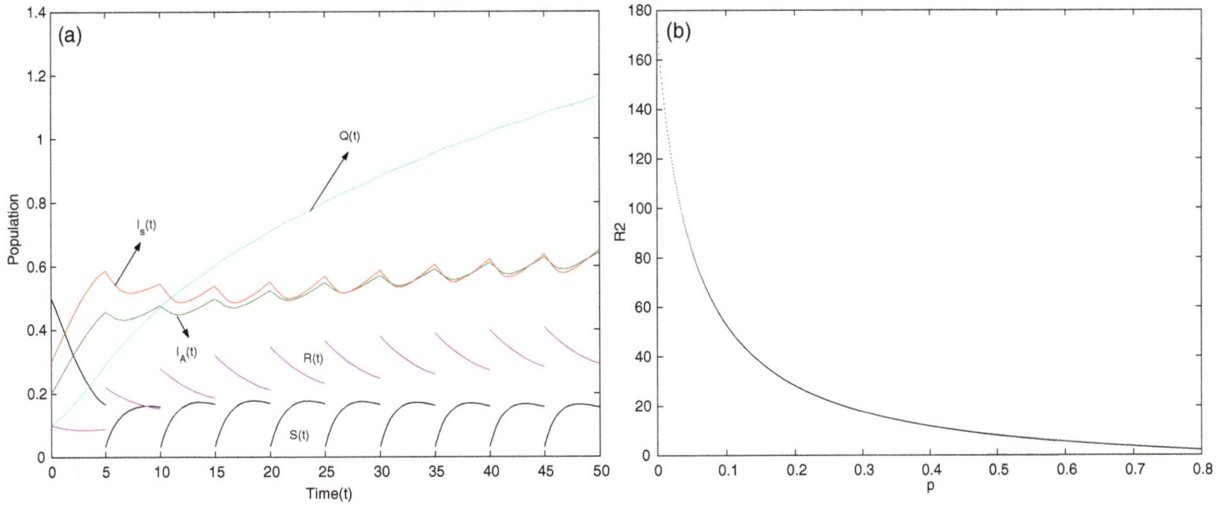

Figure 2. (a) Movement paths of $S(t), I_A(t), I_S(t), Q(t)$ and $R(t)$ for $R_2 = 2.0840 > 1$, (b) the effects of pulse vaccination (p) on the threshold value R_2, with parameter values given in Table 2.

for a sufficiently small $\epsilon_2 > 0$ (m is given by Equation (47)). Hence system (4) with initial conditions (5) is permanent and this completes the proof. ∎

5. Numerical simulations and biological interpretations

We first consider the case when $R_1 = 0.8938 < 1$ using the parameter values given in Table 1. Using these parameter values, the movement paths of $S(t), I_A(t), I_S(t), Q(t)$ and $R(t)$ are presented in Figure 1(a). This figure shows that the disease dies out when $R_1 < 1$, which supports our analytical result given in Theorem 3.1. Its epidemiological implication is that the infectious population vanishes, i.e. the disease dies out when $R_1 < 1$ (see Figure 1(a)). In Figure 1(b), the effects of pulse vaccination (p) on the threshold value R_1 is presented using the parameter values given in Table 1. It shows that the threshold values R_1 gradually decrease when the pulse vaccination rate (p) increases. This implies that the strategy of pulse vaccination is very effective to eradicate the HFMD.

Table 3. Parameter values for Figure 3.

Parameter	Values
Λ	0.1
β_1	0.6
β_2	0.8
σ_1	0.3
σ_2	0.35
μ	0.01
α	0.2
ρ	0.4
r_1	0.4
r_2	0.2
r_3	0.3
d_1	0.6
d_2	0.4
d_3	0.1
k	0.1
p	0.8
q	0.1
τ	1
T	5

Next, we consider the case when $R_2 = 2.0840 > 1$ using the parameter values given in Table 2. Using these parameter values, the movement paths of $S(t), I_A(t), I_S(t), Q(t)$ and $R(t)$ are presented in Figure 2(a). This figure shows that the disease will be permanent when $R_2 > 1$, which supports our analytical result given in Theorem 4.2. In Figure 2(b), the effects of pulse vaccination (p) on the threshold value R_2 is presented using the parameter values given in Table 2. It shows that the threshold values R_2 gradually decrease when the pulse vaccination rate (p) increases. This also implies that the strategy of pulse vaccination is very effective to eradicate the HFMD.

We also consider the case when $R_1 = 4.3601 > 1$ and $R_2 = 0.0679 < 1$ with parameter values given in Table 3. Using these parameter values, the movement paths of $S(t), I_A(t), I_S(t), Q(t)$ and $R(t)$ are presented in Figure 3(a). This figure shows that the disease dies out. For $R_1 = 4.7029 > 1$ and $R_2 = 0.9205 < 1$ where $p = 0.2$ and other parameter values are given in Table 3, the movement paths of $S(t), I_A(t), I_S(t), Q(t)$ and $R(t)$ are presented in Figure 3(b). This figure shows that the disease is still permanent though the level of disease is very low.

From the figures it is observed that a large pulse vaccination rate will lead to eradication of the HFMD.

Remark When $R_2 \leq 1 \leq R_1$, the dynamical behavior of the HFMD model (4) and (5) has not been clear.

6. Conclusions

Motivated by the recent development of the first vaccine to protect children against enterovirus 71, or EV71 (Zhu

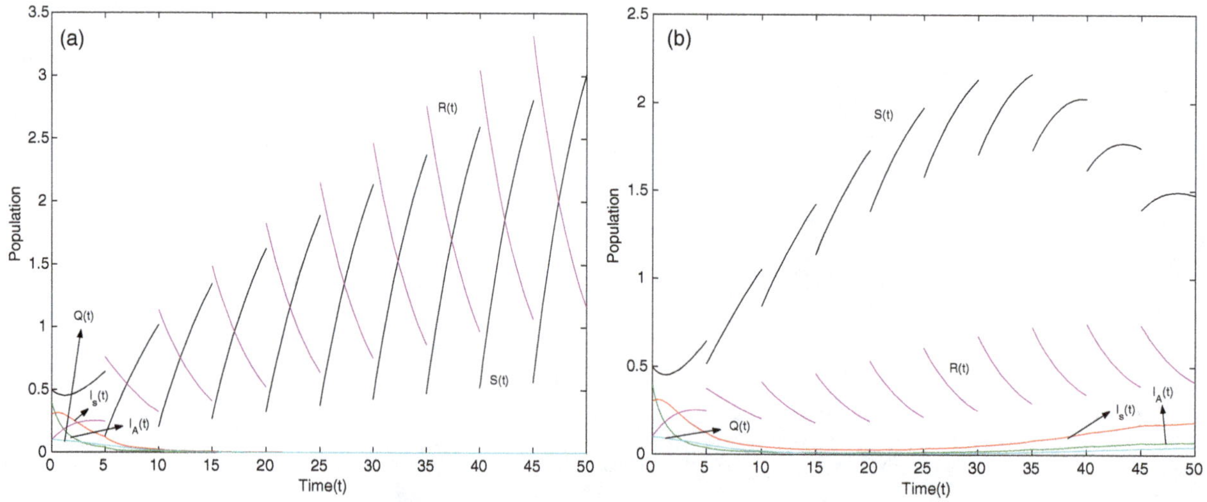

Figure 3. Movement paths of $S(t), I_c(t), I(t)$ and $R(t)$ (a) for $R_1 = 4.3601 > 1$ and $R_2 = 0.0679 < 1$ with parameter values given in Table 3, (b) for $R_1 = 4.7029 > 1$ and $R_2 = 0.9205 < 1$ where $p = 0.2$ and other parameter values are given in Table 3.

et al., 2013), in this paper we have considered a dynamical model of HFMD with discrete time delay, pulse vaccination strategy and saturation incidence rate. The entire high-risk human population is split up into six mutually exclusive epidemiological compartments (based on disease status), namely, susceptible (S), exposed (infected but not yet infectious) (E), infective in asymptomatic phase (showing no symptoms of HFMD) (I_A), infective in symptomatic phase (showing symptoms of HFMD) (I_S), infective in symptomatic phase who follow quarantine (a strict isolation imposed to prevent the spread of the disease) mechanisms and personal protection against infecting others (Q) and recovered (infectious people who have cleared (or recovered from) HFMD infection) (R). The susceptible population increases through birth (a constant influx Λ of susceptible is assumed) and from recovered hosts and decreases due to direct contact with an infectious individual (in I_A or I_S compartments), natural death and PVS. The infected classes are increased by infection of susceptible. A fraction of the exposed individuals will start to show symptoms of HFMD (and move to the class I_S), while the remaining fraction will not (but still remain capable of infecting others and move to the class I_A). Also, a fraction of the infective in symptomatic phase takes appropriate preventive measures and move to the quarantined class Q. It is assumed that there is a time lag to account for the fact that an individual infected with HFMD is not infectious until after some time (typically 3–7 days Yang et al. 2013) after exposure. A fraction of the asymptomatically infectious individuals eventually show disease symptoms and a fraction recover. The infected classes are decreased through recovery from infection, by disease-related death and by natural death. The most basic and important questions to ask for the systems in the theory of mathematical epidemiology are the persistence, extinctions, the existence of periodic solutions, global stability, etc. Here, we have established some sufficient conditions on the permanence and extinction of the disease by using inequality analytical technique. We have introduced two threshold values R_1 and R_2 and further obtained that the disease will be going to extinct when $R_1 < 1$ and the disease will be permanent when $R_2 > 1$. The important mathematical findings for the dynamical behavior of the HFMD model are also numerically verified using MATLAB. It is observed that a large pulse vaccination rate will lead to eradication of the disease and when $R_2 \leq 1 \leq R_1$, the dynamical behavior is not clear. The aim of the analysis of this model is to trace the parameters of interest for further study, with a view to informing and assisting policy-maker in targeting prevention and treatment resources for maximum effectiveness.

Acknowledgements

The author likes to thank TWAS, UNESCO and National Autonomous University of Mexico (UNAM) for financial support. He is grateful to Prof. Javier Bracho Carpizo, Prof. Marcelo Aguilar and Prof. Ricardo Gomez Aiza, Institute of Mathematics, National Autonomous University of Mexico for their helps and encouragements.

References

Agur, Z., Cojocaru, L., Mazor, G., Anderson, R. M., & Danon, Y. L. (1993). Pulse mass measles vaccination across age cohorts. *Proceedings of the National Academy of Sciences of the United States of America*, *90*, 11698–11702.

Anderson, R. M., & May, R. M. (1979). Population biology of infectious diseases. Part I. *Nature*, *180*, 361–367.

Anderson, R. M., & May, R. M. (1992). *Infectious disease of humans, dynamical and control*. Oxford: Oxford University Press.

Babiuk, L. A., Babiuk, S. L., & Baca-Estrada, M. E. (2002). Novel vaccine strategies. *Advances in Virus Research*, *58*, 29–80.

Bainov, D. D., & Simeonov, P. S. (1993). *Impulsive differential equations: Periodic solutions and applications*. New York: Longman Scientific and Technical.

Bainov, D. D., & Simeonov, P. S. (1995). *The stability theory of impulsive differential equations: Asymptotic properties of the solutions*. Singapore: World Scientific.

Bracho, M. A., González-Candelas, F., Valero, A., Córdoba, J., & Salazar, A. (2011). Enterovirus co-infections and onychomadesis after hand, foot, and mouth disease, Spain, 2008. *Emerging Infectious Diseases*, *17*(12), 2223–2231.

Brauer, F., & Castillo-Chavez, C. (2001). *Mathematical models in population biology and epidemiology*. Berlin: Springer.

Cai, L., Li, X., Ghosh, M., & Guo, B. (2009). Stability of an HIV/AIDS epidemic model with treatment. *Journal of Computational and Applied Mathematics*, *229*, 313–323.

Capasso, V. (1993). *Mathematical structures of epidemic systems, lectures notes in biomathematics, Vol. 97*. Berlin: Springer-Verlag.

Chuo, F., Tiing, S., & Labadin, J. (2008). *A simple deterministic model for the spread of hand, foot and mouth disease (HFMD) in Sarawak*. In 2008 Second Asia international conference on modelling and simulation, 947–952.3Qloc

Cooke, K. L., & van Den Driessche, P. (1996). Analysis of an SEIRS epidemic model with two delays. *Journal of Mathematical Biology*, *35*, 240–260.

Diekmann, O., & Heesterbeek, J. A. P. (2000). *Mathematical epidemiology of infectious diseases: Model building analysis, and interpretation*. Chichester: John Wiley and Sons Ltd.

Gakkhar, S., & Negi, K. (2008). Pulse vaccination in SIRS epidemic model with non-monotonic incidence rate. *Chaos, Solitons & Fractals*, *35*, 626–638.

Gao, S., Chen, L., Nieto, J. J., & Torres, A. (2006). Analysis of a delayed epidemic model with pulse vaccination and saturation incidence. *Vaccine*, *24*, 6037–6045.

Gao, S., Chen, L., & Teng, Z. (2007). Impulsive vaccination of an SEIRS model with time delay and varying total population size. *Bulletin of Mathematical Biology*, *69*, 731–745.

Gjorrgjieva, J., Smith, K., Chowell, G., Sanchez, F., Synder, J., & Castillo-Chavez, C. (2005). The role of vaccination in the control of SARS. *Mathematical Biosciences and Engineering*, *2*, 1–17.

Hethcote, H. W., & van Den Driessche, P. (1991). Some epidemiological models with nonlinear incidence. *Journal of Mathematical Biology*, *29*, 271–287.

Hui, J., & Chen, L. (2004). Impulsive vaccination of SIR epidemic models with nonlinear incidence rates. *Discrete and Continuous Dynamical Systems: Series B*, *4*, 595–605.

Kermack, W. O., & Mckendrick, A. G. (1927). Contributions to the mathematical theory of epidemics. Part I. *Proceedings of the Royal Society, A*, *115*(5), 700–721.

Lakshmikantham, V., Bainov, D. D., & Simeonov, P. S. (1989). *Theory of impulsive differential equations*. Singapore: World Scientific.

Liu, J. (2011). Threshold dynamics for a HFMD epidemic model with periodic transmission rate. *Nonlinear Dynamics*, *64*, 89–95.

Ma, W., Song, M., & Takeuchi, Y. (2004). Global stability of an SIR epidemic model with time delay. *Applied Mathematics Letter*, *17*, 1141–1145.

May, R. M., & Anderson, R. M. (1978). Regulation and stability of host-parasite population interactions II: Destabilizing process. *Journal of Animal Ecology*, *47*, 219–267.

Mena-Lorca, J., & Hethcote, H. W. (1992). Dynamic models of infectious disease as regulators of population sizes. *Journal of Mathematical Biology*, *30*, 693–716.

Meng, X., Chen, L., & Cheng, H. (2007). Two profitless delays for the SEIRS epidemic disease model with nonlinear incidence and pulse vaccination. *Applied Mathematical and Computation*, *186*, 516–529.

Naresh, R., Tripathi, A., & Omar, S. (2006). Modelling of the spread of AIDS epidemic with vertical transmission. *Applied Mathematics and Computation*, *178*, 262–272.

Nokes, D. J., & Swinton, J. (1995). The control of childhood viral infections by pulse vaccination. *IMA Journal of Mathematics Applied in Medicine and Biology*, *12*, 29–53.

d'Onofrio, A. (2002a). Pulse vaccination strategy in the SIR epidemic model: Global asymptotic stable eradication in presence of vaccine failures. *Mathematical and Computer Modelling*, *36*, 473–489.

d'Onofrio, A. (2002b). Stability properties of vaccination strategy in SEIR epidemic model. *Mathematical Biosciences*, *179*, 57–72.

d'Onofrio, A. (2005). Vaccination policies and nonlinear force of infection. *Applied Mathematics and Computation*, *168*, 613–622.

Roy, N., & Halder, N. (2010). Compartmental modeling of hand, foot and mouth infectious disease (HFMD). *Research Journal of Applied Sciences*, *5*, 177–182.

Ruan, S., & Wang, W. (2003). Dynamical behavior of an epidemic model with nonlinear incidence rate. *Journal of Differential Equations*, *188*, 135–163.

Song, X. Y., & Chen, L. S. (2001). Optimal harvesting and stability with stage-structure for a two species competitive system. *Mathematical Biosciences*, *170*, 173–186.

Stone, L., Shulgin, B., & Agur, Z. (2000). Theoretical examination of the pulse vaccination policy in the SIR epidemic models. *Mathematical and Computer Modelling*, *31*, 207–215.

Takeuchi, Y., Cui, J., Rinko, M., & Saito, Y. (2006a). Permanence of delayed population model with dispersal loss. *Mathematical Biosciences*, *201*, 143–156.

Takeuchi, Y., Cui, J., Rinko, M., & Saito, Y. (2006b). Permanence of dispersal population model with time delays. *Journal of Computational and Applied Mathematics*, *192*, 417–430.

Tang, S., Xiao, Y., & Clancy, D. (2005). New modelling approach concerning integrated disease control and cost-effectivity. *Nonlinear Analysis*, *63*, 439–471.

Thieme, H. R. (2003). *Mathematics in population biology*. Princeton, NJ: Princeton University Press.

Urashima, M., Shindo, N., & Okabe, N. (2003). Seasonal models of herpangina and hand-foot-mouth disease to simulate annual fluctuations in urban warming in Tokyo. *Japanese Journal of Infectious Diseases*, *56*, 48–53.

Wang, W. (2002). Global behavior of an SEIRS epidemic model with time delays. *Applied Mathematics Letters*, *15*, 423–428.

Wang, Y. C., & Sung, F. C. (2004). *Modeling the infectious for enteroviruses in Taiwan*. Retrieved July 21, 2007, from http://gra103.aca.ntu.edu.tw/gdoc/D91844001a.pdf

Wei, C., & Chen, L. (2008). A delayed epidemic model with pulse vaccination. *Discrete Dynamics in Nature and Society*, *2008*, 12 pages. Article ID 746951, doi:10.1155/2008/746951

Wong, S. S., Yip, C. C., Lau, S. K., & Yuen, K. Y. (2010). Human enterovirus 71 and hand, foot and mouth disease. *Epidemiology & Infection*, *138*, 1071–1089.

Yang, J. Y., Chen, Y., & Zhang, F. Q. (2013). Stability analysis and optimal control of a hand-foot-mouth disease (HFMD) model. *Applied Mathematics and Computation*, *41*, 99–117.

Zhou, Y., & Liu, H. (2003). Stability of periodic solutions for an SIS model with pulse vaccination. *Mathematical and Computer Modelling*, *38*, 299–308.

Zhu, F. C., Meng, F. Y., Li, J. X., Li, X. L., Mao, A. Y., Tao, H., … Shen, X. L. (2013, May 29). Efficacy, safety, and immunology of an inactivated alum-adjuvant enterovirus 71 vaccine in children in China: A multicentre, randomised, double-blind, placebo-controlled, phase 3 trial. *The Lancet*, *381*, 2024–2032. doi:10.1016/S0140-6736(13)61049-1.

Zhu, Q., Hao, Y. T., Ma, J. Q., Yu, S. C., & Wang, Y. (2011). Surveillance of hand, foot, and mouth disease in Mainland China (2008–2009). *Biomedical and Environmental Sciences*, *24*, 349–356.

On-line estimation of ARMA models using Fisher-scoring†

Abdelhamid Ouakasse[a] and Guy Mélard[b*]

[a]*Federal Public Service Justice of Belgium, Brussels, Belgium;* [b]*Université libre de Bruxelles (ULB), ECARES, Brussels, Belgium*

Recursive estimation methods for time series models usually make use of recurrences for the vector of parameters, the model error and its derivatives with respect to the parameters, plus a recurrence for the Hessian of the model error. An alternative method is proposed in the case of an autoregressive-moving average model, where the Hessian is not updated but is replaced, at each time, by the inverse of the Fisher information matrix evaluated at the current parameter. The asymptotic properties, consistency and asymptotic normality, of the new estimator are obtained. Monte Carlo experiments indicate that the estimates may converge faster to the true values of the parameters than when the Hessian is updated. The paper is illustrated by an example on forecasting the speed of wind.

Keywords: time series; ARMA processes; recursive estimation; Fisher information matrix

AMS 2000 MSC: Primary: 62M10, 62F10; Secondary: 93E10

1. Introduction

The development of estimation methods of the parameters of statistical and econometric models was influenced by the availability of more powerful computers. Numerical calculations are lighter and faster with the increased speed of computers, and bigger data bases can be used. For nonlinear models, it is generally not possible to find the estimator analytically so numerical optimisation procedures are applied to obtain the maximum likelihood or even the least squares estimator. These procedures are iterative and make use of all the data at each iteration. They are called off-line because they are applied when all the data are available. Each time we have a new observation the whole estimation procedure has to be repeated. This is not a problem with quarterly or monthly data but availability of large capacity memory also implies that much more data are stored and more frequently. Instead of collecting data at a yearly, quarterly or monthly level, data are more and more collected in real time, starting with financial markets. Also, new fields of applications have appeared, like mobile telecommunications or fluid flow management, where quick automated decisions are required.

When the interval of time between two observations is very short, working with past, off-line, methods becomes inefficient if all data need to be used at high frequency rates and doing huge calculations, because of the expensive computation power needed as well as the memory space. Instead of being used by humans on their desks, the work should be done 'on the spot' by computer systems and in an automated way. The idea is to use on-line or recursive methods. They make use of a very small subset of data at each time. These methods appeared first in linear models (Plackett, 1950, who referred to Gauss) when computation was a major annoyance. In statistics they reappeared later (Brown, Durbin, & Evans, 1975) as a way to check the stability of model specification with respect to time. In the discussion of that paper, the influence of Kalman (1960) is clear. Recursive methods became particularly interesting in the context of time series models, see Young (1985). These methods were indeed developed mainly in engineering, under the name of Recursive Identification, for data available on-line in telecommunications, transmissions, management of fluids, etc. For some recent contributions to recursive estimation methods, see Benveniste, Métivier, and Priouret (1990), Guo (1994), Kushner and Yin (1997), Chen (2002), Moulines, Priouret, and Roueff (2005), Dahlhaus and Subba Rao (2007), Gerencsér and Prokaj (2010), Kirshner, Maggio, and Unser (2011) and Marelli, You, and Fu (2013).

Among these recursive methods there is the RML (recursive maximum likelihood) method which was introduced by Söderström (1973), see also Young (1984). We know that, under general conditions, the (off-line) maximum likelihood method gives an estimator which is asymptotically efficient, i.e. it is distributed asymptotically like a normal law whose asymptotic variance-covariance matrix is equal

*Corresponding author. Email: gmelard@ulb.ac.be

†This work was done while the first author was with Université libre de Bruxelles, Département de Mathématique.

to the Cramér–Rao upper bound. Under certain conditions, Ljung and Söderström (1983) have shown that the RML estimator has the same asymptotic properties as the maximum likelihood estimator. But they noticed that for a finite series, $\{y_1, \ldots, y_N\}$, the maximum likelihood estimator is always better than the RML estimator. The RML estimator is based on a first-order approximation of the Taylor expansion of the sum of squares of the errors. Let β be the vector of parameters of the model. As we will see in Section 2, the estimate at time t, $\hat{\beta}_t$, makes use of the value at the previous time, $\hat{\beta}_{t-1}$, but also of a matrix R_t which is an approximation of the Hessian of the sum of squares of errors. A recurrence for the residual is used but also a recurrence for the derivative of the error with respect to the parameters and an updating recurrence for the Hessian.

Mélard (1989) and Zahaf (1999) observed that the latter recurrence, with highly variable successive values of R_t, is often the cause for wild variations in the estimates and proposed a modified RML estimator for autoregressive-moving average (ARMA) models. While keeping the spirit of the algorithm, instead of the recurrence for the Hessian R_t, Zahaf (1999) proposed to use the evaluation of the asymptotic Fisher information at the current value of the estimator, $\beta = \hat{\beta}_{t-1}$. This is a recursive analogue of the well-known Fisher-scoring algorithm, e.g. Kennedy and Gentle (1980, p. 450), Sen and Singer (1993, p. 205). Intuitively this should not change the asymptotic properties.

Zahaf (1999) noticed that the asymptotic theory developed by Ljung (1977), Solo (1981) and Ljung and Söderström (1983) no longer applies. He outlined an asymptotic theory based on the stochastic approximation of Robbins–Monro following Duflo (1997) but it was not complete. Moreover convergence in law of the estimator rested on a conjecture which was later proved to be wrong. For these reasons, after vain attempts including with the alternative approach of Kushner and Huang (1979), we preferred to adapt the approach of Ljung and Söderström (1983). An alternative which is discussed later should be to use Chen (2002) stochastic approximation theory with expanding truncations.

In Section 2, we remind the necessary concepts of RML estimation in order to be able to introduce our version at the beginning of Section 3. The remaining of Section 3 is devoted to the main theorems in order to establish consistency and asymptotic normality of the new estimator. This is done under the very general condition that fourth-order moments are finite instead of assuming that the observations are bounded, like Ljung and Söderström (1983) or that moments of order $4/(1-\delta)$, for some strictly positive δ, are finite, like Solo (1981). This is a clear improvement with respect to the literature. In Section 4, we show small samples results obtained by Monte Carlo simulations. They indicate that the new estimator can be an improvement over the classical RML estimator. Section 5 will present an example of wind forecasting.

2. RML estimation

Let us first describe the RML estimator before introducing how we have modified it. The algorithm for that estimator is derived from the off-line maximum likelihood estimator, see Ljung (1978) and Åström (1980). We assume for simplicity that the observations $\{y_t; t = 1, \ldots, N\}$ follow a univariate ARMA(p, q) model defined by the equation:

$$y_t - \phi_1 y_{t-1} - \phi_2 y_{t-2} - \cdots - \phi_p y_{t-p} = e_t - \theta_1 e_{t-1} - \theta_2 e_{t-2} - \cdots - \theta_q e_{t-q}, \quad (1)$$

where the roots of the autoregressive and moving average polynomials $\Phi(B) = 1 - \phi_1 B - \phi_2 B^2 - \cdots - \phi_p B^p$ and $\Theta(B) = 1 - \theta_1 B - \theta_2 B^2 - \cdots - \theta_q B^q$ are outside of the unit circle, $\phi_p \neq 0$ and $\theta_q \neq 0$, and e_t's are i.i.d. random variables with $E(e_t) = 0$ and $E(e_t^2) = \sigma_e^2 > 0$. Let $\beta = (\phi_1, \ldots, \phi_p, \theta_1, \ldots, \theta_q)^{\mathrm{T}}$ be the vector of the parameters of interest, where ^{T} denotes transposition, and let β^* be the true value of β. Let also $\Phi^*(B)$ and $\Theta^*(B)$, respectively, the polynomials $\Phi(B)$ and $\Theta(B)$ when $\beta = \beta^*$. The estimator at time t will be denoted $\hat{\beta}_t = (\hat{\phi}_{1,t}, \ldots, \hat{\phi}_{p,t}, \hat{\theta}_{1,t}, \ldots, \hat{\theta}_{q,t})^{\mathrm{T}}$. For a given β, the forecast $\hat{y}_{t|t-1}(\beta)$ for time t can be computed at time $t-1$, provided we replace the true errors e_s, $s < t$, also called innovations, by the residuals $\varepsilon_s(\beta) = y_s - \hat{y}_{s|s-1}(\beta)$, computed by recurrence. This requires suitable initial values whose effect can be neglected because of the assumption on the polynomials. In off-line estimation, under the Gaussian assumption on innovations e_t's, the maximum likelihood estimator is obtained, for large N, by minimising the sum of squares of the residuals

$$V_N(\beta) = \frac{1}{2} \sum_{t=1}^{N} \varepsilon_t^2(\beta). \quad (2)$$

Example 1 Specific parts will be illustrated with the ARMA(1,1) model defined by

$$y_t - \phi y_{t-1} = e_t - \theta e_{t-1}, \quad (3)$$

with $\beta^{\mathrm{T}} = (\phi, \theta)$. Note that Equation (3) implies

$$\hat{y}_{t|t-1}(\beta) = \phi y_{t-1} - \theta \varepsilon_{t-1}(\beta) \quad (4)$$

and

$$y_t - \hat{y}_{t|t-1}(\beta) = y_t - \phi y_{t-1} + \theta(y_{t-1} - \hat{y}_{t-1|t-2}(\beta)), \quad (5)$$

where the starting value $\hat{y}_{1|0}(\beta)$ can be taken equal to 0. Indeed the effect of a starting value decreases like $|\theta|^{t-1}$, and the assumption made implies that $|\theta| < 1$. This recurrence allows computing $\varepsilon_t(\beta)$.

For ARMA models, $V_N(\beta)$ is a nonlinear function of β, so $V_N(\beta)$ cannot be minimised analytically but well using numerical procedures, requiring many iterations on basis of the data from $t = 1$ to $t = N$. An on-line or recursive algorithm requires a vector of fixed size, preferably small with respect to N. Therefore we want an approximation of the off-line maximum likelihood estimator $\hat{\beta}_N$ that can be obtained by recurrences.

Given $\hat{\beta}_{t-1}$, we want to obtain $\hat{\beta}_t$ close to the minimum of $V_t(\beta)$. By a Taylor expansion of $V_t(\beta)$ around $\hat{\beta}_{t-1}$ limited to the second order we obtain

$$V_t(\beta) \simeq V_t(\hat{\beta}_{t-1}) + \left(\frac{\partial V_t(\beta)}{\partial \beta^{\mathrm{T}}}\right)_{\beta=\hat{\beta}_{t-1}} [\beta - \hat{\beta}_{t-1}] + \frac{1}{2}[\beta - \hat{\beta}_{t-1}]^{\mathrm{T}} \left(\frac{\partial^2 V_t(\beta)}{\partial \beta \partial \beta^{\mathrm{T}}}\right)_{\beta=\hat{\beta}_{t-1}} [\beta - \hat{\beta}_{t-1}]. \quad (6)$$

Minimising the right-hand side with respect to β leads to

$$\hat{\beta}_t = \hat{\beta}_{t-1} - \left(\frac{\partial^2 V_t(\beta)}{\partial \beta \partial \beta^{\mathrm{T}}}\right)^{-1}_{\beta=\hat{\beta}_{t-1}} \left(\frac{\partial V_t(\beta)}{\partial \beta^{\mathrm{T}}}\right)^{\mathrm{T}}_{\beta=\hat{\beta}_{t-1}}. \quad (7)$$

Denoting $\psi_t(\beta) = -[\partial \varepsilon_t(\beta)/\partial \beta^{\mathrm{T}}]^{\mathrm{T}}$, the opposite of the derivative of $\varepsilon_t(\beta)$ with respect to β, we have

$$\left[\frac{\partial V_t(\beta)}{\partial \beta^{\mathrm{T}}}\right]^{\mathrm{T}} = -\sum_{k=1}^{t} \psi_k(\beta)\varepsilon_k(\beta) = \left[\frac{\partial V_{t-1}(\beta)}{\partial \beta^{\mathrm{T}}}\right]^{\mathrm{T}} - \psi_t(\beta)\varepsilon_t(\beta), \quad (8)$$

and a further differentiation yields the Hessian:

$$\frac{\partial^2 V_t(\beta)}{\partial \beta \partial \beta^{\mathrm{T}}} = \frac{\partial^2 V_{t-1}(\beta)}{\partial \beta \partial \beta^{\mathrm{T}}} + \psi_t(\beta)\psi_t^{\mathrm{T}}(\beta) + \frac{\partial^2 \varepsilon_t(\beta)}{\partial \beta\, \partial \beta^{\mathrm{T}}}\varepsilon_t(\beta). \quad (9)$$

In order to evaluate Equation (7), the following approximations are made.

(1) We assume that $\hat{\beta}_t$ is close to $\hat{\beta}_{t-1}$, a quite reasonable approximation for large t, justifying Equation (6) and

$$\left(\frac{\partial^2 V_t(\beta)}{\partial \beta \partial \beta^{\mathrm{T}}}\right)_{\beta=\hat{\beta}_t} \simeq \left(\frac{\partial^2 V_t(\beta)}{\partial \beta \partial \beta^{\mathrm{T}}}\right)_{\beta=\hat{\beta}_{t-1}}. \quad (10)$$

(2) We proceed as if $\hat{\beta}_{t-1}$ were optimal at time $t-1$, i.e.

$$\left(\frac{\partial V_{t-1}(\beta)}{\partial \beta^{\mathrm{T}}}\right)_{\beta=\hat{\beta}_{t-1}} \simeq 0. \quad (11)$$

(3) Since, for β close to β^*, $\{\varepsilon_t(\beta)\}$ will almost behave like a white noise process, i.e. $\varepsilon_t(\beta)$ will have a mean close to 0 and be nearly independent from the observations and residuals before time t, allowing to neglect the last term of Equation (9).

Then, inserting Equation (10) in Equation (9) evaluated at $\beta = \hat{\beta}_{t-1}$, we have an approximation of the Hessian, $\bar{R}_t$, which can be computed recursively by

$$\bar{R}_t = \bar{R}_{t-1} + \psi_t(\hat{\beta}_{t-1})\psi_t^{\mathrm{T}}(\hat{\beta}_{t-1}). \quad (12)$$

Insertion of Equation (11) in Equation (8) evaluated at $\beta = \hat{\beta}_{t-1}$, yields

$$\left(\frac{\partial V_t(\beta)}{\partial \beta^{\mathrm{T}}}\right)^{\mathrm{T}}_{\beta=\hat{\beta}_{t-1}} = -\psi_t(\hat{\beta}_{t-1})\varepsilon_t(\hat{\beta}_{t-1}).$$

Using the approximation $\bar{R}_t$ in Equation (7), we have

$$\hat{\beta}_t = \hat{\beta}_{t-1} + \bar{R}_t^{-1}\psi_t(\hat{\beta}_{t-1})\varepsilon_t(\hat{\beta}_{t-1}). \quad (13)$$

Denoting $tR_t = \bar{R}_t$ we have the two equations

$$\begin{aligned} R_t &= R_{t-1} + \frac{1}{t}\{\psi_t(\hat{\beta}_{t-1})\psi_t^{\mathrm{T}}(\hat{\beta}_{t-1}) - R_{t-1}\} \\ \hat{\beta}_t &= \hat{\beta}_{t-1} + \frac{1}{t}R_t^{-1}\psi_t(\hat{\beta}_{t-1})\varepsilon_t(\hat{\beta}_{t-1}). \end{aligned} \quad (14)$$

There remains to derive equations for computing $\varepsilon_t(\hat{\beta}_{t-1})$ and $\psi_t(\hat{\beta}_{t-1})$. Let us first look at the ARMA(1,1) example (3).

Example 2 We have $\psi_t^{\mathrm{T}}(\beta) = \partial \hat{y}_{t|t-1}(\beta)/\partial \beta^{\mathrm{T}}$ and differentiation of $\hat{y}_{t|t-1}(\beta) - \theta\hat{y}_{t-1|t-2}(\beta) = (\phi - \theta)y_{t-1}$, which is also deduced from Equation (3), gives the two equations:

$$\frac{\partial \hat{y}_{t|t-1}(\beta)}{\partial \phi} - \theta\frac{\partial \hat{y}_{t-1|t-2}(\beta)}{\partial \phi} = y_{t-1}, \quad (15)$$

$$\frac{\partial \hat{y}_{t|t-1}(\beta)}{\partial \theta} - \hat{y}_{t-1|t-2}(\beta) - \theta\frac{\partial \hat{y}_{t-1|t-2}(\beta)}{\partial \theta} = -y_{t-1}. \quad (16)$$

The latter can also be written

$$\frac{\partial \hat{y}_{t|t-1}(\beta)}{\partial \theta} - \theta\frac{\partial \hat{y}_{t-1|t-2}(\beta)}{\partial \theta} = -\varepsilon_{t-1}(\beta). \quad (17)$$

Grouping Equations (15) and (17) gives

$$\psi_t(\beta) - \theta\psi_{t-1}(\beta) = \begin{pmatrix} y_{t-1} \\ -\varepsilon_{t-1}(\beta) \end{pmatrix}. \quad (18)$$

We can compute $\varepsilon_t(\hat{\beta}_{t-1})$ and $\psi_t(\hat{\beta}_{t-1})$ by using equations like (4) and (18) but this requires all the observations y_s, $s = 1, \ldots, t-1$. Let us derive approximations of $\varepsilon_t(\hat{\beta}_{t-1})$ and $\psi_t(\hat{\beta}_{t-1})$ that can be computed by recurrence using additional approximations. A natural approximation consists in using only the current estimator and $\max(p, q)$ previous values of ε, y and ψ as initial values.

Example 3 In the case of Equation (3), $\varepsilon_t(\hat{\beta}_{t-1})$ is approached by ε_t, computed by

$$\varepsilon_t = y_t - \hat{y}_{t|t-1} = y_t - \hat{\phi}_{t-1}y_{t-1} + \hat{\theta}_{t-1}(y_{t-1} - \hat{y}_{t-1|t-2}).$$

Let us introduce $\varphi_{t-1}^{\mathrm{T}} = (y_{t-1}, -\varepsilon_{t-1})$. Using Equation (4), we can write

$$\varepsilon_t = y_t - \hat{\beta}_{t-1}^{\mathrm{T}}\varphi_{t-1}. \tag{19}$$

Similarly, Equation (18) leads to a natural approximation ψ_t of $\psi_t(\hat{\beta}_{t-1})$

$$\psi_t = \hat{\theta}_{t-1}\psi_{t-1} + \varphi_{t-1}. \tag{20}$$

At time t we only need to know φ_{t-1}, ψ_{t-1} et $\hat{\beta}_{t-1}$. Adding these equations to those of Equation (14) and performing substitutions, we obtain the system

$$\begin{aligned} \psi_t &= \hat{\theta}_{t-1}\psi_{t-1} + \varphi_{t-1}, \\ \bar{R}_t &= \bar{R}_{t-1} + \psi_t\psi_t^{\mathrm{T}}, \\ \varepsilon_t &= y_t - \hat{\beta}_{t-1}^{\mathrm{T}}\varphi_{t-1}, \\ \hat{\beta}_t &= \hat{\beta}_{t-1} + \bar{R}_t^{-1}\psi_t\varepsilon_t. \end{aligned} \tag{21}$$

Note that Equation (21) is not computationally efficient because of the need to invert $\bar{R}_t$. There exists a more computationally efficient algorithm (Ljung & Söderström, 1983, Chap. 2, p. 19), where $P_t = \bar{R}_t^{-1}$ is updated instead of $\bar{R}_t$. The equation makes use of the inversion lemma and can be written

$$P_t = P_{t-1} - \frac{P_{t-1}\psi_t\psi_t^{\mathrm{T}}P_{t-1}}{1 + \psi_t^{\mathrm{T}}P_{t-1}\psi_t}. \tag{22}$$

Let us now go back to the general case (1). To improve the behaviour of the algorithm, we replace the factor $1/t$ by a sequence γ_t of positive scalars decreasing to 0 such that $\sum \gamma_t$ is divergent. If we now denote $\varphi_t^{\mathrm{T}} = (y_t, \ldots, y_{t-p+1}, -\varepsilon_t, \ldots, -\varepsilon_{t-q+1})$, with a due generalisation of Equation (20), the RML algorithm can now be written:

$$\begin{aligned} \psi_t &= \sum_{k=1}^{q} \hat{\theta}_{k,t-1}\psi_{t-k} + \varphi_{t-1}, \\ R_t &= R_{t-1} + \gamma_t(\psi_t\psi_t^{\mathrm{T}} - R_{t-1}), \\ \varepsilon_t &= y_t - \hat{\beta}_{t-1}^{\mathrm{T}}\varphi_{t-1}, \\ \hat{\beta}_t &= \hat{\beta}_{t-1} + \gamma_t R_t^{-1}\psi_t\varepsilon_t. \end{aligned} \tag{23}$$

3. Estimation by the RML$_{\mathrm{MZ}}$ method

Let us now consider a modification of the method of Section 2 called the RML$_{\mathrm{MZ}}$ method. From a theoretical point of view, under some assumptions, the RML algorithm (23) provides a consistent estimator with a rate of convergence $\sqrt{t}$. However, Mélard (1989) and Zahaf (1999) have observed huge variations of R_t with respect to time, which produce disturbances in the RML estimator. While keeping the recursive nature of the algorithm, they have tried to improve its accuracy by replacing the central recurrence (12) for the Hessian $\partial^2 V(\beta)/\partial\beta\partial\beta^{\mathrm{T}}$, by the computation of its expectation at the current value of the estimator. Indeed, $\sigma_e^2 R_t^{-1}$ is an approximation of the asymptotic covariance matrix $\Gamma(\beta^*)$ of the maximum likelihood estimator. But, β^* being unknown, they suggest to replace $\Gamma(\beta^*)$ by the asymptotic covariance matrix evaluated at the last value of the estimator, $\Gamma(\hat{\beta}_{t-1})$. If $\hat{\beta}_t$ converges to β^*, which will be shown later, then $\Gamma(\hat{\beta}_{t-1})$ converges to $\Gamma(\beta^*)$. Moreover, $\Gamma(\hat{\beta}_{t-1})$ is the inverse $F^{-1}(\hat{\beta}_{t-1})$ of the Fisher information matrix $F(\beta)$ computed at $\beta = \hat{\beta}_{t-1}$. At each time, we will compute $\sigma_e^2 F(\hat{\beta}_{t-1})$ and then its inverse $\sigma_e^{-2}F^{-1}(\hat{\beta}_{t-1})$ which will replace R_t^{-1} in Equation (23). This is thus a recursive analogue of the well-known Fisher-scoring algorithm, e.g. Kennedy and Gentle (1980, p. 450) and Sen and Singer (1993, p. 205). For a given σ_e^2, the algorithm is written:

$$\begin{aligned} \psi_t &= \sum_{k=1}^{q} \hat{\theta}_{k,t-1}\psi_{t-k} + \varphi_{t-1}, \\ \varepsilon_t &= y_t - \hat{\beta}_{t-1}^{\mathrm{T}}\varphi_{t-1}, \\ \hat{\beta}_t &= \hat{\beta}_{t-1} + \gamma_t\sigma_e^{-2}F^{-1}(\hat{\beta}_{t-1})\psi_t\varepsilon_t, \end{aligned} \tag{24}$$

where φ_t is like before. Therefore the recurrence for R_t in Equation (23) will no longer be needed. Note that

$$F(\beta) = \sigma_e^{-2}E\{\psi_t^1(\beta)\psi_t^{1\mathrm{T}}(\beta)\}, \tag{25}$$

where

$$\psi_t^1(\beta) = \sum_{k=1}^{q} \theta_k\psi_{t-k}^1(\beta) + \varphi_{t-1}^1,$$

and $\varphi_t^1 = (y_t, \ldots, y_{t-p+1}, -e_t, \ldots, -e_{t-q+1})$. Note also that, under a stationarity assumption, $F(\beta)$ does not depend on t. For simple models, an analytic expression does exist for $F^{-1}(\beta)$, see Box, Jenkins, and Reinsel (2008). Otherwise, there are simple algorithms for computing $F(\beta)$, see e.g. Klein and Mélard (1989).

But σ_e^2 is generally unknown so the algorithm (24) is modified as follows

$$\psi_t = \sum_{k=1}^{q} \hat{\theta}_{k,t-1}\psi_{t-k} + \varphi_{t-1}, \tag{26}$$

$$\varepsilon_t = y_t - \hat{\beta}_{t-1}^{\mathrm{T}}\varphi_{t-1}, \tag{27}$$

$$\hat{\beta}_t = \hat{\beta}_{t-1} + \gamma_t\hat{\sigma}_t^{-2}F^{-1}(\hat{\beta}_{t-1})\psi_t\varepsilon_t, \tag{28}$$

$$\hat{\sigma}_{t+1}^2 = \hat{\sigma}_t^2 + \gamma_t(\varepsilon_t^2 - \hat{\sigma}_t^2). \tag{29}$$

In order to avoid problems with uncontrolled recursions we will later change the estimator by a projection mechanism.

Remark 1 Since $F^{-1}(\hat{\beta}_{t-1})$ is to be computed at each time t, a matrix inversion is needed at each time. Consequently the RML_{MZ} method is not as computationally efficient as the RML method using Equation (22). In principle it should be possible to exploit the properties of the Fisher information matrix $F(\beta)$ to improve efficiency. In the pure AR(p) model, $F(\beta)$ is a Toeplitz matrix. Hence there are algorithms like Trench (1964) for inverting it in a number of operations proportional to p^2 instead of p^3. For an ARMA(p, p) model, $F(\beta)$ can be put under the form of a block Toeplitz matrix with $p \times p$ blocks of size 2, hence an inversion algorithm can also exploit that structure, e.g. Akaike (1973). For other ARMA(p, q) models with p different from q and p and q different from 0, some computational improvements can be found, for example by considering matrices of order $2(p+q)$. We will not further discuss improved algorithms in this paper.

To show convergence of the algorithm to the optimal value, we make two assumptions; the first one is about the true value of the vector of parameters β^* and the second one is about the errors.

Assumption 1 (on the model) The autoregressive and moving average polynomials have no common root and their roots are all outside of the unit circle (satisfying the causality or stationarity condition and the invertibility condition of the process), and $\phi_p \neq 0$ and $\theta_q \neq 0$.

Assumption 2 (on the errors) The errors $\{e_t\}$ have finite fourth-order moment: $\forall t, E(e_t^4) < \infty$.

Instead of Assumption 2, it is usually supposed, see e.g. Ljung and Söderström (1983), that the sequence of observations $\{y_t\}$ has a uniform upper bound in absolute value, more precisely $|y_t| < Y$, where Y is a random variable with a finite variance. That supposition is not really on the data but rather both on the parameter set and on the probability distribution of the errors. Our assumption is clearer on that respect. Also Ljung and Söderström (1983, p. 191) explicitly exclude errors which are not bounded, such as errors with a normal distribution, which is not the case here. Solo (1981) has slightly stronger assumptions than ours, by assuming that moments of order $4/(1-\delta)$, for some strictly positive δ, are finite.

Let $D_S = \{(\beta^{\mathrm{T}}, \sigma^2)^{\mathrm{T}} \in \mathbb{R}^{p+q+1}/$the eigenvalues of $A(\beta)$ are in the unit circle$\}$, hence $D_S = \{(\beta^{\mathrm{T}}, \sigma^2)^{\mathrm{T}} \in \mathbb{R}^{p+q+1}/$the roots of the moving average polynomial are outside of the unit circle$\}$. Because of the Fisher information matrix, the definition of the set D_R is also different:

$D_B = \{\beta \in \mathbb{R}^{p+q}/$the roots of the autoregressive and moving average polynomials are outside of the unit circle, $F(\beta)$ is invertible, $\|F^{-1}(\beta)\| < k\}$ for some constant $k > 0$ large enough.

$D_R = \{(\beta^{\mathrm{T}}, \sigma^2)^{\mathrm{T}} \in \mathbb{R}^{p+q+1}/\beta \in D_B$ and $\sigma^2 \geqslant \delta$ and $\sigma^2 \leqslant \delta'\}$ for some constant $\delta > 0$ small enough and $\delta' > \delta$ large enough.

More care is needed than in the original RML algorithm because $F^{-1}(\beta)$ can become very large. This is especially serious in the analysis where we will need a suitable projection mechanism built into the recursion, simpler than the one used in practice. As will be seen that projection mechanism makes use of β^*. This is quite disturbing for a statistician but it follows the suggestion of Benveniste et al. (1990, p. 56) to analyse a simplified algorithm. Example 4 below will show that the analysis is nevertheless very interesting. In spite of that, the study the statistical properties of the RML_{MZ} recursive estimator is very technical so most of the details will be given in Appendices 1 (Lemmas A.1–A.8) and 2 (Lemmas A.9–A.24). In Section 4, we will discuss small sample results obtained by Monte Carlo simulation, this time with a more realistic projection mechanism, and show a comparison with the original RML algorithm.

3.1. *Almost sure convergence*

Zahaf (1999) has used results from Duflo (1997) about Robbins-Monro stochastic approximation in order to obtain asymptotic properties for a Newton approximation to the RML_{MZ} estimator, called the RML_{NE} estimator. The algorithm has the form $\hat{\beta}_{t+1} = \hat{\beta}_t + \gamma_t Y_{t+1}$, where the conditional expectation of Y_{t+1} given the past information fulfils $E[Y_{t+1}/F_t]$ is a measurable function of $\hat{\beta}_t$. But here $E[Y_{t+1}/F_t]$ depends on both $\hat{\beta}_t$ and t, and it is even difficult to deduce convergence of the RML_{MZ} estimator from its Newton version. Also Kushner and Yin (1997, p. 94) cannot be applied because their assumption (A2.2) is not valid in our case.

The theory contained in Ljung and Söderström (1983) is based on writing the algorithm under the following form

$$\begin{aligned} h_t &= A(\hat{x}_{t-1})h_{t-1} + B(\hat{x}_{t-1})z_t, \\ \hat{x}_t &= \hat{x}_{t-1} + \gamma_t Q(t, \hat{x}_{t-1}, h_t), \end{aligned} \tag{30}$$

where $A(\cdot)$, $B(\cdot)$, and $Q(\cdot,\cdot,\cdot)$ are functions, γ_t is like in Section 2 and z_t makes use of the data. Like in Ljung (1977), the idea is to associate an ordinary differential equation (ODE) to the algorithm and obtain the attraction domain of an invariant set of that ODE. For the original RML estimator, $\hat{x}_t = (\hat{\beta}_t^{\mathrm{T}}, \mathrm{vec}(R_t)^{\mathrm{T}})^{\mathrm{T}}$ and it appears that $A(\cdot)$ and $B(\cdot)$ depend only on $\hat{\beta}_t$. Here we have to consider the same but where $\hat{x}_t = (\hat{\beta}_t^{\mathrm{T}}, \hat{\sigma}_{t+1}^2)^{\mathrm{T}}$

$$\begin{aligned} h_t &= A(\hat{\beta}_{t-1})\, h_{t-1} + B(\hat{\beta}_{t-1})z_t, \\ \begin{pmatrix} \hat{\beta}_t \\ \hat{\sigma}_{t+1}^2 \end{pmatrix} &= \begin{pmatrix} \hat{\beta}_{t-1} \\ \hat{\sigma}_t^2 \end{pmatrix} + \frac{1}{t} Q(t, \hat{\beta}_{t-1}, \hat{\sigma}_t^2, h_t), \end{aligned} \tag{31}$$

where h_t is $q(p+q+1) \times 1$, $Q(t, \beta, \sigma^2, \hbar)$ is $(p+q+1) \times 1$, and

$$h_t = (\varepsilon_t, \varepsilon_{t-1}, \ldots, \varepsilon_{t-q+1}, \psi_t^{\mathrm{T}}, \psi_{t-1}^{\mathrm{T}}, \ldots, \psi_{t-q+1}^{\mathrm{T}})^{\mathrm{T}},$$

$$z_t = (y_t, \ldots, y_{t-p})^{\mathrm{T}}, \tag{32}$$

$$Q(t,\beta,\sigma^2,\hbar) = [\sigma^{-2}\{F^{-1}(\beta)(\hbar_{q+1},\hbar_{q+2},\ldots,\hbar_{2q+p})^{\mathrm{T}}\}^{\mathrm{T}}\hbar_1,\ \hbar_1^2 - \sigma^2]^{\mathrm{T}}, \quad (33)$$

so that $\hbar_1$ represents ε_t, $(\hbar_{q+1},\hbar_{q+2},\ldots,\hbar_{2q+p})^{\mathrm{T}}$ represents ψ_t, and

$$Q(t,\hat{\beta}_{t-1},\hat{\sigma}_t^2,h_t) = [\hat{\sigma}_t^{-2}\{F_t^{-1}(\hat{\beta}_{t-1})\psi_t\}^{\mathrm{T}}\varepsilon_t, \varepsilon_t^2 - \hat{\sigma}_t^2]^{\mathrm{T}}.$$

Notice that R_t, obtained by the Fisher information matrix evaluated at $\beta = \hat{\beta}_{t-1}$, appears in the second term of the right-hand side of the second equation of (31), making derivations very different from Ljung and Söderström (1983). Their theory cannot be applied directly for the $\mathrm{RML_{MZ}}$ algorithm. However, the first equation of (31) still holds with the same choice for the matrices A and B as in Equation (30). For an ARMA(p,q) model, it can be seen that

$$\det(A(\beta) - \lambda I) = (-1)^{q(q-1)(p+2q-1)/2}(-(\lambda^q - \lambda^{q-1}\theta_1 - \lambda^{q-2}\theta_2 - \cdots - \theta_q))^{p+q+1}. \quad (34)$$

In the following, $x = (\beta^{\mathrm{T}},\sigma^2)^{\mathrm{T}} \in D_R$ and we write sometimes Q with three arguments instead of four. We will make use of Ljung (1977, Theorems 1 and 4), summed up as Lemma A.8 in Appendix 1. Here is the third subset of his conditions, denoted by C, without C7 which is not needed:

C1: $Q(t,x,\hbar)$ is Lipschitz continuous in x and $\hbar$:

$$\|Q(t,x_1,\hbar_1) - Q(t,x_2,\hbar_2)\| < \mathcal{K}_1(x,\hbar,\rho,\upsilon)\{\|x_1 - x_2\| + \|\hbar_1 - \hbar_2\|\}$$

for $x_i \in \mathcal{B}(x,\rho)$, an open ball of centre x and diameter ρ, for $\rho = \rho(x) > 0$, where $x \in D_R$, $\hbar_i \in \mathcal{B}(\hbar,\upsilon)$ for $\upsilon \geq 0$;

C2: matrices $A(\cdot)$and $B(\cdot)$ are Lipschitz continuous functions over D_R;

C3: $H(\bar{x}) = \lim_{t\to\infty}(1/t)\sum_{k=1}^{t} Q(k,\bar{x},\bar{h}_k(\bar{x}))$ does exist for all $\bar{x} \in D_R$;

C4: for all $\bar{x} \in D_R$, $0 < \lambda < 1$ and $c < \infty$, the random variable $k_\upsilon(t,\bar{x},\lambda,c)$ defined by

$$k_\upsilon(t,\bar{x},\lambda,c) = k_\upsilon(t-1,\bar{x},\lambda,c) + \gamma_t[\mathcal{K}_1(\bar{x},h,\rho(\bar{x}),\upsilon(t,\lambda,c))(1+\upsilon(t,\lambda,c)) - k_\upsilon(t-1,\bar{x},\lambda,c)]$$

with $k_\upsilon(0,\bar{x},\lambda,c) = 0$ and $\upsilon(t,\lambda,c) = c\sum_{k=1}^{t}\lambda^{t-k}|z(k)|$, converges to a finite limit when $t\to\infty$;

C5: $\sum_{t=1}^{\infty}\gamma_t = \infty$;

C6: $\lim_{t\to\infty}\gamma_t = 0$.

According to Ljung (1977), these conditions are used in the deterministic case, but the results are valid with probability 1 as far as z_t defined by Equation (32) is such that the conditions C3 and C4 are satisfied with probability 1.

On the basis of Equations (27) and (29), let us define $\varepsilon_t(\beta) = y_t - \beta^{\mathrm{T}}\varphi_{t-1}(\beta)$, where $\varphi_{t-1}(\beta) = (y_{t-1},\ldots,y_{t-p},-\varepsilon_{t-1}(\beta),\ldots,-\varepsilon_{t-q}(\beta))^{\mathrm{T}}$, and $\sigma_t^2(\beta) = \sigma_{t-1}^2(\beta) + (1/t)(\varepsilon_{t-1}^2(\beta) - \sigma_{t-1}^2(\beta))$. Hence

$$\sigma_t^2(\beta) = \frac{1}{t}\sum_{k=1}^{t}\varepsilon_{k-1}^2(\beta). \quad (35)$$

Define also

$$\psi_t(\beta) = \sum_{k=1}^{q}\theta_k\psi_{t-k}(\beta) + \varphi_{t-1}(\beta), \quad (36)$$

$$R_t(\beta) = R_{t-1}(\beta) + \frac{1}{t}(\psi_t(\beta)\psi_t^{\mathrm{T}}(\beta) - R_{t-1}(\beta)) = \frac{1}{t}\sum_{k=1}^{t}\psi_k(\beta)\psi_k^{\mathrm{T}}(\beta). \quad (37)$$

THEOREM 1 *Under Assumptions* 1 *and* 2, *conditions* C1–C6 *of Ljung* (1977, *Theorem* 4) *are satisfied.*

The proof is given in Appendix 1 on the basis of Lemmas A.1–A.3.

We will consider the following ODE

$$\frac{\partial x(t)}{\partial t} = \frac{\partial(\beta^{\mathrm{T}}(t),\sigma^2(t))^{\mathrm{T}}}{\partial t} = H(\beta(t),\sigma^2(t)),$$

where

$$H(\beta,\sigma^2) = [\sigma^{-2}\{F^{-1}(\beta)E(\psi(\beta)\varepsilon(\beta))\}^{\mathrm{T}}, E\{\varepsilon^2(\beta)\} - \sigma^2]^{\mathrm{T}}.$$

Letting $f(\beta) = E\{\psi(\beta)\varepsilon(\beta)\}$ and $V(\beta) = E\{\varepsilon^2(\beta)\}$, the ODE can be put under the form

$$\frac{\partial\beta(t)}{\partial t} = \sigma^{-2}(t)F^{-1}(\beta(t))f(\beta(t)), \quad (38)$$

$$\frac{\partial\sigma^2(t)}{\partial t} = V(\beta(t)) - \sigma^2(t). \quad (39)$$

THEOREM 2 *Let*

$$c(\beta^*) = \sup_{D\in K_D}\inf_{\beta\in\mathrm{Fr}(D)} V(\beta),$$

where K_D is a set of connex parts of D_B containing β^* and Fr(D) *is the frontier of D. Let $D_2 = \{(\beta^{\mathrm{T}},\sigma^2)^{\mathrm{T}} \in D_R / V(\beta) \leqslant c(\beta^*) - \varrho\}$ with a very small positive constant ϱ. Under Assumptions* 1 *and* 2, *let the recursive* $\mathrm{RML_{MZ}}$

estimator (26)–(28) *replaced by the following recurrences*

$$\begin{pmatrix} \hat{\beta}_t \\ \hat{\sigma}^2_{t+1} \end{pmatrix} = \left[\begin{pmatrix} \hat{\beta}_{t-1} \\ \hat{\sigma}^2_t \end{pmatrix} + \frac{1}{t} \begin{pmatrix} \hat{\sigma}_t^{-2} F^{-1}(\hat{\beta}_{t-1}) \psi_t \varepsilon_t \\ \varepsilon_t^2 - \hat{\sigma}_t^2 \end{pmatrix} \right]_{D_R, D_2}, \tag{40}$$

where

$$[z]_{D_R,D_2} = \begin{cases} z & \text{if } z \in D_R \\ \text{a point in } D_2 & \text{if } z \notin D_R, \end{cases}$$

and

$$\begin{cases} \left. \begin{array}{l} \varepsilon_t = y_t - \hat{\beta}^{\mathrm{T}}_{t-1}\varphi_{t-1}, \\ \psi_t = \sum_{k=1}^{q} \hat{\theta}_{k,t-1}\psi_{t-k} + \varphi_{t-1}, \end{array} \right\} & \text{if } (\hat{\beta}^{\mathrm{T}}_{t-1}, \hat{\sigma}^2_t)^{\mathrm{T}} \in D_R, \\ (\varepsilon_t, \psi_t)^{\mathrm{T}} \text{ a point in } K & \text{if } (\hat{\beta}^{\mathrm{T}}_{t-1}, \hat{\sigma}^2_t)^{\mathrm{T}} \notin D_R, \end{cases}$$

where K is a compact subset of $\mathbb{R}^{p+q+1}$ defined in advance. Then $\hat{\beta}_t$ converges to β^* almost surely when $t \to \infty$.

Proof Given that

$$\lim_{t\to\infty} \frac{1}{t} \sum_{k=1}^{t} Q(k, \beta, \sigma^2, h_k(\beta))$$
$$= E((\sigma^{-2}F^{-1}(\beta)\psi_t(\beta)\varepsilon_t(\beta))^{\mathrm{T}}, \varepsilon_t^2(\beta) - \sigma^2)^{\mathrm{T}} \tag{41}$$

(proved in Lemma A.3 when checking C3, see Appendix 1), we have to analyse (38)–(39). We need to check some assumptions on that differential equation. We have by Lemma A.4

$$V(\beta(t)) = E\{\varepsilon^2(\beta(t))\} \geqslant \sigma_e^2 > 0,$$

and

$$\frac{\partial V(\beta(t))}{\partial t} = \frac{\partial V(\beta(t))}{\partial \beta^{\mathrm{T}}(t)} \frac{\partial \beta(t)}{\partial t}$$
$$= -2f(\beta(t))^{\mathrm{T}} \sigma^{-2}(t) F^{-1}(\beta(t)) f(\beta(t)) \leqslant 0,$$

since $F^{-1}(\beta(t))$ is a symmetric positive definite matrix in D_R and $\sigma^{-2}(t)$ is positive. Let $\dot{V}(\beta(t)) = \partial V(\beta(t))/\partial t$. Hence an invariant set of the ODE is $E = \{(\beta^{\mathrm{T}}, \sigma^2)^{\mathrm{T}} \in D_R / \dot{V}(\beta) = 0\} = \{(\beta^{\mathrm{T}}, \sigma^2)^{\mathrm{T}} \in D_R / f(\beta) = 0\} = \{\beta^*\} \times [\delta; \delta']$. By Lemma A.5(a), there is a solution of the ODE (38)–(39) over some interval $[t_0, t_1]$. Let $(\beta(t_0)^{\mathrm{T}} = \beta_0^{\mathrm{T}}, \sigma^2(t_0) = \sigma_0^2)^{\mathrm{T}} \in D_2$, since $V(\beta(t))$ is decreasing in t then, for all $t > t_0$, $V(\beta(t)) < V(\beta(t_0)) \leqslant c(\beta^*) - \varrho$, by Lemma A.6, $(\beta(t)^{\mathrm{T}}, \sigma^2(t))^{\mathrm{T}} \in D_2$, $\forall t \in [t_0, t_1]$. Hence by Lemma A.5(b), $\forall t > t_0$, $(\beta(t)^{\mathrm{T}}, \sigma^2(t))^{\mathrm{T}} \in D_2$. Finally, like in Ljung and Söderström (1983), D_2 is a part of the attraction domain for the invariant set $E = \{\beta^*\} \times [\delta; \delta']$ because it fulfils the conditions of Lemma A.7. Then, we can apply Lemma A.8 which summarises Theorems 1 and 4 of Ljung (1977), by letting $D_1 = D_R$. ■

Example 4 Let the ARMA(1,1) model defined by Equation (3). Let $\beta^* = (\phi^*, \theta^*)^{\mathrm{T}}$ and assume that $\phi^* \neq \theta^*$. We know that

$$F^{-1}(\beta) = \left\{ E\left[\frac{\partial \varepsilon_t(\beta)}{\partial \beta} \frac{\partial \varepsilon_t(\beta)}{\partial \beta^{\mathrm{T}}} \right] \right\}^{-1} = \frac{1 - \phi\theta}{(\phi - \theta)^2}$$
$$\times \begin{bmatrix} (1-\phi^2)(1-\phi\theta) & (1-\phi^2)(1-\theta^2) \\ (1-\phi^2)(1-\theta^2) & (1-\theta^2)(1-\phi\theta) \end{bmatrix}.$$

The $\mathrm{RML_{MZ}}$ algorithm can be written

$$\begin{cases} h_t = \begin{pmatrix} \hat{\theta}_{t-1} & 0 & 0 \\ 0 & \hat{\theta}_{t-1} & 0 \\ -1 & 0 & \hat{\theta}_{t-1} \end{pmatrix} h_{t-1} + \begin{pmatrix} 1 & -\hat{\phi}_{t-1} \\ 0 & 1 \\ 0 & 0 \end{pmatrix} \begin{pmatrix} y_t \\ y_{t-1} \end{pmatrix}, \\ \begin{pmatrix} \hat{\beta}_t \\ \hat{\sigma}^2_{t+1} \end{pmatrix} = \begin{pmatrix} \hat{\beta}_{t-1} \\ \hat{\sigma}^2_t \end{pmatrix} + \frac{1}{t} Q(t, \hat{\beta}_{t-1}, \hat{\sigma}^2_t, h_t), \end{cases}$$

with $Q(t, \hat{\beta}_{t-1}, \hat{\sigma}_t^2, h_t) = (\hat{\sigma}_t^2 \psi_t^{\mathrm{T}} F^{-1}(\hat{\beta}_{t-1})\varepsilon_t, \varepsilon_t^2 - \hat{\sigma}_t^2)^{\mathrm{T}}$ and $h_t^{\mathrm{T}} = (\varepsilon_t, \psi_t^{\mathrm{T}})$. Hence

$$A(\beta) = \begin{pmatrix} \theta & 0 & 0 \\ 0 & \theta & 0 \\ -1 & 0 & \theta \end{pmatrix}, \quad \det(A(\beta) - \lambda I) = (\theta - \lambda)^3.$$

Let $U = \{\beta = (\phi, \theta) \in]-1, 1[\times]-1, 1[\}$. For that model, we have
$D_S = \{(\beta^{\mathrm{T}}, \sigma^2) \in \mathbb{R}^3 / \theta \in]-1, 1[\}$, $D_B = \{\beta = (\phi, \theta) \in U / \|F^{-1}(\beta)\| < k\}$,
$D_R = \{\beta \in D_B, \sigma^2 \in \mathbb{R} / \sigma^2 > \delta\}$, hence $D_R \subset D_S$ and $D_R = (U \backslash \{(\phi, \theta) \in U / |\phi - \theta| < \kappa\}) \times \{\sigma^2 > \delta\}$, where κ is a very small positive real number.

Let us compute $E(\varepsilon_t^2(\beta))$. We have $\varepsilon_t = \Theta_1^{-1}(B)\Phi_1(B)y_t = \Theta_1^{-1}(B)\Phi_1(B)\ \Phi_1^{*-1}(B)\Theta_1^*(B)e_t$. Let $\phi(\omega)$ the spectral density of ε_t. We have

$$\phi(\omega) = \frac{1}{2\pi} |\Theta_1(\mathrm{e}^{\mathrm{i}\omega})|^{-2} |\Phi_1(\mathrm{e}^{\mathrm{i}\omega})|\, |\Phi_1^*(\mathrm{e}^{\mathrm{i}\omega})|^{-2} |\Theta_1^*(\mathrm{e}^{\mathrm{i}\omega})|$$

hence

$$E(\varepsilon_t^2(\beta)) = \int_{-\pi}^{\pi} \phi(\omega)\, \mathrm{d}\omega$$
$$= \frac{1}{2\pi i} \oint \Theta_1^{-1}(z) \Theta_1^{-1}\left(\frac{1}{z}\right) \Phi_1(z) \Phi_1\left(\frac{1}{z}\right)$$
$$\times \Phi_1^{*-1}(z) \Phi_1^{*-1}\left(\frac{1}{z}\right) \Theta_1^*(z) \Theta_1^*\left(\frac{1}{z}\right) \frac{\mathrm{d}z}{z}$$
$$= \frac{1}{2\pi i} \oint \frac{(1-\phi z)(z-\phi)(1-\theta^* z)(z-\theta^*)}{(1-\theta z)(z-\theta)(1-\phi^* z)(z-\phi^*)} \frac{\mathrm{d}z}{z}.$$

For all β such that $\phi = \theta$,

$$E(\varepsilon_t^2(\beta)) = \frac{(1-\theta^*\phi^*)(\phi^* - \theta^*)}{(1-\phi^{*2})} + \frac{\theta^*}{\phi^*},$$

and the Fisher information matrix is not invertible. It is obvious that when θ comes close to 1 or -1, $E(\varepsilon_t^2(\beta))$ converges to infinity except when $\phi = \theta$. Let us consider the

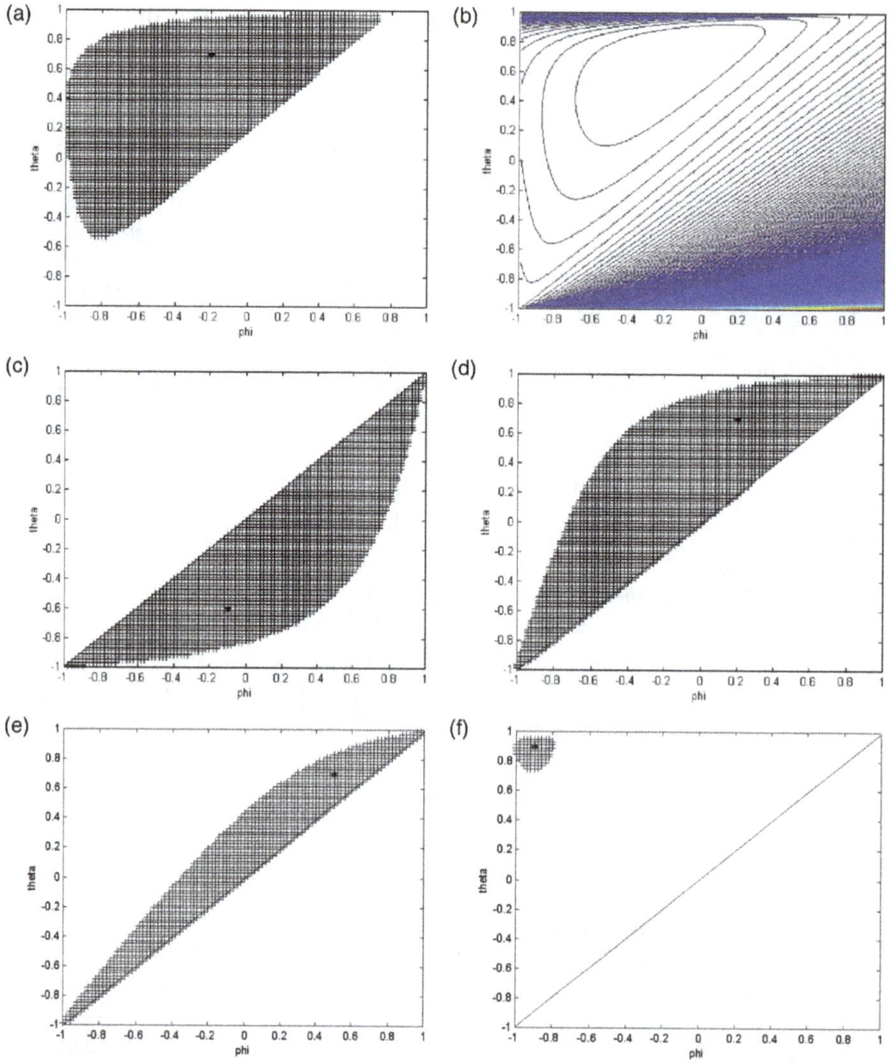

Figure 1. For several ARMA(1, 1) processes characterised by (ϕ^*, θ^*) values of ρ part D_2 is shown where we can project $\hat{\beta}_t$ in order to achieve convergence. For graph (b), level curves for several values of $V(\beta)$ are shown instead. (a) $(\phi^*, \theta^*) = (-0.2, 0.7)$, (b) $(\phi^*, \theta^*) = (-0.2, 0.7)$, (c) $(\phi^*, \theta^*) = (-0.1, -0.6)$, (d) $(\phi^*, \theta^*) = (0.2, 0.7)$, (e) $(\phi^*, \theta^*) = (0.5, 0.7)$ and (f) $(\phi^*, \theta^*) = (-0.9, 0.9)$.

ODE

$$\dot{\beta} = \sigma^{-2}(t)F^{-1}(\beta)f(\beta)$$
$$\dot{\sigma}^2(t) = V(\beta) - \sigma^2(t)$$

where

$$f(\beta) = E[\varepsilon_t(\beta)\psi_t(\beta)] \quad \text{and} \quad V(\beta) = E\{\varepsilon_t^2(\beta)\}.$$

Let

$$c(\phi^*, \theta^*) = \sup_{D \in K_D} \inf_{\beta \in \mathrm{Fr}(D)} E(V(\beta)),$$

where K_D is a set of connex parts of D_B containing β^*.

Let $D_2 = \{V(\beta) \leqslant c(\beta^*) - \varrho\}$ with a very small positive ϱ. If we know a value in D_2, we can use it for estimation by the algorithm of Ljung (1977, Theorem 4). In that case we have almost sure convergence.

In the cases shown in Figure 1, the part D_2 where we can project the estimator to achieve convergence is the crossed surface.

As said before, the admissible region is the square U except the diagonal joining the points $(-1, -1)$ and $(1, 1)$ so it is composed of two half squares. For each of the six cases $V(\beta)$ has a unique minimum located in one of the half squares. The part D_2, shown in Figure 1 except in case (b), is always in the half square where the minimum is located. If the initial value of the ODE is in that half square, and better in D_2, the solutions will turn towards that minimum when t goes to infinity, hence also the estimator $(\hat{\phi}_t, \hat{\theta}_t)$. Convergence is faster in D_2. If the initial value is in the other half square, the solutions of the ODE will turn towards the frontier formed by the diagonal joining the points $(-1, -1)$ and $(1, 1)$ but will stop before reaching it. Similarly, the estimator $(\hat{\phi}_t, \hat{\theta}_t)$ will turn towards the minimum but will have to jump over the diagonal since we may not have $\hat{\phi}_t = \hat{\theta}_t$, because the Fisher information matrix is not invertible there. Convergence will also be slower than in the other half square and much slower than in D_2. This is well illustrated by (b) which shows the contour levels of $V(\beta)$ for the same parameter values as (a). The level corresponding to D_2 can be seen and even smaller areas where convergence will be still faster. The other levels are higher in the upper half square and much higher in the lower half square.

Figure 1 shows that D_2 is sometimes lenticular, like in (c, d, e) but not always. Its size depends on the true values of the parameters and is smaller when they are close one from the other. Case (c) shows a situation where the point corresponding to the true values of the parameters is in the lower half square. The surface of D_2 is small when the point is close to the boundary, like in (e) and (f).

Remark 2 A possible alternative would have been to apply Chen (2002) stochastic approximation theory with expanding truncations. Indeed the assumptions made there can be verified. But Chen (2002, p. 291) assumption A6.1.3 that the norm of $\{z_t\}$, defined by Equation (32), is bounded by a random variable with a finite variance implies here that the observations $\{y_t\}$ are bounded by a random variable with a finite variance, ruling out again a Gaussian process. Therefore, we have preferred analysing a slightly simplified algorithm.

3.2. *Convergence in law*

Fabian (1968) has studied asymptotic normality of the algorithm

$$\tilde{\beta}_{t+1} = (I - t^{-\alpha}\Gamma_t)\tilde{\beta}_t + t^{-(\alpha+\delta)/2}\Phi_t V_t + t^{-\alpha-\delta/2}T_t,$$

where $\tilde{\beta}_t = \hat{\beta}_t - \beta^*$ in our case, Γ_t, Φ_t are matrices, V_t and T_t are vectors, by letting conditions on the components of that algorithm. He has shown that $t^{\delta/2}\tilde{\beta}_t$ converges in law to the normal distribution. In our case, we let $\alpha = \delta = 1$, $\Gamma_t = 0$, $T_t = 0$, $\Phi_t V_t = F^{-1}(\hat{\beta}_t)\psi_t\varepsilon_t$, but one of the conditions of Fabian (1968) is that Γ_t is definite positive. Ljung, Pflug, and Walk (1992) have studied a special case of that algorithm by letting $T_t = T$, $\alpha = 1$ and $\Phi_n = I$. They have shown convergence in law under other conditions. We have tried to verify the conditions of Kushner and Huang (1979) which are more general but they are not satisfied for the $\mathrm{RML_{MZ}}$ estimator. Zahaf (1999) has tried to show convergence in law of the $\mathrm{RML_{MZ}}$ estimator by using a theorem from Duflo (1997, p. 52). To achieve this goal, Zahaf (1999) makes use of an unproven conjecture which is only valid in some special cases and it is also supposed that $F^{-1}(\beta)\psi_t(\beta)\psi_t^{\mathrm{T}}(\beta)$ is (strictly) positive definite which is not true. Therefore, we have preferred to adapt the approach of Ljung and Söderström (1983).

THEOREM 3 *Consider an ARMA model defined by Equation (1) and the algorithm (40), according to Assumptions 1 and 2 of Section 3.1. Then, $\sqrt{t}(\hat{\beta}_t - \beta^*)$ converges in law to a normal distribution $N(0, F^{-1}(\beta^*))$ when $t \to \infty$.*

Proof By Theorem 2, we know that the algorithm (40), which includes a projection mechanism, leads to convergence. Hence, for $t > t_0$ for some finite t_0, that algorithm behaves like the algorithm (26)–(28) without projection. To simplify, we change the origin of time at t_0 and omit the contribution of terms before t_0 which are asymptotically negligible. We know that the algorithm (26)–(28) can be written under the form (31). We have already shown in Theorem 2 that, under the same assumptions, the estimator converges almost surely to the true value of the parameter.

Let $\varepsilon_0 = \varepsilon_0(\beta^*) = e_0 = 0$, and consider $t \geqslant 1$. Denote $\bar{\sigma}_t = t\hat{\sigma}^2_{t+1}$. Using Equations (29) and (35), we have $\bar{\sigma}_t = \bar{\sigma}_{t-1} + \varepsilon_t^2$ and

$$\hat{\sigma}^2_{t+1} = \frac{1}{t}\sum_{k=1}^{t}\varepsilon_k^2, \quad \sigma^2_{t+1}(\beta^*) = \frac{1}{t}\sum_{k=1}^{t}\varepsilon_k^2(\beta^*) = \frac{1}{t}\sum_{k=1}^{t}e_k^2 \tag{42}$$

because $\forall k > 0, \varepsilon_k(\beta^*) = e_k$. Denote

$$\bar{R}_t(\beta^*) = tR_t(\beta^*) = \bar{R}_{t-1}(\beta^*) + \psi_t(\beta^*)\psi_t^{\mathrm{T}}(\beta^*), \tag{43}$$

and let $\tilde{\beta}_t = \hat{\beta}_t - \beta^*$. From Equation (28), we have

$$\tilde{\beta}_t = \tilde{\beta}_{t-1} + \frac{1}{t}\hat{\sigma}_t^{-2}F^{-1}(\hat{\beta}_{t-1})\psi_t\varepsilon_t. \tag{44}$$

According to Lemma A.9, $K_t = \bar{\sigma}_t\tilde{\beta}_t$ can be decomposed in a sum of terms (A3). Using that decomposition, we need Lemma A.20 to show that $\forall\delta > 0, t^{1/2-\delta}\|\tilde{\beta}_t\| \to 0$ a.s. when $t \to \infty$. The proof of that Lemma A.20 makes use of Lemmas A.11–A.19. We will use that result in Lemmas A.21 and A.24. All these lemmas are in Appendix 2.

From Equations (43) and (44), we can write

$$\begin{aligned}\bar{R}_t(\beta^*)\tilde{\beta}_t &= \bar{R}_t(\beta^*)\tilde{\beta}_{t-1} + \frac{1}{t}\bar{R}_t(\beta^*)\hat{\sigma}_t^{-2}F^{-1}(\hat{\beta}_{t-1})\psi_t\varepsilon_t,\\ &= \bar{R}_{t-1}(\beta^*)\tilde{\beta}_{t-1} + \psi_t(\beta^*)\psi_t^{\mathrm{T}}(\beta^*)\tilde{\beta}_{t-1}\\ &\quad + R_t(\beta^*)\hat{\sigma}_t^{-2}F^{-1}(\hat{\beta}_{t-1})\psi_t\varepsilon_t.\end{aligned}$$

But $\psi_t\varepsilon_t$ is equal to

$$\begin{aligned}&\psi_t(\varepsilon_t - \varepsilon_t(\hat{\beta}_{t-1})) + (\psi_t - \psi_t(\hat{\beta}_{t-1}))\varepsilon_t(\hat{\beta}_{t-1})\\ &\quad + \psi_t(\hat{\beta}_{t-1})(\varepsilon_t(\hat{\beta}_{t-1}) - e_t) + \psi_t(\hat{\beta}_{t-1})e_t,\end{aligned}$$

and, using a Taylor expansion,

$$\begin{aligned}\varepsilon_t(\hat{\beta}_{t-1}) - e_t &= -\psi_t^{\mathrm{T}}(\beta^*)\tilde{\beta}_{t-1}\\ &\quad - \frac{1}{2}\tilde{\beta}_{t-1}^{\mathrm{T}}\left(\frac{\partial\psi_t(\beta)}{\partial\beta^{\mathrm{T}}}\right)_{\beta=\varkappa_t}\tilde{\beta}_{t-1},\end{aligned} \tag{45}$$

where $\varkappa_t$ is a point on the segment joining $\hat{\beta}_{t-1}$ and β^*. Letting $U_t = R_t(\beta^*)\hat{\sigma}_t^{-2}F^{-1}(\hat{\beta}_{t-1})$, we have

$$\begin{aligned}\bar{R}_t(\beta^*)\tilde{\beta}_t &= \bar{R}_{t-1}(\beta^*)\tilde{\beta}_{t-1} + \psi_t(\beta^*)\psi_t^{\mathrm{T}}(\beta^*)\tilde{\beta}_{t-1} \\ &- U_t\psi_t(\beta^*)\psi_t^{\mathrm{T}}(\beta^*)\tilde{\beta}_{t-1} \\ &- U_t(\psi_t(\hat{\beta}_{t-1}) - \psi_t(\beta^*))\psi_t^{\mathrm{T}}(\beta^*)\tilde{\beta}_{t-1} \\ &- \frac{1}{2}U_t\psi_t(\hat{\beta}_{t-1})\tilde{\beta}_{t-1}^{\mathrm{T}}\left(\frac{\partial\psi_t(\beta)}{\partial\beta^{\mathrm{T}}}\right)_{\beta=\varkappa_t}\tilde{\beta}_{t-1} \\ &+ U_t(\psi_t - (\hat{\beta}_{t-1}))\varepsilon_t(\hat{\beta}_{t-1}) \\ &+ U_t\psi_t(\varepsilon_t - \varepsilon_t(\hat{\beta}_{t-1})) + U_t\psi_t(\hat{\beta}_{t-1})e_t.\end{aligned}$$

We can write

$$\begin{aligned}&(I_{p+q} - U_t)\psi_t(\beta^*)\psi_t^{\mathrm{T}}(\beta^*)\tilde{\beta}_{t-1} \\ &= (\hat{\sigma}_t^2F(\beta^*) - R_t(\beta^*))\hat{\sigma}_t^{-2}F^{-1}(\hat{\beta}_{t-1})\psi_t(\beta^*)\psi_t^{\mathrm{T}}(\beta^*)\tilde{\beta}_{t-1} \\ &\quad + (F(\hat{\beta}_{t-1}) - F(\beta^*))F^{-1}(\hat{\beta}_{t-1})\psi_t(\beta^*)\psi_t^{\mathrm{T}}(\beta^*)\tilde{\beta}_{t-1}.\end{aligned}$$

Letting

$$\begin{aligned}B_{1,t} &= (F(\hat{\beta}_{t-1}) - F(\beta^*))F^{-1}(\hat{\beta}_{t-1})\psi_t(\beta^*)\psi_t^{\mathrm{T}}(\beta^*)\tilde{\beta}_{t-1} \\ &- U_t(\psi_t(\hat{\beta}_{t-1}) - \psi_t(\beta^*))\psi_t^{\mathrm{T}}(\beta^*)\tilde{\beta}_{t-1} \\ &- \frac{1}{2}U_t\psi_t(\hat{\beta}_{t-1})\tilde{\beta}_{t-1}^{\mathrm{T}}\left(\frac{\partial\psi_t(\beta)}{\partial\beta^{\mathrm{T}}}\right)_{\beta=\psi_t}\tilde{\beta}_{t-1}, \end{aligned} \tag{46}$$

and

$$B_{2,t} = U_t(\psi_t - \psi_t(\hat{\beta}_{t-1}))\varepsilon_t(\hat{\beta}_{t-1}) + U_t\psi_t(\varepsilon_t - \varepsilon_t(\hat{\beta}_{t-1})), \tag{47}$$

we have

$$\begin{aligned}\bar{R}_t(\beta^*)\tilde{\beta}_t &= \bar{R}_{t-1}(\beta^*)\tilde{\beta}_{t-1} + (\hat{\sigma}_t^2F(\beta^*) \\ &- R_t(\beta^*))\hat{\sigma}_k^{-2}F^{-1}(\hat{\beta}_{t-1})\psi_t(\beta^*)\psi_t^{\mathrm{T}}(\beta^*)\tilde{\beta}_{t-1} \\ &+ B_{1,t} + B_{2,t} + U_t\psi_t(\hat{\beta}_{t-1})e_t,\end{aligned}$$

hence

$$\begin{aligned}tR_t(\beta^*)\tilde{\beta}_t &= \sum_{k=1}^{t}(\hat{\sigma}_t^2F(\beta^*) \\ &- R_k(\beta^*))\hat{\sigma}_k^{-2}F^{-1}(\hat{\beta}_{k-1})\psi_k(\beta^*)\psi_k^{\mathrm{T}}(\beta^*)\tilde{\beta}_{k-1} \\ &+ \sum_{k=1}^{t}B_{1,k} + \sum_{k=1}^{t}B_{2,k} + \sum_{k=1}^{t}U_k\psi_k(\hat{\beta}_{k-1})e_k.\end{aligned} \tag{48}$$

Hence

$$\begin{aligned}\sqrt{t}\tilde{\beta}_t &= H_t + L_t + R_t^{-1}(\beta^*)\frac{1}{\sqrt{t}} \\ &\times \sum_{k=1}^{t}R_k(\beta^*)\hat{\sigma}_k^{-2}F^{-1}(\beta^*)\psi_k(\beta^*)e_k,\end{aligned} \tag{49}$$

where

$$\begin{aligned}H_t &= R_t^{-1}(\beta^*)\frac{1}{\sqrt{t}}\sum_{k=1}^{t}(\sigma_e^2F(\beta^*) \\ &- R_k(\beta^*))\hat{\sigma}_k^{-2}F^{-1}(\hat{\beta}_{k-1})\psi_k(\beta^*)\psi_k(\beta^*)\tilde{\beta}_{k-1} \\ &+ R_t^{-1}(\beta^*)\frac{1}{\sqrt{t}}\sum_{k=1}^{t}(\hat{\sigma}_k^2 - \sigma_e^2)F(\beta^*)\hat{\sigma}_k^{-2}F^{-1}(\hat{\beta}_{k-1}) \\ &\times \psi_k(\beta^*)\psi_k(\beta^*)\tilde{\beta}_{k-1} + R_t^{-1}(\beta^*) \\ &\times \frac{1}{\sqrt{t}}\sum_{k=1}^{t}B_{1,k} + R_t^{-1}(\beta^*)\frac{1}{\sqrt{t}}\sum_{k=1}^{t}B_{2,k},\end{aligned} \tag{50}$$

and

$$\begin{aligned}L_t &= R_t^{-1}(\beta^*)\frac{1}{\sqrt{t}}\sum_{k=1}^{t}R_k(\beta^*)\hat{\sigma}_k^{-2}\{F^{-1}(\hat{\beta}_{k-1}) \\ &- F^{-1}(\beta^*)\}\psi_k(\hat{\beta}_{k-1})e_k + R_t^{-1}(\beta^*)\frac{1}{\sqrt{t}} \\ &\times \sum_{k=1}^{t}R_k(\beta^*)\hat{\sigma}_k^{-2}F^{-1}(\beta^*)(\psi_k(\hat{\beta}_{k-1}) - \psi_k(\beta^*))e_k.\end{aligned} \tag{51}$$

In Lemma A.24, based on Lemma A.21, we show convergence a.s. to 0 of H_t and L_t.

From Equation (49) and according to Lemma A.23, based on Lemmas A.21 and A.22, we have that $\sqrt{t}\tilde{\beta}_t$ converges in law to a normal distribution with mean 0 and variance

$$\begin{aligned}V &= E(\psi_1(\beta^*)\psi_1^{\mathrm{T}}(\beta^*))^{-1}\sigma_e^4F(\beta^*)E(\psi_1(\beta^*)\psi_1^{\mathrm{T}}(\beta^*))^{-1} \\ &= \sigma_e^{-2}F^{-1}(\beta^*)\sigma_e^4F(\beta^*)\sigma_e^{-2}F^{-1}(\beta^*) = F^{-1}(\beta^*),\end{aligned}$$

when $t \to \infty$ since $R_t(\beta^*)$ converges a.s. to $E(\psi_1(\beta^*)\psi_1^{\mathrm{T}}(\beta^*))$ which is equal to $\sigma_e^2F(\beta^*)$ by Equation (25). ■

4. Finite sample properties

A computer program in Fortran 90 was written in order to experiment with the new method. It is a part of a bigger project described in Ouakasse and Mélard (2014) for general single input single output models. We will compare the results of our algorithm with Ljung (2007) System Identification Toolbox in Matlab version 7.0 (R2007a), and more specifically function RARMAX. Before that, let us discuss some implementation aspects that are essential for the interpretation.

First let us note that inspection of the RARMAX code reveals that the effective iterations are slightly different from Equation (23). At each iteration, only the elements ε_t, $\psi_t(1)$

and $\psi_t(p_1+1)$ of ψ_t are computed

$$\psi_t(1) = \sum_{k=1}^{q} c_{kt}\psi_t(1+k) + y_{t-1},$$
$$\psi_t(p_1+1) = \sum_{k=1}^{q} c_{kt}\psi_t(p_1+1+k) + \varepsilon_{t-1}, \tag{52}$$

with $p_1 = \max(p, q)$. After that, a sliding operation is performed by

$$\psi_{t+1}(2) = \psi_t(1), \ldots, \psi_{t+1}(p_1) = \psi_t(p_1-1),$$
$$\psi_{t+1}(p_1+2) = \psi_t(p_1+1), \ldots, \psi_{t+1}(p_1+q) = \psi_t(p_1+q-1).$$

Our program follows the same sliding operations.

Like Ljung (2007), we introduce a second estimate of the forecast error $\bar{\varepsilon}_t$, so that the algorithm (24) becomes

$$\psi_t = \sum_{k=1}^{q} \hat{\theta}_{k,t-1}\psi_{t-k} + \bar{\varphi}_{t-1}, \tag{53}$$

$$\varepsilon_t = y_t - \hat{\beta}_{t-1}^{\mathrm{T}}\bar{\varphi}_{t-1},$$
$$\hat{\beta}_t = \hat{\beta}_{t-1} + \gamma_t\hat{\sigma}_t^{-2}F^{-1}(\hat{\beta}_{t-1})\psi_t\varepsilon_t, \tag{54}$$
$$\hat{\sigma}_{t+1}^2 = \hat{\sigma}_t^2 + \gamma_t(\varepsilon_t^2 - \hat{\sigma}_t^2),$$
$$\bar{\varepsilon}_t = y_t - \hat{\beta}_t^{\mathrm{T}}\bar{\varphi}_{t-1},$$

where this time $\bar{\varphi}_t^{\mathrm{T}} = (y_t, \ldots, y_{t-p+1}, -\bar{\varepsilon}_t, \ldots, -\bar{\varepsilon}_{t-q+1})$.

Also by code inspection, the method in RARMAX makes use of a projection using the function 'FSTAB' in Matlab but, as can be seen in the code, only for the parameters of the moving average polynomial $\theta(B) = 1 - \theta_1 B - \theta_2 B^2 - \cdots - \theta_q B^q$ as follows. Let $\hat{\theta}_t(B) = 1 - \hat{\theta}_{1,t}B - \cdots - \hat{\theta}_{q,t}B^q$ be the polynomial estimated at time t. The roots should be outside of the unit circle. Therefore, those roots which are inside of the unit circle are inverted and the others are unchanged. Then, the polynomial is computed again. Because computation of the Fisher information matrix requires that the roots of the autoregressive polynomial are also outside of the unit circle, we make use of a projection mechanism for the two polynomials. Also, we have preferred a different, smoother, projection mechanism. Let us illustrate the case of an AR polynomial. If $(\hat{\phi}_{t,1}, \ldots, \hat{\phi}_{t,p})$ is not admissible, let $\rho < 1$, consider instead $(\rho\hat{\phi}_{t,1}, \rho^2\hat{\phi}_{t,2}, \ldots, \rho^p\hat{\phi}_{t,p})$ and iterate by using successive powers of ρ until the subset of parameters becomes admissible.

Another implementation aspect is about the initial values for $\hat{\beta}_0$ and other variables subject to recurrences. Initial ψ_t and ε_t are zeroes. For the RML method, an initial matrix R_0 is also needed and Ljung and Söderström (1983) recommend to use $R_0 = 10000\,I$, expressing thereby a large amount of uncertainty. For our method, we have to choose σ_0^2 instead. Clearly it should depend on the dispersion of the data. Intuitively, it is better to use a bigger initial value than needed with the hope that σ_t^2 will converge to the (supposedly constant) innovation variance. In the present illustrations, we have taken $\hat{\sigma}_0^2 = 10$ or $\hat{\sigma}_0^2 = 1000$.

Finally, we have used a factor γ_t in Equations (23) or (26)–(28) although this was taken as $1/t$ in the theory. In practice it should be selected in order to improve convergence. It is often based on the forgetting factor defined by $\lambda_t = \{\gamma_{t-1}(1-\gamma_t)\}/\gamma_t$, which corresponds to $\gamma_t = \gamma_{t-1}/\{\lambda_t + \gamma_{t-1}\}$ and $\lambda_t = 1, \forall t$, corresponds to $\gamma_t = 1/t$. We have followed the recommendation

$$\lambda_t = \lambda^0\lambda_{t-1} + (1-\lambda^0), \tag{55}$$

where typically $\lambda^0 \approx 0.95, 0.99$, or 1. We have often chosen $\lambda^0 = 1$ and $\gamma_0 = 1$.

Since Ljung (2007) makes use of the standard notation in engineering, i.e. $-\phi_j$ and $-\theta_j$ instead of ϕ_j and θ_j. we will report the results for $-\hat{\phi}_j$ and $-\hat{\theta}_j$. We will compare our estimator (solid line) with the RML estimator of Ljung as implemented in Matlab with a forgetting factor (dashed line: $\lambda = 1$, or dot-dashed line: $\lambda = 0.99$) on the same Gaussian time series. They were produced in Matlab using simple recurrences and omitting the first 150 observations. They were immediately treated with the RARMAX procedure and then exported and treated by the Fortran program. For our $\mathrm{RML_{MZ}}$ algorithm, we have used a different forgetting factor for the variance, denoted with a subscript σ, characterised by $\lambda_\sigma^0 = 1$ and $\gamma_{0\sigma} = 1$. It will be noticed that, even for RML and $t = 1000$, estimates are quite different according to λ.

4.1. *ARMA(1,1) model*

Let us consider the ARMA(1,1) model with Equation (3) with $\phi^* = -0.5$ and $\theta^* = 0.5$, with $\sigma^2 = 1$. We have generated 10,000 series of length 1000 for which we have computed the estimates of ϕ and θ, for each time $t = 1, \ldots, 1000$. The following initial values were used: $\hat{\sigma}_0^2 = 10$, $\hat{\phi}_0 = -0.25$, $\hat{\theta}_0 = 0.25$, and constant forgetting factors $\lambda_0 = 1$, $\lambda_{0\sigma} = 0.99$. The averages and standard deviations across the experiments are shown in function of time. For each plot, the true value of the parameter is given. It is even displayed in the plot of the averages as a dotted horizontal line. The averages should be as close as possible of the true value and the standard deviations should be close to 0.

The plots for averages indicate that the new estimator seems to converge faster than the RML estimator. On the plot for standard deviations, we observe that those of the RML estimator decrease more slowly than ours.

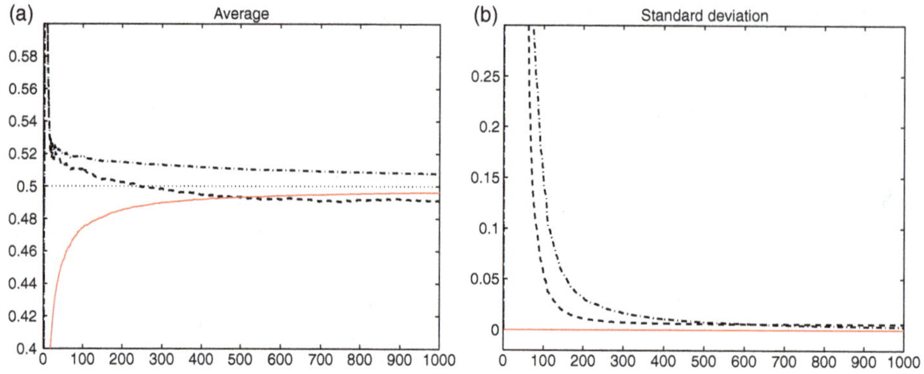

Figure 2. ARMA(1,1) with $\phi^* = -0.5$. Averages (left) and standard deviations (right) over the simulations in function of time for three estimates of $-\phi$. Solid line: our estimator, dashed line: Ljung/Matlab with $\lambda = 1$, dot-dashed line: same with $\lambda = 0.99$.

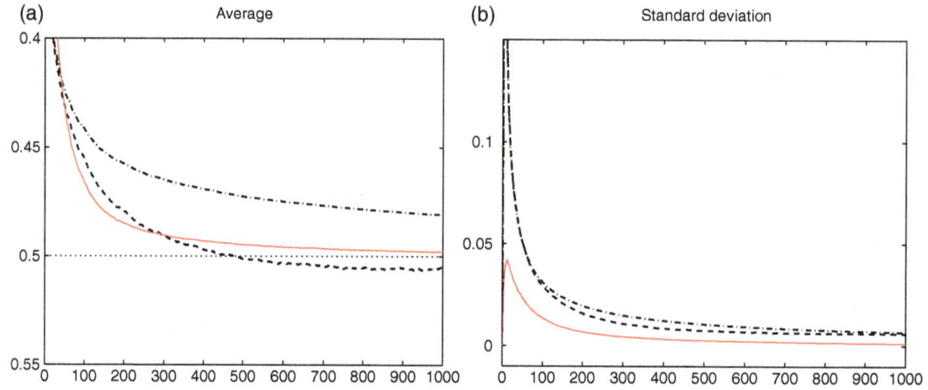

Figure 3. ARMA(1,1) with $\theta^* = 0.5$. Averages (left) and standard deviations (right) over the simulations in function of time for three estimates of $-\theta$. Solid line: our estimator, dashed line: Ljung/Matlab with $\lambda = 1$, dot-dashed line: same with $\lambda = 0.99$.

4.2. *ARMA(2, 2) model*

Let us consider the ARMA(2, 2) model with equation

$$(1 + 0.8B + 0.25B^2)y_t = (1 + 1.378B + 0.5B^2)e_t,$$

with $\sigma^2 = 1$. We have generated 10,000 series of length 1000 for which we have computed the estimates of $-\phi_1$, $-\phi_2$ and $-\theta_1$, $-\theta_2$ for each time $t = 1, \ldots, 1000$. The following initial values were used: $\hat{\sigma}_0^2 = 1000$, $\hat{\phi}_{1,0} = -0.5$, $\hat{\phi}_{2,0} = -0.8$, $\hat{\theta}_{1,0} = -0.69$, $\hat{\theta}_{2,0} = -0.14$, $\lambda_0 = 1$, $\gamma_0 = 1$, $\lambda_{0\sigma} = 0.9$, $\gamma_{0\sigma} = 1$. The results that were obtained, presented like for the previous example, are shown in Figures 4–7.

The graphs show that the averages for our method converge faster than for the RML estimator, and also that the dispersion across simulations is smaller.

4.3. *AR(1) model*

To consider a more critical situation, with a parameter close to the unit root, let us consider the AR(1) model with Equation (3) with $\phi^* = 0.9$, and $\sigma^2 = 1$. We have generated 10,000 series of length 1000 for which we have computed the estimates of ϕ, for each time $t = 1, \ldots, 1000$. But, instead of using Gaussian errors, we have used a Student distribution with five degrees of freedom, with fatter tails than the normal distribution. Note that Assumption 2 is therefore fulfilled since the fourth-order moments of the process are finite. The following initial values were used: $\hat{\sigma}_0^2 = 10$, $\hat{\phi}_0 = 0.25$, with variable forgetting factors determined by $\lambda_0 = 0.95$, $\lambda^0 = 0.99$, $\gamma_0 = 100$, and the same for the variance. Figure 8 shows the results which are very satisfactory for the RML_{MZ} method.

5. An example

We will illustrate the procedure on the following example. Windmills produce electricity in a way which is cleaner for the environment than with thermal or nuclear power stations. Electricity is however irregular because it depends of wind irregularity. When the wind is strong, more electricity is produced. Conversely, when the wind is weak, the quantity of electricity is very small. In order to maintain the offer of electricity at the level of demand, it is required to adapt production from traditional power stations in function of the amount of electricity produced by

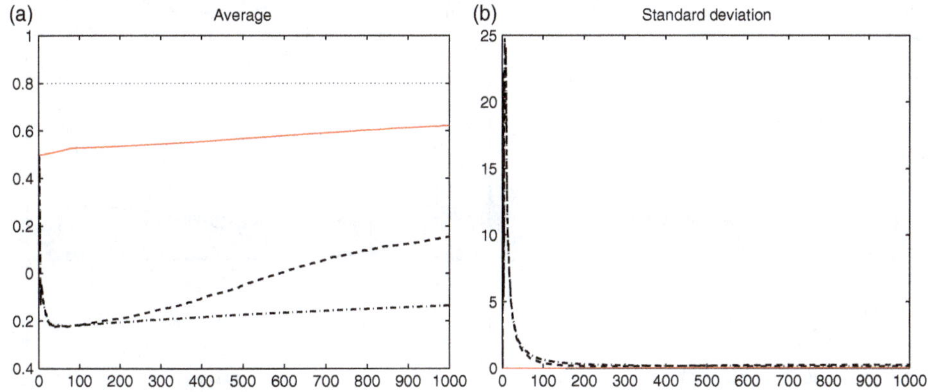

Figure 4. ARMA(2,2) with $\phi_1^* = -0.8$. Averages (left) and standard deviations (right) over the simulations in function of time for three estimates of $-\phi_1$. Solid line: our estimator, dashed line: Ljung/Matlab with $\lambda = 1$, dot-dashed line: same with $\lambda = 0.99$.

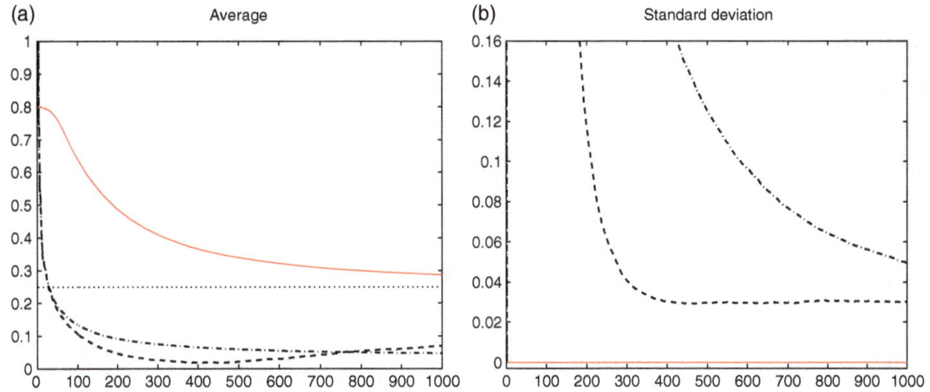

Figure 5. ARMA(2,2) with $\phi_2^* = -0.25$. Averages (left) and standard deviations (right) over the simulations in function of time for three estimates of $-\phi_2$. Solid line: our estimator, dashed line: Ljung/Matlab with $\lambda = 1$, dot-dashed line: same with $\lambda = 0.99$.

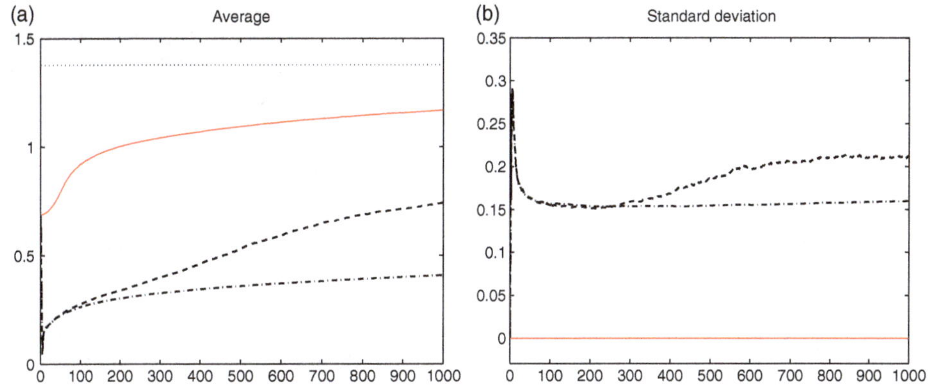

Figure 6. ARMA(2,2) with $\theta_1^* = -1.375$. Averages (left) and standard deviations (right) over the simulations in function of time for three estimates of $-\theta_1$. Solid line: our estimator, dashed line: Ljung/Matlab with $\lambda = 1$, dot-dashed line: same with $\lambda = 0.99$.

a park of windmills. Response time of a power station can go from a few minutes to several hours according to the technology being used. It is therefore useful to forecast wind speed a few hours in advance. The data come from speed of wind measurements at the top of a windmill. They are available every 10 minutes, hence 144 observations per day. We have used about 12 days of measurements, more precisely 1728 observations. The data are shown in Figure 9.

We have specified an ARMA(1,2) model with a constant, described by the equation:

$$(1 - \phi_1 B)(y_t - \mu) = (1 - \theta_1 B - \theta_2 B^2)e_t.$$

Here the vector of parameters is composed of $\beta = (\phi_1, \theta_1, \theta_2, \mu)$. A statistician or an econometrician would probably select a model with a unit root. For that reason, we have used both forgetting factors equal to 1 and an initial

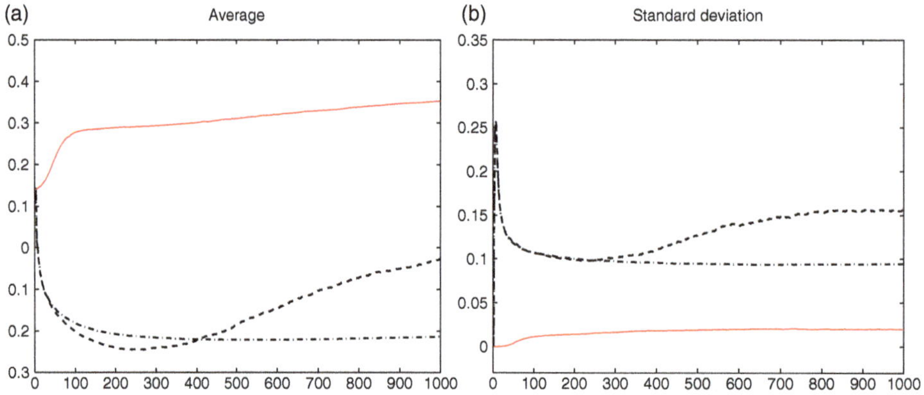

Figure 7. ARMA(2,2) with $\theta_2^* = -0.5$. Averages (left) and standard deviations (right) over the simulations in function of time for three estimates of $-\theta_2$. Solid line: our estimator, dashed line: Ljung/Matlab with $\lambda = 1$, dot-dashed line: same with $\lambda = 0.99$.

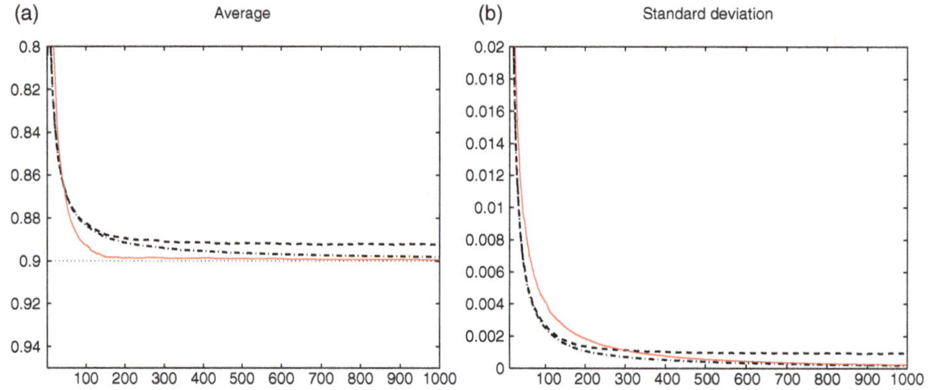

Figure 8. AR(1) with $\phi^* = 0.9$. Averages (left) and standard deviations (right) over the simulations in function of time for three estimates of $-\phi$. Solid line: our estimator, dashed line: Ljung/Matlab with $\lambda = 1$, dot-dashed line: same with $\lambda = 0.99$.

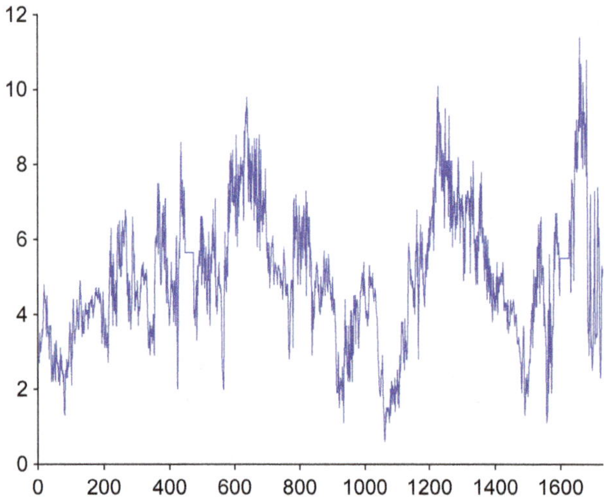

Figure 9. Speed of wind on the top of a windmill. One observation every 10 minutes during 12 days.

value of the variance which is not too large, equal to 500, given the dispersion of the data. The estimates are shown in Figure 10, where the initial values of the four parameters can be seen.

The estimates at the end of the series are $\hat{\beta} = (0.925, 0.112, 0.066, 4.933)$ and the final value of the innovation variance is 0.739. Note that the exact maximum likelihood method (Mëlard, 1984) of SPSS gives the following model using the whole data set:

$$(1 - 0.976B)(y_t - 4.894) = (1 - 0.216B - 0.196B^2)e_t,$$

with an estimate of the innovation variance equal to 0.400.

To illustrate the effect of having parameters which are not inside the stability region, we have restarted the on-line estimation but with an autoregressive parameter ϕ_1 set to 1.9 (instead of 0.5 in the case of Figure 10, top left plot). The projection described at the beginning of Section 4 is performed immediately. Since we have taken $\rho = 0.5$, the initial ϕ_1 becomes 0.95 which is inside the stability region. It can be seen that the successive values of ϕ_1 stay in the stability region. We have obtained the following estimates at the end of the series $\hat{\beta} = (0.957, 0.014, -0.009, 4.939)$ and 0.740 for the final value of the innovation variance. Of course there is little chance that using another set of initial values will give the same estimates. It is not guaranteed even with off-line method but it is more likely to happen.

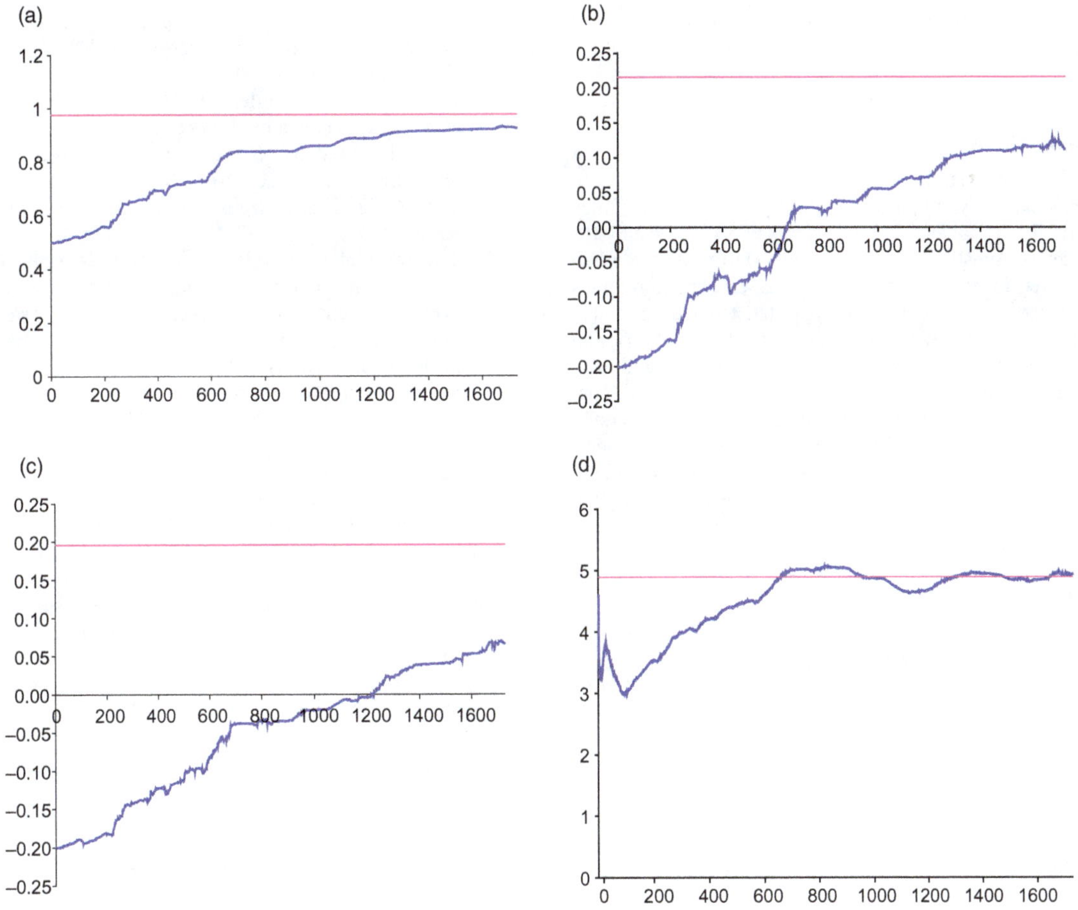

Figure 10. Estimates by the RML_{MZ} method on the data of Figure 9 in function of time: ϕ_1 (top left), θ_1 (top right), θ_2 (bottom left), μ (bottom right).

6. Conclusion

In Section 2, we have recalled the RML method proposed by Ljung (1977) and Ljung and Söderström (1983). That method provides recursive estimates using a system of equations. In one of the equations, the Hessian matrix of the error is updated. An improved RML method called RML_{MZ} is the subject of the present paper. It has been described in Section 3. It is based on using the Fisher information matrix, evaluated at the current value of the estimator, in order to update the estimator, instead of updating the Hessian. The asymptotic statistical properties of the new method have been studied in Sections 3.1 and 3.2. Under fairly general assumptions, it was proved that the RML_{MZ} estimator is consistent in the almost sure sense and also asymptotically normally distributed. This is done by following Ljung (1977) but the details are very different from those of the Ljung and Söderström (1983) approach. It is based on a result that the mathematical expectation of the errors, $E(\varepsilon^2)$, has an absolute minimum obtained at the true value of the parameter. We have obtained a part of the attraction domain around that minimum of the differential equation associated to the algorithm. In Section 4 we have shown Monte Carlo simulations (for some ARMA models and using 10,000 series of length 1000) for the comparison between the RML_{MZ} estimator and the original RML method. This suggests that indeed the RML_{MZ} estimator often does converge more quickly in practice. There is however no guarantee that it will always give a better performance. In Section 5 we show an example on real data which shows the usefulness of the new method.

One important capability of recursive estimation is when dealing with a possibly time invariant process. In that case, of course, asymptotic properties are less crucial than tracking ability. According to Ljung (1985), the latter can be achieved by using a sequence $\gamma(t)$ which converges to a strictly positive value instead of 0. Alternatively $\lambda(t)$, defined in Equation (55), is chosen as a constant λ_0.

Of course the requirement to be able to compute the Fisher information matrix implies restrictions. It has been shown in Section 5 that working outside of the stability region is not really a problem provided the projection mechanism is appropriate. If there is a unit root equal to one exactly, it is of course possible to handle the differenced series, $y_t - y_{t-1}$. But, clearly the RML_{MZ} method will not work if there is a root equal to one exactly on parts of the series and different from one on other parts. Cases where autoregressive and moving average roots come close are particularly challenging as discussed in Section 3.1.

Acknowledgements

This paper has benefited from an IAP-network in Statistics, contract P5/24, financially supported by the Belgian Federal Office for Scientific, Technical and Cultural Affairs (OSTC) and a FNRS grant. We thank Anestis Antoniadis, Marc Hallin, André Klein and Bas Werker, for their useful suggestions, and an anonymous referee of a previous version who suggested Chen's approach. We also thank Tractebel which kindly provided the data used in Section 5, and Nicolas Standaert, Denis Devleeschouwer and Bogdan Petrescu who performed an off-line analysis of these data. Finally, we would like to thank the two referees on this version whose remarks have improved the paper, like recent references, remarks on computational efficiency, discussion on Solo (1981) assumptions, and examples close or outside of the stability region.

References

Akaike, H. (1973). Block Toeplitz matrix inversion. *SIAM Journal of Applied Mathematics, 24*, 234–241.

Åström, K. J. (1980). Maximum likelihood and prediction error methods. *Automatica, 16*, 457–476.

Åström, K. J., & Söderström, T. (1974). Uniqueness of the maximum likelihood estimates of the parameters of an ARMA Model. *IEEE Transactions on Automatic Control AC, 19*, 769–773.

Benveniste, A., Métivier, M., & Priouret, P. (1990). *Adaptative algorithms and stochastic approximation*. Berlin: Springer-Verlag.

Box, G. E. P., Jenkins, G. M., & Reinsel, G. C. (2008). *Time series analysis, forecasting and control* (4th ed.). New York, NY: Wiley.

Brown, B. M. (1971). Martingale central limit theorems. *Annals of Statistics, 42*, 59–66.

Brown, R. L., Durbin, J., & Evans, J. M. (1975). Techniques for testing the constancy of regression relationships over time. *Journal of the Royal Statistical Society Series B, 37*, 149–163 (with discussion: 163–192).

Cartan, M. H. (1967). *Calcul différentiel*. Paris: Hermann.

Chen, H.-F. (2002). *Stochastic approximation and its applications*. Dordrecht: Kluwer.

Chung, K. L. (1968). *A course in probability theory*. New York, NY: Harcourt, Brace and World.

Dahlhaus, R., & Subba Rao, S. (2007). A recursive online algorithm for the estimation of time-varying ARCH parameters. *Bernoulli, 13*, 389–422.

Duflo, M. (1997). *Random iterative models*. Berlin: Springer-Verlag.

Fabian, V. (1968). On asymptotic normality in stochastic approximation. *Annals of Statistics, 39*, 1327–1332.

Gerencsér, L., & Prokaj, V. (2010). Recursive identification of continuous-time linear stochastic systems – an off-line approximation. In A. Edelmayer (Ed.), *Proceedings of the 19th international symposium on mathematical theory of networks and systems, MTNS 2010*, Budapest, Hungary, 5–9 July, pp. 375–380. Retrieved from http://www.conferences.hu/mtns2010/proceedings/start.pdf

Guo, L. (1994). Stability of recursive stochastic tracking algorithms. *SIAM Journal of Control and Optimization, 32*, 1195–1225.

Harville, D. A. (1997). *Matrix algebra from statistician's perspective*. New York, NY: Springer.

Kalman, R. E. (1960). A new approach to general filtering and prediction problems. *Transactions ASME Series D. Basic Engineering, 83*, 95–108.

Kennedy, W. J. Jr., & Gentle, J. E. (1980). *Statistical computing*. New York, NY: Marcel Dekker.

Kirshner, H., Maggio, S., & Unser, M. (2011). A sampling theory approach for continuous ARMA identification. *IEEE Transactions on Signal Processing SP, 59*, 4620–4634.

Klein, A., & Mélard, G. (1989). On algorithms for computing the covariance matrix of estimates in autoregressive-moving average models. *Computational Statistics Quarterly, 5*, 1–9.

Klein, A., & Spreij, P. J. C. (1997). On Fisher's information matrix of an ARMA process. In I. Csiszár & Gy. Michaletzky (Eds.), *Stochastic differential and difference equations. Progress in systems and control theory* (Vol. 23, pp. 273–284). Boston: Birkhäuser.

Kushner, H. J., & Huang, H. (1979). Rates of convergence for stochastic approximation type algorithms. *SIAM Journal of Control and Optimization, 17*, 607–617.

Kushner, H. J., & Yin, G. (1997). *Stochastic approximation algorithms and applications*. Berlin: Springer-Verlag.

Ljung, L. (1977). Analysis of recursive stochastic algorithms. *IEEE Transactions on Automatic Control AC, 22*, 551–575.

Ljung, L. (1978). Convergence analysis of parametric identification methods. *IEEE Transactions on Automatic Control AC, 23*, 770–783.

Ljung, L. (1985). Estimation of parameters in dynamical systems. In E. J. Hannan, P. R. Krishnaiah, & M. M. Rao (Eds.), *Handbook of statistics, Vol 5: Time series in the time domain* (pp. 189–211). Amsterdam: North Holland.

Ljung, L. (2007). *System identification toolbox user's guide: For use with MATLAB version 7.0 (Release 2007a)*. Natick, MA: The MathWorks Inc.

Ljung, L., Pflug, G., & Walk, H. (1992). *Stochastic approximation and optimisation of random systems*. Boston: Birkhäuser.

Ljung, L., & Söderström, T. (1983). *Theory and practice of recursive identification*. Cambridge, MA: MIT Press.

Loève, M. (1977). *Probability theory I* (4th ed.). New York, NY: Springer.

Lukacs, E. (1975). *Stochastic convergence* (2nd ed.). New York, NY: Academic Press.

Marelli, D., You, K., & Fu, M. (2013). Identification of ARMA models using intermittent and quantized output observations. *Automatica, 49*, 360–369.

Mélard, G. (1984). Algorithm AS197: A fast algorithm for the exact likelihood of autoregressive-moving average models. *Journal of the Royal Statistical Society Series C Applied Statistics, 33*, 104–114.

Mélard, G. (1989). Méthode d'identification récurrente pour la modélisation des données chronologiques économiques. In *Actes du Salon International de l'Informatique de la Télématique* (Séance III: Intelligence artificielle et génie logiciel, 10 pp.). Rabat, Maroc: Association Marocaine des utilisateurs de l'informatique.

Moulines, E., Priouret, P., & Roueff, F. (2005). On recursive estimation for locally stationary time varying autoregressive processes. *Annals of Statistics, 33*, 2610–2654.

Ouakasse, A., & Mélard, G. (2014). A new recursive estimation method for single input single output models. In preparation.

Plackett, R. L. (1950). Some theorems in least squares. *Biometrika, 37*, 149–157.

Protter, M. H., & Morrey, C. B. (1977). *A first course in real analysis*. New York, NY: Springer-Verlag.

Rouche, N., & Mawhin, J. (1980). *Ordinary differential equations. Stability and periodical solutions*. London: Pitman.

Sen, P. K., & Singer, J. M. (1993). *Large sample methods in statistics: An introduction with applications*. New York, NY: Chapman & Hall.
Söderström, T. (1973). *An on-line algorithm for approximate maximum likelihood identification of linear dynamic systems* (Report 7308). Lund: Department of Automatic Control, Lund Institute of Technology.
Solo, V. (1981). The second order properties of a time series recursion. *Annals of Statistics, 9*, 307–317.
Taniguchi, M., & Kakizawa, Y. (2000). *Asymptotic theory of statistical inference for time series*. New York, NY: Springer-Verlag.
Trench, W. F. (1964). An algorithm for the inversion of finite Toeplitz matrices. *Journal of the Society for Industrial and Applied Mathematics, 12*, 515–522.
van der Waerden, B. L. (1937). *Moderne algebra I*. Berlin: Springer.
Young, P. (1984). *Recursive estimation and time series analysis*. Berlin: Springer-Verlag.
Young, P. (1985). Recursive identification, estimation and control. In E. J. Hannan, P. R. Krishnaiah, & M. M. Rao (Eds.), *Handbook of statistics, Vol 5: Time series in the time domain* (pp. 213–255). Amsterdam: North Holland.
Zahaf, T. (1999). *Contributions à l'estimation des paramètres de modèles de séries chronologiques* (Unpublished doctoral thesis). Institut de Statistique et de Recherche Opérationnelle, Université libre de Bruxelles, Belgium.

Appendix 1

Here are a few lemmas needed for the proofs in Section 3.1. Invertibility of the Fisher information matrix is satisfied by Assumption 1, given the following lemma.

LEMMA A.1 (Klein & Spreij, 1997) *The Fisher information matrix $F(\beta)$ is invertible if and only if the autoregressive and moving average polynomials have no common root.*

LEMMA A.2 (Harville, 1997, p. 307) *Let F be a matrix function of $\mathbb{R}^m \to \mathbb{R}^{(n,n)}$, let $x \in \mathbb{R}^m$ be a point where matrix F is invertible and continuously differentiable, then*

$$\frac{\partial F^{-1}(x)}{\partial x_i} = -F^{-1}(x)\frac{\partial F(x)}{\partial x_i}F^{-1}(x).$$

LEMMA A.3 *The limit* (41), *see Section* (3.1) *needed in condition C3 of Ljung does exist.*

Proof On the basis of Equation (33), let $Q(t,\beta,\sigma^2,h_t(\beta)) = (\psi_t^{\mathrm{T}}(\beta)F^{-1}(\beta)\varepsilon_t(\beta), \varepsilon_t^2(\beta)-\sigma^2)^{\mathrm{T}}$. We have to prove that $(1/t)\sum_{k=1}^t \psi_k(\beta)$
$\varepsilon_k(\beta)$ converges to $E(\psi_t(\beta)\varepsilon_t(\beta))$ and that $(1/t)\sum_{k=1}^t \varepsilon_k^2(\beta)$ converges to $E(\varepsilon_t^2(\beta))$. By Lemma A.11, we have that $(1/t)\sum_{k=1}^t \psi_k(\beta)\varepsilon_k(\beta) - E\{\psi_t(\beta)\varepsilon_t(\beta)\}$ and $(1/t)\sum_{k=1}^t \varepsilon_k^2(\beta) - E\{\varepsilon_t^2(\beta)\}$ converge a.s. to 0. ■

Proof of Theorem 1 We have to prove the conditions C for the algorithm (31). Condition C2 on matrices A and B which are the same as in the RML method is of course valid. Condition C3 is proved in Lemma A.3. Of course conditions C5 and C6 are satisfied since $\gamma_t = 1/t$. There remains to check conditions C1 and C4.

Let us first check condition C1. Suppose $(\beta_1^{\mathrm{T}},\sigma_1^2)^{\mathrm{T}}$ and $(\beta_2^{\mathrm{T}},\sigma_2^2)^{\mathrm{T}}$ in a ball $\mathcal{B}((\beta^{\mathrm{T}},\sigma^2)^{\mathrm{T}},\rho(\beta,\sigma^2))$ with $\rho(\beta,\sigma^2)$ small enough such that (β_1,σ_1^2) and (β_2,σ_2^2) belong to D_R. Let $\hbar$ and $\hbar'$ be two vectors in a ball $\mathcal{B}(\hbar^0,v)$ of $\mathbb{R}^{q(p+q+1)}$ with $\hbar^0 = (\hbar_1^0,\ldots,\hbar_{q(p+q+1)}^0)^{\mathrm{T}}$, $\hbar = (\hbar_1,\ldots,\hbar_{q(p+q+1)})^{\mathrm{T}}$ and $\hbar' = (\hbar_1',\ldots,\hbar_{q(p+q+1)}')^{\mathrm{T}}$. Let $k_1^0 = (\hbar_{q+1}^0,\ldots,\hbar_{p+2q}^0)$, $k_1 = (\hbar_{q+1},\ldots,\hbar_{p+2q})$ and $k_1' = (\hbar_{q+1}',\ldots,\hbar_{p+2q}')$. By Equation (33) we have

$$\begin{aligned}
&Q(t,\beta_1,\sigma_1^2,\hbar) - Q(t,\beta_2,\sigma_2^2,\hbar')\\
&\quad = [\{(\hbar_1-\hbar_1^0)(k_1^{\mathrm{T}}-k_1'^{\mathrm{T}})\\
&\qquad + (\hbar_1-\hbar_1')(k_1'^{\mathrm{T}}-k_1^{0\mathrm{T}})\}\sigma_1^{-2}F^{-1}(\beta_1)\\
&\qquad + \{\hbar_1^0(k_1^{\mathrm{T}}-k_1'^{\mathrm{T}}) + (\hbar_1-\hbar_1')k_1^{0\mathrm{T}}\}\sigma_1^{-2}F^{-1}(\beta_1)\\
&\qquad + (\hbar_1'-\hbar_1^0)(k_1'^{\mathrm{T}}-k_1^{0\mathrm{T}})\sigma_1^{-2}(F^{-1}(\beta_1)-F^{-1}(\beta_2))\\
&\qquad + \hbar_1^0(k_1'^{\mathrm{T}}-k_1^{0\mathrm{T}})\sigma_1^{-2}(F^{-1}(\beta_1)-F^{-1}(\beta_2))\\
&\qquad + ((\hbar_1'-\hbar_1^0)k_1^{0\mathrm{T}} + \hbar_1^0k_1^{0\mathrm{T}})\sigma_1^{-2}(F^{-1}(\beta_1)-F^{-1}(\beta_2))\\
&\qquad + (\hbar_1'-\hbar_1^0)(k_1'^{\mathrm{T}}-k_1^{0\mathrm{T}})(\sigma_2^2-\sigma_1^2)\sigma_1^{-2}\sigma_2^{-2}F^{-1}(\beta_2)\\
&\qquad + \hbar_1^0(k_1'^{\mathrm{T}}-k_1^{0\mathrm{T}})(\sigma_2^2-\sigma_1^2)\sigma_1^{-2}\sigma_2^{-2}F^{-1}(\beta_2)\\
&\qquad + ((\hbar_1'-\hbar_1^0)k_1^{0\mathrm{T}} + \hbar_1^0k_1^{0\mathrm{T}})(\sigma_2^2-\sigma_1^2)\sigma_1^{-2}\sigma_2^{-2}F^{-1}(\beta_2),\\
&\qquad \times (\hbar_1-\hbar_1')(\hbar_1-\hbar^0+\hbar_1'-\hbar^0) + 2\hbar^0(\hbar_1-\hbar_1')\\
&\qquad + \sigma_2^2-\sigma_1^2]^{\mathrm{T}}.
\end{aligned}$$

Since $\sigma_1^{-2} \leqslant \delta^{-1}, \sigma_2^{-2} \leqslant \delta^{-1}, (\hbar_1-\hbar_1^0) \leqslant v$ and $(k_1-k_1^0) \leqslant v$, there exists a constant C such that

$$\begin{aligned}
&\|Q(t,\beta_1,\sigma_1^2,\hbar) - Q(t,\beta_2,\sigma_2^2,\hbar')\| \leqslant C(\|\hbar^0\|^2 + \|\hbar^0\|\\
&\quad + v^2 + v)\{\|\hbar-\hbar'\| + \|F^{-1}(\beta_1)-F^{-1}(\beta_2)\| + |\sigma_2^2-\sigma_1^2|\}.
\end{aligned}$$

According to Lemma A.2, $\partial F^{-1}(\beta)/\partial\beta$ is continuous on the ball $\mathcal{B}((\beta^{\mathrm{T}},\sigma^2)^{\mathrm{T}},\rho(\beta,\sigma^2))$ hence it is bounded, then there exists a constant $C_1 > 0$ such that $\|F^{-1}(\beta_1)-F^{-1}(\beta_2)\| \leqslant C_1\|\beta_1-\beta_2\|$, and the preceding expression can be written

$$\begin{aligned}
&\|Q(t,\beta_1,\sigma_1^2,\hbar) - Q(t,\beta_2,\sigma_2^2,\hbar')\| \leqslant C(\|\hbar^0\|^2 + \|\hbar^0\|\\
&\quad + v^2 + v + C_1)(\|(\beta_1^{\mathrm{T}},\sigma_1^2)^{\mathrm{T}} - (\beta_2^{\mathrm{T}},\sigma_2^2)^{\mathrm{T}}\| + \|\hbar-\hbar'\|).
\end{aligned}$$

Showing condition C4 means to prove that, for all $\bar{x} \in D_R$, $0 < \lambda < 1$ and $c < \infty$, the random variable $k_v(t,\bar{x},\lambda,c)$ defined by

$$\begin{aligned}
&\sum_{t=1}^{N} C(\|h_t(\beta)\|^2 + \|h_t(\beta)\| + v(t,\lambda,c)^2 + v(t,\lambda,c)\\
&\quad + C_1)(1 + v(t,\lambda,c))
\end{aligned}$$

with $k_v(0,\bar{x},\lambda,c) = 0$ and $v(t,\lambda,c) = c\sum_{k=1}^t \lambda^{t-k}\|z_k\|$, converges to a finite limit when $t \longrightarrow \infty$. In Lemma A.12. we will show that

$$\|h_t(\beta)\| = \left\|\sum_{i=1}^{t} A(\beta)^i B(\beta) z_{t-i}\right\| \leqslant C_1\sum_{i=1}^{t}\lambda_1^{t-i}\|z_i\|.$$

According to Ljung and Söderström (1983) there exists a constant C_3 such that

$$\begin{aligned}
&\sum_{t=1}^{N} C(\|h_t(\beta)\|^2 + \|h_t(\beta)\| + v(t,\lambda,c)^2 + v(t,\lambda,c)\\
&\quad + C_1)(1 + v(t,\lambda,c)) \leqslant C_3\frac{1}{N}\sum_{t=1}^{N}[1 + \|z_t\|]^3.
\end{aligned}$$

Hence we need only to prove that $(1/N)\sum_{t=1}^N[1+\|z_t\|]^3$ converges when $N \longrightarrow \infty$. Let take the norm $\|z_t\| = |y_t| + |y_{t-1}| +$

$\cdots + |y_{t-p}|$. We need to prove that $(1/N)\sum_{t=1}^{N}|y_t|^3$ is finite which is true for an ARMA process. Indeed we can write $y_t = \sum_{i=1}^{\infty}\eta_i e_{t-i}$, with $\sum_{i=1}^{\infty}|\eta_i| < \infty$, so by using Lemma A.10, part 7, $E|y_t|^3$ exists and ergodicity of the process $\{|y_t|\}$ (Taniguchi & Kakizawa, 2000) implies the result. ■

LEMMA A.4 (Åström & Söderström, 1974) *For the ARMA(p,q) model defined by Equation (1), β^* is the unique solution of $E[\varepsilon(\beta)\psi(\beta)] = 0$.*

LEMMA A.5 (Cartan, 1967, p. 122) (a) *Let $g: U \longrightarrow \mathbb{R}^n : (t,x) \longrightarrow g(t,x)$, where $U \subset \mathbb{R}\times\mathbb{R}^n$, be a continuous locally Lipschitz function. If $(t_0,x_0) \in \overset{o}{U}$, an open set included in U, then there exists a largest interval J such that $t_0 \in J$ in which there exists a unique differentiable solution $x(t)$ of $\dot{x}(t) = g(t,x(t))$, with the initial condition $x(t_0) = x_0$. That function $x(t)$ is called the maximal solution.* (b) *Let $g(x)$ be a class C_1 function $\omega \longrightarrow \mathbb{R}^n$, ω being an open set of $\mathbb{R}^n$. Let A_ω be a compact of ω. We suppose that any solution of $\dot{x}(t) = g(x(t))$, with the initial condition $x(t_0) = x_0$, defined over $[t_0,t_1]$ is such that $\forall t \in [t_0,t_1]$, $x(t) \in A_\omega$. Then the upper bound of the maximal interval of existence of the ODE is $+\infty$.*

LEMMA A.6 (Rouche & Mawhin, 1980, p. 12) *Consider the ODE in Lemma A.5 where g is a continuous locally Lipschitz function: $g: I\times B_\rho \to \mathbb{R}^n$, $B_\rho = B(0,\rho) \subset \mathbb{R}^n$. Let Γ be a part in $\mathbb{R}^n$ such that $\bar{\Gamma} \subset B_\rho$. Let $V: I\times B_\rho \longrightarrow \mathbb{R}^+$ a function of class C^1, and a, a positive constant. If*

(a) $x_0 \in \Gamma$, $t_0 \in I$,
(b) $V(t_0,x_0) < a$,
(c) $\forall (t,x) \in I\times \mathrm{Fr}(\Gamma)$, $V(t,x) \geqslant a$,
(d) $\forall (t,x) \in I\times\Gamma$, $\dot{V}(t,x) \leqslant 0$,

then the solution of the ODE is such that $\forall t \geqslant t_0$, $x(t) \in \Gamma$.

Consider the following ODE: $\dot{x} = g(x)$, $x(t_0) = x_0$ where $g: \Omega \to \mathbb{R}^n$ is a continuous locally Lipschitz function. Let $\gamma^+(x_0) = \{x(z,x_0), z \geqslant 0\}$ be the trajectory of $x(z,x_0)$.

LEMMA A.7 (Rouche & Mawhin, 1980, p. 50) *Let Ψ a compact of Ω, an open set of $\mathbb{R}^n$, and $V: \Omega \longrightarrow \mathbb{R}^+$ a function of class C^1 such that $\forall x \in \Psi$, $\dot{V}(x) \leqslant 0$. Let $E_\Psi = \{x \in \Psi / \dot{V}(x) = 0\}$ and M the largest invariant subset of E. Then for any x_0 such that $\gamma^+(x_0) \subset \Psi$, $x(z,x_0) \underset{z\to\infty}{\longrightarrow} M$.*

LEMMA A.8 (Ljung, 1977, Theorems 1 and 4) *Under conditions C, let us consider the algorithm* (30) *modified as follows:*

$$h_t = \begin{cases} A(\hat{x}_{t-1})\,h_{t-1} + B(\hat{x}_{t-1})z_t & \text{if } \hat{x}_{t-1} \in D_1 \\ \text{a point in } D_3 & \text{if } \hat{x}_{t-1} \notin D_1, \end{cases} \tag{A1}$$

$$\hat{x}_t = \left[\hat{x}_{t-1} + \gamma_t Q(t,\hat{x}_{t-1},h_t)\right]_{D_1,D_2}, \tag{A2}$$

where $D_1 \subset D_R \subset \mathbb{R}^m$ is a bounded open part containing the compact D_2, D_3 is a compact of $\mathbb{R}^m$, and m is the dimension of $\hat{x}_t$, and

$$[z]_{D_1,D_2} = \begin{cases} z & \text{if } z \in D_1, \\ \text{a point in } D_2 & \text{if } z \notin D_1. \end{cases}$$

Let $\bar{D}$ be a compact part of D_R such that the trajectories of the following ODE $\partial x(t)/\partial t = f(x(t))$ starting from a point in $\bar{D}$, stay in a closed part $\bar{D}_R$ of D_R. Suppose that the ODE possesses an invariant set D_c with its domain of attraction D_A such that $\bar{D} \subset D_A$. Let $\tilde{D} = D_1 \backslash D_2$ and suppose there exists a twice differentiable function $U(x) \geqslant 0$ defined over a neighbourhood of $\tilde{D}$ and such that:

$$\sup_{x\in\tilde{D}} U'(x)f(x) < 0,$$
$$U(x) \geqslant c_1 \quad \text{for } x \notin D_1,$$
$$U(x) \leqslant c_2 < c_1 \quad \text{for } x \in D_2.$$

Then $\hat{x}_t \to D_c$ almost surely when $t \to \infty$.

Appendix 2

LEMMA A.9 *Consider $K_t = \bar{\sigma}_t\tilde{\beta}_t$ defined in Section* 3.2. *Then*

$$\begin{aligned} K_t = {} & \sum_{k=1}^{t} A_{1,k} + \sum_{k=1}^{t} A_{2,k} + \sum_{k=2}^{t} F^{-1}(\hat{\beta}_{k-1})\psi_k e_k + T_t\tilde{\beta}_{t-1} \\ & - \sum_{k=1}^{t-1} T_k(\tilde{\beta}_k - \tilde{\beta}_{k-1}) - F^{-1}(\beta^*)S_t\tilde{\beta}_{t-1} + F^{-1}(\beta^*) \\ & \times \sum_{k=1}^{t-1} S_k(\tilde{\beta}_k - \tilde{\beta}_{k-1}) + \sum_{k=1}^{t}\frac{1}{k}(\varepsilon_k^2 - \hat{\sigma}_k^2)F^{-1}(\hat{\beta}_{k-1})\psi_k\varepsilon_k, \end{aligned} \tag{A3}$$

where

$$\begin{aligned} A_{1,t} = {} & (\varepsilon_t^2(\hat{\beta}_{t-1}) - e_t^2)\tilde{\beta}_{t-1} - F^{-1}(\hat{\beta}_{t-1})(\psi_t(\hat{\beta}_{t-1}) \\ & - \psi_t(\beta^*))\psi_t^{\mathrm{T}}(\beta^*)\tilde{\beta}_{t-1} - (F^{-1}(\hat{\beta}_{t-1}) \\ & - F^{-1}(\beta^*))\psi_t(\beta^*)\psi_t^{\mathrm{T}}(\beta^*)\tilde{\beta}_{t-1} \\ & - \frac{1}{2}F^{-1}(\hat{\beta}_{t-1})\psi_t\tilde{\beta}_{t-1}^{\mathrm{T}}\left(\frac{\partial\psi_t(\beta)}{\partial\beta^{\mathrm{T}}}\right)_{\beta=\varkappa_t}\tilde{\beta}_{t-1}, \end{aligned} \tag{A4}$$

$\varkappa_t$ is a point on the segment joining $\hat{\beta}_{t-1}$ and β^*, and

$$\begin{aligned} A_{2,t} = {} & (\varepsilon_t^2 - \varepsilon_t^2(\hat{\beta}_{t-1}))\tilde{\beta}_{t-1} + F^{-1}(\hat{\beta}_{t-1})\psi_t(\varepsilon_t - \varepsilon_t(\hat{\beta}_{t-1})) \\ & - F^{-1}(\hat{\beta}_{t-1})(\psi_t - \psi_t^{\mathrm{T}}(\hat{\beta}_{t-1}))\psi_t^{\mathrm{T}}(\beta^*)\tilde{\beta}_{t-1}, \end{aligned} \tag{A5}$$

$$S_k = \sum_{j=1}^{k}[\psi_j(\beta^*)\psi_j^{\mathrm{T}}(\beta^*) - \sigma_e^2 F(\beta^*)] \quad \text{with } S_0 = 0, \tag{A6}$$

$$T_k = \sum_{j=1}^{k}(e_j^2 - \sigma_e^2) \quad \text{with } T_0 = 0. \tag{A7}$$

Proof By Equation (44), the definition of $\bar{\sigma}_t = t\hat{\sigma}_t^2$ and Equation (35),

$$\begin{aligned} K_t = \bar{\sigma}_t\tilde{\beta}_t = {} & \bar{\sigma}_{t-1}\tilde{\beta}_{t-1} + \varepsilon_t^2\tilde{\beta}_{t-1} + F^{-1}(\hat{\beta}_{t-1})\psi_t\varepsilon_t \\ & + \frac{1}{t}(\varepsilon_t^2 - \hat{\sigma}_t^2)F^{-1}(\hat{\beta}_{t-1})\psi_t\varepsilon_t, \end{aligned}$$

and we have

$$\begin{aligned} K_t = {} & K_{t-1} + (\varepsilon_t^2 - \varepsilon_t^2(\hat{\beta}_{t-1}))\tilde{\beta}_{t-1} + (\varepsilon_t^2(\hat{\beta}_{t-1}) - e_t^2)\tilde{\beta}_{t-1} \\ & + e_t^2\tilde{\beta}_{t-1} + F^{-1}(\hat{\beta}_{t-1})\psi_t(\varepsilon_t - \varepsilon_t(\hat{\beta}_{t-1})) \\ & + F^{-1}(\hat{\beta}_{t-1})\psi_t(\varepsilon_t(\hat{\beta}_{t-1}) - e_t) + F^{-1}(\hat{\beta}_{t-1})\psi_t e_t \\ & + \frac{1}{t}(\varepsilon_t^2 - \hat{\sigma}_t^2)F^{-1}(\hat{\beta}_{t-1})\psi_t\varepsilon_t. \end{aligned}$$

We can use Equation (45) so that

$$K_t = K_{t-1} + (\varepsilon_t^2 - \varepsilon_t^2(\hat\beta_{t-1}))\bar\beta_{t-1} + (\varepsilon_t^2(\hat\beta_{t-1}) - e_t^2)\tilde\beta_{t-1}$$
$$+ e_t^2\tilde\beta_{t-1} + F^{-1}(\hat\beta_{t-1})\psi_t(\varepsilon_t - \varepsilon_t(\hat\beta_{t-1}))$$
$$- F^{-1}(\hat\beta_{t-1})(\psi_t - \psi_t(\hat\beta_{t-1}))\psi_t^{\mathrm{T}}(\beta^*)\tilde\beta_{t-1}$$
$$- F^{-1}(\hat\beta_{t-1})(\psi_t(\hat\beta_{t-1}) - \psi_t(\beta^*))\psi_t^{\mathrm{T}}(\beta^*)\tilde\beta_{t-1}$$
$$- (F^{-1}(\hat\beta_{t-1}) - F^{-1}(\beta^*))\psi_t(\beta^*)\psi_t^{\mathrm{T}}(\beta^*)\tilde\beta_{t-1}$$
$$- F^{-1}(\beta^*)\psi_t(\beta^*)\psi_t^{\mathrm{T}}(\beta^*)\tilde\beta_{t-1}$$
$$- \frac{1}{2}F^{-1}(\hat\beta_{t-1})\psi_t\tilde\beta_{t-1}^{\mathrm{T}}\left(\frac{\partial\psi_t(\beta)}{\partial\beta^{\mathrm{T}}}\right)_{\beta=\varkappa_t}\tilde\beta_{t-1}$$
$$+ F^{-1}(\hat\beta_{t-1})\psi_t e_t$$
$$+ \frac{1}{t}(\varepsilon_t^2 - \hat\sigma_t^2)F^{-1}(\hat\beta_{t-1})\psi_t\varepsilon_t.$$

Moving some terms leads to

$$K_t = K_{t-1} + A_{1,t} + A_{2,t} + F^{-1}(\hat\beta_{t-1})\psi_t e_t + (e_t^2 - \sigma_e^2)\tilde\beta_{t-1}$$
$$- F^{-1}(\beta^*)[\psi_t(\beta^*)\psi_t^{\mathrm{T}}(\beta^*) - \sigma_e^2 F(\beta^*)]\tilde\beta_{t-1}$$
$$= \sum_{k=1}^{t} A_{1,k} + \sum_{k=1}^{t} A_{2,k} + \sum_{k=2}^{t} F^{-1}(\hat\beta_{k-1})\psi_k e_k$$
$$+ \sum_{k=1}^{t}(e_k^2 - \sigma_e^2)\tilde\beta_{k-1} - F^{-1}(\beta^*)$$
$$\times \sum_{k=1}^{t}[\psi_k(\beta^*)\psi_k^{\mathrm{T}}(\beta^*) - \sigma_e^2 F(\beta^*)]\tilde\beta_{k-1}$$
$$+ \sum_{k=1}^{t}\frac{1}{k}(\varepsilon_k^2 - \hat\sigma_k^2)F^{-1}(\hat\beta_{k-1})\psi_k\varepsilon_k,$$

since $K_0 = 0$. Introducing S_t, defined by Equation (A6), yields after some algebra

$$\sum_{k=1}^{t}[\psi_k(\beta^*)\psi_k^{\mathrm{T}}(\beta^*) - \sigma^2 F(\beta^*)]\tilde\beta_{k-1}$$
$$= \sum_{k=1}^{t}(S_k - S_{k-1})\tilde\beta_{k-1} = S_t\tilde\beta_{t-1} - \sum_{k=1}^{t-1} S_k(\tilde\beta_k - \tilde\beta_{k-1}).$$

Similarly using T_k defined by Equation (A7),

$$\sum_{k=1}^{t}(e_k^2 - \sigma^2)\tilde\beta_{k-1} = \sum_{k=1}^{t}(T_k - T_{k-1})\tilde\beta_{k-1}$$
$$= T_t\tilde\beta_{t-1} - \sum_{k=1}^{t-1} T_k(\tilde\beta_k - \tilde\beta_{k-1}). \tag{A8}$$

■

We need several classical lemmas which are collected here for convenience, see e.g. Loéve (1977, pp. 250–251), Lukacs (1975, p. 80) and Chung (1968, p. 307), plus two other lemmas.

LEMMA A.10

(1) (*Kolmogorov's proposition*) *Let x_n be independent random variables, $b_n \nearrow \infty$, and $S_n = \sum_{k=1}^n x_k$. Suppose that $\sum_{k=1}^n \mathrm{Var}(x_k)/b_k^2 < \infty$, then $(S_n - E(S_n))/b_n \longrightarrow 0$ a.s. when $n \longrightarrow \infty$.*

(2) (*Kronecker*) *Let x_k be a sequence of real numbers, a_k a sequence of positive numbers which converges to ∞. Then $\sum_{k=1}^{\infty}(x_k/a_k) < \infty$ implies $(1/a_n)\sum_{k=1}^n x_k \to 0$ when $n \longrightarrow \infty$.*

(3) (*Toeplitz*) *Let $a_{nk}, k = 1,2,\ldots,$ be numbers such that for every fixed k, $a_{nk} \to 0$ when $n \longrightarrow \infty$ and, for all n, $\sum_{k=1}^{\infty}|a_{nk}| \le c < \infty$. Let x_k be a sequence of real numbers. If $x_k < \infty$, then $\sum_{k=1}^n a_{nk}x_k < \infty$ when $n \longrightarrow \infty$.*

(4) (*Lukacs*) *Let $\{x_n\}$ be a sequence of random variables with finite expectations and suppose that $\sum_{n=1}^{\infty} E(|x_n|)$ is finite. Then the infinite series $\sum_{n=1}^{\infty}|x_n|$ is a.s. convergent.*

(5) (*Chung*) *Let $\{x_n\}$ be a submartingale satisfying the condition $\lim_{n\longrightarrow\infty} E(\max\{x_n, 0\}) < \infty$. Then $\{x_n\}$ converges a.s. to a finite limit when $n \longrightarrow \infty$.*

(6) *Let $\{x_t; t \in Z\}$ and $\{z_t; t \in Z\}$ be ARMA(p_1, q_1) and ARMA(p_2, q_2) stationary processes, respectively, in terms of the same errors. Then, for all $\varepsilon > 0$ and all $s \in Z$, $t^{1/2-\varepsilon}((1/t)\sum_{k=1}^t x_k z_{k-s} - E\{x_k z_{k-s}\})$ converges a. s. to 0.*

(7) *Let $\{x_t^{(i)}; t \in Z\}$, $i = 1,2,3,4$, be four processes defined in terms of a sequence of i.i.d. random variables $\{e_t; t \in Z\}$, with finite fourth-order moment, by $x_t^{(i)} = \sum_{v=1}^{\infty}\xi_v^{(i)}|e_{t-v}|$ with $\sum_{v=1}^{\infty}|\xi_v^{(i)}| < \infty$. Then $E\{x_t^{(1)}x_t^{(2)}x_t^{(3)}x_t^{(4)}\}$ is finite.*

Proof To prove part 6, let

$$x_k - \phi_{1,1}x_{k-1} - \cdots - \phi_{1,p_1}x_{k-p_1}$$
$$= e_k - \theta_{1,1}e_{k-1} - \cdots - \theta_{1,q_1}e_{k-q_1}$$

and

$$z_k - \phi_{2,1}z_{k-1} - \cdots - \phi_{2,p_2}z_{k-p_2}$$
$$= e_k - \theta_{2,1}e_{k-1} - \cdots - \theta_{2,q_2}e_{k-q_2}.$$

We have

$$x_k z_{k-s-i} - \phi_{1,1}x_{k-1}z_{k-s-i} - \cdots - \phi_{1,p_1}x_{k-p_1}z_{k-s-i}$$
$$= (e_k - \theta_{1,1}e_{k-1} - \cdots - \theta_{1,q_1}e_{k-q_1})z_{k-s-i} \tag{A9}$$

with $i = 0, 1, \ldots, p_2 - 1$ and

$$z_k x_{k+s-j} - \phi_{2,1}z_{k-1}x_{k+s-j} - \cdots - \phi_{2,p_2}z_{k-p_2}x_{k+s-j}$$
$$= (e_k - \theta_{2,1}e_{k-1} - \cdots - \theta_{2,q_2}e_{k-q_2})x_{k+s-j} \tag{A10}$$

with $j = 1, \ldots, p_1$. Denote $\gamma_s = \lim_{t\to\infty} t^{-1/2-\varepsilon}(\sum_{k=1}^t (x_k z_{k-s} - E\{x_k z_{k-s}\}))$ hence the equation of (A9) for $i = 0$ yields

$$\gamma_s - \phi_{1,1}\gamma_{s-1} - \cdots - \phi_{1,p_1}\gamma_{s-p_1}$$
$$= \lim_{t\to\infty} t^{-1/2-\varepsilon}\sum_{k=1}^{t}[(e_k - \theta_{1,1}e_{k-1} - \cdots - \theta_{1,q_1}e_{k-q_1})z_{k-s}$$
$$- E\{(e_k - \theta_{1,1}e_{k-1} - \cdots - \theta_{1,q_1}e_{k-q_1})z_{k-s}\}]. \tag{A11}$$

Writing z_{k-s} as function of $e_{k-s}, \ldots, e_{k-2q_1}, z_{k-q_1-1}, \ldots, z_{k-q_1-p_2}$, we have in the right-hand side a finite sum of expressions of the form

$$\lim_{t\to\infty} t^{-1/2-\varepsilon}\left(\sum_{k=1}^{t}(e_k z_{k-h} - E\{e_k z_{k-h}\})\right)$$

$$= \lim_{t\to\infty} t^{-1/2-\varepsilon}\sum_{k=1}^{t} e_k z_{k-h} \quad \text{(A12)}$$

for $h > 0$, or similar expressions with $e_k e_{k-h}$ instead of $e_k z_{k-h}$, and also an expression with e_k^2. The latter provides a sum of independent variables

$$\lim_{t\to\infty} t^{-1/2-\varepsilon}\sum_{k=1}^{t}(e_k^2 - E\{e_k^2\}).$$

Since $E\{e_k^4\}$ exists, this limit is 0 by applying Lemma A.10 (Kolmogorov's proposition). Let us take for example (A12) and prove that it equals 0. Consider $r_t = \sum_{k=1}^{t} k^{-1/2-\varepsilon} e_k z_{k-h}$ and the σ-algebra $\mathcal{F}_{t-1}$, spanned by the e_i, $i \leqslant t-1$. For $h > 0$, we have

$$E(r_t \mid \mathcal{F}_{t-1}) = E\left(\sum_{k=1}^{t} k^{-1/2-\varepsilon} e_k z_{k-h} \,\middle|\, \mathcal{F}_{t-1}\right) = r_{t-1},$$

hence $\{r_t\}$ is a martingale with respect to $\mathcal{F}_{t-1}$. Moreover

$$E|r_t|^2 \leqslant \sum_{k=1}^{t} k^{-1-2\varepsilon} E|z_{k-h}|^2 E|e_k|^2 \leqslant C\sum_{k=1}^{t} k^{-1-2\varepsilon} < \infty,$$

where C denotes a constant. Hence r_t is a martingale with a bounded variance, and according to Lemma A.10 (Chung, 1968), r_t converges a.s. to a finite limit r_∞. Hence by Lemma A.10 (Kronecker's lemma), $t^{-1/2-\varepsilon}\sum_{k=1}^{t} e_k z_{k-h} \to 0$ a.s. when $t \to \infty$, for $h > 0$, with a similar reasoning for terms in $e_k e_{k-h}$. Hence the right-hand side of Equation (A11) converges to 0 a.s. Proceeding in the same way for all the equations of (A9) and (A10), and denoting the $(p_2 + p_1) \times (p_2 + p_1)$ matrix

$$A = \begin{pmatrix} 1 & -\phi_{1,1} & \dots & -\phi_{1,p_1} & \dots & \dots & 0 \\ 0 & 1 & -\phi_{1,1} & \dots & -\phi_{1,p_1} & \dots & 0 \\ \dots & \dots & \dots & \dots & \dots & \dots & \dots \\ 0 & 0 & \dots & 1 & -\phi_{1,1} & \dots & -\phi_{1,p_1} \\ -- & -- & -- & -- & -- & -- & -- \\ -\phi_{2,p_2} & -\phi_{2,p_2-1} & \dots & 1 & \dots & \dots & 0 \\ 0 & -\phi_{2,p_2} & -\phi_{2,p_2-1} & \dots & 1 & \dots & 0 \\ \dots & \dots & \dots & \dots & \dots & \dots & \dots \\ 0 & 0 & 0 & -\phi_{2,p_2} & -\phi_{2,p_2-1} & \dots & 1 \end{pmatrix},$$

and

$$X' = (\gamma_{s+p_2-1} \quad \dots \quad \gamma_s \quad \dots \quad \gamma_{s-p_1}),$$

we obtain

$$AX = 0.$$

From van der Waerden (1937), A is invertible because the zeroes of the polynomials $1 - \phi_{1,1}B - \dots - \phi_{1,p_1}B^{p_1}$ and $1 - \phi_{2,1}B - \dots - \phi_{2,p_2}B^{p_2}$ are outside the unit circle so the polynomials $1 - \phi_{1,1}B - \dots - \phi_{1,p_1}B^{p_1}$ and $B^{p_2} - \phi_{2,1}B^{p_2-1} - \dots - \phi_{2,p_2}$ have no common zeroes. Hence $X = 0$, which implies the result.

To prove part 7, we write

$$\begin{aligned} E\{x_t^{(1)}x_t^{(2)}x_t^{(3)}x_t^{(4)}\} = {} & \sum_{v=1}^{\infty} \xi_v^{(1)}\xi_v^{(2)}\xi_v^{(3)}\xi_v^{(4)} E\{|e_{t-v}|^4\} \\ & + \sum_{v_1 \neq v_2} \sum_{v_1=1}^{\infty} \sum_{v_2=1}^{\infty} \{\xi_{v_1}^{(1)}\xi_{v_1}^{(2)}\xi_{v_2}^{(3)}\xi_{v_2}^{(4)} \\ & + \xi_{v_1}^{(1)}\xi_{v_2}^{(2)}\xi_{v_1}^{(3)}\xi_{v_2}^{(4)} \\ & + \xi_{v_1}^{(1)}\xi_{v_2}^{(2)}\xi_{v_2}^{(3)}\xi_{v_1}^{(4)}\} E\{|e_{t-v_1}|^2\} E\{|e_{t-v_2}|^2\} \\ & + \sum_{v_1 \neq v_2 \neq v_3} \sum_{v_1=1}^{\infty} \sum_{v_2=1}^{\infty} \sum_{v_3=1}^{\infty} \{\xi_{v_1}^{(1)}\xi_{v_1}^{(2)}\xi_{v_2}^{(3)}\xi_{v_3}^{(4)} \\ & + \xi_{v_1}^{(1)}\xi_{v_2}^{(2)}\xi_{v_1}^{(3)}\xi_{v_3}^{(4)} + \xi_{v_1}^{(1)}\xi_{v_2}^{(2)}\xi_{v_3}^{(3)}\xi_{v_1}^{(4)}\} \\ & \times E\{|e_{t-v_1}|^2\} E\{|e_{t-v_2}|\} E\{|e_{t-v_3}|\} \\ & + \sum_{\substack{v_1=1 \\ v_1 \neq v_2 \neq v_3 \neq v_4}}^{\infty} \sum_{v_2=1}^{\infty} \sum_{v_3=1}^{\infty} \sum_{v_4=1}^{\infty} \xi_{v_1}^{(1)}\xi_{v_2}^{(2)}\xi_{v_3}^{(3)}\xi_{v_4}^{(4)} \\ & \times E\{|e_{t-v_1}|\} E\{|e_{t-v_2}|\} E\{|e_{t-v_3}|\} E\{|e_{t-v_4}|\} \end{aligned}$$

Denoting $m_j = E(|e_t|^j)$, the absolute value of the left-hand side is bounded by the maximum of $m_4, m_2^2, m_2 m_1^2$ and m_1^4, which is finite, multiplied by the $\prod_{i=1}^{4}\sum_{v=1}^{\infty}|\xi_v^{(i)}|$, which is also finite. ■

LEMMA A.11 *Let $\beta \in D_B$ and consider $R_t(\beta)$ defined by Equation (37). Then for all $\varepsilon > 0$, $t^{1/2-\varepsilon}(R_t(\beta) - \sigma_e^2 F(\beta))$ converges to 0 a.s. when $t \to \infty$ as well as $t^{1/2-\varepsilon}[(1/t)\sum_{k=1}^{t}\psi_k(\beta)\varepsilon_k(\beta) - E\{\psi_t(\beta)\varepsilon_t(\beta)\}]$ and $t^{1/2-\varepsilon}(\sigma_t^2(\beta) - E\{\varepsilon_t^2(\beta)\})$, where $\sigma_t^2(\beta)$ is defined by Equation (35). Similarly, considering S_t and T_t defined by Equations (A6) and (A7), respectively, we have that, for all $\varepsilon > 0$, $t^{-1/2-\varepsilon}S_t$ and $t^{-1/2-\varepsilon}T_t$ converge to 0 a.s. when $t \to \infty$.*

Proof Let $\psi_t(\beta) = (\psi_{1,t}(\beta), \dots, \psi_{p+q,t}(\beta))^{\mathrm{T}}$. From Equation (36) we have $\Theta(B)\psi_t(\beta) = \varphi_{t-1}(\beta)$, hence

$$\psi_{i,t}(\beta) = \Theta(B)^{-1}y_{t-i}, \quad i = 1, \dots, p,$$

$$\psi_{p+i,t}(\beta) = -\Theta(B)^{-1}\varepsilon_{t-i}(\beta), \quad i = 1, \dots, q,$$

but, since $\Phi^*(B)y_t = \Theta^*(B)e_t$, we know that $y_{t-i} = \Phi^*(B)^{-1}\Theta^*(B)e_{t-i}$, and that $\varepsilon_{t-i}(\beta) = \Theta(B)^{-1}\Phi(B)\Phi^*(B)^{-1}\Theta^*(B)e_{t-i}$, hence

$$\psi_{i,t}(\beta) = \Theta(B)^{-1}\Phi^*(B)^{-1}\Theta^*(B)e_{t-i}, \quad i = 1, \dots, p,$$

$$\psi_{p+i,t}(\beta) = -\Theta(B)^{-1}\Theta(B)^{-1}\Phi(B)\Phi^*(B)^{-1}\Theta^*(B)e_{t-i},$$
$$i = 1, \dots, q.$$

Consequently the $\varepsilon_t(\beta)$ and $\{\psi_{i,t}(\beta),\ i = 1, 2, \dots, p\}$ are ARMA processes in terms of the errors e_t of the original process. The proof is completed by using three times Lemma A.10, part 6. The last part of the lemma is immediate given the definition of S_t and T_t with $\beta \in D_B$ replaced by β^*. ■

A part of the following lemma is inspired by Ljung and Söderström (1983, pp. 441–442), but is stated under Assumption 2, without using that the observations are bounded. In the latter case, in particular, it is enough to state that $\|h_t\|$, $\|h_t(\beta)\|$, and $\|\partial h_t(\beta)/\partial\beta^{\mathrm{T}}\|$ are bounded and that $\|h_t - h_t(\hat{\beta}_t)\| \leqslant M/t$.

LEMMA A.12 *Let us consider the algorithm (31). First* $\|h_t\|$, $\|h_t(\beta)\|$, *and* $\|\partial h_t(\beta)/\partial\beta^{\mathrm{T}}\|$, *where* $\beta = \varkappa_t$ *is sufficiently close to* β^*, *are bounded by a process* $\sum_{i=1}^{\infty}|\nu_i|\,|e_{t-i}|$ *where* $\sum_{i=1}^{\infty}|\nu_i|$ *is finite. Then, for all random variable* b_t *such that,* $b_t = \sum_{v=1}^{\infty}\xi_v|e_{t-v}|$ *with* $\sum_{v=1}^{\infty}|\xi_v| < \infty$, *there exists a positive constant* M *such that* $E(\|h_t - h_t(\hat\beta_t)\|b_t) \leqslant M/t$ *where* $\forall\beta \in D_B$, $h_t(\beta) = A(\beta)h_{t-1}(\beta) + B(\beta)z_t$. *Moreover* $h_t - h_t(\beta^*) \to 0$ *a.s, when* $n \to \infty$.

Proof From Equation (1) we have the infinite moving average representation

$$y_t = \sum_{i=1}^{\infty}\mu_i e_{t-i}, \tag{A13}$$

where $\sum_{i=1}^{\infty}|\mu_i| < \infty$. $\forall\beta \in D_B$ we may write $h_t(\beta)$ under the form

$$h_t(\beta) = \sum_{i=0}^{t-1}A(\beta)^iB(\beta)z_{t-i}(\beta) + A(\beta)^th_0(\beta), \tag{A14}$$

We have also

$$\frac{\partial h_t(\beta)}{\partial\beta^{\mathrm{T}}} = A(\beta)\frac{\partial h_{t-1}(\beta)}{\partial\beta^{\mathrm{T}}} + (h_{t-1}(\beta)\otimes I_v)\frac{\partial\,\mathrm{vec}A(\beta)}{\partial\beta^{\mathrm{T}}} + (z_t\otimes I_v)\frac{\partial\,\mathrm{vec}B(\beta)}{\partial\beta^{\mathrm{T}}}.$$

where $v = q(p+q+1)$. Let

$$G_t(\beta) = (h_{t-1}(\beta)\otimes I_v)\frac{\partial\,\mathrm{vec}A(\beta)}{\partial\beta^{\mathrm{T}}} + (z_t\otimes I_v)\frac{\partial\,\mathrm{vec}B(\beta)}{\partial\beta^{\mathrm{T}}}. \tag{A15}$$

We can write $\partial h_t(\beta)/\partial\beta^{\mathrm{T}}$ under the form

$$\frac{\partial h_t(\beta)}{\partial\beta^{\mathrm{T}}} = \sum_{i=0}^{\infty}A(\beta)^iG_{t-i}(\beta). \tag{A16}$$

We know that $\beta^* \in D_B$, so $\|A(\beta^*)^t\| \leqslant C\lambda^t$ for some $\lambda < 1$. Furthermore, for β_k belonging to a neighbourhood of β^* small enough, we have also $\|\prod_{k=1}^{t}A(\beta_k)\| \leqslant C\lambda_1^t$ for some $\lambda_1 < 1$ since $\prod_{k=1}^{t}A(\beta)$ is a continuous function of β. In Section 3.1, we have proved that $\hat\beta_t \to \beta^*$ a.s. when $t \to \infty$, then for a large enough t, $\exists T > 0$, such as $\forall s > T$, $\|A(\hat\beta_s)\| < \lambda_1$, so $\forall t > T$,

$$\left\|\prod_{k=1}^{t}A(\hat\beta_k)\right\| \leqslant \left\|\prod_{k=1}^{T-1}A(\hat\beta_k)\right\|\left\|\prod_{k=T}^{t}A(\hat\beta_k)\right\| \leqslant C_0C\lambda_1^{t-T} = C_2\lambda_1^t, \tag{A17}$$

where C_0 can be taken as $(C_1)^{\mathrm{T}}$, for example, where $C_1 = \sup_{\beta\in D_R}\|A(\beta)\|$.

According to Equation (31) we have

$$h_t = A(\hat\beta_{t-1})h_{t-1} + B(\hat\beta_{t-1})z_t.$$

Since h_t contains ε_t and $\psi_t, \psi_{t-1}, \ldots, \psi_{t-q}$, we can suppose that $h_0 = 0$ hence

$$\begin{aligned}h_t &= A(\hat\beta_{t-1})A(\hat\beta_{t-2})h_{t-2} + A(\hat\beta_{t-1})B(\hat\beta_{t-2})z_{t-1} + B(\hat\beta_{t-1})z_t = \cdots\\ &= \sum_{k=1}^{t}\left[\prod_{j=k}^{t-1}A(\hat\beta_j)\right]B(\hat\beta_{k-1})z_k + \left[\prod_{j=k}^{t}A(\hat\beta_j)\right]h_0,\end{aligned} \tag{A18}$$

with the convention $\prod_{j=t}^{t-1}A(\hat\beta_j) = I_{p+q}$. Since $\forall\beta \in D_B$, there exists a positive real C such that $\|B(\beta)\| < C$, so Equations (A17) and (A18) imply

$$\|h_t\| \leqslant C\sum_{k=1}^{t}\lambda_1^{t-k}\|z_k\|. \tag{A19}$$

To conclude the proof of the first statement, let us first show that $\|h_t\|$ is bounded by some process. Indeed, we deduce from Equation (A19) that

$$\|h_t\| \leqslant C\sum_{k=0}^{t-1}\lambda_1^k\|z_{t-k}\| \leqslant C\sum_{k=0}^{\infty}\lambda_1^k\sum_{i=1}^{\infty}|\alpha_i||e_{t-k-i}|,$$

where the coefficients α_i are related to the μ_i in Equation (A13). Hence $\|h_t\|$ is dominated by a process such as $\sum_{i=1}^{\infty}|\nu_i||e_{t-i}|$ where $\sum_{i=1}^{\infty}|\nu_i|$ is finite. The derivations are similar for $\|h_t(\beta)\|$ and $\|\partial h_t(\beta)/\partial\beta^{\mathrm{T}}\|$, where $\beta = \varkappa_t$ is sufficiently close to β^*, using Equations (A14) and (A16), respectively, with appropriate λ_1, α_i and ν_i.

For the second part of the proof, let $\tilde h_k(\beta) = h_k - h_k(\beta)$, $\tilde A(\hat\beta_k, \beta) = A(\hat\beta_k) - A(\beta)$, $\tilde B(\hat\beta_k, \beta) = B(\hat\beta_k) - B(\beta)$. For $k \leqslant t$, we have similarly to Equation (A18)

$$\tilde h_t(\beta) = \sum_{k=1}^{t}\left(\prod_{j=k}^{t-1}A(\hat\beta_j)\right)[\tilde A(\hat\beta_{k-1}, \beta)h_{k-1}(\beta) + \tilde B(\hat\beta_{k-1}, \beta)z_k]$$

since $\tilde h_0(\beta) = 0$. Hence for $\beta = \hat\beta_{t-1}$, we have

$$\tilde h_t(\hat\beta_{t-1}) = \sum_{k=1}^{t}\left(\prod_{j=k}^{t-1}A(\hat\beta_j)\right)[\tilde A(\hat\beta_{k-1}, \hat\beta_{t-1})h_{k-1}(\hat\beta_{t-1}) + \tilde B(\hat\beta_{k-1}, \hat\beta_{t-1})z_k]. \tag{A20}$$

thus

$$\|\tilde h_t(\hat\beta_{t-1})\| \leqslant \sum_{k=1}^{t}\lambda_1^{t-k}\|\tilde A(\hat\beta_{k-1}, \hat\beta_{t-1})h_{k-1}(\hat\beta_{t-1}) + \tilde B(\hat\beta_{k-1}, \hat\beta_{t-1})z_k\|.$$

Since $A(\beta)$ and $B(\beta)$ are Lipschitz continuous, there exists a constant C_{AB} such that

$$\|\tilde A(\hat\beta_{k-1}, \hat\beta_{t-1})\| \leqslant C_{AB}\|\hat\beta_{k-1} - \hat\beta_{t-1}\|,$$

$$\|\tilde B(\hat\beta_{k-1}, \hat\beta_{t-1})\| \leqslant C_{AB}\|\hat\beta_{k-1} - \hat\beta_{t-1}\|.$$

We have

$$\begin{aligned}t\|\tilde h_t(\hat\beta_{t-1})\| &\leqslant tC_{AB}\sum_{k=1}^{t}\lambda_1^{t-k}\|\hat\beta_{k-1} - \hat\beta_{t-1}\|[\|h_{k-1}(\hat\beta_{t-1})\| + \|z_k\|],\\ &\leqslant tC\sum_{k=1}^{t}\lambda_1^{t-k}\left(\sum_{l=k}^{t}\frac{1}{l}\|\psi_l\varepsilon_l\|\right)\times[\|h_{k-1}(\hat\beta_{t-1})\| + \|z_k\|].\\ &\leqslant Ct\sum_{k=1}^{t}\sum_{l=k}^{t}\lambda_1^{t-k}\frac{1}{l}\|\psi_l\varepsilon_l\|[\|h_{k-1}(\hat\beta_{t-1})\| + \|z_k\|].\\ &\leqslant tC\sum_{l=1}^{t}\frac{\lambda_1^{t-l}}{l}\|\psi_l\varepsilon_l\|\sum_{k=1}^{l}\lambda_1^{l-k}\times[\|h_{k-1}(\hat\beta_{t-1})\| + \|z_k\|].\end{aligned}$$

Let us prove that, for all random variable b_t such that $b_t = \sum_{u=1}^{\infty}\xi_v|e_{t-v}|$ with $\sum_{v=1}^{\infty}|\xi_v| < \infty$, $tE(\|\tilde h_t(\hat\beta_{t-1})\|b_t)$ is bounded. Indeed, by using two times Lemma A.10, part 7,

$$\begin{aligned} tE(\|\tilde{h}_t(\hat{\beta}_{t-1})\|b_t) &\leqslant tC\sum_{l=1}^{t}\frac{\lambda_1^{t-l}}{l}\sum_{k=1}^{l}\lambda_1^{l-k} \\ &\quad\times E[b_t\|\psi_l\varepsilon_l\|\|h_{k-1}(\hat{\beta}_{t-1})\| + b_t\|\psi_l\varepsilon_l\|\|z_k\|] \\ &\leqslant tC_1\sum_{l=1}^{t}\frac{\lambda_1^{t-l}}{l}\sum_{k=1}^{l}\lambda_1^{l-k} \\ &= tC_1\sum_{l=1}^{t}\frac{\lambda_1^{t-l}}{l}\frac{\lambda_1-\lambda_1^l}{1-\lambda_1} \\ &= \frac{\lambda_1}{1-\lambda_1}tC_1\left(\sum_{l=1}^{t}\frac{\lambda_1^{t-l}}{l}-\sum_{l=1}^{t}\frac{\lambda_1^{t-1}}{l}\right) \end{aligned}$$

which is bounded

For the third part of the proof, let $\tilde{h}_k(\beta^*) = h_k - h_k(\beta^*)$, $\tilde{A}(\hat{\beta}_k,\beta^*) = A(\hat{\beta}_k) - A(\beta^*)$, $\tilde{B}(\hat{\beta}_k,\beta^*) = B(\hat{\beta}_k) - B(\beta^*)$. We have

$$\tilde{h}_t(\beta^*) = \sum_{k=1}^{t}\left(\prod_{j=k}^{t-1}A(\hat{\beta}_j)\right)[\tilde{A}(\hat{\beta}_{k-1},\beta^*)h_k(\beta^*) + \tilde{B}(\hat{\beta}_{k-1},\beta^*)z_k].$$

Because $\tilde{A}(\hat{\beta}_{k-1},\beta^*)$, $\tilde{B}(\hat{\beta}_{k-1},\beta^*)$ converge to 0, hence $\tilde{h}_k \longrightarrow 0$ a.s., when $n \longrightarrow \infty$. ■

Lemma A.13 *There exists a process a_t such that $(1/t)\sum_{k=1}^{t} a_k$ is bounded and $\forall t \geqslant 1$, $\|A_{1,t}\| < a_t\|\tilde{\beta}_{t-1}\|^2$, where $A_{1,t}$ is defined by* (A4).

Proof We know that $\hat{\beta}_t \to \beta^*$ a.s., and $F^{-1}(\beta)$ being continuous over D_B, $F^{-1}(\hat{\beta}_{t-1}) \to F^{-1}(\beta^*)$ a.s., so there exists a positive constant C_1 such that

$$\begin{aligned} &\left\|\frac{1}{2}F^{-1}(\hat{\beta}_{t-1})\psi_t\tilde{\beta}_{t-1}^{\mathrm{T}}\left(\frac{\partial\psi_t(\beta)}{\partial\beta^{\mathrm{T}}}\right)_{\beta=\varkappa_t}\tilde{\beta}_{t-1}\right\| \\ &\quad< C_1\|\psi_t\|\left\|\left(\frac{\partial\psi_t(\beta)}{\partial\beta^{\mathrm{T}}}\right)_{\beta=\varkappa_t}\right\|\|\tilde{\beta}_{t-1}\|^2 \end{aligned}$$

where $\varkappa_t$ is a point on the segment joining $\hat{\beta}_{t-1}$ and β^* near to β^*. Also, $\partial F(\beta)/\partial\beta$ and $F^{-1}(\beta)$ are bounded over D_B so that, by Lemma A.2, $\partial F^{-1}(\beta)/\partial\beta$ is bounded, there exist positive constants C_2 and C_3 such that

$$\begin{aligned} &\|(F^{-1}(\hat{\beta}_{t-1}) - F^{-1}(\beta^*))\psi_t(\beta^*)\psi_t^{\mathrm{T}}(\beta^*)\tilde{\beta}_{t-1}\| \\ &\quad< C_2\|\psi_t(\beta^*)\|^2\|\tilde{\beta}_{t-1}\|^2 \end{aligned}$$

and

$$\begin{aligned} &\|F^{-1}(\hat{\beta}_{t-1})\psi_t(\psi_t^{\mathrm{T}}(\hat{\beta}_{t-1}) - \psi_t^{\mathrm{T}}(\beta^*))\tilde{\beta}_{t-1}\| \\ &\quad< C_3\|\psi_t\|\left\|\left(\frac{\partial\psi_t(\beta)}{\partial\beta^{\mathrm{T}}}\right)_{\beta=\varkappa_t}\right\|\|\tilde{\beta}_{t-1}\|^2, \end{aligned}$$

and there exists a constant C_4 such that

$$\begin{aligned} \|(\varepsilon_t^2(\hat{\beta}_{t-1}) - e_t^2)\tilde{\beta}_{t-1}\| &= \|(\varepsilon_t^2(\hat{\beta}_{t-1}) - \varepsilon_t^2(\beta^*))\tilde{\beta}_{t-1}\| \\ &< C_4\|\psi_t(\varkappa_t)\|\|\varepsilon_t(\varkappa_t)\|\|\tilde{\beta}_{t-1}\|^2. \end{aligned}$$

We have to prove that

$$\begin{aligned} &\frac{1}{t}\sum_{k=1}^{t}\left(\|\psi_k\|\left\|\left(\frac{\partial\psi_k(\beta)}{\partial\beta^{\mathrm{T}}}\right)_{\beta=\varkappa_k}\right\| + \|\psi_k(\beta^*)\|^2 + \|\psi_k\|\right. \\ &\quad\left.\times\left\|\left(\frac{\partial\psi_k(\beta)}{\partial\beta^{\mathrm{T}}}\right)_{\beta=\varkappa_k}\right\| + \|\psi_k(\varkappa_k)\|\|\varepsilon_{k-1}(\varkappa_k)\|\right) \end{aligned} \tag{A21}$$

is finite. Since ψ_t, both $\psi_t(\beta)$ and $\varepsilon_t(\beta)$, and $\partial\psi_t(\beta)/\partial\beta^{\mathrm{T}}$ are components of h_t, $h_t(\beta)$, and $\partial h_t(\beta)/\partial\beta^{\mathrm{T}}$, respectively, from Lemma A.12, since $\varkappa_t$ is near β^*, we have for each term of (A21) an upper bound of the form $\sum_{i=1}^{\infty}|v_i||e_{k-i}|\sum_{j=1}^{\infty}|\eta_j||e_{k-j}|$ which is an ergodic process, hence we have the result. ■

Lemma A.14 *For any $\vartheta < 1$, and any a_t such that $(1/t)\sum_{k=1}^{t} a_k$ is finite, $t^{-1+\vartheta}\sum_{k=1}^{t} k^{-\vartheta}a_k$ is finite. If $\vartheta > 1$ then $\sum_{k=1}^{t} k^{-\vartheta}a_k$ is finite.*

Proof We have

$$\begin{aligned} \frac{1}{t^{1-\vartheta}}\sum_{k=1}^{t}k^{-\vartheta}a_k &= \frac{t^{-\vartheta}}{t^{1-\vartheta}}\sum_{k=1}^{t}a_k - \frac{1}{t^{1-\vartheta}} \\ &\quad\times\sum_{k=1}^{t-1}((k+1)^{-\vartheta} - k^{-\vartheta})\sum_{l=1}^{k}a_l \\ &= \frac{1}{t}\sum_{k=1}^{t}a_k + \frac{1}{t^{1-\vartheta}}\sum_{k=1}^{t-1}k(k^{-\vartheta} - (k+1)^{-\vartheta}) \\ &\quad\times\frac{1}{k}\sum_{l=1}^{k}a_l. \end{aligned}$$

Let us prove that $t^{-1+\vartheta}\sum_{k=1}^{t-1} k(k^{-\vartheta} - (k+1)^{-\vartheta})$ is finite when $t \to \infty$ in order to apply Toeplitz lemma (see Lemma A.10). Denote

$$\Sigma_t = \sum_{k=1}^{t}k(k^{-\vartheta} - (k+1)^{-\vartheta}) \quad\text{and}$$

$$\Sigma(x) = x(x^{-\vartheta} - (x+1)^{-\vartheta}).$$

Since $\Sigma(x)$ is a decreasing function of x, we have by using Cauchy theorem on comparisons between series and integrals (e.g. Protter & Morrey, 1977)

$$\frac{\Sigma_t - (1-(2)^{-\vartheta})}{t^{1-\vartheta}} \leqslant \frac{1}{t^{1-\vartheta}}\int_1^t\Sigma(x)\,\mathrm{d}x \leqslant \frac{1}{t^{1-\vartheta}}\Sigma_{t-1}$$

and, integrating by parts and rearranging terms,

$$\begin{aligned} \int_1^t\Sigma(x)\,\mathrm{d}x &= \frac{1}{1-\vartheta}t\left(t^{1-\vartheta}\left(1-\left(1+\frac{1}{t}\right)^{1-\vartheta}\right)\right) \\ &\quad- \frac{1}{1-\vartheta}(1-2^{1-\vartheta}) - \frac{1}{(1-\vartheta)(2-\vartheta)} \\ &\quad\times\left[t^{2-\vartheta}\left(1-\left(1+\frac{1}{t}\right)^{2-\vartheta}\right) - (1-(2)^{2-\vartheta})\right]. \end{aligned}$$

Hence $\Sigma_t/t^{1-\vartheta}$ is bounded by above by the integral divided by $t^{1-\vartheta}$ which is finite. If $\vartheta > 1$, we will have $\int_1^t\Sigma(x)\,\mathrm{d}x < \infty$. ■

LEMMA A.15 *Consider $A_{2,t}$ defined by Equation* (A5). $\forall\delta > 0$, $t^{-1/2-\delta}\sum_{k=1}^{t} A_{2,k}$ *and* $t^{-1/2-\delta}\sum_{k=1}^{t}(1/k)(\varepsilon_k^2 - \hat{\sigma}_k^2)F^{-1}(\hat{\beta}_{k-1})\psi_k\varepsilon_k$, *converge to* 0 *a.s. when* $t \to \infty$.

Proof A sketch of the proof is given by Ljung and Söderström (1983, p. 444), Lemma 4.B.4 but the proof given here differs because of Assumption 2. Let us consider the series $Z_t = \sum_{k=1}^{t} k^{-1/2-\delta}A_{2,k}$ obtained by

$$Z_t = \sum_{k=1}^{t} k^{-1/2-\delta}(\varepsilon_k - \varepsilon_k(\hat{\beta}_{k-1}))(\varepsilon_k + \varepsilon_k(\hat{\beta}_{k-1}))\tilde{\beta}_{k-1}$$
$$+ \sum_{k=1}^{t} k^{-1/2-\delta}F^{-1}(\hat{\beta}_{k-1})\psi_k(\varepsilon_k - \varepsilon_k(\hat{\beta}_{k-1}))$$
$$- \sum_{k=1}^{t} k^{-1/2-\delta}F^{-1}(\hat{\beta}_{k-1})(\psi_k - \psi_k(\hat{\beta}_{k-1}))\psi_k^{\mathrm{T}}(\beta^*)\tilde{\beta}_{k-1}.$$

By using four times Lemma A.12, we know that

$$kE(|\varepsilon_k - \varepsilon_k(\hat{\beta}_{k-1})||\varepsilon_k + \varepsilon_k(\hat{\beta}_{k-1})|)$$
$$\leq kE(|\varepsilon_k - \varepsilon_k(\hat{\beta}_{k-1})||\varepsilon_k|) + kE(|\varepsilon_k - \varepsilon_k(\hat{\beta}_{k-1})||\varepsilon_k(\hat{\beta}_{k-1})|),$$

$kE(\|\psi_k\||\varepsilon_k - \varepsilon_k(\hat{\beta}_{k-1})|)$ and $kE(\|(\psi_k - \psi_k(\hat{\beta}_{k-1}))\psi_k^{\mathrm{T}}(\beta^*)\|)$ are bounded, as well as $F^{-1}(\beta)$ over D_B, hence by using Lemma A.10 (Lukacs, 1975) we prove that Z_t converges a.s. and by Lemma A.10 (Kronecker's lemma) $t^{-1/2-\delta}\sum_{k=1}^{t} A_{2,k} \to 0$ a.s. when $t \to \infty$. Similarly, this time with $Z_t = \sum_{k=1}^{t} k^{-1/2-\delta}(1/k)(\varepsilon_k^2 - \hat{\sigma}_k^2)F^{-1}(\hat{\beta}_{k-1})\psi_k\varepsilon_k$, since $E(\|(\varepsilon_k^2 - \hat{\sigma}_k^2)F^{-1}(\hat{\beta}_{k-1})\psi_k\varepsilon_k\|)$ is bounded, then Z_t is finite and the proof proceeds as before. ■

LEMMA A.16 $\forall\delta > 0$, $t^{-1/2-\delta}\sum_{k=1}^{t} F^{-1}(\hat{\beta}_{k-1})\psi_k e_k \to 0$ *a.s. when* $t \to \infty$.

Proof The proof is given by Ljung and Söderström (1983, p. 442), Lemma 4.B.3 and is repeated here for completeness. Consider the series

$$s_t = \sum_{k=1}^{t} k^{-1/2-\delta}F^{-1}(\hat{\beta}_{k-1})\psi_k e_k.$$

That random vector is a martingale with respect to the σ-algebra $\mathcal{F}_{t-1}$, spanned by the e_i, $i \leqslant t-1$. Indeed

$$E(s_t \mid \mathcal{F}_{t-1}) = s_{t-1} + E(t^{-1/2-\delta}F^{-1}(\hat{\beta}_{k-1})\psi_k e_k \mid \mathcal{F}_{t-1})$$
$$= s_{t-1} + t^{-1/2-\delta}F^{-1}(\hat{\beta}_{k-1})\psi_k E(e_t|\mathcal{F}_{t-1}) = s_{t-1},$$

since $F^{-1}(\hat{\beta}_{t-1})$ and ψ_t do not depend of e_i, $i \leqslant t-1$. Moreover

$$E\|s_t\|^2 \leqslant \sum_{k=1}^{t} k^{-1-2\delta}E\|F^{-1}(\hat{\beta}_{k-1})\psi_k\|^2 E|e_k|^2$$
$$\leqslant C\sum_{k=1}^{t} k^{-1-2\delta} < \infty,$$

where C denotes a constant. Hence s_t is a martingale with a bounded variance, and according to Lemma A.10 (Chung, 1968), s_t converges a.s. to a finite limit s_∞. Hence by Lemma A.10 (Kronecker's lemma), $t^{-1/2-\delta}\sum_{k=1}^{t} F^{-1}(\hat{\beta}_{k-1})\psi_k e_k \to 0$ a.s. when $t \to \infty$. ■

LEMMA A.17 *Let S_t defined by Equation* (A6) *and T_k defined by Equation* (A7), *then* $\forall\delta > 0$,

$$G_t = t^{-1/2-\delta}\left\{T_t\tilde{\beta}_{t-1} - \sum_{k=1}^{t-1} T_k(\tilde{\beta}_k - \tilde{\beta}_{k-1}) - F^{-1}(\beta^*)S_t\tilde{\beta}_{t-1} + F^{-1}(\beta^*)\sum_{k=1}^{t-1} S_k(\tilde{\beta}_k - \tilde{\beta}_{k-1})\right\}$$

converges to 0 *a.s. when* $t \to \infty$.

Proof Replacing β by β^* in Lemma A.11 implies that, for all $\delta > 0$, $t^{-1/2-\delta}S_t$ and $t^{-1/2-\delta}T_t$ converge to 0 a.s. when $t \to \infty$, which implies that $t^{-1/2-\delta}T_t\tilde{\beta}_{t-1}$ and $t^{-1/2-\delta}F^{-1}(\beta^*)S_t\tilde{\beta}_{t-1}$ converge to 0 a.s. when $t \to \infty$. Let us now prove that

$$t^{-1/2-\delta}F^{-1}(\beta^*)\sum_{k=1}^{t-1} S_k(\tilde{\beta}_k - \tilde{\beta}_{k-1}) \to 0, \quad \text{a.s. when } t \to \infty.$$

Consider the series

$$F^{-1}(\beta^*)\sum_{k=1}^{t-1} k^{-1/2-\delta}S_k(\tilde{\beta}_k - \tilde{\beta}_{k-1}).$$

From Equation (44), we know that $(\tilde{\beta}_k - \tilde{\beta}_{k-1}) = (1/k)\hat{\sigma}_t^{-2}F^{-1}(\hat{\beta}_{t-1})\psi_t\varepsilon_t$ and $\forall\mu > 0$, $k^{-1/2-\mu}S_k = o(1)$. Let $0 < \mu < \delta$, so

$$k^{-1/2-\delta}\|S_k(\tilde{\beta}_k - \tilde{\beta}_{k-1})\| = k^{-1/2-\mu}k^{\mu-\delta}\|S_k(\tilde{\beta}_k - \tilde{\beta}_{k-1})\|$$
$$\leqslant Ck^{-1+\mu-\delta}\|\psi_t\varepsilon_t\|$$

hence, by Lemmas A.12 and A.14, we conclude that $F^{-1}(\beta^*)\sum_{k=1}^{t} k^{-1/2-\delta}S_k(\tilde{\beta}_k - \tilde{\beta}_{k-1})$ is finite, and by Lemma A.15, when $t \to \infty$, $t^{-1/2-\delta}F^{-1}(\beta^*)\sum_{k=1}^{t} S_k(\tilde{\beta}_k - \tilde{\beta}_{k-1}) \to 0$ a.s. Similarly, we show that $t^{-1/2-\delta}\sum_{k=1}^{t} T_k(\tilde{\beta}_k - \tilde{\beta}_{k-1}) \to 0$, a.s. ■

LEMMA A.18 *Using notations in the proof of Theorem* 3 *and the sequence a_k from Lemma* A.13, *there exist two constants positive M_1 and M_2 such that*

$$\forall t > 0, \quad t\|\tilde{\beta}_t\| < M_1\sum_{k=1}^{t-1} a_k\|\tilde{\beta}_k\|^2 + M_2 t^{1/2+\delta} \quad \textit{a.s.} \tag{A22}$$

Proof From Equation (A3) and Lemma A.9, we have

$$t\hat{\sigma}_{t+1}^2\tilde{\beta}_t = \sum_{k=1}^{t} A_{1,k} + \sum_{k=1}^{t} A_{2,k} + \sum_{k=2}^{t} F^{-1}(\hat{\beta}_{k-1})\psi_k e_k$$
$$+ T_t\tilde{\beta}_{t-1} - \sum_{k=1}^{t-1} T_k(\tilde{\beta}_k - \tilde{\beta}_{k-1})$$
$$- F^{-1}(\beta^*)S_t\tilde{\beta}_{t-1} + F^{-1}(\beta^*)\sum_{k=1}^{t-1} S_k(\tilde{\beta}_k - \tilde{\beta}_{k-1})$$
$$+ \sum_{k=1}^{t} \frac{1}{k}(\varepsilon_k^2 - \hat{\sigma}_k^2)F^{-1}(\hat{\beta}_{k-1})\psi_k\varepsilon_k \tag{A23}$$

hence

$$t\tilde{\beta}_t = \hat{\sigma}_{t+1}^{-2}\sum_{k=1}^{t} A_{1,k} + \hat{\sigma}_{t+1}^{-2}\sum_{k=1}^{t} A_{2,k} + \hat{\sigma}_{t+1}^{-2}\sum_{k=2}^{t} F^{-1}(\hat{\beta}_{k-1})\psi_k e_k + \hat{\sigma}_{t+1}^{-2}T_t\tilde{\beta}_{t-1} - \hat{\sigma}_{t+1}^{-2}\sum_{k=1}^{t-1} T_k(\tilde{\beta}_k - \tilde{\beta}_{k-1}) - \hat{\sigma}_{t+1}^{-2}F^{-1}(\beta^*)S_t\tilde{\beta}_{t-1} + \hat{\sigma}_{t+1}^{-2}F^{-1}(\beta^*)\sum_{k=1}^{t-1} S_k(\tilde{\beta}_k - \tilde{\beta}_{k-1}) + \hat{\sigma}_{t+1}^{-2}\sum_{k=1}^{t}\frac{1}{k}(\varepsilon_k^2 - \hat{\sigma}_k^2)F^{-1}(\hat{\beta}_{k-1})\psi_k\varepsilon_k,$$

which implies

$$\|t\tilde{\beta}_t\| \leqslant |\hat{\sigma}_{t+1}^{-2}|\sum_{k=1}^{t}\|A_{1,k}\| + |\hat{\sigma}_{t+1}^{-2}|\left\|\sum_{k=1}^{t} A_{2,k}\right\| + |\hat{\sigma}_{t+1}^{-2}|\left\|\sum_{k=1}^{t} F^{-1}(\hat{\beta}_{k-1})\psi_k e_k\right\| + |\hat{\sigma}_{t+1}^{-2}|\left\| T_t\tilde{\beta}_{t-1} + \sum_{k=1}^{t-1} T_k(\tilde{\beta}_k - \tilde{\beta}_{k-1}) + F^{-1}(\beta^*)S_t\tilde{\beta}_{t-1} + F^{-1}(\beta^*)\sum_{k=1}^{t-1} S_k(\tilde{\beta}_k - \tilde{\beta}_{k-1})\right\| + |\hat{\sigma}_{t+1}^{-2}|\left\|\sum_{k=1}^{t}\frac{1}{k}(\varepsilon_k^2 - \hat{\sigma}_k^2)F^{-1}(\hat{\beta}_{k-1})\psi_k\varepsilon_k\right\| \quad \text{(A24)}$$

By Lemma A.13, there exists a constant C such that $\forall k \geqslant 1$, $\|A_{1,k}\| < Ca_k\|\tilde{\beta}_{k-1}\|^2$, hence $\sum_{k=1}^{t}\|A_{1,k}\| < C\sum_{k=1}^{t} a_k \|\tilde{\beta}_{k-1}\|^2$, so there exists a constant $C_1 > 0$ such that $\forall k > 1$,

$$|\hat{\sigma}_{t+1}^{-2}|\sum_{k=1}^{t}\|A_{1,k}\| < C_1\sum_{k=1}^{t} a_k\|\tilde{\beta}_{k-1}\|^2.$$

By Lemma A.16, we know that for all $\delta > 0$, $t^{-1/2-\delta}\sum_{k=1}^{t} F^{-1}(\hat{\beta}_{k-1})\psi_k e_k \to 0$, a.s when $t \to \infty$, so there exists a constant $C_2 > 0$ such that $\forall k \geqslant 1$,

$$|\hat{\sigma}_{t+1}^{-2}|\left\|\sum_{k=1}^{t} F^{-1}(\hat{\beta}_{k-1})\psi_k e_k\right\| < C_2 t^{1/2+\delta}.$$

By Lemma A.17, we have that

$$G_t = t^{-1/2-\delta}\left\{T_t\tilde{\beta}_{t-1} - \sum_{k=1}^{t} T_k(\tilde{\beta}_k - \tilde{\beta}_{k-1}) - F^{-1}(\beta^*)S_t\tilde{\beta}_{t-1} + F^{-1}(\beta^*)\sum_{k=1}^{t} S_k(\tilde{\beta}_k - \tilde{\beta}_{k-1})\right\}$$

converges to 0 a.s when $t \to \infty$ and, by Lemma A.15, $t^{-1/2-\delta}\sum_{k=1}^{t} A_{2,k} \to 0$ a.s. when $t \to \infty$, so that there exists a constant $C_3 > 0$ such that $\forall k \geqslant 1$,

$$|\hat{\sigma}_{t+1}^{-2}|\left\|\sum_{k=1}^{t} A_{2,k}\right\| + |\hat{\sigma}_{t+1}^{-2}|\left\| T_t\tilde{\beta}_{t-1} + \sum_{k=1}^{t} T_k(\tilde{\beta}_{k-1} - \tilde{\beta}_{k-2}) + F^{-1}(\beta^*)S_t\tilde{\beta}_{t-1} + F^{-1}(\beta^*)\sum_{k=1}^{t} S_k(\tilde{\beta}_{k-1} - \tilde{\beta}_{k-2})\right\| + |\hat{\sigma}_{t+1}^{-2}|\left\|\sum_{k=1}^{t}\frac{1}{k}(\varepsilon_k^2 - \hat{\sigma}_k^2)F^{-1}(\hat{\beta}_{k-1})\psi_k\varepsilon_k\right\| < C_3 t^{1/2+\delta}.$$

From Equation (A24), we can conclude that for each $\delta > 0$, there exists a constant $M > 0$, such that

$$t\|\tilde{\beta}_t\| \leqslant M_1\sum_{k=1}^{t} a_k\|\tilde{\beta}_{k-1}\|^2 + M_2 t^{1/2+\delta}. \quad \text{(A25)}$$

■

The following lemma is more general than a result of Ljung and Söderström (1983, p. 445) and the proof is very different.

LEMMA A.19 *Let b_t and a_t sequences of positive numbers such that b_t converges to 0, and $(1/t)\sum_{k=1}^{t} a_k$ converges. If for all t*

$$tb_t < M_1\sum_{k=1}^{t-1} a_k b_k^2 + M_2 t^{1/2+\delta} \quad \text{(A26)}$$

then for all $\gamma < \frac{1}{2} - \delta$, $t^{\gamma} b_t$ converges to 0.

Proof Let us take $\vartheta < (1-2\delta)/4$ and suppose that $t^{\vartheta} b_t$ is not bounded. From Lemma A.14, $t^{-1+\vartheta}\sum_{k=1}^{t} k^{-\vartheta} a_k$ is finite. Let

$$\varepsilon < \left(2M_1\sup_{t>1}(t^{-1+2\vartheta}\sum_{k=1}^{t} a_k k^{-2\vartheta})\right)^{-1}. \quad \text{(A27)}$$

There exists t_0 such that $b_t < \varepsilon$ for all $t \geqslant t_0$. Let M a positive real number such that $\sup_{t<t_0} t^{\vartheta} b_t < M$, hence for all $t < t_0$, $t^{\vartheta} b_t < M$. Let $t_1 \geqslant t_0$ such that $t_1^{\vartheta} b_{t_1} > M$ and for all $t < t_1$, $t^{\vartheta} b_t < M$ thus $t_1^{\vartheta} > M/\varepsilon$. To have t_1 larger it suffices to take M larger. We have

$$t_1 b_{t_1} < M_1 M^2\sum_{k=1}^{t_1-1} a_k k^{-2\vartheta} + M_2 t_1^{1/2+\delta} < M_1 t_1^{2\vartheta} b_{t_1}^2\sum_{k=1}^{t_1-1} a_k k^{-2\vartheta} + M_2 t_1^{1/2+\delta}$$

hence

$$b_{t_1} < M_1 b_{t_1}^2 t_1^{-1+2\vartheta}\sum_{k=1}^{t_1-1} a_k k^{-2\vartheta} + M_2 t_1^{-1/2+\delta}. \quad \text{(A28)}$$

Let us consider the corresponding second degree equation in b_{t_1} and its roots

$$b_{t_1}^{\pm} = \frac{1 \pm \sqrt{1 - 4M_1M_2(t_1^{-1+2\vartheta}\sum_{k=1}^{t_1-1} a_k k^{-2\vartheta})t_1^{-1/2+\delta}}}{2M_1(t_1^{-1+2\vartheta}\sum_{k=1}^{t_1-1} a_k k^{-2\vartheta})}. \quad \text{(A29)}$$

These roots are real for a large enough t_1, hence for large values of M. Inequality (A28) implies either $b_{t_1} > b_{t_1}^{+}$ or $b_{t_1} < b_{t_1}^{-}$. The former is impossible because we would have $b_{t_1} > \varepsilon$ in contradiction with the choice of $t_1 \geqslant t_0$. Hence $b_{t_1} < b_{t_1}^-$ which implies

$$t_1^{\vartheta} b_{t_1} < \frac{t_1^{\vartheta} - \sqrt{t_1^{2\vartheta} - 4M_1M_2(t_1^{-1+2\vartheta}\sum_{k=1}^{t_1-1} a_k k^{-2\vartheta})t_1^{-1/2+\delta+2\vartheta}}}{2M_1(t_1^{-1+2\vartheta}\sum_{k=1}^{t_1-1} a_k k^{-2\vartheta})}.$$

Since $-\frac{1}{2}+\delta+2\vartheta < 0$, when M is large the right-hand side becomes smaller, consequently we will have $t_1^{\vartheta} b_{t_1} < M$ which is absurd since $t_1^{\vartheta} b_{t_1} > M$. Hence for all $\vartheta < (1-2\delta)/4$, $t^{\vartheta} b_t$ is bounded.

Let now take $\gamma = 1/2 - \delta - \varepsilon_0$, where ε_0 is small. From Equation (A26), we have

$$tb_t < M_1 \sum_{k=1}^{t_1-1} a_k k^{-\gamma-\varepsilon_0/2}(k^{\gamma/2+\varepsilon_0/4} b_k)^2 + M_2 t^{1/2+\delta}$$

$$< M_1 C \sum_{k=1}^{t_1-1} a_k k^{-\gamma-\varepsilon_0/2} + M_2 t^{1/2+\delta}$$

since $\gamma/2+\varepsilon_0/4 < (1-2\delta)/4$ and where C is a constant, hence

$$t^{\gamma} b_t < M_1 C t^{-1+\gamma} \sum_{k=1}^{t_1-1} a_k k^{-\gamma-\varepsilon_0/2} + M_2 t^{-1/2+\delta+\gamma}$$

$$= t^{-\varepsilon_0/2} M_1 C t^{-1+\gamma+\varepsilon_0/2} \sum_{k=1}^{t_1-1} a_k k^{-\gamma-\varepsilon_0/2} + M_2 t^{-\varepsilon_0}$$

which implies the result. ■

LEMMA A.20 *For all $\gamma < \frac{1}{2}$, $t^{\gamma}\|\tilde{\beta}_t\|$ converges to 0 almost surely.*

Proof It is an immediate consequence of Lemmas A.18 and A.19 ■

LEMMA A.21 *For all γ such that $0 < \gamma < \frac{1}{2}$, then $t^{\gamma}(\hat{\sigma}_t^2 - \sigma_e^2) \longrightarrow 0$ a.s. when $t \longrightarrow \infty$.*

Proof From Equation (35) we have

$$t^{\gamma}(\hat{\sigma}_t^2 - \sigma_e^2) = \frac{1}{t^{1-\gamma}} \sum_{k=1}^{t} (\varepsilon_{k-1} - \varepsilon_{k-1}(\hat{\beta}_{k-2})) \times (\varepsilon_{k-1} + \varepsilon_{k-1}(\hat{\beta}_{k-2})) \quad \text{(A30)}$$

$$+ \frac{1}{t^{1-\gamma}} \sum_{k=1}^{t} (\varepsilon_{k-1}(\hat{\beta}_{k-2}) - \varepsilon_{k-1}(\beta^*)) \times (\varepsilon_{k-1}(\hat{\beta}_{k-2}) + \varepsilon_{k-1}(\beta^*)) \quad \text{(A31)}$$

$$+ \frac{1}{t^{1-\gamma}} \sum_{k=1}^{t} (\varepsilon_{k-1}^2(\beta^*) - \sigma_e^2). \quad \text{(A32)}$$

By Lemma A.12 we know that $kE(|\varepsilon_k - \varepsilon_k(\hat{\beta}_{k-1})||\varepsilon_{k-1} + \varepsilon_{k-1}(\hat{\beta}_{k-2})|)$ is bounded, hence

$$\sum_{k=1}^{t} k^{-1+\gamma} E|(\varepsilon_{k-1} - \varepsilon_{k-1}(\hat{\beta}_{k-2}))(\varepsilon_{k-1} + \varepsilon_{k-1}(\hat{\beta}_{k-2}))| \leqslant C \sum_{k=1}^{t} k^{-2+\gamma}$$

and, by Lemma A.10 (Lukacs, 1975),

$$\sum_{k=1}^{t} k^{-1+\gamma} |(\varepsilon_{k-1} - \varepsilon_{k-1}(\hat{\beta}_{k-2}))(\varepsilon_{k-1} + \varepsilon_{k-1}(\hat{\beta}_{k-2}))|$$

converges, and, by Lemma A.10 (Kronecker's Lemma), Equation (A30) $\rightarrow$ 0.

There exists a constant C such that

$$\sum_{k=1}^{t} k^{-1+\gamma} |(\varepsilon_{k-1}(\hat{\beta}_{k-2}) - \varepsilon_{k-1}(\beta^*))(\varepsilon_{k-1}(\hat{\beta}_{k-2}) + \varepsilon_{k-1}(\beta^*))|$$

$$\leqslant C \sum_{k=1}^{t} k^{-1+\gamma} \|\psi_{k-1}(\chi_{k-2})\| |\varepsilon_{k-1}(\hat{\beta}_{k-2}) + \varepsilon_{k-1}(\beta^*)| \|\tilde{\beta}_{k-1}\|.$$

By Equation (A22) and Lemma A.19, we have, $\forall \gamma_1$ such that $\gamma < \gamma_1 < \frac{1}{2} - \delta$, $\|t^{\gamma_1}\tilde{\beta}_t\|$ converges to 0 a.s.,

$$C \sum_{k=1}^{t} k^{-1+\gamma} \|\psi_{k-1}(\chi_{k-2})\| |\varepsilon_{k-1}(\hat{\beta}_{k-2}) + \varepsilon_{k-1}(\beta^*)| \|\tilde{\beta}_{k-1}\|$$

$$\leqslant C_1 \sum_{k=1}^{t} k^{-1+\gamma-\gamma_1} \|\psi_{k-1}(\chi_{k-2})\| |\varepsilon_{k-1}(\hat{\beta}_{k-2}) + \varepsilon_{k-1}(\beta^*)|,$$

from which by Lemmas A.12 and A.14,

$$\sum_{k=1}^{t} k^{-1+\gamma} (\varepsilon_{k-1}(\hat{\beta}_{k-2}) - \varepsilon_{k-1}(\beta^*))(\varepsilon_{k-1}(\hat{\beta}_{k-2}) + \varepsilon_{k-1}(\beta^*))$$

converges to a finite limit, and, by Kronecker's Lemma A.10, Equation (A31) $\rightarrow$ 0 a.s. By Lemma A.11, Equation (A32) $\rightarrow$ 0 a.s. when $t \rightarrow \infty$. ■

LEMMA A.22 (Brown, 1971, Theorem 1) *Let $\{S_t, \mathcal{F}_t, t = 1, \ldots\}$ be a martingale. Let*

$$X_t = S_t - S_{t-1}, \quad V_t^2 = \sum_{k=1}^{t} E(X_k^2 \mid \mathcal{F}_{k-1}), \quad s_t^2 = EV_t^2 = ES_t^2.$$

Suppose that $V_t^2 s_t^{-2}$ converges with probability 1 to 0 when $t \rightarrow \infty$, and that the following Lindeberg condition is satisfied: $\forall \delta > 0$, $s_t^{-2}\sum_{j=1}^{t} EX_j^2 I(X_j \geqslant \delta s_t)$ converges in probability to 1 when $t \rightarrow \infty$. Then S_t/s_t converges in law to the normal distribution with mean 0 and variance 1.

LEMMA A.23

$$\left(\frac{1}{\sqrt{t}}\right) \sum_{k=1}^{t} R_k(\beta^*)\hat{\sigma}_k^{-2} F^{-1}(\beta^*)\psi_k(\beta^*)e_k \quad \text{(A33)}$$

converges in law to the normal distribution $N(0, \sigma_e^4 F(\beta^*))$.

Proof A sketch of the proof is given by Ljung and Söderström (1983, p. 448), Lemma 4.B.7 but there are again differences here because of Assumption 2.

We use Lemma A.22 with the Cramér–Wold device. Let

$$Y_t^2 = \sum_{k=1}^{t} E[R_k(\beta^*)\hat{\sigma}_k^{-2}F^{-1}(\beta^*)\psi_k(\beta^*)e_k \times (R_k(\beta^*)\hat{\sigma}_k^{-2}F^{-1}(\beta^*)\psi_k(\beta^*)e_k)^{\mathrm{T}}|\mathcal{F}_{k-1}]$$

and $M_t^2 = EY_t^2$. We have

$$\begin{aligned} Y_t^2 =& \sum_{k=1}^{t}\{R_k(\beta^*) - \sigma_e^2F(\beta^*)\}\hat{\sigma}_k^{-4}F^{-1}(\beta^*)\psi_k(\beta^*) \\ &\times \psi_k^{\mathrm{T}}(\beta^*)F^{-1}(\beta^*)R_k(\beta^*)\sigma_e^2 \\ &+ \sum_{k=1}^{t}\sigma_e^4\hat{\sigma}_k^{-4}\psi_k(\beta^*)\psi_k^{\mathrm{T}}(\beta^*)F^{-1}(\beta^*)\{R_k(\beta^*) - \sigma_e^2F(\beta^*)\} \\ &+ \sum_{k=1}^{t}\sigma_e^6(\sigma_e^2 - \hat{\sigma}_k^2)(\sigma_e^2 + \hat{\sigma}_k^2)\hat{\sigma}_k^{-4}\sigma_e^{-4}\psi_k(\beta^*)(\beta^*) \\ &+ \sum_{k=1}^{t}\sigma_e^2\psi_k(\beta^*)\psi_k^{\mathrm{T}}(\beta^*). \end{aligned} \quad \text{(A34)}$$

Let x_k be the first component of $R_k(\beta^*)\hat{\sigma}_k^{-2}F^{-1}(\beta^*)\psi_k(\beta^*)$. To prove that $S_t = \sum_{k=1}^{t} x_k e_k$ converges in law to the normal distribution, we check the two conditions of Lemma A.22, with $V_t^2 = \sum_{k=1}^{t} E(|x_k e_k|^2|\mathcal{F}_{k-1})$ and $s_t^2 = E(V_t^2)$.

First, by Lemma A.21, we have $(\sigma_e^2 - \hat{\sigma}_k^2) = o(t^{-\gamma})$, hence

$$\frac{1}{t}\sum_{k=1}^{t}\sigma_e^6(\sigma_e^2 - \hat{\sigma}_k^2)(\sigma_e^2 + \hat{\sigma}_k^2)\hat{\sigma}_k^{-4}\sigma_e^{-4}\|\psi_k(\beta^*)\psi_k^{\mathrm{T}}(\beta^*)\| \leqslant C\frac{1}{t}\sum_{k=1}^{t}k^{-\gamma}\|\psi_k(\beta^*)\psi_k^{\mathrm{T}}(\beta^*)\|$$

and by Lemma A.11,

$$(R_t(\beta^*) - \sigma_e^2F(\beta^*)) = o(t^{-1/2+\delta}),$$

so by applying Kronecker's Lemma A.10 to Equation (A34), the first three terms of Y_t^2/t converge to 0 a.s. when $t \to \infty$, hence $Y_t^2/t \to \sigma_e^4F(\beta^*)$. It is obvious that $M_t^2/t \to \sigma_e^4F(\beta^*)$ so $V_t^2/s_t^2 \to 1$ in probability when $t \to \infty$, where $s_t/\sqrt{t}$ is the square root of the element (1,1) of M_t^2/t.

To prove the Lindeberg condition, let us consider

$$W_t = \sum_{k=1}^{t}\left(\frac{1}{s_k^2}\right)E|x_ke_k|^2I(|x_ke_k| > \delta s_t).$$

We have $s_k^2 = E(V_k^2) = kC$, where C is a constant. By Tchebychev inequality we obtain

$$P(|x_ke_k| > \delta s_k) \leqslant \frac{1}{|\delta s_k|^2}E|x_ke_k|^2 \leqslant \frac{1}{\delta^2kC}E|x_ke_k|^2.$$

Therefore, since for all $k < t$, $s_k < s_t$

$$W_t \leqslant \sum_{k=1}^{t}\left(\frac{1}{s_k^2}\right)E|x_ke_k|^2E(I(|x_ke_k| > \delta s_k)) \leqslant \frac{1}{\delta^2k^2C^2}(E|x_ke_k|^2)^2,$$

hence, by Lemma A.10 (Lukacs, 1975) and Kronecker's Lemma A.10, $W_t \to 0$ when $t \to \infty$. Hence Lemma A.22 implies convergence in law of $(S_t/\sqrt{t})/(s_t/\sqrt{t})$. In a similar way, we can show that any linear combination of the components of Equation (A33) converges in law to a normal distribution. ■

Lemma A.24 *The series H_t and L_t defined by Equations (50) and (51) converge to 0 a.s. when $t \to \infty$.*

Proof Let us show that L_t converges to 0 a.s. when $t \to \infty$. Consider the series

$$L_t^1 = \sum_{k=1}^{t}\frac{1}{\sqrt{k}}R_k(\beta^*)\hat{\sigma}_k^{-2}(F^{-1}(\hat{\beta}_{k-1}) - F^{-1}(\beta^*))\psi_k(\hat{\beta}_{k-1})e_k + \sum_{k=1}^{t}\frac{1}{\sqrt{k}}R_k(\beta^*)\hat{\sigma}_k^{-2}F^{-1}(\beta^*)(\psi_k(\hat{\beta}_{k-1}) - \psi_k(\beta^*))e_k.$$

L_t^1 is a martingale and

$$\begin{aligned} E\|L_t^1\|^2 \leqslant& \sum_{k=1}^{t}\frac{1}{k}\|R_k(\beta^*)\hat{\sigma}_k^{-2}(F^{-1}(\hat{\beta}_{k-1}) - F^{-1}(\beta^*))\psi_k(\hat{\beta}_{k-1})e_k\|^2 \\ &+ \sum_{k=1}^{t}\frac{1}{k}\|R_k(\beta^*)\hat{\sigma}_k^{-2}F^{-1}(\beta^*)(\psi_k(\hat{\beta}_{k-1}) - \psi_k(\beta^*))e_k\|^2. \end{aligned}$$

We know that $F^{-1}(\beta)$ and $\partial F(\beta)/\partial\beta$ are bounded, then by Lemma A.2, we have that $\partial F^{-1}(\beta)/\partial\beta$ is bounded, then there exists a constant C such that

$$E\|L_t^1\|^2 \leqslant C\sum_{k=1}^{t}\frac{1}{k}E\|\tilde{\beta}_{k-1}\|^2E|e_k|^2.$$

Using Equation (A22) and Lemma A.20, for all γ such that $0 < \gamma < 1/2 - \delta$, $\|t^\gamma\tilde{\beta}_t\|^2 = t^{2\gamma}\|\tilde{\beta}_t\|^2$ converges to 0, hence $E\|L_t^1\|^2$ is bounded. Then L_t^1 is a martingale with a bounded variance, and using Chung (1968, p. 310) L_t^1 converges a.s. to a finite limit when $t \to \infty$, hence $R_t^{-1}(\beta^*)L_t^1$ converges, and by Kronecker's Lemma A.10, L_t converges a.s. to 0 when $t \to \infty$.

Let us now show that $H_t \to 0$ a.s. when $t \to \infty$. Let

$$\begin{aligned} H_t^1 =& R_t^{-1}(\beta^*)\sum_{k=1}^{t}\frac{1}{\sqrt{k}}(\sigma_e^2F(\beta^*) - R_k(\beta^*))\hat{\sigma}_k^{-2}F^{-1}(\hat{\beta}_{k-1})\psi_k\beta^*\psi_k^{\mathrm{T}}(\beta^*)\tilde{\beta}_k \\ &+ R_t^{-1}(\beta^*)\sum_{k=1}^{t}\frac{1}{\sqrt{k}}(\hat{\sigma}_k^2 - \sigma_e^2)F(\beta^*)\hat{\sigma}_k^{-2}F^{-1}(\hat{\beta}_{k-1}) \times \psi_k(\beta^*)\psi_k^{\mathrm{T}}(\beta^*)\tilde{\beta}_{k-1} \\ &+ R_t^{-1}(\beta^*)\sum_{k=1}^{t}\frac{1}{\sqrt{k}}B_{1,k} + R_t^{-1}(\beta^*)\sum_{k=1}^{t}\frac{1}{\sqrt{k}}B_{2,k}, \end{aligned}$$

By Lemma A.11, we know that $\forall\mu > 0$,

$$t^{1/2-\mu}(R_t(\beta^*) - \sigma_e^2F(\beta^*)) = o(1),$$

and since for every γ positive such that $0 < \gamma < \frac{1}{2}$, $t^{\gamma}\tilde{\beta}_k$ converges to 0, then

$$\|k^{-1/2}(\sigma_e^2 F(\beta^*) - R_k(\beta^*))\hat{\sigma}_k^{-2}F^{-1}(\hat{\beta}_{k-1})(\beta^*)\psi_k^{\mathrm{T}}(\beta^*)\tilde{\beta}_k\| \leqslant Ck^{-1+\mu-\gamma}\|\psi_k^{\mathrm{T}}(\beta^*)(\beta^*)\|$$

hence with $\mu < \gamma$ by Lemmas A.12 and A.14

$$R_t^{-1}(\beta^*)\sum_{k=1}^{t}\frac{1}{\sqrt{k}}(\sigma_e^2 F(\beta^*) - R_k(\beta^*))\hat{\sigma}_k^{-2}F^{-1}(\hat{\beta}_{k-1})\psi_k^{\mathrm{T}}(\beta^*)(\beta^*)\tilde{\beta}_k$$

is convergent. In the same manner, we have

$$\frac{1}{\sqrt{k}}\|(\hat{\sigma}_k^2 - \sigma_e^2)F(\beta^*)\hat{\sigma}_k^{-2}F^{-1}(\hat{\beta}_{k-1})\psi_k(\beta^*)\psi_k^{\mathrm{T}}(\beta^*)\tilde{\beta}_{k-1}\| \leqslant Ck^{-1/2}|\hat{\sigma}_k^2 - \sigma_e^2|\|\psi_k(\beta^*)\psi_k^{\mathrm{T}}(\beta^*)\|.$$

by Lemma A.21, $t^{1/2-\mu}(\hat{\sigma}_t^2 - \sigma_e^2) \longrightarrow 0$ a.s. when $t \longrightarrow \infty$, hence

$$\frac{1}{\sqrt{k}}\|(\hat{\sigma}_k^2 - \sigma_e^2)F(\beta^*)\hat{\sigma}_k^{-2}F^{-1}(\hat{\beta}_{k-1})(\beta^*)\psi_k^{\mathrm{T}}(\beta^*)\tilde{\beta}_{k-1}\| \leqslant Ck^{-1+\mu-\gamma}\|\psi_k(\beta^*)\psi_k^{\mathrm{T}}(\beta^*)\|$$

hence

$$R_t^{-1}(\beta^*)\sum_{k=1}^{t}\frac{1}{\sqrt{k}}(\hat{\sigma}_k^2 - \sigma_e^2)F(\beta^*)\hat{\sigma}_k^{-2}F^{-1}(\hat{\beta}_{k-1}) \times \psi_k(\beta^*)\psi_k^{\mathrm{T}}(\beta^*)\tilde{\beta}_{k-1}$$

is convergent.

In the same way as in Lemma A.13, we can show from Equation (46) that $\|B_{1,t}\| < Ca_t\|\tilde{\beta}_t\|^2$, where $(1/t)\sum_{k=1}^{t} a_t$ is finite a.s., and since $\forall\gamma < 1/2$, $t^{2\gamma}\|\tilde{\beta}_t\|^2$ is bounded, then with γ such that $\gamma > \frac{1}{4}$,

$$\sum_{k=1}^{t}\frac{1}{\sqrt{k}}B_{1,k} \leqslant C\sum_{k=1}^{t} a_t k^{-1/2-2\gamma},$$

thus $R_t^{-1}(\beta^*)\sum_{k=1}^{t} B_{1,k}/\sqrt{k}$ converges a.s. Let us now consider $B_{2,t}$ defined by Equation (47).

From Lemma A.12 we know that $tE(\|\psi_t - \psi_t(\hat{\beta}_{t-1})\| \|\varepsilon_t(\hat{\beta}_{t-1})\|)$ and $tE(|\varepsilon_t - \varepsilon_t(\hat{\beta}_{t-1})|\|\psi_t\|)$ are bounded and since $\hat{\sigma}_t^{-2}$ is bounded, then $R_t^{-1}(\beta^*)\sum_{k=1}^{t} E\|B_{2,k}\|/\sqrt{k}$ converges a.s. hence by Lukacs' Lemma A.10 $R_t^{-1}(\beta^*)\sum_{k=1}^{t} B_{2,k}/\sqrt{k}$ converges a.s. Finally H_t^1 is finite. According to Kronecker's Lemma A.10, $H_t \to 0$ a.s. when $t \to \infty$. ■

26

Design and implementation of an open circuit voltage prediction mechanism for lithium-ion battery systems

T. Stockley*, K. Thanapalan, M. Bowkett and J. Williams

Centre for Automotive & Power System Engineering (CAPSE), Faculty of Computing, Engineering and Science, University of South Wales, Treforest, UK

This paper describes an open circuit voltage (OCV) prediction technique for lithium cells. The work contains an investigation to examine the charge and mixed state relaxation voltage curves, to analyse the potential for the OCV prediction technique in a practical system. The underlying principal of the technique described in this paper employs a simple equation paired with a polynomial to predict the equilibrated cell voltage after a small rest period. The polynomial coefficients are devised by the use of curve fitting and system identification techniques. The practical work detailed in this paper was conducted at the Centre for Automotive and Power System Engineering (CAPSE) battery laboratories at the University of South Wales. The results indicate that the proposed OCV prediction technique is highly effective and may be implemented with a simple battery management system.

Keywords: OCV; cell relaxation; lithium technology; battery system performance; prediction mechanism

1. Introduction

This paper investigates the effect of a constant current charge and mixed state duty cycle, and how the cell voltage reacts when in a state of suspension. Previous research work relating to this study proved that a simple equation is adequate for an improved accuracy open circuit voltage (OCV) measurement. The study also provided a detailed explanation of how constant current discharges affect such lithium cell OCV measurements (Stockley, Thanapalan, Bowkett, & Williams, 2013).

The reliance that the world's consumer markets have placed on battery technologies can be seen in the vast amount of applications required today. Typically used in portable electronic devices such as cell phones and small items of equipment (Zhang & Harb, 2013), lithium battery technologies are now becoming more common in the automotive sector (Weinert, Burke, & Wei, 2007) in its quest to power the next generation of clean, green vehicles (Affanni, Bellini, Franceschini, Guglielmi, & Tassoni, 2005). A less obvious application is their integration into many stationary applications. The industrial applications include backup power systems for the telecommunications industry (Lu, Han, Li, Hua, & Ouyang, 2013; Suzuki, Shizuki, & Nishiyama, 2003), and energy storage for renewable energy systems (Li et al., 2012).

There are several large battery technologies being used in today's industry. The following advantages are gained through using lithium technologies over other cell chemistries: (i) a higher cell voltage, which is key to the high energy density; (ii) greatly improved cycle count, with typical figures of 300–400 cycles; (iii) a more consistent manufacturing process between cells of the same type, resulting in more balanced battery modules and (iv) an improved specific energy and energy density. However, the benefits of lithium batteries are counterbalanced by several drawbacks: (i) high initial cost – although prices are reducing with increased high volume production; (ii) costly electronics for the battery management system (BMS) to protect the cells; (iii) increased risk of overheating and fire due to the high energy, albeit this is mitigated by the use of a BMS and safer cell chemistries. Failure to implement a BMS or the consequences of error can result in overheating, fire or explosion such as the well-publicised incidents in 2006 (Sima, 2006).

The function of the BMS extends beyond preventing damage to the lithium cells. BMSs also contribute to the advantages of using the lithium technology. The BMS monitors the voltage level of each of the cells in the battery pack to ensure that the cells are well balanced, which increases the performance of the battery pack (Bowkett, Thanapalan, Stockley, Hathway, & Williams, 2013). A key element of the BMS is estimation of the state of charge (SoC) of the battery. The SoC indicates what percentage of the battery capacity has been used so that the user can recharge the cells when necessary. The most common SoC estimation techniques are referred to as follows.

*Corresponding author. Email: thomas.stockley@southwales.ac.uk

Coulomb counting is the most basic capacity measurement technique and is typically implemented in the majority of BMSs available to date. The method works by counting the Amp-hours (Ah) in and out of a cell/battery and then calculating how many Ah are left as a percentage of the initial capacity (Ng, Moo, Chen, & Hsieh, 2009). Although extremely simple to implement and requiring very little processing space and time, this method does have two major disadvantages: (i) the initial SoC of the cell/battery must be known or calculation errors occur and (ii) the capacity of the cell must be known, and this value changes with both temperature and age.

A spectrum of frequencies is used by electrochemical impedance spectroscopy (EIS) methods to measure the internal resistance and the impedance of the cell. By measuring these factors, the SoC of the cell can be obtained as (Salkind, Fennie, Singh, Atwater, & Reisner, 1999) has for NiMH cells. Whilst this method is very accurate, the cell must be fully equilibrated, requiring an unused state of between 2 and 6 hours for lithium cells prior to measurement. It is, therefore, impractical for real-time use in many applications.

The OCV of a cell can be used to find the SoC of the cell due to its relationship (Chiang, Sean, & Ke, 2011). This method can accurately estimate the SoC; however, research indicates that the OCV is temperature reliant (Pop et al., 2006). It has been shown by Roscher and Sauer (2011) that OCV is prevented from being the same after a charge and discharge for the same SoC by hysteresis. The introduction of a recovery factor to decide whether the cell has performed a charge or discharge has reduced the error in OCV–SoC measurements (Roscher & Sauer, 2011). As with the EIS method, a long rest period is required because the cell needs to be equilibrated for an accurate measurement to be taken. Work has been completed by Aylor, Thieme, and Johnson (1992) on an OCV prediction technique which estimates what the OCV will be after 3 hours, from a single measurement taken at 30 minutes, allowing the OCV–SoC method to be used practically.

Aylor et al. (1992) provided two methods of OCV prediction for lead acid batteries. The techniques estimate what the OCV will be after 3 hours, from a single measurement taken at 30 minutes, allowing the OCV–SoC method to be used practically. In their work two methods are identified: (i) a simple summing method is used to quickly calculate the SoC of the battery. This method was disregarded due to the varying performance of lead acid batteries even from the same production batch. The simple summing method provided disappointing results for Aylor et al. when used on lead acid batteries. However, Stockley et al. (2013) showed very promising results when the summing method was applied to two different chemistry-based lithium ion cells. (ii) A more complex method is provided to eliminate the problems encountered in the first method. This method uses asymptotes on a logarithmic scale and proves viable with errors in SoC estimation of less than 5% and a prediction time of just 6.6 minutes.

The asymptote method has been proven to work on lithium cells as well as lead acid batteries (Pop, Bergveld, Danilov, Regiten, & Notten, 2008). This method was used to validate the model proposed by Pop et al. (2008), but in doing so was effectively utilised when adapted to lithium cells. Three different methods were used throughout the paper, with the asymptote method giving an error of just 0.92% SoC, an improvement from the 20.19% SoC provided by a combination of a voltage change model and a temperature model. However, the asymptote method was trumped in accuracy by the model developed by Pop et al., which works on the voltage relaxation accounting for temperature, SoC, charge/discharge rate and age of the cell. The accuracy of the model by Pop et al. achieved an error of 0.19% SoC.

The work conducted by Weng, Sun, and Peng (2013) uses a new sigmoid function approach to provide a model capable of predicting the OCV of a lithium cell. The author compared the proposed model to several polynomial based models to good effect with an error of 2.5, 4.8 mV lower than the most accurate polynomial model. The SoC comparison and also SoH comparison are proven to be effective when combined with an extended Kalman filter (EKF).

Pei, Wang, Lu, and Zhu (2014) uses a combination of practical tests on a lithium iron phosphate cell and a theoretical second-order resistor capacitor equivalent circuit to show that the long relaxation time is due to the diffusion process. The diffusion process was then proven to have a linear relationship with the cell OCV. By using this information, a linear regression model has been created to determine the cell conditions before a voltage relaxation model can be used for the OCV prediction.

Polynomials have been used to reduce the relaxation time before OCV can be measured for both lithium iron phosphate and lithium manganese based cells in the work by Hu, Li, Peng, and Sun (2012). In their work the polynomials are used to model the curve of a lithium cell relaxation and then an EKF is used for the estimation of the SoC. This work paves the way for the work conducted in the remaining sections of this research work.

The remaining sections of this paper are as follows: Section 2 provides a brief explanation of the cell modelling and the simple equation used for the OCV prediction tests, and Section 3 explains how the tests were conducted. A summary of the background work is also presented in Section 3, followed by the constant current charge test results. Section 4 presents the mixed state relaxation test and implementation of the prediction mechanism. Section 4 also hosts a brief discussion of the polynomial which will allow rapid OCV prediction. Finally, a discussion of the results and the conclusions are presented at the end of the paper.

2. Modelling of a lithium cell

The OCV analysis carried out in the previous work has indicated that in order to obtain an accurate OCV measurement of a lithium cell, the cell needs to be in a fully equilibrated state. This is because the OCV is influenced by several factors, including cell chemistry, age of cell, cell temperature, charge/discharge rate and cell voltage characteristics. As has been previously mentioned, waiting for a cell to become equilibrated is a large disadvantage for the OCV–SoC estimation method because real world applications can rarely be halted for up to 3 hours for a measurement to be taken. Although several methods of OCV prediction were presented in the Introduction, it is thought that these methods are complicated and could prove too intense for simple BMS applications. Therefore, this work aims to develop a robust yet simple strategy to address the problems caused by the uncertainty that is associated with OCV prediction.

The proposed method comprises two sections: the first is to determine the OCV of the battery during relaxation, by using probabilistic methods such as what-if predictions. The second part is to determine the equilibrated OCV. Accurate prediction of OCV after a 30-minute relaxation period has been proven by Stockley et al. (2013), and will be incorporated in this work.

A mathematical model will be developed and incorporated to the original model developed by Stockley et al. (2013), resulting in a cascade-connected nonlinear model with a complex probabilistic component and a deterministic component.

Thus, the model will have the following form:

$$f = f_1 + f_2, \quad (1)$$

$$f_1 = \phi(t, \mathrm{soc}, c, i), \quad (2)$$

where $\phi = \delta x^2 + \gamma x + \vartheta$, ϑ is the relaxation start voltage, and δ and γ are the polynomial coefficients. The function f_1 is dependent upon the temperature (t), the SoC, the age of the cell (c) and the charge/discharge rate (i). All of the aforementioned parameters can be identified by monitoring the charge/discharge curve and by extrapolation of the battery voltage sampled during initial relaxation, or by the use of estimation algorithms. For example, Hu et al. (2012) used the linear regression model to predict the OCV. Function 1 is explored further in Section 4.

$$f_2 = V_{\mathrm{tr}} \pm K_{\mathrm{v}}, \quad (3)$$

where V_{tr} is the voltage at a known measurement interval (8 minutes in this work, 30 minutes in Stockley et al. (2013)) and K_{v} is a predefined constant derived from the equation $V_{\mathrm{OC}} - V_{\mathrm{tr}}$. Function 2 is the focus of this work and is found in Sections 3 and 4.

3. Measurement and monitoring

This section describes the test set-up, results and analysis of two types of lithium cell. It aims to prove that Equation (3) will be effective for both pouch and cylindrical lithium cells. The two types of cells that have been chosen for the tests in this section of work are the $LiNiMgCO_2 \cdot 20$ Ah pouch cell by Energy Innovation Group (EIG) and the $LiFePO_4 \cdot 8$ Ah cylindrical cell by Lifebatt.

3.1. *Experimental set-up*

To gain a greater knowledge of the relaxation curves of lithium cells, tests were conducted on two of the most common types of lithium cells, cylindrical and pouch type. Figures 1 and 2 show the pouch cell on test and a block diagram of the test set-up, respectively. As previously stated, the tests were conducted in the Centre for Automotive and Power System Engineering (CAPSE) laboratories at the University of South Wales. An industry standard cell tester unit was used to charge/discharge the cell with a temperature logger to monitor the ambient temperature.

3.2. *Similarity between cells*

For the implementation of the OCV prediction mechanism, the cells had to be proven to have a "clone-like" similarity between two cells of the same chemistry and type. To achieve this, tests were carried out (these were a 0.5C capacity test and a 0.3C relaxation test) which could be used to compare the two pouch cells. The cells used for these tests were $LiNiMgCO_2$ pouch cells. Figure 3 shows the results of the 0.5C capacity tests.

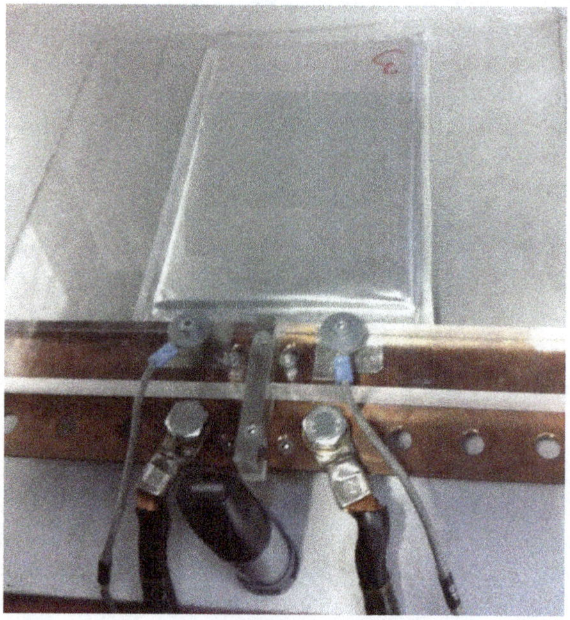

Figure 1. Testing of lithium pouch cell.

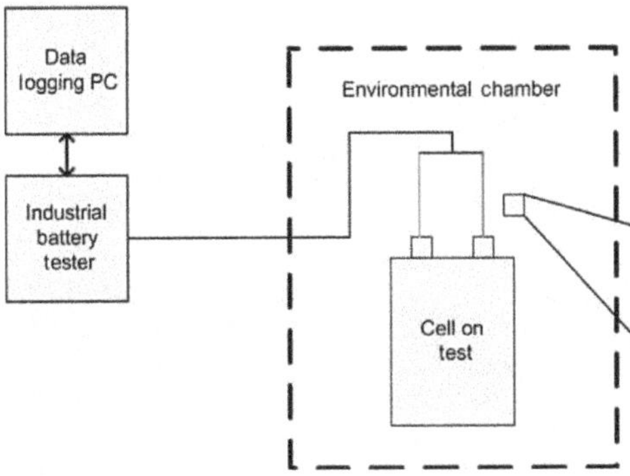

Figure 2. Block diagram of test equipment.

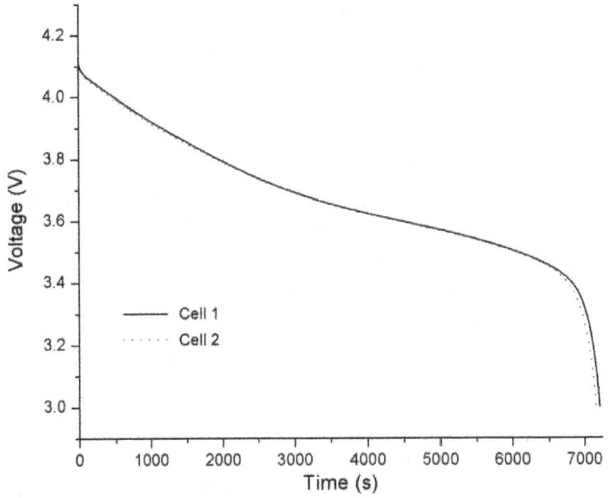

Figure 3. 0.5C discharge curves for pouch cell comparison.

As it can clearly be seen in Figure 3, the cells are almost a clone of each other throughout the discharge until the "knee" is reached. As Tremblay and Dessaint (2009) shows, the "knee" occurs during the final 20% of the discharge when the cell is placed under the most stress, and therefore, the cycle life is reduced. In Figure 4 the discharge voltage and the capacity of the cell are displayed, showing that the "knee" occurs within the last 10% SoC (less than 2 Ah). This means that even though there is a slight discrepancy between the two cells' data, use of the cell after this point should be avoided and, therefore, the results are largely irrelevant. Guena and Leblanc (2006) concur with Shim and Striebel (2003) and the concept that a cell's cycle life can be greatly improved by only using the cell within 70% of capacity.

Figure 5 shows a comparison of the relaxation voltages between the two $LiNiMgCO_2$ pouch cells to further highlight the clone-like effect. The relaxation tests were carried out at 80%, 60%, 40% and 20% SoC. This was so that a broad spectrum of results could be obtained that would represent the full capacity range of the cell. The relaxation curves were not measured at 0% SoC because the cells are rarely used below 20% SoC as mentioned earlier.

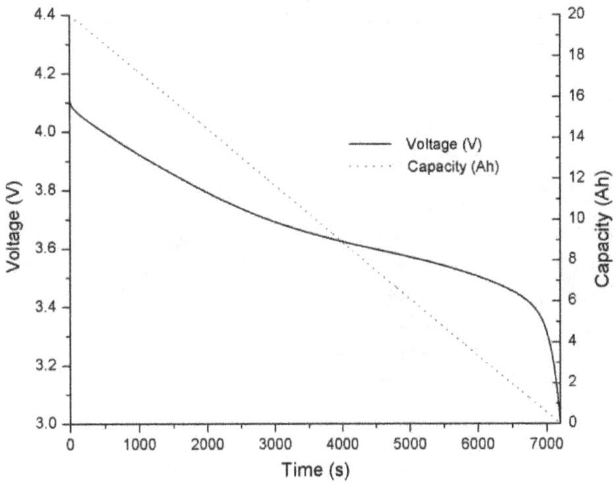

Figure 4. Cell capacity vs. voltage.

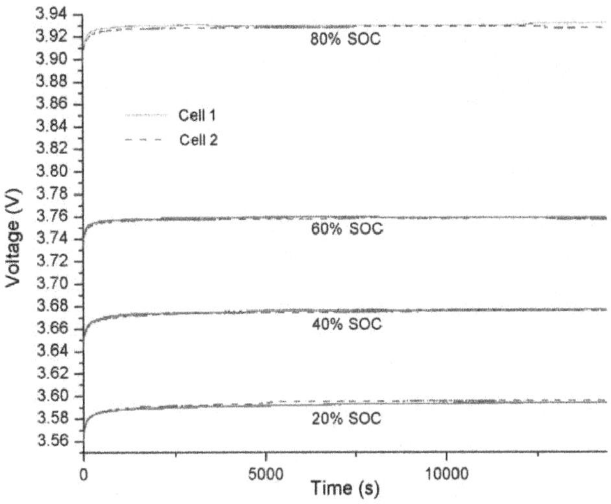

Figure 5. Relaxation curves following a 0.3C discharge for pouch cell comparison.

3.3. Cell relaxation testing discharge state

The cells' relaxation performance following a discharge state can be seen in Stockley et al. (2013), which also highlights the potential for an improved OCV prediction mechanism with very low prediction errors. This section presents a brief summary of the cell relaxation testing following a constant current discharge state.

The relaxation tests were conducted by discharging the cell at 20% SoC intervals, followed by a 4-hour open circuit period. The OCV was measured during the 4-hour rest period at a sample rate of 0.1 s to ensure the accuracy of the relaxation curves. The relaxation curves were measured at 80%, 60%, 40% and 20%, following a 0.3C, 1C and 3C discharge.

To ensure that the proposed OCV prediction technique was not limited to a particular type of cell, the tests were carried out on two types of cells. The two types of cells were not only two different package styles (pouch and

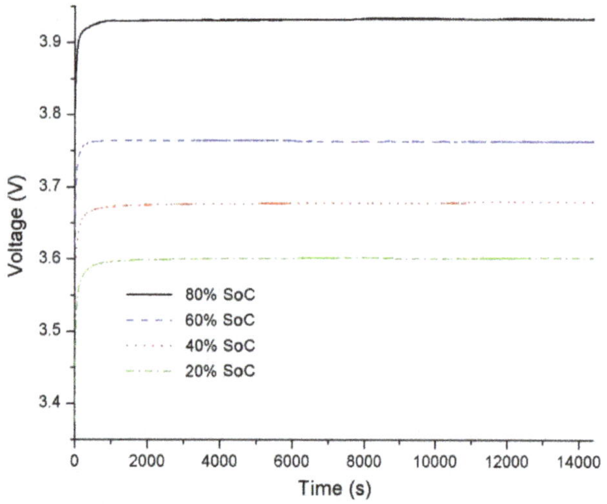

Figure 6. Relaxation curves of lithium pouch cell following a 1C discharge.

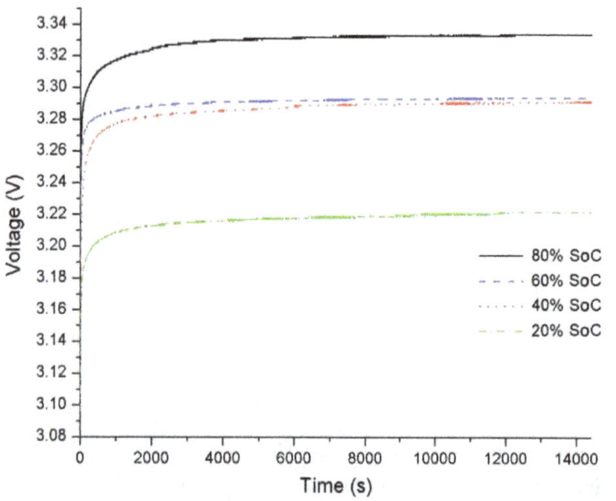

Figure 7. Relaxation curves of lithium cylindrical cell following a 1C discharge.

cylindrical), but also two different types of cell chemistries. The two different types of cell chemistries allowed the prediction technique to be conducted on different capacities and different operating voltages. Figures 6 and 7 show the discharge relaxation curves obtained for the pouch cell and the cylindrical cell, respectively.

The differences in the shape of the curves between Figures 6 and 7 were noted as being caused by the characteristic discharge curve from each type of cell. The difference in these discharge curves can be seen in Figure 8. As can be observed in Figure 8, the $LiNiMgCO_2$ cell has a very steep curve initially with a flat profile towards the end of discharge, as can be highlighted by the large gap between the 80% and 60% SoC curves in Figure 6. The $LiFePO_4$ cell has the opposite characteristics, as can be noted by the large difference between the 40% and 20% SoC curves in Figure 7.

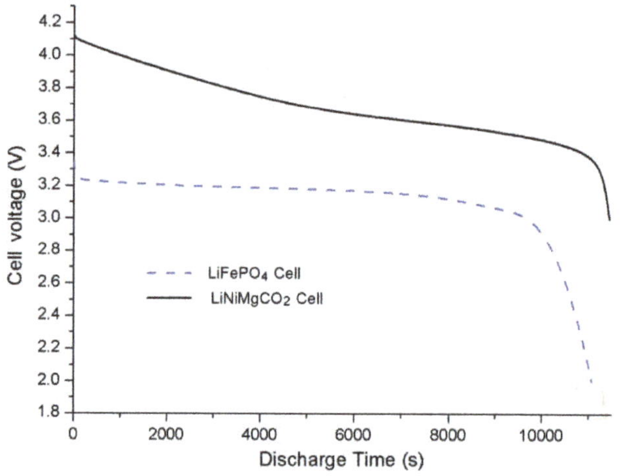

Figure 8. Characteristic discharge curves of $LiNiMgCO_2$ (pouch) and $LiFePO_4$ (cylindrical) cells.

Table 1 is a summary of the results for the discharge tests. The table shows the measured OCV voltage after a 3-hour rest state for each of the test curves. In addition it shows the calculated 3-hour OCV from Equation (3) and the error between the calculated OCV and the measured OCV. These data are supplied for both the pouch cell and the cylindrical cell.

As the test uses constant discharges, the relaxation curve will be positive. This means that Equation (3) had to be used with the addition of the constant K_v. The constant K_v was derived from calculating the average difference between the 30-minute OCV measurement and the 3-hour OCV measurement for each of the test curves. Therefore, the chosen K_v values were 0.002 and 0.01 for the pouch and cylindrical cells, respectively. The OCV prediction technique proved to be successful with a maximum error of 3 mV and 5 mV for the pouch and cylindrical cells, respectively. As is derived from alternate tests, the pouch cells have a SoC–OCV relationship of 1% SoC equals 9.9 mV. Therefore the pouch cell resulted in a SoC error of less than 1%. This is a vast improvement on the 5% error recorded by Aylor et al. (1992).

3.4. *Cell relaxation testing charge state*

The success of the OCV prediction mechanism following a constant current discharge has led to a continuation of research into the relaxation curves following a constant current charge state, and also into a quicker OCV prediction to make the system more practical in real world applications. To conduct the research into the charge relaxation curves, the cell was discharged down to 0% SoC and then charged in 20% steps to 80% SoC as can be seen in Figure 9.

As with the discharge tests in Section 3.3, the tests were conducted at 0.3C, 1C and 3C. The $LiNiMgCO_2$ pouch cell

Table 1. Comparison of measured OCV and calculated OCV after a 3-hour rest period for discharge tests.

	Pouch cell			Cylindrical cell		
	180 minute (real)	180 minute (calc)	Error (mV)	180 minute (real)	180 minute (calc)	Error (mV)
0.3C 80% SoC	3.929	3.929	0	3.328	3.332	4
0.3C 60% SoC	3.759	3.759	0	3.293	3.296	3
0.3C 40% SoC	3.675	3.676	1	3.287	3.288	1
0.3C 20% SoC	3.593	3.595	2	3.21	3.212	2
1C 80% SoC	3.934	3.937	3	3.333	3.333	0
1C 60% SoC	3.766	3.766	0	3.293	3.298	5
1C 40% SoC	3.679	3.681	3	3.289	3.29	1
1C 20% SoC	3.599	3.601	2	3.219	3.22	1
3C 80% SoC	3.932	3.933	1	3.333	3.333	0
3C 60% SoC	3.766	3.763	3	3.294	3.298	4
3C 40% SoC	3.677	3.678	1	3.291	3.291	0
3C 20% SoC	3.601	3.601	0	3.221	3.222	1

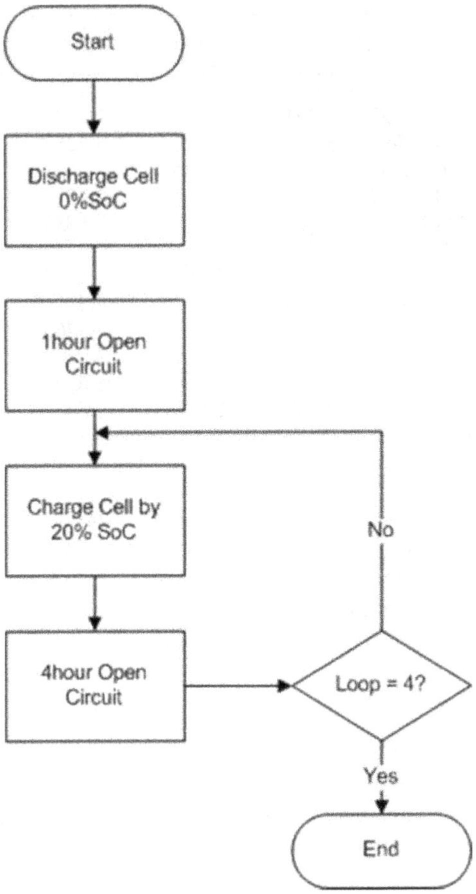

Figure 9. Charge test programme flowchart.

results can be seen in Table 2. Figure 10 shows the four relaxation curves for the 3C charge relaxation test.

The most notable difference between the discharge and charge curves (Figures 6 and 10) is the fact that the charge results in a declining relaxation curve. This is as expected from first principles, as the charge voltage needs to be higher than the cell voltage to reverse the electron flow. Therefore, the voltage falls from the charging voltage to the OCV when the cell charger is removed. The larger voltage difference between 80% and 60% curves in comparison to that of the other curves is expected from the OCV–SoC relationship, as can be seen in Figure 4.

From the relaxation curves in Figure 10 an observation was made that the curve became linear for the $LiNiMgCO_2$ cells after 8 minutes. Therefore, during analysis the voltage measurement taken for the value V_{tr} in Equation (3) was the 8-minute measurement. As the voltage is falling after a charge state, the constant K_v is subtracted from the measurement V_{tr}. To ensure that the 8-minute prediction interval was optimum for the type of cell, the relationship between the error and the prediction time was calculated. Figure 11 shows the error vs. relaxation time for the 0.3C charge test at 80%, 60%, 40% and 20% SoC.

Similarly to the relaxation tests in Section 3.3, the constant K_v was derived from the average of the difference between the 8-minute and 180-minute measurements for each of the relaxation curves. Therefore, for the charging relaxation test, the constant K_v was calculated as 0.0087. The following worked example shows the calculation for a relaxation curve following a 0.3C charge at 60% SoC.

$$V_{OC} = V_{tr} - K_v$$

$$V_{OC} = 3.737V - 0.0087$$

$V_{OC} = 3.728V$ is obtained. From the experimental tests conducted,

$V_{real} = 3.729V$ and therefore the error is 0.7 mV, as shown later.

Table 3 provides a comparison of the voltage measurements at the 180-minute interval and the calculated voltage at 180 minutes for all charge relaxation curves. It can be noted that the maximum calculated error for the pouch cells during a charge test was 8.3 mV. As mentioned in Section 3.3, it has been established that for each of the $LiNiMgCO_2$ cells, 9.9 mV of the OCV represents a SoC of

Table 2. Voltage measurements from pouch cell charge relaxation tests.

Cell test	0 (minute)	8 (minute)	30 (minute)	60 (minute)	120 (minute)	180 (minute)
0.5C 80% SoC	3.92	3.891	3.889	3.888	3.887	3.887
0.5C 60% SoC	3.775	3.737	3.732	3.731	3.73	3.729
0.5C 40% SoC	3.702	3.671	3.668	3.667	3.666	3.665
0.5C 20% SoC	3.632	3.594	3.588	3.586	3.585	3.584
1C 80% SoC	3.948	3.892	3.889	3.888	3.888	3.888
1C 60% SoC	3.786	3.73	3.72	3.719	3.718	3.718
1C 40% SoC	3.72	3.665	3.661	3.66	3.658	3.658
1C 20% SoC	3.625	3.586	3.578	3.576	3.573	3.572
3C 80% SoC	4.027	3.885	3.882	3.883	3.883	3.883
3C 60% SoC	3.904	3.737	3.724	3.721	3.72	3.72
3C 40% SoC	3.815	3.668	3.661	3.66	3.66	3.66
3C 20% SoC	3.736	3.587	3.578	3.576	3.574	3.574

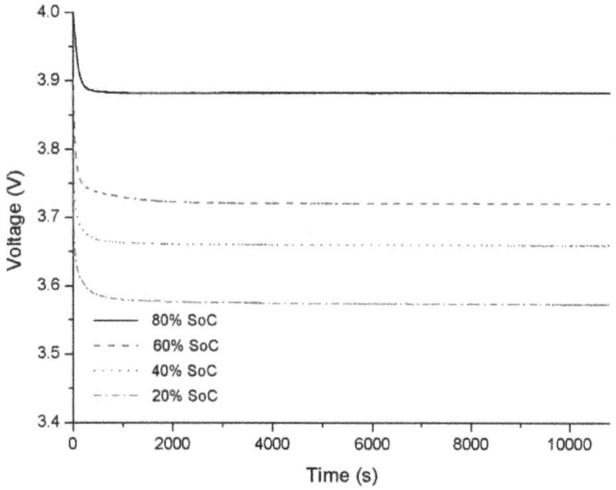

Figure 10. Relaxation curves of lithium pouch cell following a 3C charge.

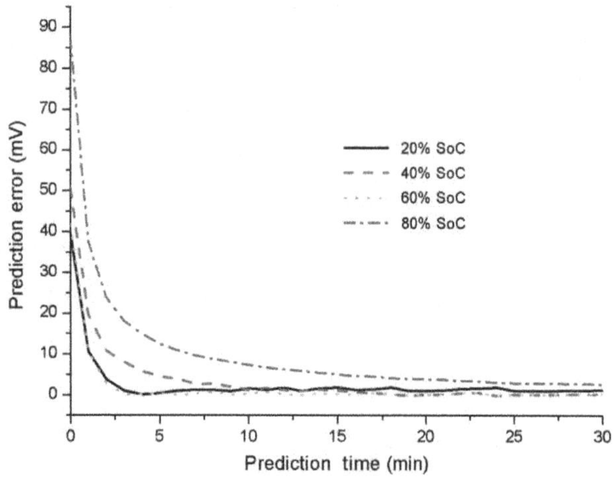

Figure 11. Prediction error vs. relaxation time.

1%. Therefore, a maximum error of just 8.3 mV equals an error of less than 1% SoC at a prediction time of 8 minutes.

This SoC estimation technique also works on the cylindrical $LiFePO_4$ cells, as can be seen by the comparison of

Table 3. Calculation results of Equation (3) for pouch cells during a constant current charge test.

Cell test	8 minute	180 minute (real)	180 minute (calc)	Error (mV)
0.5C 80% SoC	3.891	3.887	3.8823	4.7
0.5C 60% SoC	3.737	3.729	3.7283	0.7
0.5C 40% SoC	3.671	3.665	3.6623	2.7
0.5C 20% SoC	3.594	3.584	3.5853	1.3
1C 80% SoC	3.892	3.888	3.8833	4.7
1C 60% SoC	3.73	3.718	3.7213	3.3
1C 40% SoC	3.665	3.658	3.6563	1.7
1C 20% SoC	3.586	3.572	3.5773	5.3
3C 80% SoC	3.885	3.883	3.8763	6.7
3C 60% SoC	3.737	3.72	3.7283	8.3
3C 40% SoC	3.668	3.66	3.6593	0.7
3C 20% SoC	3.587	3.574	3.5783	4.3

the measured and calculated results in Table 5. The same procedure was followed as in the pouch cell tests, and the results of the test can be seen in Table 4.

As Table 5 shows, the charge relaxation tests proved to be effective, with a maximum error of just 11 mV.

4. Mixed state results and implementation

With the success of the discharge and charge tests, a test plan was conducted to allow the pouch cell to be used in a set of mixed states. This test ensured that the OCV prediction technique could be used in a practical system and not just be reserved for charge or discharge only cycles. The test profile can be seen in Figure 12. A positive current represents that the cell is in a charge state and a negative current represents the discharge state.

A notable feature of the profile in Figure 12 is the long open circuit states (periods of 0A). These periods are where the relaxation voltage was measured, and represent the cell at 80%, 60%, 40% and 20% SoC. The relaxation curves measured during these periods can be seen in Figure 13. The charge/discharge current used for this test was ±10A (0.5C).

Table 4. Voltage measurements from cylindrical cell charge relaxation tests.

Cell test	0 minute	8 minute	30 minute	60 minute	120 minute	180 minute
0.5C 80% SoC	3.468	3.344	3.333	3.332	3.331	3.331
0.5C 60% SoC	3.414	3.315	3.307	3.304	3.303	3.302
0.5C 40% SoC	3.385	3.312	3.302	3.3	3.299	3.228
0.5C 20% SoC	3.326	3.247	3.324	3.232	3.23	3.228
1C 80% SoC	3.409	3.341	3.334	3.332	3.332	3.331
1C 60% SoC	3.373	3.313	3.308	3.305	3.304	3.301
1C 40% SoC	3.351	3.309	3.304	3.302	3.3	3.299
1C 20% SoC	3.289	3.244	3.237	3.233	3.231	3.23
3C 80% SoC	3.394	3.339	3.337	3.336	3.332	3.332
3C 60% SoC	3.357	3.314	3.309	3.307	3.305	3.304
3C 40% SoC	3.342	3.307	3.306	3.304	3.302	3.3
3C 20% SoC	3.279	3.242	3.24	3.237	3.235	3.234

Table 5. Calculation results of Equation (3) for cylindrical cells during a constant current charge test.

Cell test	8 minute	180 minute (real)	180 minute (calculated)	Error (mV)
3C 80% SoC	3.344	3.331	3.331	0
3C 60% SoC	3.315	3.302	3.302	0
3C 40% SoC	3.312	3.288	3.299	11
3C 20% SoC	3.247	3.228	3.234	6
1C 80% SoC	3.341	3.331	3.328	3
1C 60% SoC	3.313	3.301	3.3	1
1C 40% SoC	3.309	3.299	3.296	3
1C 20% SoC	3.244	3.23	3.231	1
0.3C 80% SoC	3.339	3.332	3.326	6
0.3C 60% SoC	3.314	3.304	3.301	3
0.3C 40% SoC	3.307	3.3	3.294	6
0.3C 20% SoC	3.242	3.234	3.229	5

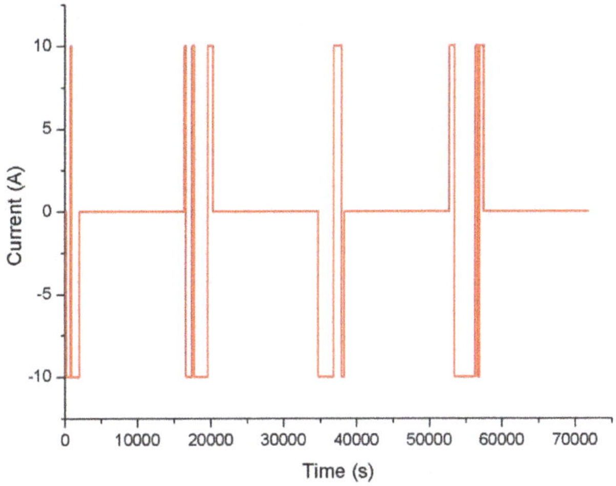

Figure 12. Mixed state test profile.

From both Figures 12 and 13, it can be seen that the profile is split into two relaxation curves following a brief charge state (60% and 20%) and two relaxation curves following a brief discharge state (80% and 40%). This is why two of the relaxation curves fall and two rise.

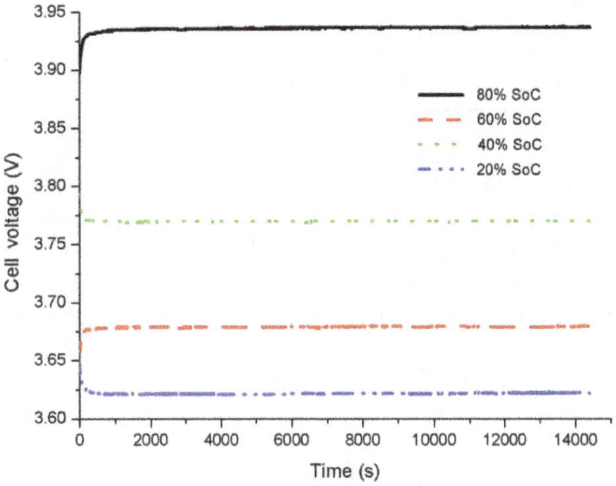

Figure 13. Relaxation curves of lithium pouch cell following a mixed state test profile.

The measurement values from the relaxation curves in Figure 13 are summarised in Table 6.

To test whether Equation (3) (function 2) worked with a mixed state profile, the 8-minute charge and discharge constants (K_v) that were calculated for the charge and discharge relaxation tests were used. Therefore the constants used were 0.0083 if the cell is relaxing from a discharge and 0.0087 if the cell is recovering from a charge state. A comparison of the measured and calculated cell voltages at the 180-minute interval is provided in Table 7.

Table 7 shows that the OCV prediction mechanism can be successfully applied to a cell during a mixed state test. This is an important finding as constant charge/discharge curves are rarely seen in practical applications. The maximum error of 13.3 mV shows that the constants (K_v) used for the constant discharge and charge tests can be applied to both constant states and mixed states with a SoC error of less than 1.5%.

Figure 14 shows a proposed design for the implementation of the OCV prediction technique into a real world BMS. BMSs currently monitor both cell voltage and the current. This enables the block diagram in Figure 14 to be

Table 6. Relaxation voltage measurements for pouch cells during mixed state test.

	OCV (V)					
Cell test	0 minute	8 minute	30 minute	60 minute	120 minute	180 minute
0.5C 80% SoC	3.897	3.903	3.935	3.936	3.937	3.937
0.5C 60% SoC	3.794	3.789	3.77	3.77	3.77	3.77
0.5C 40% SoC	3.656	3.66	3.679	3.679	3.679	3.68
0.5C 20% SoC	3.649	3.644	3.621	3.621	3.622	3.622

Table 7. Comparison of calculated 3-hour voltage value and measured 3-hour voltage value for mixed state test.

Cell test	8 minute	180 minute (real)	180 minute (calc)	Error (mV)
0.3C 80% SoC	3.935	3.937	3.9433	6.3
0.3C 60% SoC	3.789	3.77	3.7803	10.3
0.3C 40% SoC	3.66	3.68	3.6683	11.7
0.3C 20% SoC	3.644	3.622	3.6353	13.3

incorporated without major hardware additions. The design can be simply programmed into a small microcontroller.

As aforementioned, the cell voltage and cell current are already known by the BMS as they are both used for alternative functions and safety and are performance based. The cell voltage can easily be stored after an open circuit state of 8 or 30 minutes (OCV measurement block). The start of the open circuit state can be derived from the cell current. The cell current is also monitored by a constant selector. This block works out if the cell has just finished a charge or discharge state, and then chooses the required constant value. For example if the 30-minute measurement was required for higher accuracy, in the case of the pouch cell this would be 0.0026 for the charge state and −0.002 for the discharge state. The K_v constant value would then be summed to the 30-minute OCV measurement to calculate the predicted OCV value after a 3-hour rest state.

Although an OCV prediction time of less than 8 minutes would be practically acceptable in an application such as stand-alone photo-voltaic systems where the battery would not be in constant use, work was conducted to reduce the prediction time to make the system transferable to alternative applications. For this reason, the relaxation curves were studied immediately following the charge or discharge state.

As mentioned in Section 2, a second-order polynomial function has been derived from the results, and takes the following form:

$$\phi = \delta x^2 + \gamma x + \vartheta,$$

where

$$-0.0000467 < \delta < -0.0005$$

$$-0.00865 < \gamma < 0.0242.$$

From the analysis of the polynomials and by the use of simultaneous equations, it can be noted that as the temperature is increased from 25°C to 45°C, the coefficient altered to

$$-0.000046 < \delta < -0.0005$$

$$-0.0085 < \gamma < 0.0235.$$

The polynomial was also calculated for a $LiNiMgCO_2$ pouch cell which has been aged to a capacity of 17.2 Ah from 20.8 Ah. The coefficient values for the aged cell are calculated as follows:

$$-0.000046 < \delta < -0.0005$$

$$-0.0085 < \gamma < 0.0236.$$

However, the coefficients calculated for the temperature change and cell ageing still fall within the range calculated for a new cell at 25°C.

5. Performance analysis

The ability to estimate the equilibrated OCV at an interval as little as 8 minutes allows the OCV–SoC method to be used practically in BMSs (Aylor et al., 1992). Furthermore, it is important to note that due to the simplicity of the method described in this work, the equation used in this paper to calculate the equilibrated OCV is so simple that it can be easily implemented into a small microcontroller, as described in the previous section. The results of this paper show that the prediction mechanism works exceptionally well with both charge and discharge states, with a maximum error of 8.3 and 8.25 mV error for the discharge and charge tests, respectively. With a voltage of 9.9 mV resulting in a SoC change of 1%, the simple summation method of function 2 (Equation (3)) is a significant improvement on the ±5% achieved by Aylor et al. (1992). Furthermore, the author of Aylor et al. (1992) used a more complicated mathematical model because of the high errors incurred using Equation (3). The results also show that although the error is very small, there is a difference between the pouch cells and the cylindrical cells, with the cylindrical cells providing a larger error for both the charge and discharge tests. This shows that the constant K_v needs to be calculated for each type of cell prior to cell use. When applied to a mixed

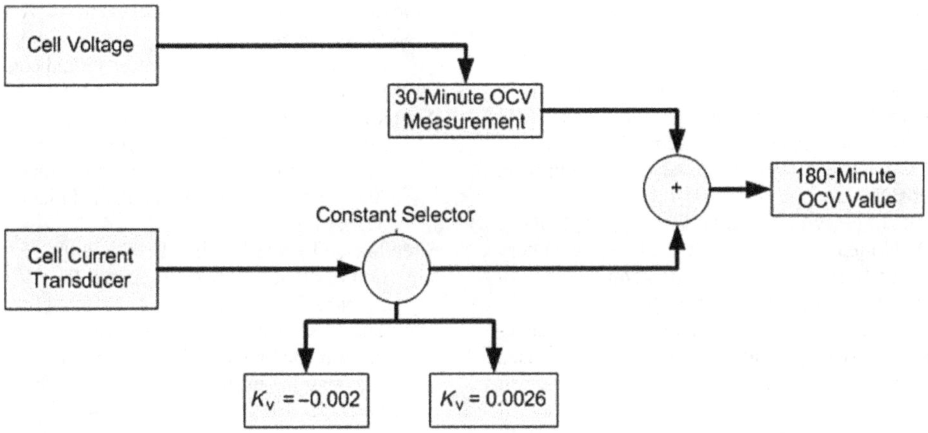

Figure 14. Block diagram of the proposed OCV predictor implementation.

state duty cycle in Section 4, the error was slightly greater but still resulted in a SoC error of less than 1.5%, further proving the function's effectiveness.

A notable point is that although the error is very low at 8 minutes, an OCV prediction at the 30-minute interval as in the original work in Stockley et al. (2013) results in even lower error rates. The maximum error of 8.25 mV for the discharge test and 8.3 mV for the charge test reduces to 3 and 6 mV, respectively. Although the use of the polynomial reduces the possible prediction interval, the error is increased. This is a case for future BMSs in the field of stand-alone PV–lithium hybrid systems to make use of all 3 prediction points. By using all the three methods in a single BMS, the SoC can always be calculated using its most accurate (longest prediction time) value.

6. Conclusion

This paper presents a simple but effective methodology to predict the OCV after a small rest period. It used a simple equation to predict the OCV. Previous work had proved that lithium cells had a unique "clone-like" similarity, which is the key for the use of Equation (3), and the high success that occurred during the testing with constant current discharging.

By performing the relaxation tests discussed in Sections 3 and 4, the relaxation curves of several $LiNiMgCO_2$ pouch cells were recorded, so that the equation could be applied following a constant current charge and also during the rest periods of a mixed state duty cycle. By using the OCV at the 8-minute interval, the equilibrated voltage was successfully estimated with a maximum error of just 8.3 mV for the charge tests and 8.25 mV for the mixed state test. The constant current charge tests were also conducted on a set of $LiFePO_4$ cylindrical cells, proving that Equation (3) could be used on a wide variety of lithium cells with a maximum error of 11 mV. In addition to this, possible implementation of this system into a BMS is also described.

Acknowledgements

The first author would like to acknowledge the financial support from KESS, RUMM Ltd. and the University of South Wales during this research project. The authors would also like to thank the staff of CAPSE for their assistance.

References

Affanni, A., Bellini, A., Franceschini, G., Guglielmi, P., & Tassoni, C. (2005). Battery choice and management for new-generation electric vehicles. *IEEE Transactions on Industrial Electronics, 52* (5), 1343–1349.

Aylor, J. H., Thieme, A., & Johnson, B. W. (1992). A battery state-of-charge indicator for electric wheelchairs. *IEEE Transactions on Industrial Electronics, 39* (5), 398–409.

Bowkett, M., Thanapalan, K., Stockley, T., Hathway, M., & Williams, J. (2013). *Design and implementation of an optimal battery management system for hybrid electric vehicles.* 19th international conference on automation and computing (ICAC), London, UK (pp. 213–217).

Chiang, Y. H., Sean, W. Y., & Ke, J. C. (2011). Online estimation of internal resistance and open-circuit voltage of lithium-ion batteries in electric vehicles. *Journal of Power Sources, 196* (8), 3921–3932.

Guena, T., & Leblanc, P. (2006). *How depth of discharge affects the cycle life of lithium-metal-polymer batteries.* Telecommunications Energy Conference, INTELEC '06. 28th Annual International, Providence, RI, USA (pp. 1–8).

Hu, X., Li, S., Peng, H., & Sun, F. (2012). Robustness analysis of state-of-charge estimation methods for two types of Li-ion batteries. *Journal of Power Sources, 217*, 209–219.

Li, X., Hui, D., Xu, M., Wang, L., Guo, G., & Zhang, L. (2012). *Integration and energy management of large-scale lithium-ion battery energy storage station.* 2012 15th international conference on electrical machines and systems (ICEMS), Sapporo, Japan (pp. 1–6).

Lu, L., Han, X., Li, J., Hua, J., & Ouyang, M. (2013). A review on the key issues for lithium-ion battery management in electric vehicles. *Journal of Power Sources, 226*, 272–288.

Ng, K. S., Moo, C.-S., Chen, Y.-P., & Hsieh, Y.-C. (2009). Enhanced coulomb counting method for estimating state-of-charge and state-of-health of lithium-ion batteries. *Applied Energy, 86* (9), 1506–1511.

Pei, L., Wang, T., Lu, R., & Zhu, C. (2014). Developement of a voltage relaxation model for rapid open-circuit voltage prediction in lithium-ion batteries. *Journal of Power Sources, 253*, 412–418.

Pop, V., Bergveld, H. J., Danilov, D., Regiten, P. P. L., & Notten, P. H. L. (2008). Battery Management Systems. In V. Pop, H. J. Bergveld, D. Danilov, P. P. L. Regiten, & P. H. L. Notten, (Eds.), *Methods for measuring and modelling a battery's electro-motive force* (vol. 9, pp. 63–94). Eindhoven, The Netherlands: Springer.

Pop, V., Bergveld, H. J., het Veld, J. H. G. O., Regtien, P. P. L., Danilov, D., & Notten, P. H. L. (2006). Modeling battery behavior for accurate state-of-charge indication. *Journal of the Electrochemical. Society, 153* (11), A2013–A2022.

Roscher, M. A., & Sauer, D. U. (2011). Dynamic electric behavior and open-circuit-voltage modeling of LiFePO4-based lithium ion secondary batteries. *Journal of Power Sources, 196* (1), 331–336.

Salkind, A. J., Fennie, C., Singh, P., Atwater, T., & Reisner, D. E. (1999). Determination of state-of-charge and state-of-health of batteries by fuzzy logic methodology. *Journal of Power Sources, 80* (1–2), 293–300.

Shim, J., & Striebel, K. A. (2003). Characterization of high-power lithium-ion cells during constant current cycling: Part I. Cycle performance and electrochemical diagnostics. *Journal of Power Sources, 122* (2), 188–194.

Sima, A. (2006, October 13). Sony exploding batteries – The chronicles. *Softpedia.* Retrieved from http://news.softpedia.com/news/Sony-Exploding-Batteries-Chronicles-37848.shtml

Stockley, T., Thanapalan, K., Bowkett, M., & Williams, J. (2013). *Development of an OCV prediction mechanism for lithium-ion battery system.* 19th international conference on automation and computing (ICAC), London, UK (pp. 48–53).

Suzuki, I., Shizuki, T., & Nishiyama, K. (2003). *High power and long life lithium-ion battery for backup power sources.* Telecommunications energy conference, 2003. INTELEC '03. The 25th International, Yokohama, Japan (pp. 317–322).

Tremblay, O., & Dessaint, L. A. (2009). Experimental validation of a battery dynamic model for EV applications. *World Electric Vehicle Journal, 3*, 1–10.

Weinert, J. X., Burke, A. F., & Wei, X. (2007). Lead-acid and lithium-ion batteries for the Chinese electric bike market and implications on future technology advancement. *Journal of Power Sources, 172* (2), 938–945.

Weng, C., Sun, J., & Peng, H. (2013, October 21–23). *An open-circuit-voltage model of lithium-ion batteries for effective incremental capacity analysis.* Proceedings of the ASME 2013 Dynamic Systems Control Conference, California, USA.

Zhang, Y., & Harb, J. N. (2013). Performance characteristics of lithium coin cells for use in wireless sensing systems: Transient behavior during pulse discharge. *Journal of Power Sources, 229*, 299–307.

27

Estimating parameters of S-systems by an auxiliary function guided coordinate descent method

Li-Zhi Liu, Fang-Xiang Wu* and Wen-Jun Zhang

Department of Mechanical Engineering, Division of Biomedical Engineering, University of Saskatchewan, Saskatoon, SK, Canada

The S-system, a set of nonlinear ordinary differential equations and derived from the generalized mass action law, is an effective model to describe various biological systems. Parameters in S-systems have significant biological meanings, yet difficult to be estimated because of the nonlinearity and complexity of the model. Given time series biological data, its parameter estimation turns out to be a nonlinear optimization problem. A novel method, auxiliary function guided coordinate descent, is proposed in this paper to solve the optimization problem by cyclically optimizing every parameter. In each iteration, only one parameter value is updated and it proves that the objective function keeps nonincreasing during the iterations. The updating rules in each iteration is simple and efficient. Based on this idea, two algorithms are developed to estimate the S-systems for two different constraint situations. The performances of algorithms are studied in several simulation examples. The results demonstrate the effectiveness of the proposed method.

Keywords: parameter estimation; nonlinear programming; optimization; S-system; coordinate descent

1. Introduction

There are many various molecules and productions in a cell. Some of them can regulate others via some mechanisms to achieve specific cellular functions, based on which the cells adapt to the changing environments. These components and their interactions constitute biological systems, such as metabolic pathways and genetic regulatory networks. One task of systems biology is to reveal the interactions and the biological functions those interactions may result in Klipp, Herwig, Kowald, Wierling, and Lehrach (2005). Instead of focusing on individual components, systems biology applies system engineering methods and principles to studying all components and their interactions as parts of a biological system. Such a systematic view provides an insight into the control and optimization of parts of the systems while taking the effects those may have on the whole system into account. It is of help to the discovery of new properties of biological systems and may provide valuable clues and new ideas in practical areas such as disease treatment and drug design (Gardner, di Bernardo, Lorenz, and Collins 2003).

One effective way to study the biological system is using mathematical or computational methods. Many mathematical models have been developed to describe the molecular biological systems based on biochemical principles. Most of them are nonlinear in both parameters and state variables (Klipp et al. 2005; Voit 2000). Nonlinear optimization problems are usually formulated for estimating the parameters in those models. However, analytical solutions to those problems are hardly available. One of the most popular models is the S-system which is highly nonlinear and derived from the generalized mass action law (Voit, 2000).

The S-system is an effective mathematical framework to characterize and analyse the molecular biological systems. An S-system consisting of N components is type of power-law formalism and typically a group of nonlinear ordinary differential equations (ODEs),

$$\dot{X}_i = \alpha_i \prod_{j=1}^{N} X_j^{g_{ij}} - \beta_i \prod_{j=1}^{N} X_j^{h_{ij}}, \quad i = 1, \ldots, N, \qquad (1)$$

where X_i represents the concentration of molecular species i measured at time t, whose changes are the difference between production and degradation. α_i's and β_i's are nonnegative rate constants, and g_{ij}'s and h_{ij}'s are real-valued kinetic orders that reflect the interaction intensity of X_j to X_i. If $g_{ij} > 0, X_j$ activates the production of X_i; if $g_{ij} < 0, X_j$ inhibits the production of X_i. h_{ij} has the same effects but on the degradation. A zero-valued kinetic order indicates that X_j has no such effect on X_i. The representation of this model maps the dynamical and topological information of the biological system onto its parameters.

The parameter estimation and structure identification of the S-systems are extremely difficult tasks. The parameter estimation usually occurs after or in the process of structure identification. As the parameter estimation of the S-systems is essentially a nonlinear problem, in principle, all nonlinear

*Corresponding author. Email: faw341@mail.usask.ca

optimization algorithms can be used, e.g. Gauss–Newton method and its variants, such as Box–Kanemasu interpolation method, Levenberg damped least-squares method, and Marquardt's method (Beck & Arnold, 1977). However, most of these methods are initial-sensitive and need to calculate the inverse of the Hessian which is computational expensive.

Several numerical methods have been proposed to estimate the parameters in S-systems. Most of them are based on heuristic methods. For example, Kikuchi, Tominaga, Arita, Takahashi, and Tomita (2003) employ a genetic algorithm to infer the S-systems. The effectiveness of the simulated annealing is studied in Gonzalez, Kper, Jung, Naval, and Mendoza (2007). Voit and Almeida (2004) develop an artificial neural network (ANN)-based method to identify and estimate the parameters of S-systems. Ho, Hsieh, Yu, and Huang (2007) develop an intelligent two-stage evolutionary algorithm that combines the genetic algorithm and the simulated annealing. An unified approach has been proposed in Wang, Qian, and Dougherty (2010). Most of these methods are computational expensive and do not sufficiently take advantage of special model structure of the S-systems.

Several methods taking the model structure of S-systems into account have been proposed. Wu and Mu (2009) introduce a separable parameter estimation method which divides the parameters into two groups: one group is linear in model while the other nonlinear. This method has been extended with a genetic algorithm to the case when system topology is unavailable in Liu, Wu, and Zhang (2012b). One can observe that if parameters in one term on the right-hand side of Equation (1) is known, moving it to the left and taking logarithm of both sides, a linear model is obtained. Using this observation, an alternating regression method is proposed by Chou, Martens, and Voit (2006), which reduces the nonlinear optimization into iterative procedures of linear regression. However, the convergence of the iterations cannot be guaranteed. Similarly, an alternating weighted least-squares method is proposed by Liu, Wu, and Zhang (2012a), in which the objective function to be optimized is approximated by a weighted linear regression in each iteration. However, this approximation may fail in some cases.

This study focus on the parameter estimation of S-systems when system topology is known. A novel method, auxiliary function guided coordinate descent (AFGCD), is proposed. After decoupling the S-systems by replacing the derivatives with numerical slops, the parameter estimation is formulated as a nonlinear optimization problem. AFGCD iteratively solves this problem and in each iteration only one parameter is optimized with other parameters fixed. Instead of directly optimizing the objective function, AFGCD takes advantage of a property of the exponential function and constructs an auxiliary function for each kinetic order parameter to optimize. It shows that optimizing the auxiliary function makes the objective function keep nonincreasing. Because the auxiliary function is simple and its optimization has an analytical solution, the parameter updating in each iteration is simple and efficient. Based on this idea, two algorithms are developed to estimate the parameters in S-systems under two different situations. One algorithm is developed for the case when the range of each parameter is known and the other algorithm for the case that the only constraint is the non-negativity of the rate constants.

The remaining of this paper is organized as follows. In Section 2, two algorithms are developed based on the idea of AFGCD and their descent properties are proven. In Section 3, the effectivenesses of the proposed algorithms are studied by several simulation examples. Finally, Section 4 concludes this study and points out some future works along this research.

2. Auxiliary function guided coordinate descent

2.1. *Problem statement*

Consider a biological system of N molecular species, described by an S-system as Equation (1), and assume that for each molecular species X_i, a time series concentration data measured at n equally spaced time points, $x_{i1}, \ldots, x_{in}$ are obtained. This study assumes that the topology of the system is available, i.e. the zero-values kinetic orders are known. The purpose is to estimate the parameters of nonzero-valued kinetic orders and rate constants. We substitute the derivative of X_i at time t with the estimated slop, S_{it}, so that the original coupled ordinary differential equations are decoupled into $n \times N$ uncoupled algebraic equations (Vilela et al., 2008; Voit & Almeida, 2004):

$$S_{it} = \alpha_i \prod_{j=1}^{N} x_{jt}^{g_{ij}} - \beta_i \prod_{j=1}^{N} x_{jt}^{h_{ij}} + \epsilon_{it}, \tag{2}$$

where $i = 1, \ldots, N, t = 1, \ldots, n$ and ϵ_{it}'s are errors or noises. The estimation of slopes is a crucial step and may affect the final results. To increase the accuracy, the five-point numerical derivative method is used in this study, i.e.

$$S_{it} = \frac{-x_{i,t+2} + 8x_{i,t+1} - 8x_{i,t-1} + x_{i,t-2}}{12\Delta t}, \tag{3}$$

where Δt is the length of the sampling step.

To determine the parameter values, the sum of least squares is usually employed as an objective function to be minimized, i.e. the parameters of each ODE equation i in Equation (1) is obtained by solving

$$\begin{aligned} &\underset{\alpha_i,\beta_i,g_i,h_i}{\text{minimize}} \quad J_i = \frac{1}{2}\sum_{t=1}^{n}\left[S_{it} - \alpha_i \prod_{j\in G_i} x_{ij}^{g_{ij}} + \beta_i \prod_{j\in H_i} x_{ij}^{h_{ij}}\right]^2 \\ &\text{subject to} \quad \alpha_i \geq \delta, \quad \beta_i \geq \delta \end{aligned} \tag{4}$$

where δ is a predefined small positive number; G_i and H_i are the sets of indices of molecular species which have effects on the production and degradation of molecular species i, respectively; $g_i = \{g_{ij}, j \in G_i\}$ and $h_i = \{h_{ij}, j \in H_i\}$. The minimization of Equation (4) is non-trivial, as it is highly nonlinear and contains a lot of parameters.

2.2. *Proposed method*

The optimization problem (4) has no analytical solutions and therefore, iterative methods are considered. A coordinate descent strategy is applied in this study, i.e. we cyclically optimize one parameter in one iteration with other parameters fixed such that the objective function keeps nonincreasing in every iteration. More specifically, denote by $\boldsymbol{\theta}^{(\ell)} = (\theta_1^{(\ell)}, \ldots, \theta_p^{(\ell)})$ the parameters of an optimization problem (e.g. for Equation (4), $\boldsymbol{\theta} = (\alpha_i, g_i, \beta_i, h_i)$) at iteration ℓ. After one iteration, the first parameter is updated, $\boldsymbol{\theta}_1^{(\ell+)} = (\theta_1^{(\ell+1)}, \theta_2^{(\ell)}, \ldots, \theta_p^{(\ell)})$ and after k iterations, the first k parameters are updated, $\boldsymbol{\theta}_k^{(\ell+)} = (\theta_1^{(\ell+1)}, \ldots, \theta_k^{(\ell+1)}, \theta_{k+1}^{(\ell)}, \ldots, \theta_p^{(\ell)})$.

Consider the problem (4) and suppose the parameter h_{ik}, for some $k \in H_i$, needs to be updated at the current iteration, one simple way to make sure that the objective function keeps nonincreasing is to use the classical gradient descent method:

$$h_{ik}^{(\ell+1)} = h_{ik}^{(\ell)} - d_\ell \frac{\partial J_i(h_{ik}^{(\ell)})}{\partial h_{ik}}, \tag{5}$$

where values of other parameters are fixed and d_ℓ is the stepsize taken along the negative gradient direction. Some methods can be used to search and select the stepsize, such as minimization rule and Armijo rule (Bazaraa, Sherali, & Shetty, 2006; Bertsekas, 1999). However, these methods may either depend on a line search algorithm or introduce an extra iteration loop.

In this paper, a smart stepsize is derived. The introduced stepsize is computational efficient and the non-increase of the objective function in each iteration is guaranteed, which is proved in the following subsection. The algorithm for solving the problem (4) is illustrated in Algorithm 1, in which the updates for each parameter are shown in Equations (6)–(9). To check the convergence, the following stopping criteria is used

$$\frac{\|\boldsymbol{\theta}^{(\ell+1)} - \boldsymbol{\theta}^{(\ell)}\|_2}{\|\boldsymbol{\theta}^{(\ell)}\|_2} < \eta,$$

here η is a preset threshold. In this paper, we set $\eta = 10^{-5}$.

In some situations, from experiments, the range of each parameter is known. Then, more constraints can be added to the problem (4) as follows:

$$\underset{\alpha_i,\beta_i,g_i,h_i}{\text{minimize}} \quad J_i = \frac{1}{2}\sum_{t=1}^{n}\left[S_{it} - \alpha_i \prod_{j\in G_i} x_{ij}^{g_{ij}} + \beta_i \prod_{j\in H_i} x_{ij}^{h_{ij}}\right]^2$$

$$\begin{aligned} \text{subject to} \quad & r_{\min} \le \alpha_i \le r_{\max}, \quad r_{\min} \le \beta_i \le r_{\max}, \\ & k_{\min} \le g_{ij} \le k_{\max}, \quad j \in G_i, \\ & k_{\min} \le h_{ij} \le k_{\max}, \quad j \in H_i, \end{aligned} \tag{10}$$

that is, we require the rate constants stay in the range $[r_{\min}, r_{\max}]$ and kinetic orders in the range $[k_{\min}, k_{\max}]$. Simple adaptations to Algorithm 1 lead to Algorithm 2 for solving the problem (10). The update rules (11) and (13) are slight modifications to Equations (6) and (8). They efficiently make the objective function keep nonincreasing, which is proved in the following subsection.

2.3. *Descent property*

Similar to Seung and Lee (2001) and Li, Wu, Zhang, and Wu (2013), we make use of an auxiliary function to prove the nonincrease of the objective function in each iteration.

Definition 1 *$F(h, h')$ is an auxiliary function of $J(h)$ if there exists $\sigma(h')$, such that, if $|h - h'| \le \sigma(h')$*

$$F(h, h') \ge J(h) \quad \text{and} \quad F(h, h) = J(h) \tag{15}$$

are satisfied.

The auxiliary function is useful because of the following lemma.

Lemma 1 *If F is an auxiliary function, then J is nonincreasing under the update*

$$h^{(\ell+1)} = \underset{h\in\{h:|h-h^{(\ell)}|\le\sigma(h^{(\ell)})\}}{\arg\min} F(h, h^{(\ell)}). \tag{16}$$

Proof

$$J(h^{(\ell+1)}) \le F(h^{(\ell+1)}, h^{(\ell)}) \le F(h^{(\ell)}, h^{(\ell)}) = J(h^{(\ell)})$$

■

Lemma 2 shows a property of the exponential function, which is useful for the following discussions.

Lemma 2 *For some point h', the exponential function e^h satisfies*

$$e^h \le e^{h'} + e^{h'}(h - h') + e^{h'}(h - h')^2 \Delta(\sigma) \quad \text{if } |h - h'| \le \sigma, \tag{17}$$

where

$$\Delta(\sigma) = \frac{1}{\sigma^2}(e^\sigma - 1 - \sigma). \tag{18}$$

Algorithm 1 AFGCD for problem (4).

(1) Assign initial values to $\boldsymbol{\theta}^{(0)} = (\alpha_i^{(0)}, g_i^{(0)}, \beta_i^{(0)}, h_i^{(0)})$ and set $\ell = 0$.

(2) Let $y_t = -S_{it} - \beta_i^{(\ell)} \prod_{j \in H_i} x_{jt}^{h_{ij}^{(\ell)}}$, $t = 1, \ldots, n$ and cyclically update $g_{ik}^{(\ell)}$ to $g_{ik}^{(\ell+1)}$ for each $k \in G_i$ as follows:

$$g_{ik}^{(\ell+1)} = g_{ik}^{(\ell)} - \frac{1}{2\tau_g(\sigma_g^{(\ell)})} \frac{\partial J_i(g_{ik}^{(\ell)})}{\partial g_{ik}}, \tag{6}$$

where

$$\tau_g(\sigma_g^{(\ell)}) = \sum_t a_t |y_t| x_{kt}^{g_{ik}^{(\ell)}} (\log^2 x_{kt}) \Delta(\sigma_g^{(\ell)} | \log x_{kt}|) + 2 \sum_t a_t^2 x_{kt}^{2g_{ik}^{(\ell)}} (\log^2 x_{kt}) \Delta(2\sigma_g^{(\ell)} | \log x_{kt}|)$$

$$\sigma_g^{(\ell)} = \frac{|\partial J_i(g_{ik}^{(\ell)}) / \partial g_{ik}|}{\sum_t a_t |y_t| x_{kt}^{g_{ik}^{(\ell)}} \log^2 x_{kt} + 2 \sum_t a_t^2 x_{kt}^{2g_{ik}^{(\ell)}} \log^2 x_{kt}},$$

$$a_t = \alpha_i^{(\ell)} \prod_{j \in G_i, j \neq k} x_{jt}^{g_{ij}^{(\ell+1)I(j<k)+(\ell)I(j \geq k)}}$$

and the function $\Delta(\cdot)$ is defined in Equation (18).

(3) Compute

$$\alpha_i^{(\ell+1)} = \max\left\{ -\frac{\sum_t y_t p_t}{\sum_t p_t^2}, \delta \right\}, \quad \text{where } p_t = \prod_{j \in G_i} x_{jt}^{g_{ij}^{(\ell+1)}} \tag{7}$$

(4) Let $y_t = S_{it} - \alpha_i^{(\ell+1)} \prod_{j \in G_i} x_{jt}^{g_{ij}^{(\ell+1)}}$, $t = 1, \ldots, n$ and cyclically update $h_{ik}^{(\ell)}$ to $h_{ik}^{(\ell+1)}$ for each $k \in H_i$ as follows

$$h_{ik}^{(\ell+1)} = h_{ik}^{(\ell)} - \frac{1}{2\tau_h(\sigma_h^{(\ell)})} \frac{\partial J_i(h_{ik}^{(\ell)})}{\partial h_{ik}}, \tag{8}$$

where

$$\tau_h(\sigma_h^{(\ell)}) = \sum_t b_t |y_t| x_{kt}^{h_{ik}^{(\ell)}} (\log^2 x_{kt}) \Delta(\sigma_h^{(\ell)} | \log x_{kt}|) + 2 \sum_t b_t^2 x_{kt}^{2h_{ik}^{(\ell)}} (\log^2 x_{kt}) \Delta(2\sigma_h^{(\ell)} | \log x_{kt}|)$$

$$\sigma_h^{(\ell)} = \frac{|\partial J_i(h_{ik}^{(\ell)}) / \partial h_{ik}|}{\sum_t b_t |y_t| x_{kt}^{h_{ik}^{(\ell)}} \log^2 x_{kt} + 2 \sum_t b_t^2 x_{kt}^{2h_{ik}^{(\ell)}} \log^2 x_{kt}},$$

and

$$b_t = \beta_i^{(\ell)} \prod_{j \in H_i, j \neq k} x_{jt}^{h_{ij}^{(\ell+1)I(j<k)+(\ell)I(j \geq k)}}$$

(5) Compute

$$\beta_i^{(\ell+1)} = \max\left\{ -\frac{\sum_t y_t d_t}{\sum_t d_t^2}, \delta \right\}, \quad \text{where } d_t = \prod_{j \in H_i} x_{jt}^{h_{ij}^{(\ell+1)}} \tag{9}$$

(6) Let $\boldsymbol{\theta}^{(\ell+1)} = (\alpha_i^{(\ell+1)}, g_i^{(\ell+1)}, \beta_i^{(\ell+1)}, h_i^{(\ell+1)})$ and check the convergence. If converged, output $\boldsymbol{\theta}^{(\ell+1)}$, otherwise, $\ell \leftarrow \ell + 1$ and go to 2.

Algorithm 2 AFGCD for problem (10)

(1) Assign initial values to $\boldsymbol{\theta}^{(0)} = (\alpha_i^{(0)}, g_i^{(0)}, \beta_i^{(0)}, h_i^{(0)})$ and set $\ell = 0$.

(2) Let $y_t = -S_{it} - \beta_i^{(\ell)} \prod_{j\in H_i} x_{jt}^{h_{ij}^{(\ell)}}$, $t = 1, \ldots, n$ and cyclically update $g_{ik}^{(\ell)}$ to $g_{ik}^{(\ell+1)}$ for each $k \in G_i$ as follows

$$g_{ik}^{(\ell+1)} = \min\left(\max\left(g_{ik}^{(\ell)} - \frac{1}{2\tau_g(\sigma_g^{(\ell)})}\frac{\partial J_i(g_{ik}^{(\ell)})}{\partial g_{ik}}, L\right), R\right), \tag{11}$$

where $L = \max(g_{ik}^{(\ell)} - \sigma_g^{(\ell)}, k_{\min})$, $R = \min(g_{ik}^{(\ell)} + \sigma_g^{(\ell)}, k_{\max})$ and $\tau_g(\sigma_g^{(\ell)})$ is calculated in the same way as in Algorithm 1

(3) Compute

$$\alpha_i^{(\ell+1)} = \max\left\{-\frac{\sum_t y_t p_t}{\sum_t p_t^2}, \delta\right\}, \quad \text{where } p_t = \prod_{j\in G_i} x_{jt}^{g_{ij}^{(\ell+1)}} \tag{12}$$

(4) Let $y_t = S_{it} - \alpha_i^{(\ell+1)} \prod_{j\in G_i} x_{jt}^{g_{ij}^{(\ell+1)}}$, $t = 1, \ldots, n$ and cyclically update $h_{ik}^{(\ell)}$ to $h_{ik}^{(\ell+1)}$ for each $k \in H_i$ as follows

$$h_{ik}^{(\ell+1)} = \min\left(\max\left(h_{ik}^{(\ell)} - \frac{1}{2\tau_h(\sigma_h^{(\ell)})}\frac{\partial J_i(h_{ik}^{(\ell)})}{\partial h_{ik}}, L\right), R\right), \tag{13}$$

where $L = \max(h_{ik}^{(\ell)} - \sigma_h^{(\ell)}, k_{\min})$ and $R = \min(h_{ik}^{(\ell)} + \sigma_h^{(\ell)}, k_{\max})$.

(5) Compute

$$\beta_i^{(\ell+1)} = \max\left\{-\frac{\sum_t y_t d_t}{\sum_t d_t^2}, \delta\right\}, \quad \text{where } d_t = \prod_{j\in H_i} x_{jt}^{h_{ij}^{(\ell+1)}} \tag{14}$$

(6) Let $\boldsymbol{\theta}^{(\ell+1)} = (\alpha_i^{(\ell+1)}, g_i^{(\ell+1)}, \beta_i^{(\ell+1)}, h_i^{(\ell+1)})$ and check the convergence. If converged, output $\boldsymbol{\theta}^{(\ell+1)}$, otherwise, $\ell \leftarrow \ell + 1$ and go to Step 2 .

Proof

$$\begin{aligned}
e^h &= e^{h'} e^{h-h'} \\
&= e^{h'}\left(1 + (h-h') + \frac{1}{2!}(h-h')^2 + \cdots \right. \\
&\quad \left. + \frac{1}{n!}(h-h')^n + \cdots\right) \\
&= e^{h'} + e^{h'}(h-h') + e^{h'}(h-h')^2 \\
&\quad \left(\frac{1}{2!} + \cdots + \frac{1}{n!}(h-h')^{n-2} + \cdots\right) \\
&\le e^{h'} + e^{h'}(h-h') + e^{h'}(h-h')^2 \\
&\quad \left(\frac{1}{2!} + \cdots + \frac{1}{n!}\sigma^{n-2} + \cdots\right) \\
&= e^{h'} + e^{h'}(h-h') + e^{h'}(h-h')^2 \frac{1}{\sigma^2}(e^\sigma - 1 - \sigma).
\end{aligned}$$

■

Then, the descent property of Algorithm 1 is shown in the following theorem.

THEOREM 1 *The updating rules (6)–(9) in Algorithm 1 keep the objective function in problem (4) nonincreasing.*

Proof First, we show that the updating rules (6) and (8) can be obtained by minimizing constructed auxiliary functions. Here, we only prove that the rule (8) and the rule (6) can be proved similarly.

Suppose in one iteration, we need to update the parameter $h_{ik}^{(\ell)}$ to $h_{ik}^{(\ell+1)}$ with all the other parameters fixed. Using the notations in Algorithm 1, the objective function is

$$J_i(h_{ik}) = \frac{1}{2}\sum_t (y_t + b_t x_{kt}^{h_{ik}})^2. \tag{19}$$

Because other parameters except h_{ik} are fixed, J_i can be considered only depending on h_{ik}. The following auxiliary function is proposed:

$$\begin{aligned}
F(h_{ik}, h_{ik}^{(\ell)}) &= J_i(h_{ik}^{(\ell)}) + J_i'(h_{ik}^{(\ell)})(h - h_{ik}^{(\ell)}) \\
&\quad + \tau_h(\sigma_h^{(\ell)})(h_{ik} - h_{ik}^{(\ell)})^2.
\end{aligned} \tag{20}$$

The definitions of $\tau_h(\cdot)$ and $\sigma_h^{(\ell)}$ can be found in Algorithm 1.

Obviously, $F(h_{ik}, h_{ik}) = J_i(h_{ik})$. We only need to verify the first condition in Definition 1. Using Lemma 2, we have, if $|h_{ik} - h_{ik}^{(\ell)}| \le \sigma_h^{(\ell)}$

$$\begin{aligned}J_i(h_{ik}) &= \frac{1}{2}\sum_t y_t^2 + \sum_t b_t y_t e^{h_{ik}\log x_{kt}} + \frac{1}{2}\sum_t b_t^2 e^{2h_{ik}\log x_{kt}} \\ &\le \frac{1}{2}\sum_t y_t^2 + \sum_t b_t y_t e^{h_{ik}^{(\ell)}\log x_{kt}} + \frac{1}{2}\sum_t b_t^2 e^{2h_{ik}^{(\ell)}\log x_{kt}} \\ &\quad + \sum_t b_t y_t e^{h_{ik}^{(\ell)}\log x_{kt}}(\log x_{kt})(h_{ik} - h_{ik}^{(\ell)}) \\ &\quad + \sum_t b_t^2 e^{2h_{ik}^{(\ell)}\log x_{kt}}(\log x_{kt})(h_{ik} - h_{ik}^{(\ell)}) \\ &\quad + \sum_t b_t |y_t| e^{h_{ik}^{(\ell)}\log x_{kt}}(\log^2 x_{kt})(h_{ik} - h_{ik}^{(\ell)})^2 \\ &\quad\quad \Delta(\sigma_h^{(\ell)}|\log x_{kt}|) \\ &\quad + 2\sum_t b_t^2 e^{2h_{ik}^{(\ell)}\log x_{kt}}(\log^2 x_{kt})(h_{ik} - h_{ik}^{(\ell)})^2 \\ &\quad\quad \Delta(2\sigma_h^{(\ell)}|\log x_{kt}|) \\ &= J_i(h_{ik}^{(\ell)}) + J_i'(h_{ik}^{(\ell)})(h - h_{ik}^{(\ell)}) + \tau_h(\sigma_h^{(\ell)}) \\ &\quad\quad (h_{ik} - h_{ik}^{(\ell)})^2,\end{aligned}$$

where the function $\Delta(\cdot)$ is defined in Equation (18). Therefore, F defined in Equation (20) is an auxiliary function. By Lemma 1, the minimization of F with respect to h_{ik} in the neighbourhood of $h_{ik}^{(\ell)}$ makes the objective function J keep nonincreasing. Thus, the updating rule is obtained by solving

$$\underset{h_{ik}\in\{h_{ik}:|h_{ik}-h_{ik}^{(\ell)}|\le\sigma_h^{(\ell)}\}}{\arg\min} \quad F(h_{ik}, h_{ik}^{(\ell)}). \tag{21}$$

Note that the auxiliary function (20) is quadratic and its global minimum is attained at

$$h_{ik}^* = h_{ik}^{(\ell)} - \frac{1}{2\tau_h(\sigma_h^{(\ell)})} J_i'(h_{ik}^{(\ell)}), \tag{22}$$

Also note that

$$\begin{aligned}|h_{ik}^* - h_{ik}^{(\ell)}| &= \frac{|J_i'(h_{ik}^{(\ell)})|}{2\tau_h(\sigma_h^{(\ell)})} \\ &\le \frac{|J_i'(h_{ik}^{(\ell)})|}{\sum_t b_t|y_t| x_{kt}^{h_{ik}^{(\ell)}} \log^2 x_{kt} + 2\sum_t b_t^2 x_{kt}^{2h_{ik}^{(\ell)}} \log^2 x_{kt}} \\ &= \sigma_h^{(\ell)},\end{aligned} \tag{23}$$

the inequality is due to that $\Delta(\sigma) \ge \frac{1}{2}$ for all $\sigma > 0$. Therefore, the updating rule is (22) or (8) and the objective function keeps nonincreasing during the iterations.

The updating steps (7) and (9) are obtained by simply solving an univariate least-squares problem with a simple constraint, whose objective function is quadratic and thus has an analytical solution. The derivation is simple and we omit it here. Therefore, the objective function keeps nonincreasing in every iterations of Algorithm 1. ■

For Algorithm 2, we have the same descent property.

THEOREM 2 *The updating rules (11)–(14) in Algorithm 2 keep the objective function in problem (10) nonincreasing.*

Proof Similar to Theorem 1, considering the updating of $h_{ik}^{(\ell)}$, the auxiliary function is defined as in Equation (20). The updating rule is obtained by solving

$$\begin{aligned}&\underset{h_{ik}}{\text{argmin}} \quad F(h_{ik}, h_{ik}^{(\ell)}) \\ &\text{subject to} \quad h_{ik}^{(\ell)} - \sigma_h^{(\ell)} \le h_{ik} \le h_{ik}^{(\ell)} + \sigma_h^{(\ell)} \\ &\qquad\qquad k_{\min} \le h_{ik} \le k_{\max}.\end{aligned} \tag{24}$$

Since F is quadratic in h_{ik} and the constraints are very simple, analytical solution (13) exists. The remaining proof is similar to Theorem 1 and we omit it here. ■

3. Simulation examples

To study the performances of the proposed methods, we apply the algorithms to simulated data and compare the estimated parameters with the corresponding true ones.

3.1. *Performances of Algorithm 1*

3.1.1. *Four-dimensional model*

Consider the following S-system of four molecular species (Voit, 2000):

$$\begin{aligned}\dot{X}_1 &= 12X_3^{-0.8} - 10X_1^{0.5}, \\ \dot{X}_2 &= 8X_1^{0.5} - 3X_2^{0.75}, \\ \dot{X}_3 &= 3X_2^{0.75} - 5X_3^{0.5}X_4^{0.2}, \\ \dot{X}_4 &= 2X_1^{0.5} - 6X_4^{0.8}.\end{aligned} \tag{25}$$

The noise-free time series data are obtained by numerically solving the S-system with an initial condition $X(0) = [x_{10}, x_{20}, x_{30}, x_{40}]^{\mathrm{T}}$. The data are sampled at time points in the interval $[0, 5]$ with $\Delta t = 0.1$.

In this example, the data are generated with $X(0) = [10, 1, 2, 3]^{\mathrm{T}}$. The time series data are shown in Figure 1, from which we can see all states of X_i's are eventually in the steady states. Algorithm 1 is applied to estimating the parameters from these data with initial values for all parameters chosen by

$$\theta^{\text{initial}} = \theta^{\text{true}}(1 + s\epsilon) \quad \text{for any} \theta \in \{\alpha_i, g_i, \beta_i, h_i\}, \tag{26}$$

where ϵ is a standard Gaussian random variable and s is a positive constant. Since Equation (25) is a nonlinear optimization problem, to avoid falling into the local optimum,

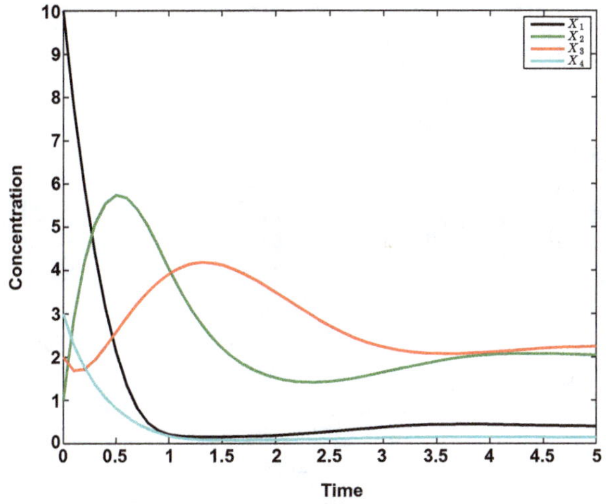

Figure 1. Time series data of the four-dimensional model.

we apply Algorithm 1 for 100 times initiated with different values and choose the one with minimum objective value as the final solution. Here, we set $s = 90\%$. The objective values J_i of the first 100 iterations in one run are illustrated in Figure 2. It can be seen that the objective values decrease with the increase of iteration steps. Table 1 shows the estimated results, from which we can see that the estimated values are quite close to their true values and the optimal objective values are all very small.

3.1.2. *Five-dimensional model*

A benchmark five-dimensional model (Vilela et al., 2008; Yang, Dent, & Nardini, 2012) is considered,

$$\begin{aligned}\dot{X}_1 &= 5X_3X_5^{-1} - 10X_1^2,\\ \dot{X}_2 &= 10X_1^2 - 10X_2^2,\\ \dot{X}_3 &= 10X_2^{-1} - 10X_2^{-1}X_3^2,\\ \dot{X}_4 &= 8X_3^2X_5^{-1} - 10X_4^2,\\ \dot{X}_5 &= 10X_4^2 - 10X_5^2.\end{aligned} \tag{27}$$

The data used in this example are generated with the initial condition $X(0) = [0.1, 0.7, 0.7, 0.16, 0.18]^{\mathrm{T}}$, the same as in Yang et al. (2012). Figure 3 shows the time series data that

Table 1. Estimated results of the four-dimensional model from Algorithm 1.

Parameter	True value	Estimation	Relative error (%)	Objective value
α_1	12	11.9856	0.12	6.809×10^{-4}
β_1	10	9.9819	0.18	
g_{13}	−0.8	−0.8022	0.27	
h_{11}	0.5	0.5013	0.25	
α_2	8	7.9857	0.18	3.650×10^{-4}
β_2	3	2.9912	0.29	
g_{21}	0.5	0.5005	0.11	
h_{22}	0.75	0.7511	0.14	
α_3	3	3.0377	1.26	5.8635×10^{-5}
β_3	5	5.0375	0.75	
g_{32}	0.75	0.7444	0.74	
h_{33}	0.5	0.4957	0.86	
h_{34}	0.2	0.1978	1.11	
α_4	2	1.9965	0.17	4.0889×10^{-5}
β_4	6	5.9977	0.04	
g_{41}	0.5	0.5014	0.29	
h_{44}	0.8	0.8014	0.18	

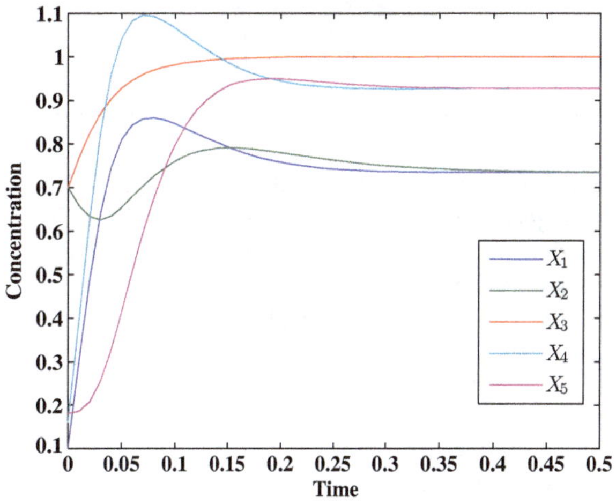

Figure 3. Time series data of the five-dimensional model.

are sampled in the interval $[0, 0.5]$ with $\Delta t = 0.01$. Note that the states of all variables quickly converge to the steady state. Hence, only limited information on the dynamics of the system is contained in the data. We run Algorithm 1 for 100 times with different initial values obtained from

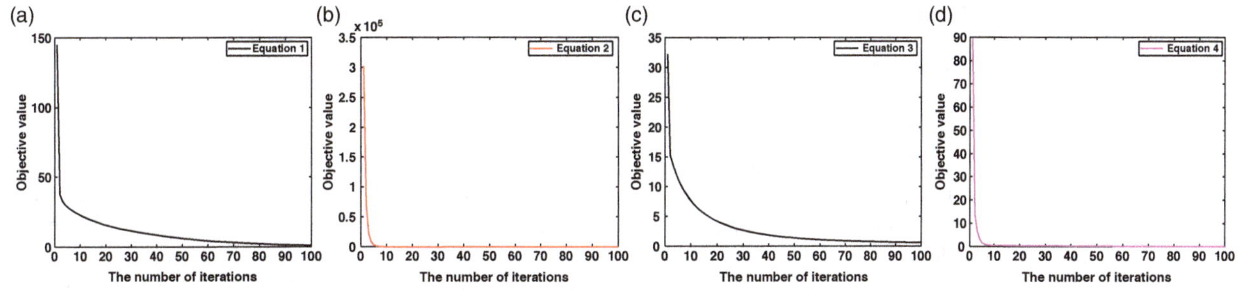

Figure 2. Objective values over various numbers of iterations in Algorithm 1.

Table 2. Estimated results of the five-dimensional model from Algorithm 1.

Parameter	True value	Estimation	Relative error (%)	Objective value
α_1	5	4.7238	5.52	3.7096×10^{-4}
β_1	10	9.7188	2.81	
g_{13}	1	1.1732	17.32	
g_{15}	−1	−1.0500	4.99	
h_{11}	2	2.0783	3.92	
α_2	10	10.0280	0.28	3.8752×10^{-4}
β_2	10	9.9796	0.20	
g_{21}	2	1.9953	0.24	
h_{22}	2	1.9778	1.11	
α_3	10	9.0097	9.90	4.2896×10^{-4}
β_3	10	9.0220	9.78	
g_{32}	−1	−1.1133	11.33	
h_{32}	−1	−1.1082	10.82	
h_{33}	2	2.1220	6.10	
α_4	8	7.5761	5.30	6.5891×10^{-4}
β_4	10	9.5605	4.40	
g_{43}	2	2.1641	8.21	
g_{45}	−1	−1.0457	4.57	
h_{44}	2	2.0790	3.95	
α_5	10	9.9333	0.67	6.6801×10^{-4}
β_5	10	9.9402	0.60	
g_{54}	2	2.0138	0.69	
h_{55}	2	2.0240	1.20	

Equation (26) with $s = 80\%$ and select the one with minimum objective value as the solution. The results in Table 2 show the effectiveness of this algorithm: estimated values of parameters are close to the true values and the optimal objective values are all very small.

3.1.3. *Six-dimensional model*

In this example, Algorithm 1 is applied to estimate the parameters in the following six-dimensional S-system Yang et al. (2012):

$$\begin{aligned} \dot{X}_1 &= 10X_3^{-2}X_5 - 5X_1^{0.5}, \\ \dot{X}_2 &= 5X_1^{0.5} - 10X_2^{0.5}, \\ \dot{X}_3 &= 2X_2^{0.5} - 1.25X_3^{0.5}, \\ \dot{X}_4 &= 8X_2^{0.5} - 5X_4^{0.5}, \\ \dot{X}_5 &= 0.5 - X_6, \\ \dot{X}_6 &= X_5 - 0.5. \end{aligned} \tag{28}$$

The noise-free time series data are generated with the initial condition $X(0) = [1.1, 0.5, 0.9, 0.75, 0.5, 0.75]^{\mathrm{T}}$ which matches that in Yang et al. (2012). Figure 4 illustrates the time series data which are sampled in the interval [0, 10] with $\Delta t = 0.1$. Figure 4 also shows the periodic oscillating behaviour of the data. The initial values are also chosen by Equation (26) with $s = 50\%$. The solution shown in Table 3 is the best one among 100 runs of Algorithm 1 with different initial values. The results in Table 3 indicate that the estimated parameters are close to their true values and the optimal objective values are all very small.

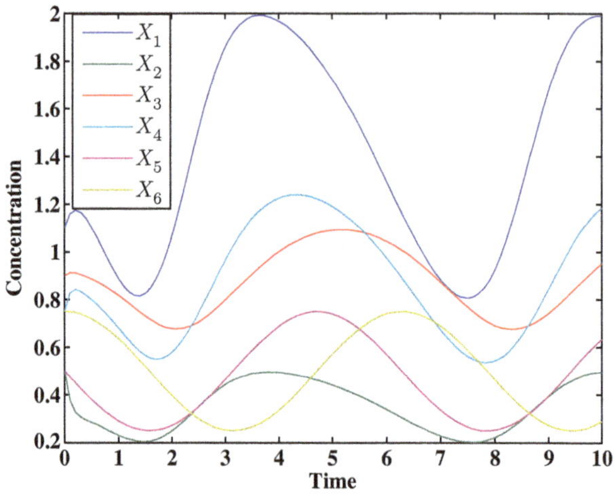

Figure 4. Time series data of the six-dimensional model.

Table 3. Estimated results of the six-dimensional model from Algorithm 1.

Parameter	True value	Estimation	Relative error (%)	Objective value
α_1	10	9.7145	2.85	4.2743×10^{-4}
β_1	5	4.8192	3.61	
g_{13}	−2	−2.0397	1.99	
g_{15}	1	1.0146	1.46	
h_{11}	0.5	0.5047	0.94	
α_2	5	4.7858	4.28	3.4432×10^{-4}
β_2	10	9.9353	0.65	
g_{21}	0.5	0.5266	5.32	
h_{22}	0.5	0.5269	5.39	
α_3	2	2.0059	0.29	4.1841×10^{-6}
β_3	1.25	1.2580	0.64	
g_{32}	0.5	0.4963	0.74	
h_{33}	0.5	0.4964	0.72	
α_4	8	8.0231	0.29	1.2605×10^{-4}
β_4	5	5.0138	0.28	
g_{42}	0.5	0.5002	0.04	
h_{44}	0.5	0.5002	0.05	
α_5	0.5	0.4962	0.76	1.9439×10^{-6}
β_5	1	0.9978	0.22	
h_{56}	1	1.0086	0.86	
α_6	1	0.9978	0.22	1.9949×10^{-6}
β_6	0.5	0.4962	0.76	
g_{65}	1	1.0087	0.87	

3.2. ***Performances of Algorithm 2***

The performances of Algorithm 2 is studied in this example. We apply the algorithm to the data generated from the four-dimensional, five-dimensional, and six-dimensional models, respectively. The time series data we use are exactly the same as above-mentioned examples. Since Algorithm 2 solves the problem (10) which includes a range constraint for each parameter, the initial values for the algorithm are randomly generated in those ranges. In this study, the rate

Table 4. Estimated results of the four-dimensional model from Algorithm 2.

Parameter	True value	Estimation	Relative error (%)	Objective value
α_1	12	11.945	0.4585	8.2968×10^{-4}
β_1	10	9.9288	0.7119	
g_{13}	−0.8	−0.807	0.8789%	
h_{11}	0.5	0.5034	0.6858	
α_2	8	7.9857	0.1787	3.65×10^{-4}
β_2	3	2.9912	0.2931	
g_{21}	0.5	0.5005	0.1093	
h_{22}	0.75	0.7511	0.1411	
α_3	3	3.0337	1.1233	7.9833×10^{-5}
β_3	5	5.0244	0.4879	
g_{32}	0.75	0.7435	0.8691	
h_{33}	0.5	0.4937	1.26	
h_{34}	0.2	0.1964	1.7893	
α_4	2	2.0051	0.2528	1.991×10^{-4}
β_4	6	5.9847	0.2551	
g_{41}	0.5	0.4927	1.457	
h_{44}	0.8	0.7938	0.7776	

Table 5. Estimated results of the five-dimensional model from Algorithm 2.

Parameter	True value	Estimation	Relative error (%)	Objective value
α_1	5	4.7704	4.5918	3.1738×10^{-4}
β_1	10	9.7686	2.3138	
g_{13}	1	1.1535	15.3495	
g_{15}	−1	−1.0424	4.2402	
h_{11}	2	2.0652	3.2612	
α_2	10	10.0133	0.1333	3.9712×10^{-4}
β_2	10	9.9644	0.3563	
g_{21}	2	2.0022	0.1122	
h_{22}	2	1.9845	0.7763	
α_3	10	8.1857	18.1435	9.5442×10^{-4}
β_3	10	8.1839	18.1609	
g_{32}	−1	−1.2361	23.6145	
h_{32}	−1	−1.2363	23.6264	
h_{33}	2	2.2322	11.6124	
α_4	8	7.5565	5.5442	6.7221×10^{-4}
β_4	10	9.5398	4.6023	
g_{43}	2	2.1688	8.4413	
g_{45}	−1	−1.0476	4.7607	
h_{44}	2	2.0827	4.1342	
α_5	10	9.9649	0.3505	5.6977×10^{-4}
β_5	10	9.958	0.4204	
g_{54}	2	2.0152	0.7598	
h_{55}	2	2.0047	0.2345	

constants are restricted in the range $[0.1, 10]$ and kinetic orders are in the range $[-2, 3]$. For each model, we run the algorithm for 100 times with different initial values and choose the one with minimum objective value as the final result. The estimated results from Algorithm 2 for each model are reported in the Tables 4–6. From the results, it can be seen that estimated parameters for the four-dimensional and six-dimensional models have very small relative errors

Table 6. Estimated results of the six-dimensional model from Algorithm 2.

Parameter	True value	Estimation	Relative error (%)	Objective value
α_1	10	9.7512	2.4881	4.3511×10^{-4}
β_1	5	4.9344	1.3118	
g_{13}	−2	−1.9891	0.5461	
g_{15}	1	0.9872	1.2845	
h_{11}	0.5	0.4899	2.0172	
α_2	5	4.6623	6.7549	3.3703×10^{-4}
β_2	10	9.6115	3.8853	
g_{21}	0.5	0.5221	4.4167	
h_{22}	0.5	0.5218	4.3684	
α_3	2	1.9983	0.084	4.0053×10^{-6}
β_3	1.25	1.2484	0.1248	
g_{32}	0.5	0.5005	0.0925	
h_{33}	0.5	0.5005	0.0912	
α_4	8	7.9848	0.1896	1.2578×10^{-4}
β_4	5	4.9926	0.1487	
g_{42}	0.5	0.4996	0.0864	
h_{44}	0.5	0.4995	0.0953	
α_5	0.5	0.4991	0.1864	1.2022×10^{-7}
β_5	1	0.9994	0.0609	
h_{56}	1	1.002	0.1989	
α_6	1	0.9977	0.2326	2.2004×10^{-6}
β_6	0.5	0.496	0.8035	
g_{65}	1	1.0091	0.9094	

compared with their corresponding true values. For the five-dimensional model, most parameters are estimated with low errors, while the estimations of the parameters in the third ODE have relatively large errors. This may be due to the limited information contained in the time series data. We can also see that the estimated objective function values for these three models are all very small. All these results show the effectiveness of the proposed algorithm.

4. Conclusions

The S-system is an effective mathematical model to characterize and analyse the molecular biological systems. The parameters in S-systems have significant biological meanings. However, the estimation of these parameters from time series biological data is non-trivial, because of the nonlinearity and complexity of this model. An novel method, the AFGCD, is proposed in this study to estimate the parameters of S-systems. To solve the nonlinear optimization problem involved in the parameter estimation, the proposed method optimizes one parameter at a time. Taking advantage of a property of the exponential function, an auxiliary function is constructed for each kinetic order parameter. We prove that updating the parameter value by optimizing the auxiliary function keeps the objective function nonincreasing. As the auxiliary function is quadratic, the analytical solution exists and therefore the updating rule for each parameter is simple and efficient. Based on this idea, two algorithms are developed: one estimates the parameters in S-system with the only non-negative constraints on rate constants; the other puts constraints both on rate constants and kinetic orders. Their performances are studied in simulation examples, which show that the estimated parameter values from the proposed algorithms are very close to the corresponding true values and the optimized objective functions has very low values. All of these results demonstrate the effectivenesses of the proposed algorithms.

In this study, the topology information of the system is assumed to be known. One direction of the future work is to extend the current method with structure identification method to infer the S-system without knowing the system topology.

Acknowledgements

This study is supported by Natural Science and Engineering Research Council of Canada (NSERC).

References

Bazaraa, M. S., Sherali, H. D., & Shetty, C. M. (2006). *Nonlinear programming: Theory and algorithms*. Hoboken, NJ: Wiley-Interscience. ISBN 0471486000.

Beck, J. V., & Arnold, K. J. (1977). *Parameter estimation in engineering and science*. New York: Wiley.

Bertsekas, D. P. (1999). *Nonlinear programming*. Belmont, MA: Athena Scientific. ISBN 1886529000.

Chou, I.-C., Martens, H., & Voit, E. (2006). Parameter estimation in biochemical systems models with alternating regression. *Theoretical Biology and Medical Modelling, 3*(1), 25. ISSN 1742-4682. doi:10.1186/1742-4682-3-25

Gardner, T. S., di Bernardo, D., Lorenz, D., & Collins, J. J. (2003). Inferring genetic networks and identifying compound mode of action via expression profiling. *Science, 301*(5629), 102–105. doi:10.1126/science.1081900

Gonzalez, O. R., Kper, C., Jung, K., Naval, P. C., & Mendoza, E. (2007). Parameter estimation using simulated annealing for S-system models of biochemical networks. *Bioinformatics, 23*(4), 480–486. doi:10.1093/bioinformatics/btl522

Ho, S.-Y., Hsieh, C.-H., Yu, F.-C., & Huang, H.-L. (2007). An intelligent two-stage evolutionary algorithm for dynamic pathway identification from gene expression profiles. *Computational Biology and Bioinformatics, IEEE/ACM Transactions on, 4*(4), 648–704. ISSN 1545-5963. doi:10.1109/tcbb.2007.1051

Kikuchi, S., Tominaga, D., Arita, M., Takahashi, K., & Tomita, M. (2003). Dynamic modeling of genetic networks using genetic algorithm and S-system. *Bioinformatics, 19*(5), 643–650. doi:10.1093/bioinformatics/btg027

Klipp, E., Herwig, R., Kowald, A., Wierling, C., & Lehrach, H. (2005, May). *Systems biology in practice: Concepts, implementation and application*. Weinheim: Wiley-VCH.

Li, L.-X., Wu, L., Zhang, H.-S., & Wu, F.-X. (2013). A fast algorithm for nonnegative matrix factorization and its convergence. *Neural Networks and Learning Systems, IEEE Transactions, PP*(99), 1. doi:10.1109/TNNLS.2013.2296627

Liu, L.-Z., Wu, F.-X., & Zhang, W.-J. (2012a). *Alternating weighted least squares parameter estimation for biological s-systems*. 2012 IEEE 6th international conference on systems biology (ISB), Xi'an, China, pp. 6–11. doi:10.1109/ISB.2012.6314104

Liu, L.-Z., Wu, F.-X., & Zhang, W. J. (2012b). Inference of biological s-system using the separable estimation method and the genetic algorithm. *IEEE/ACM Transactions on Computational Biology and Bioinformatics, 9*(4), 955–965. ISSN 1545-5963. doi:10.1109/TCBB.2011.126

Seung, D., & Lee, L. (2001). Algorithms for non-negative matrix factorization. *Advances in Neural Information Processing Systems, 13*, 556–562.

Vilela, M., Chou, I.-C., Vinga, S., Vasconcelos, A., Voit, E., & Almeida, J. (2008). Parameter optimization in S-system models. *BMC Systems Biology, 2*(1), 35. ISSN 1752-0509. doi:10.1186/1752-0509-2-35

Voit, E. O. (2000, September). *Computational analysis of biochemical systems: A practical guide for biochemists and molecular biologists*. Cambridge, UK: Cambridge University Press. ISBN 0521785790.

Voit, E. O., & Almeida, J. (2004). Decoupling dynamical systems for pathway identification from metabolic profiles. *Bioinformatics, 20*(11), 1670–1681. doi:10.1093/bioinformatics/bth140

Wang, H., Qian, L., & Dougherty, E. (2010, March). Inference of gene regulatory networks using S-system: A unified approach. *Systems Biology, IET, 4*(2), 145–156. ISSN 1751-8849. doi:10.1049/iet-syb.2008.0175

Wu, F.-X., & Mu, L. (2009, June). *Separable parameter estimation method for nonlinear biological systems*. ICBBE 2009. 3rd international conference on bioinformatics and biomedical engineering, Beijing, China, 2009, pp. 1–4. doi:10.1109/ICBBE.2009.5163379

Yang, X., Dent, J. E., & Nardini, C. (2012). An S-system parameter estimation method (SPEM) for biological networks. *Journal of Computational Biology, 19*(2), 175–187. doi:10.1089/cmb.2011.0269

Permissions

All chapters in this book were first published in SSCE, by Taylor & Francis; hereby published with permission under the Creative Commons Attribution License or equivalent. Every chapter published in this book has been scrutinized by our experts. Their significance has been extensively debated. The topics covered herein carry significant findings which will fuel the growth of the discipline. They may even be implemented as practical applications or may be referred to as a beginning point for another development.

The contributors of this book come from diverse backgrounds, making this book a truly international effort. This book will bring forth new frontiers with its revolutionizing research information and detailed analysis of the nascent developments around the world.

We would like to thank all the contributing authors for lending their expertise to make the book truly unique. They have played a crucial role in the development of this book. Without their invaluable contributions this book wouldn't have been possible. They have made vital efforts to compile up to date information on the varied aspects of this subject to make this book a valuable addition to the collection of many professionals and students.

This book was conceptualized with the vision of imparting up-to-date information and advanced data in this field. To ensure the same, a matchless editorial board was set up. Every individual on the board went through rigorous rounds of assessment to prove their worth. After which they invested a large part of their time researching and compiling the most relevant data for our readers.

The editorial board has been involved in producing this book since its inception. They have spent rigorous hours researching and exploring the diverse topics which have resulted in the successful publishing of this book. They have passed on their knowledge of decades through this book. To expedite this challenging task, the publisher supported the team at every step. A small team of assistant editors was also appointed to further simplify the editing procedure and attain best results for the readers.

Apart from the editorial board, the designing team has also invested a significant amount of their time in understanding the subject and creating the most relevant covers. They scrutinized every image to scout for the most suitable representation of the subject and create an appropriate cover for the book.

The publishing team has been an ardent support to the editorial, designing and production team. Their endless efforts to recruit the best for this project, has resulted in the accomplishment of this book. They are a veteran in the field of academics and their pool of knowledge is as vast as their experience in printing. Their expertise and guidance has proved useful at every step. Their uncompromising quality standards have made this book an exceptional effort. Their encouragement from time to time has been an inspiration for everyone.

The publisher and the editorial board hope that this book will prove to be a valuable piece of knowledge for researchers, students, practitioners and scholars across the globe.

List of Contributors

Ruoxun Zhang and Jingbo Gong
College of Teacher Education, Xingtai University, Hebei province, 054001, People's Republic of China

Fengjuan Chen, Shujiao Jin and Liqun Zhou
College of Mathematics, Physics and Information Engineering, Zhejiang Normal University, Jinhua, Zhejiang 321004, People's Republic of China

Guoqing Huang
School of Civil Engineering and Architecture, Nanchang University, Nanchang, People's Republic of China;
School of Science, Nanchang University, Nanchang, People's Republic of China

Zuozun Cao
School of Science, Nanchang University, Nanchang, People's Republic of China

Jun Shen and James Lam
Department of Mechanical Engineering, The University of Hong Kong, Pokfulam Road, Hong Kong

Zhang Sheng-Hai, Yang Hua and Zhao Zhen-Hua
Institute of Science, Information Engineering University, Zhengzhou 450001, People's Republic of China

Zhiquan Qin, Ranchao Wu and Yanfen Lu
School of Mathematics, Anhui University, Hefei 230601, People's Republic of China

Mario Lefebvre
Département de mathématiques et de génie industriel, École Polytechnique, C.P. 6079, Succursale Centre-ville, Montréal, Québec, Canada H3C 3A7

Foued Zitouni
Département de mathématiques et de statistique, Université de Montréal, C.P. 6128, Succursale Centre-ville, Montréal, Québec, Canada H3C 3J7

Xin Wanga
Department of Electrical and Computer Energy Engineering, Southern Illinois University Edwardsville, Edwardsville, IL 62026, USA

Edwin E. Yaz
Department of Electrical and Computer Engineering, Marquette University, Milwaukee, WI 53201, USA

James Long
Computer Systems Engineering, Oregon Institute of Technology, Klamath Falls, OR 97601, USA

Jianbin He and Jianping Cai
College of Mathematics and Statistics, Minnan Normal University, Zhangzhou, People's Republic of China

T. Binazadeh and M.H. Shafiei
Department of Electrical and Electronic Engineering, Shiraz University of Technology, Modares Blvd., Shiraz, Iran

Wei Wang, Guangyi Wang and Xiaoyuan Wang
School of Electronics Information, Hangzhou Dianzi University, Hangzhou, People's Republic of China

M. Kermani and A. Sakly
Research Unit of Industrial Systems Study and Renewable Energy (ESIER), National Engineering School of Monastir (ENIM), Ibn El Jazzar, Skaness, 5019, Monastir, Tunisia

H.Y. Jia and Q. Tao
Department of Automation, Tianjin University of Science and Technology, Tianjin 300222, People's Republic of China

Z.Q. Chen
Department of Automation, Nankai University, Tianjin 300071, People's Republic of China

Jing Chen
School of IoT Engineering, Jiangnan University, Wuxi 214122, People's Republic of China

Hongfen Zou
Wuxi Professional College of Science and Technology, Wuxi 214028, People's Republic of China

Xin Wang
Department of Electrical and Computer Energy Engineering, Southern Illinois University Edwardsville, Edwardsville, IL 62026, USA

Edwin E. Yaz
Department of Electrical and Computer Engineering, Marquette University, Milwaukee, WI 53201, USA

James Long
Computer Systems Engineering, Oregon Institute of Technology, Klamath Falls, OR 97601, USA

Yang Fang and Kang Yan
School of Automation and Electronic Information, Sichuan University of Science & Engineering, Sichuan 643000, People's Republic of China;

Kelin Li
Institute of Nonlinear Science and Engineering Computing, Sichuan University of Science & Engineering, Sichuan 643000, People's Republic of China

Qingsong Liu and Yiping Lin
Department of Applied Mathematics, Kunming University of Science and Technology, Kunming, Yunnan 650500, People's Republic of China

Neha Sharma and Kailash Singh
Department of Chemical Engineering, Malaviya National Institute of Technology, JLN Marg, Malaviya Nagar, Jaipur 302017, India

Justin M. Selfridge and Gang Tao
Department of Electrical and Computer Engineering, University of Virginia, Charlottesville, VA 22904, USA

Jinling Liang and Fangbin Sun
Department of Mathematics, Southeast University, Nanjing 210096, People's Republic of China;

Xiaohui Liu
School of Computer Science and Technology, Nanjing University of Science and Technology, Nanjing 210094, People's Republic of China

R. Whalley and A. Abdul-Ameer
Faculty of Engineering, The British University in Dubai, PO Box 345015, Dubai, UAE

V. Costanza and P.S. Rivadeneira
"Grupo de Sistemas No Lineales", INTEC-Facultad de Ingeniería Química (UNL-CONICET), Güemes 3450, 3000 Santa Fe, Argentina

Kanwar Bharat Singh
Department of Mechanical Engineering, Virginia Tech, Randolph Hall (MC0238), Blacksburg, VA 24061, USA

Saied Taheri
Department of Mechanical Engineering, Virginia Tech, Randolph Hall (MC0238), Blacksburg, VA 24061, USA
NSF I/UCRC Center for Tire Research (CenTiRe), Virginia Tech, Blacksburg, VA, USA

G.P. Samanta
Institute of Mathematics, National Autonomous University of Mexico, Mexico D.F., C.P. 04510, Mexico

Abdelhamid Ouakasse
Federal Public Service Justice of Belgium, Brussels, Belgium

Guy Mélard
Université libre de Bruxelles (ULB), ECARES, Brussels, Belgium

T. Stockley, K. Thanapalan, M. Bowkett and J. Williams
Centre for Automotive & Power System Engineering (CAPSE), Faculty of Computing, Engineering and Science, University of South Wales, Treforest, UK

Li-Zhi Liu, Fang-Xiang Wu and Wen-Jun Zhang
Department of Mechanical Engineering, Division of Biomedical Engineering, University of Saskatchewan, Saskatoon, SK, Canada

www.ingramcontent.com/pod-product-compliance
Lightning Source LLC
Chambersburg PA
CBHW080500200326
41458CB00012B/4038
* 9 7 8 1 6 3 2 3 8 5 0 3 1 *